AF784545

The Role of Cell Types in Hepatocarcinogenesis

Edited by
Alphonse E. Sirica, M.S., Ph.D.
Professor of Pathology
Medical College of Virginia
Virginia Commonwealth University
Richmond, Virginia

CRC Press
Boca Raton Ann Arbor London Tokyo

Library of Congress Cataloging-in-Publication Data

The role of cell types in hepatocarcinogenesis / edited by Alphonse E. Sirica.
p. cm.
Includes bibliographical references and index.
ISBN 0-8493-4746-7
1. Liver—Cancer—Cytopathology. 2. Carcinogenesis. 3. Liver cells. I. Sirica, Alphonse E.
[DNLM: 1. Cell Transformation, Neoplastic. 2. Liver—cytology. 3. Liver Neoplasms—physiopathology. WI 735 R745]
RC280.L5R65 1992
616.99'436—dc20
DNLM/DLC
for Library of Congress 92-4798
CIP

This book represents information obtained from authentic and highly regarded sources. Reprinted material is quoted with permission, and sources are indicated. A wide variety of references are listed. Every reasonable effort has been made to give reliable data and information, but the author and the publisher cannot assume responsibility for the validity of all materials or for the consequences of their use.

Neither this book nor any part may be reproduced or transmitted in any form or by any means, electronic or mechanical, including photocopying, microfilming, and recording, or by any information storage and retrieval system, without permission in writing from the publisher.

Chapters 4 and 6 were prepared as U.S. government works and as such are not subject to copyright.

Chapter 9 was supported by USPHS grant 2R01 CA 39225 to A.E.S. from the National Cancer Institute. Its contents are solely the responsibility of the authors and do not necessarily represent the official views of the National Cancer Institute.

Plate 3A and 3C reprinted with permission from Sirica, A.E. and Williams, T.W., *Am. J. Pathol.*, 140, 129, 1992.

Direct all inquiries to CRC Press, Inc., 2000 Corporate Blvd., N.W., Boca Raton, Florida 33431.

© 1992 by CRC Press, Inc.

International Standard Book Number 0-8493-4746-7

Library of Congress Card Number 92-4798

Printed in the United States of America 1 2 3 4 5 6 7 8 9 0

Printed on acid-free paper

FOREWORD

The aim of this volume is to provide an up-to-date and detailed analysis of the different cell types, including various precursor hepatocyte populations and ''oval cells'', which are currently believed to be important in the multistage development of hepatocellular carcinoma and some other types of hepatic neoplasms. The focus is on cell lineages in both human and experimental forms of hepatic epithelial cancers and in normal liver development, as well as on hepatocyte and intrahepatic biliary cell plasticity in various noncarcinogenic and carcinogenic forms of liver injury. Relevant topics, such as the clonality of hepatic neoplasms; evidence for the existence of a ''facultative stem cell'' in liver; putative cell precursor relationships in the genesis of hepatocellular carcinoma, intrahepatic cholangiocarcinoma and hepatoblastoma; the roles of specific viral and cellular oncogenes, p53 gene, and growth factors in the neoplastic transformation of hepatocytes and other liver cell types; and important *in vivo* and *in vitro* models of hepatocarcinogenesis are also addressed in specific chapters.

In editing this volume, I wished to ensure continuity and to limit redundancy, while still preserving the unique perspective and individual views brought by the contributing authors to their respective chapters. My intent also was to provide an overall balance of views of what is now known (as well as hypothesized) about cell lineages in hepato- and cholangiocarcinogenesis and for some other types of nonmesenchymal liver tumors.

I want to express my appreciation and heartfelt thanks to all of the authors for their hard work and for the careful attention they paid to preparing their individual chapters. I am also grateful to Ms. Julie Spadaro, Project Editor with CRC Press, for the valuable help she has given me in preparing this volume. I further hope that this volume will become a valuable reference to basic and clinical researchers with interests in liver carcinogenesis, development and differentiation, as well as serve to stimulate new research aimed at furthering our understanding of the differentiation potential of the different parencyhmal and nonparenchymal cell types of liver under specific conditions of carcinogenesis and hepatic injury.

Alphonse E. Sirica

THE EDITOR

Alphonse E. Sirica, Ph.D. is a professor in the Department of Pathology at the Medical College of Virginia, Virginia Commonwealth University in Richmond.

Dr. Sirica received his B.A. degree from St. Michael's College in 1965, his M.S. in cell biology from Fordham University in 1968 and his Ph.D. degree in biomedical sciences from the University of Connecticut in 1976. He did his postdoctoral training with Dr. Henry C. Pitot at the McArdle Laboratory for Cancer Research, the University of Wisconsin, Madison, WI, and from 1979 to 1984 served on the faculty of the Medical School of the University of Wisconsin, Madison. In 1984, he joined the faculty of the Medical College of Virginia, VCU where he is currently director of the Department of Pathology's Program in Carcinogenesis and Experimental Tumor Biology.

Dr. Sirica serves as a regular member of the Metabolic Pathology Study Section, N.I.H. (1991–1995) and is a regular member of the Scientific Advisory Committee on Carcinogenesis and Nutrition of the American Cancer Society (1989–1992). In addition, he is currently a member of the Program Committee of the American Association of Pathologists, is on the editorial boards of a number of biomedical and scientific journals, and is the editor/author of *The Pathobiology of Neoplasia,* which was published in 1989 by Plenum Publishing, New York, NY.

His major research interests are in the area of liver carcinogenesis with a focus on the pathobiology of intrahepatic biliary epithelial cells.

CONTRIBUTORS

Marie-José Blouin, Ph.D.
Postdoctoral Fellow
Ludwig Institute for Cancer Research
London, United Kingdom

Sara L. Cole, B.S.
Laboratory Specialist
Department of Pathology
Medical College of Virginia
Virginia Commonwealth University
Richmond, Virginia

Valeer J. Desmet, M.D., Ph.D.
Professor in Histology and Pathology
Catholic University of Leuven
and
Chairman
Department of Pathology II
University Hospital St. Rafaël
Leuven, Belgium

Bhalchandra Diwan, M.Sc., Ph.D.
Senior Scientist
Program Resources Inc./Dyncorp
National Cancer Institute
Frederick Cancer Research and Development Center
Frederick, Maryland

Lynne W. Elmore, B.A.
Ph.D. Candidate
Department of Pathology
Medical College of Virginia
Virginia Commonwealth University
Richmond, Virginia

Ritva P. Evarts, D.V.M., Ph.D.
Veterinary Medical Officer
Laboratory of Experimental Carcinogenesis
National Cancer Institute, NIH
Bethesda, Maryland

Emmanuel Farber, M.D., Ph.D., F.R.S.C.
Professor of Pathology and Biochemistry
Department of Pathology
University of Toronto
Toronto, Ontario, Canada

Ronald Faris, Ph.D.
Assistant Professor
Department of Medicine/Pathology
Brown University
Providence, Rhode Island

Nelson Fausto, M.D., M.S.
Professor and Chairman
Department of Pathology and Laboratory Medicine
Brown University
Providence, Rhode Island

Michael A. Gerber, M.D.
Professor and Chairman
Department of Pathology and Laboratory Medicine
Tulane University School of Medicine
New Orleans, Louisiana

Stanley Goldfarb, M.D.
Professor
Department of Pathology and Laboratory Medicine
University of Wisconsin Medical School
Madison, Wisconsin

Douglas C. Hixson, Ph.D.
Associate Professor
Department of Medicine/Pathology
Brown University
Providence, Rhode Island

Philip M. Iannaccone, M.D., D.Phil.
Professor, Department of Pathology
and
Director, Markey Program in Developmental Biology
Northwestern University Medical School
Chicago, Illinois

Harriet C. Isom, Ph.D.
Professor
Department of Microbiology and Immunology
Penn State University
College of Medicine
Hershey, Pennsylvania

Brian J. Knoll, Ph.D.
Assistant Professor
Department of Internal Medicine
Baylor College of Medicine
Houston, Texas

Joan M. Lemire, Ph.D.
Postdoctoral Research Associate
Department of Pathology
University of Washington
Seattle, Washington

Anne Loranger, Ph.D.
Postdoctoral Fellow
Centre de recherche
Hôtel-Dieu de Québec
Québec, Canada

Normand Marceau, Ph.D.
Professor
Department of Medicine
Laval University, Ste-Foy
Quebec, Canada

Yuk K. Ng, Ph.D.
Postdoctoral Research Associate
BMBCB
Northwestern University
Evanston, Illinois

Micheline Noël, B.Sc.
Research Assistant
Centre de recherche
Hôtel-Dieu de Québec
Québec, Canada

Phyllis M. Novikoff, Ph.D.
Associate Professor
Department of Pathology
Albert Einstein College of Medicine
New York, New York

Thomas D. Pugh, Ph.D.
Assistant Scientist
Department of Pathology and Laboratory Medicine
University of Wisconsin Medical School
Madison, Wisconsin

Jerry M. Rice, Ph.D.
Chief
Laboratory of Comparative Carcinogenesis
National Cancer Institute
Frederick, Maryland

Stewart Sell, M.D.
Professor
Department of Pathology and Laboratory Medicine
University of Texas Health Science Center
Houston, Texas

Nobuyoshi Shiojiri, Ph.D.
Assistant Professor
Faculty of Science
Department of Biology
Shizuoka University
Shizuoka 422, Japan

Alphonse E. Sirica, M.S., Ph.D.
Professor and Director
Program in Carcinogenesis and Experimental Tumor Biology
Department of Pathology
Medical College of Virginia
Virginia Commonwealth University
Richmond, Virginia

Stephen C. Strom, Ph.D.
Associate Professor of Pathology
Medical College of Virginia
Virginia Commonwealth University
Richmond, Virginia

Bud C. Tennant, D.V.M.
James Law Professor of Comparative Medicine
Department of Clinical Sciences
Cornell University
Ithaca, New York

Snorri S. Thorgeirsson, M.D., Ph.D.
Chief
Laboratory of Experimental Carcinogenesis
National Cancer Institute, NIH
Bethesda, Maryland

Swan Nio Thung, M.D.
Professor of Pathology
Department of Pathology
Mount Sinai Medical Center
New York, New York

Natalia Török, M.D.
Postdoctoral Fellow
Centre de recherche
Hôtel-Dieu de Québec
Québec, Canada

Peter Van Eyken, M.D., Ph.D.
Assistant
Department of Pathology II
Catholic University of Leuven
Leuven, Belgium

Jerrold M. Ward, D.V.M., Ph.D.
Chief
Tumor Pathology and Pathogenesis Section
Laboratory of Comparative Carcinogenesis
National Cancer Institute
Frederick, Maryland

Wendy C. Weinberg, Ph.D.
Postdoctoral Fellow
Laboratory of Cellular Carcinogenesis and Tumor Promotion
National Cancer Institute, NIH
Bethesda, Maryland

Teresa W. Williams, B.S.
Laboratory Specialist
Department of Pathology
Medical College of Virginia
Virginia Commonwealth University
Richmond, Virginia

Li Yang, M.D.
Department of Biology/Medicine
Brown University
Providence, Rhode Island

Dedicated with love
to my wife, Annette,
and to my children,
Gabrielle and Nicholas

TABLE OF CONTENTS

Chapter 1

ON CELLS OF ORIGIN OF LIVER CELL CANCER

E. Farber

TABLE OF CONTENTS

ISBN 0-8493-4746-7
© 1992 by CRC Press, Inc.

I. INTRODUCTION

The clarification of the histogenesis of cancer is fundamental to our understanding of how cancer develops in any organ, including the liver. Even though organ and systemic influences in carcinogenesis have been studied from time to time and no doubt play important roles in the cellular evolution during the carcinogenic process, the cell as the smallest potentially autonomous biological unit remains the most useful starting point for any discussion of the origin of any cancer. Although biochemical-molecular events underlie the biological behavior of cells, "the cell is nevertheless the smallest integrating unit in biology: a pseudointelligent computer that receives, screens, changes, reacts to and adapts to a host of environmental signals, all of this activity apparently designed, through evolution, for cell survival and host survival".[1]

The liver in humans and in mammalian animals is the site of development of several different types of liver cancer.[2-4] This spectrum includes hepatoblastoma, "angiosarcoma" (hemangiosarcoma), hepatocellular ("liver cell") carcinoma, and cholangiocarcinoma. This discussion is limited almost entirely to hepatocellular carcinoma in rats and refers only incidentally and peripherally to cholangiocarcinoma. However, it must be pointed out that liver cell carcinoma often shows areas of glandular or ductular differentiation that are frequently confused with cholangiocarcinoma.

II. SOME FUNDAMENTAL PRINCIPLES OF THE STUDY OF HISTO-CYTO-GENESIS

There are at least four important and fundamental aspects that must be considered briefly as a prelude to an analysis of the histo-cyto-genesis of liver cell cancer: (1) the importance of the separate identification of the first or initial relevant cell and of each subsequent step, including the ultimate cell of origin of cancer in any multistep process; (2) the importance of *material* (or linear) *continuity* of cells[5]; (3) the unreliability of phenotype ''markers'' as a guide to cell lineage; and (4) the great difficulty, if not impossibility, of making a *reliable* and *valid* extrapolation from an *in vitro* culture system to an *in vivo* pathological process, such as carcinogenesis.

A. CELL OF ORIGIN OF EACH STEP FROM BEGINNING TO CANCER

Some biological processes appear to involve a single step change from one cell to another. Cancer in general is not known to be among such processes. During the development of liver cell cancer in humans, in young or adult rodents, and in transgenic mouse models, there is virtually universal agreement that many changes in the liver, including focal proliferative lesions of phenotypically different hepatocytes (so-called benign neoplasms) consistently precede the ultimate appearance of unequivocal hepatocellular malignancy. In at least some systems, some of these lesions participate in a regular sequence that may culminate ultimately in cancer. In any multistep process, be it chemical or biological, the only rational approach to understanding is to identify and analyze each step with respect to origin, properties, and fate. Any hope of understanding anything about histo-cyto-genesis by looking only at the initial and final events is doomed from the outset. Each step can only be understood through comparison with its *immediate precursor.* With chemical carcinogenesis, only the *first cellular response* can be directly related to the carcinogenic agent and any attempt to relate the agent with the ultimate cancer is entirely a futile exercise in guesswork. *There is no known relevance of the properties of any carcinogen to the properties of any final cancer.* Thus, the title of this chapter is rationally only appropriate for the last step in the process, the *ultimate precursor to cancer.* For all other steps, *cells of origin for cellular evolution,* rather than cells of origin for cancer, is the only scientifically acceptable way to formulate the critical question of the cellular origins of cancer.

Clearly, it becomes important to delineate the step or steps being considered, since it is theoretically possible that different types of cells might be involved at different steps or stages.

B. MATERIAL OR LINEAR CONTINUITY

Foulds,[5] in an unusually decisive analysis of cancer development in humans and animals, has emphasized repeatedly the need for material continuity to establish cell lineage. As he stated some years ago, "the statistical time relationship alone is not sufficient evidence of sequential development which requires demonstration of the *material continuity* of the suspected sequential lesions".[5] The only convincing evidence for the conversion of one cell or one lesion to another is to observe such a change grossly or microscopically. Most models of liver cell cancer do not allow for an unequivocal answer, either yes or no, but rather for a "maybe" answer. However, some models do allow such an answer.

C. PHENOTYPIC "MARKERS" AND CELL LINEAGE

The commonest approach to the problem of cell lineage is the use of phenotypic expressions or markers. This is based upon a modern expression of a very old false paradigm in biology. It is readily assumed, and even considered axiomatic, that the appearance and behavior of differentiated cells, i.e., the phenotypic expression, are quite fixed and that this expression is a reliable index of cell lineage. Innumerable studies in intact organisms and in isolated cells indicate the plasticity and enormous variability in the phenotypic expression of any type of cell depending upon wide fluctuations in the immediate cellular environment under different physiological and pathological circumstances. This is to be expected, since each cell in an organism has, presumably, the same genetic information as all the others, yet the spectrum of phenotypic "markers" among the many different cells in a single host is extremely large. The liver cells are no exception. They show great degrees of variability in phenotypic expression under different conditions, including exposure to xenobiotics, as will be discussed.

Based on an archaic and outmoded misconception in the 19th century view of cell biology, carcinogenesis, including liver carcinogenesis, and cancer are considered to arise by dedifferentiation of a mature differentiated cell or by aberrant differentiation of a pluripotential stem cell.[6,7] A third, more modern, view states that cancer can arise from a mature differentiated cell by further differentiation options that are available to many if not all mature cells capable of cell proliferation and that may be initiated or favored by many cellular environmental variations.[8-10] This latter view is more in keeping with modern cell biology and with our realization that cells, including "mature" cells, are capable of large variations in phenotype and of unusual degrees of plasticity. It is time for the concept of the fixation of the phenotype during differentiation to be discarded.

It is also important to realize that no "phenotypic markers" have been found for any so-called preneoplastic and precancerous liver cell populations or for liver cell cancer. Any so-called marker for cancer or "precancer" is repeatedly found in different cells under different physiological or pathological conditions. We would all be well advised to drop the term "phenotypic marker" for putative preneoplastic, precancerous, or malignant cells until such a property has been found.

D. EXTRAPOLATION FROM CELL CULTURE TO *IN VIVO*

It is also important to establish, if possible, the relevance of studies in isolated cells from the liver to the known or suspected steps in the intact host. This is difficult and is rarely possible.

The difficulties of lineage analyses *in vitro* are seriously compounded by the inability to follow a transformation of one single cell to another cell in any cell culture system. Given the common occurrence of phenotypic heterogeneity in *all in vitro* cultures,[8,9] it is impossible to follow visually or with any known *phenotypic* "marker" the origin of one type of cell from another. This approach is further made almost impossible in the case of the liver, where no one has yet devised a primary isolation procedure for any cell type (especially for the nonhepatocyte cells) that generates *exclusively* (100%) only one type of cell.[11] The contamination of any isolated population with even a few percent of other cells, such as small hepatocytes, makes it impossible so far to devise a definitive experiment for cell lineage *in vitro,* even if this were theoretically possible, since it is not inconceivable that many transformed lines may originate from the rare minority of "contaminating" cells. This criticism is also applicable to the cell of origin of the rat liver epithelial-like cell lines that can be readily developed from normal rat liver. The cell or cells of origin of such cell lines remain in the realm of opinion or conjecture.

In addition, the well-known great variability and plasticity in the phenotype of any single type of cell with the same apparent genotype under different environmental conditions *in vitro* (as well as *in vivo*), including isolated liver presumptive epithelial cells,[8,9] makes the measurement of a single or even of a few biochemical indices of a phenotype totally unreliable and useless for the studies of a definitive cell lineage.

III. DEVELOPMENT OF A HEPATOCELLULAR CARCINOMA AS A MULTICELLULAR, MULTISTEP PROCESS

There is direct evidence and considerable indirect evidence that the development of liver cell cancer in many instances occurs by a sequence of different cells, perhaps as many as eight or ten,[10] that can be characterized as a process of cellular evolution.[1,5,10] In a few models, it is possible to follow *visually* the appearance of new or altered cells in a discrete precursor cell population, including the appearance, in a hepatocyte nodule, of a single unequivocal hepatocellular carcinoma with metastases to the lung.[12]

It must be emphasized that there may well be more than one cellular sequence throughout or at any one step in the whole process. The clarification of one sequence under some circumstances does not rule out or preclude the existence of one or more additional sequences under other circumstances. This tolerance for more than one alternative set of alterations is essential if we are to analyze the process critically, scientifically, and rationally, especially with an eye toward the prevention of cancer and perhaps even the treatment of advanced cancer in humans.

In the analysis of the cell or cells of origin of liver cell cancer, the following *assumption* is made: epithelial cell cancers arise from epithelial cells. Since no unequivocal examples of the connective tissue-epithelial cell metaplasia are documented despite their common genome, we "assume that the assumption" is valid. I too subscribe to this prejudice for the time being and currently consider that only epithelial cells can generate epithelial cell cancer (carcinoma), including liver cell cancer.

IV. ONE KNOWN SEQUENCE OF DEVELOPMENT OF HEPATOCELLULAR CARCINOMA

Before we can discuss the question of cell or cells of origin, it is important to delineate, insofar as we can, the first product of the supposed cell of origin and the different biological steps in the sequence between the initial "altered cell" that appears during or after initiation and the first appearance of cancer. In order to do this on a solid, rational basis, it seems most appropriate to begin at the "end" and work backwards. In this way, the possible relevance to cancer should be more convincing. What is the origin and fate of each step and how might it be studied and ultimately modulated?

A. ULTIMATE PRECURSOR TO HEPATOCELLULAR CARCINOMA

There is unequivocal evidence that the late or persistent[10] hepatocyte nodule is one precursor lesion for liver cell cancer: (1) cancer has been seen to arise inside nodules in different models of hepatocarcinogenesis with aramite,[13] ethionine,[14,15] 3′-methyl-4-dimethylaminoazobenzene,[16] 2-acetylaminofluorene (2-AAF),[17] and diethylnitrosamine (DEN)[12,18] (in one report, metastasizing hepatocellular carcinoma was found in a late nodule);[12] (2) persistent hepatocyte nodules, unlike early nodules, show a high rate of development of hepatocellular carcinoma with metastases when transplanted individually to the spleen of syngeneic rats;[19] and (3) one can often see "nodules in nodules" in studies of the behavior of late hepatocyte nodules during liver carcinogenesis.[13,20,21]

In some models,[10] the number of late persistent "precancerous" nodules shows a progressive decrease such that one sees only one to three nodules at the time that unequivocal cancer can be identified. The resistant hepatocyte model[22,23] and the orotic acid (OA) model[24,25] are two such models. This progressive decrease with increasing relevance to cancer has not been seen using phenobarbital (PB) or polychlorinated biphenyls (PCB) for promotion, where the number of nodules may *increase* as a function of time. This may be a reflection of the asynchronous nature of the appearance and cellular evolution of nodules.

B. HEPATOCYTE NODULES TO ULTIMATE PRECANCEROUS NODULE

This is the least-understood phase of liver carcinogenesis in humans and animals. Since most models of liver cell cancer do not show a synchrony in the development of early or late nodules, it is not possible to delineate ''which nodule precedes or follows another nodule''. Under these conditions, the formulation of a sequence is largely based on intuition or guessing.

In the resistant hepatocytes-model, it is possible to follow a sequence because of three properties:[10,21] (1) the intensity of the selection of resistant hepatocytes to generate nodules during promotion; (2) the rapid synchronous development of nodules from the resistant hepatocytes generated during initiation, with all visible foci enlarging in unison; and (3) the brief period of exposure of the animal to ''the promoting or selecting environment''.

Of the selectable resistant hepatocytes generated during initiation, the vast majority remodel quite quickly after the first expansion phase of the nodules. Over 95% of several hundred nodules begin remodeling at week 6 and have been integrated into the liver architecture by weeks 8 to 10. Of the 10 or 15 nodules that persist after weeks 8 to 10, the majority undergo slow remodeling, such that only 1 to 3 nodules remain by the time unequivocal cancer appears in a nodule at 10 months or so.

The continuity between early nodules, the persistent nodules, and the ultimate precancerous nodules is obvious by gross and microscopic examination and is confirmed unequivocally by following the early nodules as a function of time after they have been selectively and exclusively labeled with tritiated thymidine.[26]

However, given the two major options that nodules express *in situ,* remodeling and persistence, and given the large number that are present early, there is at present no valid way to identify which individual early nodule may persist and which may remodel. Also, since as many as 10 to 15 may show persistence, but since most of these will remodel slowly, there is no known way (if one exists) to specify relevance to cancer at any time until the 1 to 3 late persistent precancerous nodules appear. Thus, the rat is much more efficient in selecting for relevance than we are.

Despite these uncertainties, it is very clear that the ultimate persistent precancerous nodule is derived from the more numerous population of early persistent nodules. These in turn are derived from a much larger number of early developing nodules, the majority of which remodel and a minority persist.

C. FOCI OF ALTERED HEPATOCYTES TO PERSISTENT NODULES

The appearance of small foci of altered hepatocytes is a regular feature of most if not all the models for liver carcinogenesis using genotoxic carcinogens or exposure to a choline-devoid diet. These have a variety of biochemical differences from the surrounding liver. In several models, the majority of foci become visible only after exposure to a promoting environment. They often appear slowly over many weeks (sometimes 10 weeks or longer) and do so asynchronously.

Their expansion to nodules during promotion is often asynchronous, so that it is impossible to delineate any regular sequence or predictable times of appearance. Again, it becomes impossible under these conditions to designate a precursor-product relationship, a prerequisite for any study of the site and cell of origin of cancer.

In the resistant hepatocytes model, the foci appear synchronously in a remarkably uniform fashion. Unlike the foci and nodules in some other models, the variations in diameters of the foci and nodules are so small that no correction is needed for this variable when enumerating the number of foci per area.

D. SITE AND CELL OF ORIGIN OF HEPATOCYTE FOCI

The earliest foci seen in liver carcinogenesis involve obvious hepatocytes. However, whether they are derived from preexisting hepatocytes or from other epithelial elements in the liver, such as the ductular or periductular system, they are topics of considerable interest, both at present and in the past.

Since the foci or islands of altered hepatocytes can be shown to develop into hepatocyte nodules and since a linear sequence of lesions with material continuity can be shown from foci, early hepatocyte nodules, persistent hepatocyte nodules, late persistent precancerous hepatocyte nodules, and unequivocal hepatocellular carcinoma, it is clear that any study of cells of origin, in this sequence, has to be addressed to several sites in the sequence — original cell and site of origin for foci, origin of hepatocyte nodules from foci, origin of persistent hepatocyte nodules from early nodules, origin of late precancerous nodules from persistent nodules, and origin of liver cell cancer from precancerous nodules.

If other sequences occur, these will have to be addressed independently. For example, speculations of cancer development from altered atypical hepatocytes without a nodule as an intermediary are not uncommon in the medical and scientific literature. Although such speculations are interesting medically, they seem to be impossible to address scientifically at this time.

There has been some periodic discussion also about other possible sequences — cancer from microscopic foci without the need for a grossly visible nodule or cancer from oval-ductular cells without a nodule as the immediate precursor for cancer. If such proposed sequences are to be studied and analyzed in detail, appropriate *analyzable models* must be developed. An obvious problem in model design that makes a negative finding most difficult to interpret is the fact that new focal proliferations usually grow more quickly than do their precursors. Naturally, under these conditions, any proliferating cell population very often destroys its precursor population. Thus, the absence of a visible precursor at any time point does *not* establish that such a precursor did *not* exist. Clearly, a *large expanded precursor population, such as a nodule,* is necessary in most instances, in order to observe a new population arising in an older one. This serious experimental complication has been overlooked in many studies that claim the absence of a precursor for any cell population acting as a step in a sequence leading to cancer. The corollary of this is the obvious bias of all studies that implicate a discrete lesion such as a nodule as the only precursor.

V. POSSIBLE CELLS OF ORIGIN OF EACH STEP IN THE SEQUENCE OF FOCI OF ALTERED HEPATOCYTES TO HEPATOCYTE NODULES TO LIVER CELL CANCER

If one assumes that *liver cell carcinoma originates from a liver epithelial cell,* there are only a few choices — hepatocyte, biliary ductal and/or ductular epithelial cells (DEC), and proliferating primary biliary DEC (so-called "oval cells"). Since the bile flow is from bile canaliculus to biliary ductule and then to ducts, we shall designate the first ductule, the canal of Hering, as the *primary biliary ductule*. Since in this entire sequence all relevant steps involve epithelial cells, emphasis is only on epithelial cells throughout the process. Most current discussion centers around the nature of the epithelial cell that generates the foci of altered hepatocytes as precursors for nodules and, to a lesser degree, the cells in the ultimate precancerous hepatocyte nodule that are the precursors for hepatocellular carcinoma.

A. HEPATOCYTE

1. As the Cell of Origin of Foci and Early Nodules

There is direct visual, incontrovertible evidence that a few original hepatocytes can be altered during initiation and that some of these altered hepatocytes are sites of origin for foci, and in turn nodules. The models in which such evidence is clear are indicated in Table 1. They include models v, viii, a subset of (i) with DEN in which no ductular-oval cell proliferation is seen, and model vi in which such proliferation is present briefly, but an origin of foci and nodules from preexisting hepatocytes is clearly visible and obvious. With models v, vi, and viii, almost 100 different chemical carcinogens have been studied as initiators, and the patterns are remarkably reproducible.[133-141]

TABLE 1
Models of Liver Cancer Development in the Rat

	Model	Ductular-cell proliferation	Cell of origin of foci-nodules	Ref.
	A. Single agent			
i)	Long-term continuous exposure to a carcinogen	+	?	5, 10, 15–21, 40, 42, 53, 92, 93
	Except diethylnitrosamine	−	Hepatocyte (?)	4, 89, 90
ii)	Intermittent chronic exposure to a carcinogen	+	?	15, 152, 153
iii)	The stop model	−	Hepatocyte(?)	4, 90
iv)	Chronic exposure to peroxisome proliferators	−	Hepatocyte(?)	142, 143
	B. Initiation-promotion-progression			
v)	Chronic enzyme induction model (i.e., phenobarbital, PCB, etc.)	−	Hepatocyte	144
vi)	Resistant hepatocyte model	+	Hepatocyte	10, 12, 21
vii)	Choline-methionine-deficient model	+	?	67, 145, 146
viii)	Orotic acid model	−	Hepatocyte	147
	C. Dietary deficiency			
ix)	Choline-devoid without added carcinogens	+	?	148–151

In the models in which initiation is induced by a brief exposure to one of many different genotoxic carcinogens and promotion is by PB or PCB or by OA, foci of altered hepatocytes are seen to arise from original altered hepatocytes without any visible proliferation or alteration of other types of epithelial cells. Here the random appearance of the smallest focus of altered hepatocytes in zones 1, 2, or 3 of the liver acinus and the uncomplicated hepatocyte composition of these areas with no ductular-oval cell proliferation is unequivocal and conclusive. Even though the foci develop and enlarge slowly over many weeks and in a nonsynchronous fashion, the interpretation of their origin appears to be quite evident.

In the resistant hepatocytes model, foci of altered hepatocytes, resistant to the inhibition of cell proliferation by the 2-AAF, begin to appear within 24 h after the partial hepatectomy (PH), well *before* ductular-oval proliferation is seen. As with the PB and OA models, the foci of altered hepatocytes appear randomly throughout the liver acinus in zones 1, 2, or 3.[12] The induction of such resistant hepatocytes is seen with almost 100 genotoxic carcinogens with quite different chemical structures.

Recently, with the use of a histochemical stain for glutathione *S*-transferase P(7-7) (GST-P), it has been found that positive isolated single hepatocytes or doublets can be seen in the liver following the administration of DEN (10 to 80 mg/kg body weight), dimethylnitrosamine (DMN, 40 mg/kg), aflatoxin B_1 (1 mg/kg), or methylazoxymethanol (35 mg/kg), mostly prior to PH.[27] The number of stained cells or doublets was proportional to the dose of DEN from 10 to 120 mg/kg.[27-29] The appearance of GST-P, first isolated from rat placenta, is probably the commonest enzyme change found in foci and nodules in the liver with several carcinogens.[30]

Although it was suggested that the stained hepatocytes represent the first step in carcinogenesis,[27,29] more recent studies have shown that only some of the cells are selectable by one promoting environment for resistant hepatocytes and many stained cells do not develop into foci or islands, but remain as single cells.[27,30-32] Whether these stainable hepatocytes not selected with one promoting environment would be selected or promoted with another promoting environment, such as with PB or OA, is an important question, so far unanswered. However, it is clear that some of the stainable *hepatocytes* are resistant to the mitoinhibitory action of 2-acetylaminofluorene and thus generate nodules, similar to the hepatocyte nodules that are central to cancer development in the RH and other models.

Since the induction of a new phenotype in a rare *hepatocyte* occurs reproducibly with so many genotoxic carcinogens, and since such hepatocytes are the cells of origin of foci and early hepatocyte nodules in some models, it would seem logical and rational to begin any detailed analysis of the histo-cyto-genesis of a foci-nodules-cancer sequence with the hypothesis that the hepatocytes with the altered phenotype are also cells of origin in the several more complex models that have so far not been analyzed to yield an unequivocal answer. Critical experiments designed to disprove the validity of this hypothesis are needed.

Consistent with the conclusion that original hepatocytes are the cells of origin of foci and nodules are the results of several studies showing their clonal origin in rats[12,28,29,33-35] and in mice.[36-38]

2. As the Cell of Origin of Early Remodeling and Persistent Nodules

The foci and early nodules, both remodeling and persistent, consist of hepatocytes as the only epithelial cell component.[5,10,12-16] The close precursor-product relationship between early nodules and later persistent ones is seen when the hepatocytes in the early nodules are selectively and exclusively labeled with tritiated thymidine and followed as a function of time.[26] All the hepatocytes in the persistent nodules are derived from the hepatocytes in some early nodules. However, at present there is no way to predict which of the early nodules will remodel and which will persist.

The cellular composition of the nodules as they develop and then exercise their options (remodeling vs. persistence) is noteworthy. Initially, as the hepatocyte nodules expand by cell proliferation, they are remarkably similar in appearance, on histological examination, to the proliferating hepatocytes during posthepatectomy or postnecrotic regeneration. They cannot be distinguished from regenerating hepatocytes except for their nodular, rather than zonal or diffuse distribution. Histochemically, these proliferating hepatocytes in the developing nodules show in part a biochemical pattern similar to that seen in regenerating hepatocytes. However, in addition, they show a pattern of gene expression that is special for nodule hepatocytes — large increases in the expression of microsomal epoxide hydrolase, glutathione *S*-transferase (7-7)(P), and γ-glutamyl transpeptidase.[10]

After the initial phase of clonal expansion by hepatocyte proliferation, the majority of the nodules become virtually quiescent and the nodules show major changes associated with remodeling by differentiation. The persistent nodules show patterns of interference with glycogen breakdown, progressive proliferation of endoplasmic reticulum, cyst formation, and other histologic-cytologic changes. Only later do they show the appearance of ducts, presumably by differentiation of hepatocytes or by the ingrowth of ducts from the surrounding liver.

3. As the Cell of Origin of Liver Cell Cancer in the Late Persistent Nodule

Histologic examination of the persistent nodules as a function of time during the progression phase reveals the new appearance of ducts lined by cuboidal epithelium in the late persistent precancerous nodules. These ducts are not seen in every section of late persistent nodules, but are not infrequent. Parenthetically, it should be pointed out that no study has appeared of how the hepatocyte nodule is integrated with the surrounding liver anatomically and functionally in regard to bile secretion.

Also, in the models with long-term continuous (model i) or intermittent (model ii) exposure to a carcinogen (azo dyes, 2-AAF, ethionine), one occasionally can see ductular proliferation inside a nodule. I have only seen this cellular response with the long-term exposures and never in the initiation-promotion-progression models, even with models (such as model vi [Table 1]) that show a short period of ductular-oval cell proliferation early in the carcinogenic process.

Theoretically, hepatocellular carcinomas might arise from the predominant hepatocyte population or in only some models, from a proliferating ductular-oval cell population. As of now, there is no definitive and convincing evidence concerning the cell or cells of origin of liver cell carcinomas. This problem is discussed in some detail in Section V.C.

It is thus evident that hepatocarcinogenesis with many genotoxic carcinogens shows one well-established sequence of events — altered rare hepatocyte to foci to early nodules to persistent nodules to late persistent precancerous nodule to liver cell cancer. The immediate *cell* of origin of the cancer is not established.

B. INTRAHEPATIC BILIARY EPITHELIAL CELLS

There are no data pro or con to either implicate or exclude the lining epithelium of the intrahepatic biliary ducts and ductules (canals of Hering) as the original cell of origin of hepatocellular carcinoma. Although these cells would logically appear to be the cell of origin of bona fide cholangiocarcinoma, direct evidence for this lineage is so far not available. Furthermore, no clearly defined sequence of biliary ductal or ductular cell to hepatocyte foci, nodules, or cancer has ever been seen.

C. PROLIFERATING DUCTULAR EPITHELIAL CELLS (SO-CALLED "OVAL CELLS")

A common finding in experimental carcinogenesis is the appearance of proliferating DEC early in the process and their persistence throughout the period of cancer development in some models in which the carcinogenic stimulus is continued until cancer develops. This includes continuation of exposure to either a chemical carcinogen or to a carcinogenic diet, such as the choline-devoid diet. This hepatic response was noted in the very first report of the induction of liver cancer in the rat with a chemical, *o*-aminoazotoluene, by Sasaki and Yoshida,[39] and subsequently by many investigators in many countries with many different carcinogens, including 4-dimethylaminoazobenzene, its 3′-methyl or 4′-fluoro derivatives, a naphthalene derivative and mixtures of azo dyes, 2-AAF, *m*-toluenediamine, dibenzcarbazole, tannic acid (see Reference 40 for references), ethionine,[40-44] thioacetamide,[45] aramite,[13] aflatoxins,[46,47] pyrrolizidine alkaloids,[48-50] and purified extracts and compounds from *Fusarium moniliforme*[51] and probably other hepatocarcinogens, either purified from natural products or synthetic.

Extensive DEC proliferation, with a resemblance to the DEC proliferation seen with carcinogens, can be induced in the rat with noncarcinogenic stimuli, such as α-naphthyl-isothiocyanate (ANIT),[52] bile duct ligation,[53] feeding a fat-free diet,[53] or lithocholic acid.[54-57] However, α-fetoprotein, albumin, and glucose-6-phosphatase activity were not found in some studies of proliferating DEC induced by bile duct ligation or by ANIT, but were found in proliferating DEC with ethionine, 2-AAF, or 3′-Me-DAB.[58-74] In one study with ANIT, some proliferating DEC did express α-fetoprotein.[75]

Since the uncertainty concerning the major origin of the "oval cells"[40] no longer exists,[4,41,42,44,76-85] it is appropriate to drop the descriptive term "oval cells" and to refer to this interesting cellular response as *ductular proliferation.* It is clearly evident from the many studies that this tissue response consists fundamentally of a *proliferation of fine bile ductules,* analogous in some respects to the proliferation of vascular capillaries seen in response to different types of tissue injury and repair. Naturally, the cells lining these fine ductules should be designated as *DEC*. The DEC do not appear to be proliferating randomly during ductular proliferation, but seem to be predominantly if not exclusively related to their organization as ductules.[41,83,84,86] There may be a fundamental difference in the genesis and nature of *ductal* and *ductular* proliferative responses.[11,86,87]

Since proliferation of ductules and DEC is seen in several models of liver carcinogenesis, the relevance of DEC to liver cancer induction in the rat has been discussed frequently since the original description in 1933 to 1935. Up until the early 1970s, the general impression among investigators of hepatocarcinogenesis was that the proliferating DEC did not seem to be major cells of origin for hepatocellular carcinoma with most carcinogens,[4,10,88-91] with the possible exception of butter yellow (4-dimethylaminoazobenzene) and its derivatives.

From the earliest studies of Kinosita,[92] it has been repeatedly and consistently observed that the ductular proliferation seen with butter yellow and its derivatives (''azo dyes'') is quite different from that seen with all the other hepatocarcinogens.[40,58-71,93] The proliferating DEC within 4 to 8 weeks after beginning the carcinogenic regimen with the azo dyes become enlarged, lose their apparent oval shape, and increasingly resemble hepatocytes. Whether or not they ever become functionally and anatomically ''normal'' hepatocytes that can participate in the usual hepatocyte functions has not been studied. So far, the pattern with DAB and derivatives is quite different than that with several other hepatocarcinogens and *cannot be used as a general model* to study the histo-cyto-genesis of hepatocarcinogenesis in the rat. It is perhaps noteworthy that the liver cancers seen with the azo dyes are frequently more varied in their histologic type than are the cancers with 2-AAF, ethionine, or DEN.

Some of the DEC induced with DAB and derivatives were found to contain an increased amount of AFP. This finding was not particularly unexpected, since some of the liver cell cancers induced by DAB and derivatives are AFP positive and the serum of animals on DAB show an early elevation of AFP during carcinogenesis.[94-97]

However, more recently, some DEC induced with 2-AAF, ethionine, or nitrosomorphine were also found to have histochemically detected AFP as well as albumin.[44,58-74] The DEC in these models do not show the apparent change to cells resembling hepatocytes, as they do with DAB.

These observations helped to focus anew on the DEC as possible original cells of origin from which hepatocellular carcinomas evolve. This orientation was strengthened by the proposal that DEC (as well as ductal cells and DEC generally) are resistant to the inhibitory effects of 2-AAF and other hepatocarcinogens on cell proliferation.[12,98] Since acquisition of resistance to the ''mitoinhibitory'' effects of hepatocarcinogens is a common property acquired by a small number of cells in the liver during initiation of hepatocarcinogenesis with many carcinogens, it was proposed that this characteristic was consistent with the hypothesis that some DEC might be the cells from which liver cell cancers evolve.[98] This hypothesis states that the selective inhibition, which *so far* is the only *established* pattern of promotion in hepatocarcinogenesis in the rat,[10] could apply just as well to preformed resistant cells, the duct-ductular epithelial cells, as to the rare hepatocytes that acquired the resistant phenotype during initiation.[98] The role or the contribution of ''initiation'' with genotoxic carcinogens to the carcinogenic process in the liver is not clear under this hypothesis. Carcinogenesis should require no initiating exposure to a carcinogen according to this hypothesis. The origin of a focal change in a rare cell is not clear.

1. As Cells of Origin of Hepatocyte Foci and Nodules

There is *no direct evidence* that any cells among the proliferating DEC are the cells of origin for foci and, in turn, of nodules. Thus, the basis for the hypothesis is all circumstantial, speculative, and indirect.

Price et al.[93] suggested that nodules of hepatocytes in animals on continuous 3′-Me-DAB or 4′-fluoro-DAB may be derived from the ''hyperplastic duct cells''. However, they could not rule out at all that '' . . . recrudescence of parenchymal cells may come from the scattered parenchymal cells trapped among the hyperplastic duct tissue''.

Even though no linear continuity has been seen or established between proliferating DEC and foci or nodules of altered hepatocytes, it has been suggested that some cells among the proliferating DEC induced by 2-AAF (or other carcinogens) may be direct precursors of hepatocytes. The proliferating DEC or the hepatocytes postulated to be derived from the proliferating DEC would then be precursors for the foci of altered hepatocytes. In models other than those using DAB or derivatives, the first suggestion came from presumed phenotypic markers in or phenotypic properties of some of the proliferating DEC.[44,59,65,74,98]

This hypothesis has been recently explored further by the use of labeled thymidine incorporation into DNA as a marker for proliferating DEC. It has been reported that proliferating DEC, labeled with tritiated thymidine, appear to become transformed into small basophilic hepatocytes in zone 1 of the liver when PH is performed on animals given 2-AAF.[73,74] Although interesting, these studies are not conclusive and at best suggestive, because of the following limitations: (a) it was not shown that the proliferation of the original hepatocytes was completely or almost completely inhibited at the time that the transformation from DEC to basophilic hepatocytes was first seen. Such an inhibition can occur in the rat liver with 2-AAF[82] and is essential in order to rule out incorporation of the labeled precursor by the original hepatocytes themselves. Also, zone 1 hepatocytes recover quickly from inhibition with 2-AAF.[99] It is noteworthy that the apparent ''transformation'' from proliferating DEC to basophilic hepatocytes was seen only with small doses of 2-AAF[74] and not with doses that were shown previously to inhibit proliferation of hepatocytes almost 100%.[82] (b) No bona fide hepatocytes were found as the unequivocal product of DEC, since the longest period of observation was 13 d posthepatectomy, a time at which the liver still contained many DEC.[73,74] Longer time studies of a duration of 10 weeks failed to find any hepatocytes that could be considered as products of the DEC.[41,82] (c) The reported studies[73,74] did not include any measurement of the *total radioactivity in liver DNA*. The reported[74] persistence for 13 d of *total liver radioactivity* after radioactive thymidine administration to the animal is not interpretable, since it could include both acid soluble (thymidine, thymidine metabolic, and breakdown products) and acid insoluble (DNA, some protein, etc.). A previous study by Rubin and co-workers[78] showed that *labeled DNA* in animals injected with radioactive thymidine during proliferation of DEC rapidly disappeared. Thus, it has yet to be shown that bona fide hepatocytes that become part of the liver can be generated from proliferating DEC in the models under study.

This doubt is further strengthened by the apparent failure of proliferating DEC to restore the loss of hepatocytes following the administration of a necrogenic dose of CCl_4[100] or the removal of 67% of the liver by PH.[12] It is difficult to assign the designation "stem cells"[7,97,101] if the cells involved have not been shown to be able to generate and restore a significant number of functionally active hepatocytes in the liver. It is, of course, possible that a "facultative stem cell" might be present which acts as a precursor or progenitor for hepatocytes under the special conditions of "extreme liver injury". Conceivably, the intensive inhibition of hepatocyte proliferation plus progressive liver cell necrosis might be such a situation.

It has been suggested repeatedly that the similarity in the phenotypic properties between some proliferating DEC and the hepatocytes in foci and nodules is closer than between original hepatocytes and the hepatocytes in the foci and nodules. This suggestion is not supported by the available experimental observations. The proliferating DEC as well as the original hepatocytes show considerable variation in phenotypic biochemical properties.[44,59,72-74,97] AFP, albumin, and glucose-6-phosphatase expressions varied much in the same cell type ("oval cells, hepatocytes, DEC, etc.") at different times during a period up to 42 d. Also, these studies failed to emphasize that the original hepatocytes vary very much under different conditions of exposure to a variety of xenobiotics, including carcinogens. It is well documented that original hepatocytes in rats and mice can undergo many reversible changes in phenotypic properties on exposure to many different xenobiotics[102] and during liver regeneration following PH.[66,68,103-111] For example, 12 different patterns of microsomal protein response in the rat were found with 12 different xenobiotics.[107,108]

This raises a more fundamental question concerning the use for cell lineage studies of any biochemical phenotypic properties, including not only those with a known property (e.g., albumin, AFP, glucose-6-phosphatase, etc.), but also those still without any known function, such as the several antigens, presumably proteins, for which a variety of monoclonal antibodies have been generated.[11,112-114] Since the biochemical components of any phenotype are assumed to have some possible relevant role in the biological behavior of any cell and cell population, it is important to determine the functions of these so-called "antigenic markers" so that they may become useful in formulating mechanistic hypotheses for our ultimate understanding of the different steps in hepatocarcinogenesis. This will be discussed briefly later under Section V.C.3.

An area of considerable research activity relates to the use of isolated liver cells and their possible role in hepatocarcinogenesis.[11,112,115-127] To date, no reported study on *in vitro* neoplastic transformation of liver cells has been able to match up, step-by-step with what is seen *in vivo* in any model, except for some similarities between some of the cancers. Transformation studies with some rat liver epithelial cell lines have generated a spectrum of malignant neoplasms, including hepatocellular carcinoma hepatoblastoma and cholangiocarcinoma.[123] One change that appears to show some similarity to hepatocarcinogenesis *in vivo* is the appearance of resistant liver cells on transformation of rat liver epithelial cells with v-H-*ras* or v-*raf*.[116]

While interesting from the point of view of *what might happen theoretically,* these *in vitro* studies in composite have been of limited value so far in clarifying *what does happen.* Because of this, their utility in the molecular, biochemical, genetic, or cell biological analysis of the multistep carcinogenic process in the rat liver leading to liver cell carcinoma has yet to be demonstrated. This regrettable circumstance is a reflection of the great variability and plasticity that the liver epithelial cells (and most other types of cells) show when the cellular environment is modified. In every study so far reported, virtually no so-called "phenotypic marker" has shown a degree of consistency that enables the observer to make an unequivocal conclusion concerning possible lineages among the cells.[11,44,112,114,120-127]

To this must be superimposed another major hurdle, the inability to rule out a rare unidentified cell as the cell from which the neoplastic transformation develops. Since cancer most frequently begins as a phenotypic change in a rare cell, both *in vivo* and *in vitro,* it becomes impossible *in vitro* to identify the original cell of origin. This problem may be circumvented *in vivo* where the liver cells begin with a fixed geography at reproducible times in the life history of the host. Careful observation in some models may readily allow the "material continuity" to be made.

A third major complication in the extrapolation from *in vitro* to *in vivo* concerns cell proliferation. The systems *in vitro* utilize a proliferating cell system with which to begin, while the systems *in vivo* in the adult rat begin with a highly controlled quiescent system in which *initiation does not turn on cell proliferation.* There is a requirement for one round of cell proliferation during initiation. Following this, *controlled, repeated* cell proliferation is seen only during the clonal expansion associated with promotion and only becomes apparently autonomous at a later step.[128] It is questionable, therefore, whether or not the step-by-step sequence *in vitro* is at all comparable to that seen *in vivo,* especially given the large number of biochemical and molecular changes in DNA, protein, and RNA that can be induced by an activated chemical carcinogen and the lack of a rational hypothesis for a selection procedure for initiated cells *in vitro.*

2. As Cells of Origin of Persistent Nodules From Early Nodules

Since the early nodules consist almost exclusively of hepatocytes with very few visible ductules or ducts, and since selectively labeled hepatocytes in early nodules are found to predominate in the persistent nodules,[26] an origin from nodular hepatocytes is evident. Whether or not any ductular elements participate is not known.

3. As Cells of Origin of Liver Cell Carcinoma in Late Persistent Nodules

As indicated under Section V.A.3, the late persistent hepatocyte nodules, the truly ''precancerous'' nodules, consist mainly of hepatocytes, but with variable numbers of ductular elements. Dunsford and colleagues[129] and Hixson and co-workers[114] have studied this step using mainly monoclonal antibodies and to a lesser degree other antibodies to presumptive ''phenotypic markers''. Given the wide variations in the biochemical properties of the cancer cells in any single cancer and among different cancers, given the considerable variations in the properties of the hepatocytes and ductular or ductal cells in the late nodules, and given the failure of the studies to concentrate on the very few late persistent nodules that have a high probability of being the lesions of origin for liver cell cancer, the data obtained remain unclear with respect to determining a definitive cell lineage for the ultimate precursor of cancer in hepatocarcinogenesis in the rat.[44,59,72-74,97,103,119] The data available further indicate a diversity and variability in histochemical including immunohistochemical properties that do not allow so far any definitive conclusion concerning linear or material continuity. As emphasized above, without a linear or material continuity, any suggestion concerning possible lineage remains conjectural.

4. Role of Proliferating Ductules in Hepatocarcinogenesis

An intriguing question in experimental hepatocarcinogenesis is the role, if any, of proliferating ductules in the process of cancer development. Since there is no direct and unequivocal evidence for the sequence-ductular proliferation (DEC) — some intermediate (hepatocytes?) — nodules — cancer or for the sequence-ductular proliferation (DEC) — cancer, what might be the role of ductular proliferation?

One attractive hypothesis proposes that the ductular proliferation is a normal liver response to an inhibition of hepatocyte proliferation when a stimulus for regenerative hyperplasia of hepatocytes is applied. As first suggested by Inaoka[58] with subsequent further evidence,[100] proliferation of fine biliary ductules is a common reaction[41,83,84] when the normally integrated restoration of liver is interfered with by inhibition of hepatocyte proliferation without coincident inhibition of DEC proliferation.

With chemical carcinogens that require metabolic activation, the hepatocytes, but not the duct-ductular epithelial cells, are by far the most active.[11,115,130] Since activated carcinogens are strongly inhibitory to hepatocyte proliferation, since cells altered during initiation of chemical carcinogenesis are *not* associated with any spontaneous or autonomous cell proliferation and are thus dependent upon a regenrative stimulus for cell proliferation, and since DEC are stimulated to proliferative as part of the restorative process and are not inhibited, physiologic ductular proliferation is the natural response to the set of local environmental perturbations created by exposure to the metabolism-dependent carcinogens. This response is seen *predominantly* or *only* when promotion by clonal expansion of chemical hepatocarcinogenesis is carried out with hepatocarcinogens with either a brief[10,12] or a much longer exposure.[40,66-68,92,93] The intensification of the process with the use of a choline-deficient diet with hepatocarcinogens is readily explained by the continuing and progressive hepatocyte cell death over several weeks associated with this diet.[131,132] This leads to a stimulus for restorative hepatocyte cell proliferation more prolonged and more intensive than with hepatocarcinogens alone.

VI. SUMMARY AND CONCLUSIONS

Rare original mature hepatocytes in zones 1, 2, or 3 of the adult rat liver appearing after initiation with genotoxic carcinogens have been shown to be the cells of origin for foci or islands of altered hepatocytes and of nodules derived from these foci. This clear-cut lineage is established in a few models of hepatocarcinogenesis with about 100 different genotoxic chemical carcinogens. The early nodules are lesions of origin for persistent nodules and the few late persistent nodules are precursor lesions for liver cell cancer.[152,153]

These models show either no ductular proliferation or a very limited ductular proliferation and these relatively simpler models allow for this definitive analysis of lineage.

When the models involve long-term exposure to carcinogens alone or to carcinogens plus a choline-deficient diet, the tissue and cellular changes are so very complex as to prevent a definitive analysis of cell lineage. At best, suggestions concerning "possible cells of origin" can be presented. In these complex models, conclusions concerning whether original hepatocytes, altered hepatocytes, or proliferating DEC are the likely cells of origin for evolution to cancer are not possible.

Since these complex models use many of the same carcinogens as have been used for the simplified models in which an origin for cancer evolution from original hepatocytes has been established, it seems prudent to consider as likely that the same type of cells, the original hepatocytes, may also be the cells of origin in the complex models. Why implicate a new possible cell of origin for the evolution to cancer until the hepatocyte has been ruled out? This approach appears to this reviewer to be scientific and rational. Since a hypothesis can be disproven, but not proven, why not disprove the hepatocyte hypothesis before a new hypothesis is seriously entertained? Such an approach seems logical and scientifically prudent.

REFERENCES

1. **Farber, E.,** Carcinogenesis — cellular evolution as a unifying thread: presidential address, *Cancer Res.,* 33, 2537, 1973.
2. **Okuda, K. and Peters, R. L.,** *Hepatocellular Carcinoma,* John Wiley & Sons, New York, 1976.
3. **Ishak, K.,** Newer developments in diagnostic liver pathology, in *Pathogenesis of Liver Disease,* Farber, E., Phillips, M. J., and Kaufman, N., Eds., Williams & Wilkins, Baltimore, 1987, chap. 11.
4. **Bannasch, P. and Zerban, H.,** Tumours of the liver, in *Pathology of Tumours in Laboratory Animals,* Vol. 1, 2nd ed., Turusov, V. and Mohr, U., Eds., International Agency for Research on Cancer, IARC Sci. Publ. 99, Lyon, 1990.
5. **Foulds, L.,** *Neoplastic Development,* Vol. 2, Academic Press, London, 1975, 11.
6. **Pierce, G. B., Shikes, R., and Fink, L. M.,** *Cancer: A Problem of Developmental Biology,* Prentice-Hall, Engelwood Cliffs, NJ, 1978.
7. **Sell, S.,** Is there a liver stem cell?, *Cancer Res.,* 50, 3811, 1990.
8. **Rubin, H.,** The significance of biological heterozygosity, *Cancer Met. Rev.,* 9, 1, 1990.
9. **Farber, E. and Rubin, H.,** Cellular adaptation in the development of cancer, *Cancer Res.,* 51, 2751, 1991.
10. **Farber, E. and Sarma, D. S. R.,** Hepatocarcinogenesis: a dynamic cellular perspective, *Lab. Invest.,* 56, 4, 1987.
11. **Sirica, A. E., Mathis, G. A., Sano, N., and Elmore, L. W.,** Isolation, culture and transplantation of intrahepatic biliary epithelial cells and oval cells, *Pathobiology,* 58, 44, 1990.
12. **Solt, D. B., Medline, A., and Farber, E.,** Rapid emergence of carcinogen-induced hyperplastic lesions in a new model for the sequential analysis of liver carcinogenesis, *Am. J. Pathol.,* 88, 595, 1977.
13. **Popper, H., Sternberg, S. S., Oser, B. C., and Oser, M.,** The carcinogenic effect of aramite in rats. A study of hepatic nodules, *Cancer,* 13, 1035, 1960.
14. **Farber, E.,** Ethionine carcinogenesis, *Adv. Cancer Res.,* 7, 383, 1963.
15. **Epstein, S., Ito, N., Merkow, L., and Farber, E.,** Cellular analysis of liver carcinogenesis: the induction of large hyperplastic nodules in the liver with 2-fluorenylacetamide or ethionine and some aspects of their morphology and glycogen metabolism, *Cancer Res.,* 27, 1702, 1967.
16. **Goldfarb, S.,** A morphological and histochemical study of carcinogenesis of the liver in rats fed 3′-methyl-4-dimethylaminoazobenzene, *Cancer Res.,* 33, 1119, 1973.

17. **Pugh, T. D. and Goldfarb, S.,** Quantitative histochemical and autoradiographic studies of hepatocarcinogenesis in rats fed 2-acetylaminofluorene followed by phenobarbital, *Cancer Res.,* 38, 4450, 1978.
18. **Scherer, E. and Emmelot, P.,** Kinetics of evolution and growth of precancerous liver-cell foci and liver tumor formation by diethylnitrosamine in the rat, *Eur. J. Cancer,* 11, 689, 1975.
19. **Tatematsu, M., Lee, G., Hayes, M. A., and Farber, E.,** Progression in hepatocarcinogenesis: differences in growth and behaviour of transplants of early and late hepatocyte nodules in the rat spleen, *Cancer Res.,* 47, 4699, 1987.
20. **Farber, E.,** The biology of carcinogen-induced hepatocyte nodules and related liver lesions in the rat, *Toxicol. Pathol.,* 10, 197, 1982.
21. **Farber, E., Chen, Z.-Y., Harris, L., Lee, G., Rinaudo, J. S., Roomi, W. M., Rotstein, J., and Semple, E.,** The biochemical-molecular pathology of the stepwise development of liver cancer: new insights and problems, in *Liver Cell Carcinoma,* Bannasch, P., Kappler, D., and Weber, G., Eds., Kluwer Academic, Dordrecht, 1989, p. 273.
22. **Solt, D. B. and Farber, E.,** New principle for the analysis of chemical carcinogenesis, *Nature,* 263, 702, 1976.
23. **Semple-Roberts, E., Hayes, M. A., Armstrong, D., Becker, R. A., Racz, W. J., and Farber, E.,** Alternative methods of selecting rat hepatocellular nodules resistant to 2-acetylaminofluorene, *Int. J. Cancer,* 40, 643, 1987.
24. **Laurier, C., Tatematsu, M., Rao, P. M., Rajalakshmi, S., and Sarma, D. S. R.,** Promotion by orotic acid of liver carcinogenesis in rats initiated by 1,2-dimethylhydrazine, *Cancer Res.,* 44, 2186, 1984.
25. **Pichiri-Coni, G., Coni, P., Laconi, E., Schwarze, P. E., Seglen, P. O., Rao, P. M., Rajalakshmi, S., and Sarma, D. S. R.,** Studies on the mitoinhibitory effect of orotic acid in hepatocytes in primary culture, *Carcinogenesis,* 11, 981, 1990.
26. **Tatematsu, M., Nagamine, Y., and Farber, E.,** Redifferentiation as a basis for remodeling of carcinogen-induced hepatocyte nodules to normal appearing liver, *Cancer Res.,* 43, 5049, 1983.
27. **Moore, M. A., Nakagawa, K., Satoh, K., Ishikawa, T., and Sato, K.,** Single GST-P positive liver cells — putative initiated hepatocytes, *Carcinogenesis,* 8, 483, 1987.
28. **Satoh, K., Hatayama, I., Tateoka, N., Tamai, K., Shimizu, T., Tatematsu, M., Ito, N., and Sato, K.,** Transient induction of single GST-P positive hepatocytes by DEN, *Carcinogenesis,* 10, 2107, 1989.
29. **Cameron, R. G.,** Identification of the putative first cellular step of chemical hepatocarcinogenesis, *Cancer Lett.,* 47, 163, 1989.
30. **Sato, K.,** Glutathione transferases as markers of preneoplasia and neoplasia, *Adv. Cancer Res.,* 52, 205, 1989.
31. **Cameron, R. G.,** Analysis of the biology of initiation using the resistant hepatocyte model, in *Chemical Carcinogenesis: Models and Mechanisms,* F. Feo, Ed., Plenum Press, New York, 1988, 293.
32. **Cameron, R. G.,** Comparison of GST-P versus GGT as markers of hepatocellular lineage during analyses of initiation of carcinogenesis, *Cancer Invest.,* 6, 725, 1988.
33. **Scherer, E. and Hoffmann, M.,** Probable clonal genesis of cellular islands induced in rat liver by diethylnitrosamine, *Eur. J. Cancer,* 7, 369, 1971.
34. **Weinberg, W. C., Berkwits, L., and Iannaccone, P. M.,** The clonal nature of carcinogen-induced altered foci of γ-glutamyltranspeptidase expression in rat liver, *Carcinogenesis,* 8, 565, 1987.
35. **Ogawa, K., Yokota, K.-I., Sonoda, T., Piao, Z. S., Mori, M., and Nagase, S.,** Albumin-negative hepatocytes in Sprague-Dawley x analbuminemic F_1 rats treated with hepatic carcinogens, *Int. J. Cancer,* 41, 727, 1988.

36. **Rabes, H. M., Bücher, T. G., Hartmann, A., Linke, I., and Dünnwald, M.,** Clonal growth of carcinogen-induced enzyme-deficient preneoplastic cell populations in mouse liver, *Cancer Res.,* 42, 3220, 1982.
37. **Williams, E. D., Wareham, K. A., and Howell, S.,** Direct evidence for the single cell origin of mouse liver cell tumours, *Br. J. Cancer,* 47, 723, 1983.
38. **Howell, S., Wareham, K. A., and Williams, E. D.,** Clonal origin of mouse liver cell tumors, *Am. J. Pathol.,* 121, 426, 1985.
39. **Sasaki, T. and Yoshida, T.,** Experimentelle Enzeugung des Lebercarcinoms durch Fütterung mit o-Amidoazooluol, *Virchows Arch. Pathol. Anat. Physiol.,* 295, 175, 1935.
40. **Farber, E.,** Similarities in the sequence of early histological changes induced in the liver of the rat by ethionine, 2-acetylaminofluorene and 3′-methyl-4-dimethylaminoazobenzene, *Cancer Res.,* 16, 142, 1956.
41. **Popper, H., Kent, G., and Stein, R.,** Ductular cell reaction in the liver in hepatic injury, *J. Mt. Sinai Hosp.,* 24, 551, 1957.
42. **Rubin, E.,** Origin and fate of proliferated bile ductular cells, *Exp. Mol. Pathol.,* 3, 279, 1964.
43. **Rubin, E., Masuko, K., Goldfarb, S., and Zak, F. G.,** Role of cell proliferation in hepatic carcinogenesis, *Proc. Soc. Exp. Biol. Med.,* 115, 381, 1964.
44. **Yokoyama, S., Satoh, M., and Lombardi, B.,** Bile ductular cells and the phenotypic heterogeneity of the population of hepatic non-parenchymal epithelial cells induced in rats by chemical carcinogens, *Carcinogenesis,* 7, 1215, 1986.
45. **Lopez, M.,** Proliferazione iperplastiche e neoplastiche delle vie bilifere intraepatiche da tioacetamide, *Experientia,* 12, 185, 1956.
46. **Newberne, P. M., Carlton, W. W., and Wogan, G. N.,** Hepatomas in rats and hepatorenal injury in ducklings fed peanut meal or *Aspergillus flavis* extracts, *Pathol. Vet.,* 1, 105, 1964.
47. **Butler, W. H. and Barnes, J. M.,** Toxic effects of groundnut meal containing aflatoxin to rats and guinea-pigs, *Br. J. Cancer,* 17, 699, 1964.
48. **Bull, L. B., Culvenor, C. C. J., and Dick, A. T.,** *The Pyrrolizidine Alkaloids,* North-Holland, Amsterdam, 1968.
49. **Jago, M. V.,** Factors affecting the chronic hepatotoxicity of pyrrolizidine alkaloids, *J. Pathol. Bacteriol.,* 105, 1, 1971.
50. **Reddy, J. K., Rao, M. S., and Jago, M. V.,** Rapid development of hyperplastic nodules and cirrhosis in liver of rats treated concurrently with thioacetamide and the pyrrolizidine alkaloid lasiocarpine, *Int. J. Cancer,* 17, 621, 1976.
51. **Marasas, W. F. O., Kriek, N. P. J., Fincham, J. E., and van Rensburg, S. J.,** Primary liver cancer and oesophageal basal cell hyperplasia in rats cased by *Fusarium Moniliforme, Int. J. Cancer,* 34, 383, 1984.
52. **Lopez, M. and Mazzanti, L.,** Experimental investigations on alpha-naphthyl-iso-thiocyanate as a hyperplastic agent of the biliary ducts in the rat, *J. Pathol. Bacteriol.,* 69, 243, 1955.
53. **Gillman, J., Gilbert, C., and Spence, J.,** Some factors regulating the structural integrity of the intrahepatic bile ducts with special reference to primary carcinoma of the liver and vitamin A, *Cancer,* 7, 1109, 1954.
54. **Hunt, R. H., Leveille, G. A., and Sauberlich, H. E.,** Dietary bile acids and lipid metabolism. II. The ductular cell reaction induced by lithocholic acid, *Proc. Soc. Exp. Biol. Med.,* 113, 139, 1963.
55. **Hunt, R. D., Leveille, G. A., and Sauberlich, H. E.,** Dietary bile acids and lipid metabolism. III. Effects of lithocholic acid in mammalian species, *Proc. Soc. Exp. Biol. Med.,* 115, 277, 1964.
56. **Hardison, W. G. M.,** Bile salts and the liver, in *Progress in Liver Disease,* Vol. 4, Popper, H. and Schaffner, F., Eds., Grune & Stratton, New York, 1972, 83.

57. **Palmer, R. H. and Hruban, Z.,** Production of bile duct hyperplasia and gallstones by lithocholic acid, *J. Clin. Invest.*, 45, 1255, 1966.
58. **Inaoka, Y.,** Significance of the so-called oval cell proliferation during azo-dye hepatocarcinogenesis, *Gann,* 58, 355, 1967.
59. **Kitagawa, T., Yokochi, T., and Sugano, H.,** α-Fetoprotein and hepatocarcinogenesis in rats fed 3′-methyl-4-dimethylaminoazobenzene or N-2-fluorenylacetamide, *Int. J. Cancer,* 10, 368, 1972.
60. **Ogawa, K. T., Minase, T., and Onoe, T.,** Demonstration of glucose-6-phosphatase activity in oval cells and the significance of oval cells in azo-dye carcinogenesis, *Cancer Res.*, 34, 3379, 1974.
61. **Dempo, K., Chisaka, N., Yoshida, Y., Kaneko, A., and Onoe, T.,** Immunofluorescent study on α-fetoprotein-producing cells in the early stage of 3′-methyl-4-dimethylaminoazobenzene carcinogenesis, *Cancer Res.*, 35, 1282, 1975.
62. **Onoe, T., Kaneko, A., Dempo, K., Ogawa, K., and Minase, T.,** α-Fetoprotein and early histological changes of hepatic tissue in DAB-hepatocarcinogenesis, *Ann. N.Y. Acad. Sci.*, 259, 168, 1975.
63. **Fujita, S., Ishizuka, H., Kamimura, N., Kaneda, H., and Ariga, K.,** The α-fetoprotein-producing cells in the early stage of experimental liver cancer, *Ann. N.Y. Acad. Sci.*, 269, 217, 1975.
64. **Onda, H.,** Immunohistological studies on alphafetoprotein and alpha-acid glycoprotein during azo dye hepatocarcinogenesis in rats, *Gann,* 67, 253, 1976.
65. **Tchipysheva, T. A., Guelstein, V. I., and Bannikov, G. A.,** α-Fetoprotein-containing cells in the early stages of liver carcinogenesis induced by 3′-methyl-4-dimethylaminoazobenzene and 2-acetylaminofluorene, *Int. J. Cancer,* 20, 388, 1977.
66. **Sell, S.,** Distribution of alphafetoprotein and albumin containing cells in the livers of Fischer rats fed four cycles of N-2-fluorenylacetamide, *Cancer Res.*, 38, 3107, 1978.
67. **Shinozuka, H., Lombardi, B., Sell, S., and Iammarino, R. M.,** Early histological and functional alterations of ethionine liver carcinogenesis in rats fed a choline-deficient diet, *Cancer Res.*, 38, 1092, 1978.
68. **Kuhlmann, W. D.,** Localization of alpha-fetoprotein and DNA-synthesis in liver cell population during experimental hepatocarcinogenesis in rats, *Int. J. Cancer,* 21, 368, 1978.
69. **Sell, S., Leffert, H. L., Shinozuka, H., Lombardi, B., and Gochman, N.,** Rapid development of large numbers of α-fetoprotein containing "oval" cells in the liver of rats fed N-2-fluorenylacetamide in a choline devoid diet, *Gann,* 72, 479, 1981.
70. **Wu, P. C., Ma, L., Gibson, J. B., Hirai, H., and Tsukada, Y.,** Serum alpha-fetoproteins in rats after ligation of the common bile duct: relation to ductular cell (oval) proliferation, *J. Pathol.*, 133, 61, 1981.
71. **Sell, S.,** A comparison of oval cells induced by feeding N-2-fluorenylacetamide in a choline devoid diet and bile duct cells induced by feeding 4,4′-diaminodiphenyl-methane, *Cancer Res.*, 43, 1761, 1983.
72. **Evarts, R. P., Nagy, P., Marsden, E., and Thorgeirsson, S. S.,** In situ hybridization studies on expression of albumin and α-fetoprotein during early stages of neoplastic transformation in rat liver, *Cancer Res.*, 47, 5469, 1987.
73. **Evarts, R. P., Nagy, P., Marsden, E., and Thorgeirsson, S. S.,** A precursor-product relationship exists between oval cells and hepatocytes in rat liver, *Carcinogenesis,* 8, 1737, 1987.
74. **Evarts, R. P., Nagy, P., Nakatsukasa, H., Marsden, E., and Thorgeirsson, S. S.,** In vivo differentiation of rat liver oval cells into hepatocytes, *Cancer Res.*, 49, 1541, 1989.
75. **Richards, W. L., Tsukada, Y., and Potter, V. R.,** γ-Glutamyl transpeptidase and α-fetoprotein expression during α-naphthylisothiocyanate-induced hepatotoxicity in rats, *Cancer Res.*, 42, 5133, 1982.

76. **Grisham, J. W. and Hartroft, W. S.,** Morphologic identification by electron microscopy of "oval cells" in experimental hepatic degeneration, *Lab. Invest.,* 10, 317, 1961.
77. **Schaffner, F. and Popper, H.,** Electron microscopic studies of normal and proliferated bile ductules, *Am. J. Pathol.,* 38, 393, 1961.
78. **Rubin, E., Hutterer, F., Danon, G., and Popper, H.,** Deoxyribonucleic acid turnover of specific cells in a heterogeneous population of life span of hepatic ductular cells, *Lab. Invest.,* 12, 657, 1963.
79. **Steiner, J. W. and Carruthers, J. S.,** Electron microscopy of hyperplastic ductular cells in α-naphthyl isothiocyanate-induced cirrhosis, *Lab. Invest.,* 12, 471, 1963.
80. **Grisham, J. W. and Porta, E. A.,** Origin and fate of proliferated ductal cells in the rat: electron microscopic and autoradiographic studies, *Exp. Mol. Pathol.,* 3, 242, 1964.
81. **Sell, S. and Salman, J.,** Light and electron-microscopic autoradiographic analysis of proliferating cells during the early stages of chemical hepatocarcinogenesis in the rat induced by feeding N-2-fluorenylacetamide in a choline-deficient diet, *Am. J. Pathol.,* 114, 287, 1984.
82. **Tatematsu, M., Ho, R. H., Kaku, T., Ekem, J. K., and Farber, E.,** Studies on the proliferation and fate of oval cells in the liver of rats treated with 2-acetylaminofluorene and partial hepatectomy, *Am. J. Pathol.,* 114, 418, 1984.
83. **Dunsford, H. S., Maset, R., Salman, J., and Sell, S.,** Connection of duct-like structures induced by a chemical hepatocarcinogen to portal bile ducts in the rat liver detected by injection of bile ducts and pigmented barium gelatin medium, *Am. J. Pathol.,* 118, 218, 1985.
84. **Makino, Y., Yamamoto, K., and Tsuji, T.,** Three-dimensional arrangement of ductular structures formed by oval cells during hepaticarcinogenesis, *Acta Med. Okayama,* 42, 143, 1988.
85. **Lenzi, R., Liu, M. H., Tarsetti, F., Slott, P. A., Alpini, G., Zhai, W.-R., Paronetto, F., Lenzen, R., and Tavoloni, N.,** Histogenesis of bile duct-like cells proliferating during ethionine hepatocarcinogenesis. Evidence for a biliary epithelial nature of oval cells, *Lab. Invest.,* 66, 390, 1992.
86. **Tavoloni, N.,** The intrahepatic biliary epithelium: an area of growing interest in hepatology, *Semin. Liver Dis.,* 7, 280, 1987.
87. **McLean, M. R. and Rees, K. R.,** Hyperplasia of bile-ducts induced by alpha-naphthyl-isothiocyanate: experimental biliary cirrhosis free from biliary obstruction, *J. Pathol. Bacteriol.,* 76, 175, 1958.
88. **Stewart, H. L. and Snell, K. C.,** The histopathology of experimental tumors of the liver of the rat: a critical review of the histopathogenesis, *Acta Unio Int. Contra Cancrum,* 13, 770, 1957.
89. **Grundmann, E. and Sieburg, H.,** Die Histogenese und Cytogenese des Lebercarcinoms der Ratte durch Diäthynitrosamin im lichtmikroskopischen Bild, *Beitr. Pathol. Anat. Allg. Pathol.,* 126, 57, 1962.
90. **Bannasch, P.,** *The Cytoplasm of Hepatocytes During Carcinogenesis. Electron and Light Microscopical Investigations of the Nitrosomorpholine Intoxicated Rat Liver,* Springer-Verlag, New York, 1968.
91. **Farber, E. and Cameron, R.,** The sequential analysis of cancer development, *Adv. Cancer Res.,* 31, 125, 1980.
92. **Kinosita, R.,** Studies on the carcinogenic chemical substances, *Jpn. Path. Soc. Trans.,* 27, 665, 1937.
93. **Price, J. M., Harman, J. W., Miller, E. C., and Miller, J. A.,** Progressive microscopic alterations in the livers of rats fed the hepatic carcinogens 3′-methyl-4-dimethylaminoazobenzene and 4′-fluoro-4-dimethylaminoazobenzene, *Cancer Res.,* 12, 192, 1952.
94. **Watabe, H.,** Early appearance of embryonic alpha-globulin in rat serum during carcinogenesis with 4-dimethylaminoazobenzene, *Cancer Res.,* 31, 1192, 1971.

95. **Kores, R., Williams, G. M., and Weisburger, J. H.**, Early appearance of serum alpha-fetoprotein during hepatocarcinogenesis as a function of age of rats and extent of treatment with 3′-methyl-4-dimethyl-aminoazobenzene, *Cancer Res.*, 32, 1526, 1972.
96. **Becker, F. F. and Sell, S.**, Early elevation of alpha-fetoprotein in N-′2-fluorenylacetamide hepatocarcinogenesis, *Cancer Res.*, 34, 2489, 1974.
97. **Sell, S. and Leffert, H. L.**, On evaluation of cellular lineages in the pathogenesis of experimental hepatocellular carcinoma, *Hepatology*, 2, 77, 1982.
98. **Lombardi, B.**, On the nature, properties and significance of oval cells, in *Recent Trends in Chemical Carcinogenesis*, Vol. 1, Pani, P., Feo, F., and Columbano, A., Eds., ESA, Cagliari, 1982, 37.
99. **Laws, J. O., Mabille, Ph., Royer, R., and Rudali, G.**, Etude histopathologique des lesions provoquees par le acetylaminofluorene dans differents organs du rat, *Bull. Assoc. Franc. Etude Cancer*, 39, 450, 1952.
100. **Ghoshal, A. K., Mullen, B., Medline, A., and Farber, E.**, Sequential analysis of hepatic carcinogenesis: regeneration of liver after carbon tetrachloride induced necrosis when hepatocyte proliferation is inhibited by 2-acetylaminofluorene, *Lab. Invest.*, 48, 224, 1983.
101. **Sell, S. and Dunsford, H. A.**, Evidence for the stem cell origin of hepatocellular carcinoma and cholangiocarcinoma, *Am. J. Pathol.*, 134, 1347, 1989.
102. **Farber, E.**, Clonal adaptation during carcinogenesis, *Biochem. Pharmacol.*, 39, 1837, 1990.
103. **Tatematsu, M., Kaku, T., Medline, A., Eriksson, L., Roomi, M. W., Sharma, R. N., Murray, R. K., and Farber, E.**, Markers of liver neoplasia — real or fictional?, in *Application of Biological Markers to Carcinogen Testing*, Milman, H. A. and Sell, S., Eds., Plenum Press, New York, 1983, 25.
104. **Engelhardt, N. V., Lazareva, M. N., Abelev, G. I., Urivaeva, I. V., Factor, I. M., and Brodsky, V. Ya.**, Detection of α-fetoprotein in mouse liver differentiated hepatocytes before their progression through S-phase, *Nature*, 263, 146, 1976.
105. **Englehardt, N. V., Baranov, V. N., Lazareva, M. N., and Gusev, A. I.**, Ultrastructural localization of alpha-fetoprotein (AFP) in regenerating mouse liver poisoned with CCl_4. 1. Reexpression of AFP in differentiated hepatocytes, *Histochemistry*, 80, 401, 1984.
106. **Baranov, V. N., Englehardt, N. V., Lazareva, M. N., Gusev, A. I., Yazova, A. K., Shakhlamov, V. A., and Abelev, G. I.**, Atypical cells synthesizing α-fetoprotein in the regenerating mouse liver (transl.), *Bull. Exp. Biol. Med.*, 94, 82, 1982.
107. **Sharma, R. N., Cameron, R. G., Farber, E., Griffin, M., Joly, J.-G., and Murray, R. K.**, Multiplicity of induction of patterns of rat liver microsomal mono-oxygenases and other polypeptides produced by administration of various xenobiotics, *Biochem. J.*, 182, 317, 1979.
108. **Cameron, R., Sharma, R. N., Murray, R. K., and Farber, E.**, Some patterns of response of liver to environmental agents, *Ann. N.Y. Acad. Sci.*, 329, 39, 1979.
109. **Sell, S. and Wepsic, H. T.**, Alpha-fetoprotein, in *The Liver: Normal and Abnormal Function*, Part B, Becker, F. F., Ed., Marcel Dekker, New York, 1975, 773.
110. **Uriel, J.**, Fetal characteristics of cancer, in *Cancer: A Comprehensive Treatise*, Vol. 3, Becker, F. F., Ed., Plenum Press, New York, 1975, 21.
111. **Tournier, I., Legres, L., Schoevaert, D., Feldmann, G., and Bernuau, D.**, Cellular analysis of α-fetoprotein gene activation during carbon tetrachloride and D-galactosamine-induced acute liver injury in rats, *Lab. Invest.*, 59, 657, 1988.
112. **Marceau, N., Blouin, M.-J., Germain, L., and Noël, M.**, Role of different epithelial cell types in liver oncogenesis, regeneration and neoplasia, *In Vitro Cell Dev. Biol.*, 25, 336, 1989.
113. **Dunsford, H. A. and Sell, S.**, Production of monoclonal antibodies to preneoplastic liver cell populations induced by chemical carcinogens in rats and to transplantable Morris hepatomas, *Cancer Res.*, 49, 4887, 1989.

114. **Hixon, D. C., Faris, R. A., and Thompson, N. L.,** An antigenic portrait of the liver during carcinogenesis, *Pathobiology,* 58, 65, 1990.
115. **Ledda, G. M., Sells, M. A., Yokoyama, S., and Lombardi, B.,** Metabolic properties of isolated rat liver cell preparations enriched in epithelial cells other than hepatocytes, *Int. J. Cancer,* 31, 231, 1983.
116. **Burt, B. K., Garfield, S., Johnson, K., and Thorgeirsson, S. S.,** Transformation of rat liver epithelial cells with v-H-*ras* or v-*raf* causes expression of MDR-1, glutathione S-transferase P and increased resistance to cytotoxic chemicals, *Carcinogenesis,* 9, 2329, 1988.
117. **Sells, M. A., Katyal, S. L., Shinozuka, H., Estes, L. W., Sell, S., and Lombardi, B.,** Isolation of oval cells and transitional cells from the liver of rats fed the carcinogen DL-ethionine, *J. Natl. Cancer Inst.,* 66, 355, 1981.
118. **Yoshimura, H., Yokoyama, S., Harris, R., Schwartz, G., Takahashi, S., Sells, M. A., Pan, S. F., and Lombardi, B.,** Anaplastic carcinomas in nude mice and in original donor strain rats inoculated with cultured oval cells, *Am. J. Pathol.,* 110, 322, 1983.
119. **Yaswen, P., Hayner, N. T., and Fausto, N.,** Isolation of oval cells by centrifugal elutriation and comparison with other cell types purified from normal and preneoplastic livers, *Cancer Res.,* 44, 324, 1986.
120. **Sirica, A. E. and Cihla, H. P.,** Isolation and partial characterization of oval and hyperplastic bile ductular cell-enriched populations from the livers of carcinogen and non-carcinogen-treated rats, *Cancer Res.,* 44, 3454, 1984.
121. **Hayner, N. T., Braun, L., Yaswen, P. et al.,** Isozyme profiles of oval cells, parenchymal cells and biliary cells isolated by centrifugal elutriation from normal and preneoplastic livers, *Cancer Res.,* 44, 332, 1984.
122. **Tsao, M.-S., Smith, J. D., Nelson, K. D. et al.,** A diploid epithelial cell line from normal adult rat liver with phenotypic properties of "oval" cells, *Exp. Cell Res.,* 154, 38, 1984.
123. **Tsao, M.-S. and Grisham, J. W.,** Hepatocarcinomas, cholangiocarcinomas and hepatoblastomas produced by chemically transformed cultured rat liver epithelial cells in a light and electron-microscopic analysis, *Am. J. Pathol.,* 127, 168, 1987.
124. **Fausto, N., Thompson, N. L., and Braun, L.,** Purification and culture of oval cells from rat liver, in *Cell Separation — Methods and Selected Applications,* Vol. 4, Pretlow, T. G., II and Pretlow, T. P., Eds. Academic Press, Orlando, FL, 1987, 45.
125. **Braun, L., Goyette, M., Yaswen, P., Thompson, M. L., and Fausto, N.,** Growth in culture and tumorigenicity after transfection with *ras* oncogene of liver epithelial cells from carcinogen-treated rats, *Cancer Res.,* 47, 4116, 1987.
126. **Germain, L., Noël, M., Gourdeau, H. et al.,** Promotion of growth and differentiation of rat ductular oval cells in primary culture, *Cancer Res.,* 48, 368, 1988.
127. **Braun, L., Mikumo, R., and Fausto, N.,** Production of hepatocellular carcinoma by oval cells: cell cycle expression of c-myc and P53 at different stages of oval cell transformation, *Cancer Res.,* 49, 1554, 1989.
128. **Farber, E.,** Hepatocyte proliferations in the stepwise development of experimental liver cell cancer, *Digest. Dis. Sci.,* 36, 973, 1991.
129. **Dunsford, H. A., Karnasuta, C., Hunt, J. et al.,** Different lineages of chemically induced hepatocellular carcinomas in rats defined by monoclonal antibodies, *Cancer Res.,* 49, 4894, 1989.
130. **Mathis, G. A., Walls, S. A., D'Amico, P. et al.,** Enzyme profile of rat bile ductular epithelial cells in reference to the resistance phenotype in hepatocarcinogenesis, *Hepatology,* 9, 477, 1989.
131. **Giambarresi, L. I., Katyal, S. L., and Lombardi, B.,** Promotion of liver carcinogenesis in the rat by a choline-devoid diet: role of liver cell necrosis and regeneration, *Br. J. Cancer,* 46, 825, 1982.

132. **Ghoshal, A. K., Ahluwalia, M., and Farber, E.,** The rapid induction of liver cell death in rats fed a choline-deficient methionine-low diet, *Am. J. Pathol.,* 113, 309, 1983.
133. **Tsuda, H., Lee, G., and Farber, E.,** Induction of resistant hepatocytes as a new principle for a possible short-term in vivo test for carcinogens, *Cancer Res.,* 40, 1157, 1980.
134. **Tsuda, H. and Farber, E.,** Resistant hepatocytes as early changes in liver induced by polycyclic aromatic hydrocarbons, *Int. J. Cancer,* 25, 137, 1980.
135. **Farber, E. and Tsuda, H.,** Induction of a resistant preneoplastic liver cell as a new principle for a short-term assay in vivo for carcinogens, in *Short-Term Test for Chemical Carcinogens,* Stich, H. F. and San, R.-H. C., Eds., Springer-Verlag, New York, 1981, 372.
136. **Tatematsu, M., Murasaki, G., Nakanishi, K., Miyata, Y., Shinohara, Y., and Ito, N.,** Sequential quantitative studies in hyperplastic nodules in the liver of rats treated with carcinogenic chemicals, *Gann,* 70, 125, 1979.
137. **Ito, N., Tatematsu, M., Nakanishi, K., Hosegawa, R., Takano, T., Imaida, K., and Ogiso, T.,** The effects of various chemicals on the development of hyperplastic liver nodules in hepatectomized rats treated with N-nitrosoethylamine or N-2-fluorenylacetamide, *Gann,* 71, 832, 1980.
138. **Van der Heijden, C. A., Dormans, J. A. M. A., and Van Nesselrooij, J. H. U.,** Short-term induction of preneoplastic nodules in the rat liver. I. Role of 2-AAF as selecting agent, *Eur. J. Cancer,* 16, 1389, 1980.
139. **Van der Heijden, C. A. and Dormans, J. A. M.,** Short-term induction of neoplastic nodules in the liver. II. Study of their development and effects of withdrawal of 2-acetylaminofluorene, *Carcinogenesis,* 2, 147, 1981.
140. **Leonard, T. B., Dent, J. G., Graichen, M. E., Lyght, O., and Popp, J. A.,** Comparison of hepatic carcinogen and initiator-promoter systems, *Carcinogenesis,* 3, 851, 1982.
141. **Möller, L., Torndal, U.-B., Eriksson, L. C., and Gustafsson, J. A.,** The air pollutant 2-nitrofluorene as initiator and prmoter in a liver model for chemical carcinogenesis, *Carcinogenesis,* 10, 435, 1989.
142. **Rao, M. S., Lalwani, N. D., and Reddy, J. K.,** Sequential histologic study of rat liver during peroxisome proliferator [4-chloro-6-(2,3-xylidino)-2-pyrimidinylthio]-acetic acid (WY-14,643)-induced carcinogenesis, *J. Natl. Cancer Inst.,* 73, 983, 1984.
143. **Greaves, P., Irisarri, E., and Monro, A. M.,** Hepatic foci of cellular and enzymatic alteration and nodules in rats treated with clofibrate or diethylnitrosamine followed by phenobarbital: their rate of onset and their reversibility, *J. Natl. Cancer Inst.,* 76, 475, 1986.
144. **Sarma, D. S. R., Rao, P. M., and Rajalakshmi, S.,** Liver tumor promotion by chemicals: models and mechanisms, *Cancer Surv.,* 5, 781, 1986.
145. **Sells, M. A., Katyal, S. L., Sell, S., Shinozuka, H., and Lombardi, B.,** Induction of foci of altered γ-glutamyltranspeptidase-positive hepatocytes in carcinogen-treated rats fed a choline-deficient diet, *Br. J. Cancer,* 40, 274, 1979.
146. **Shinozuka, H., Sells, M. A., Katyal, S. L., Sell, S., and Lombardi, B.,** Effects of a choline devoid diet on the emergence of γ-glutamyltranspeptidase-positive foci in the liver of carcinogen-treated rats, *Cancer Res.,* 39, 2515, 1979.
147. **Laconi, E.,** Studies on Rat Liver Tumor Promotion by Orotic Acid, Ph.D. thesis, University of Toronto, 1988.
148. **Mikol, Y. B., Hoover, K. L., Creasia, D., and Poirier, L. A.,** Hepatocarcinogenesis in rats fed methyl-deficient, amino acid defined diets, *Carcinogenesis,* 4, 1619, 1983.
149. **Ghoshal, A. K. and Farber, E.,** The induction of liver cancer by dietary deficiency of choline and methionine without carcinogens, *Carcinogenesis,* 5, 1367, 1984.
150. **Yokoyama, S., Sells, M. A., Reddy, T. V., and Lombardi, B.,** Hepatocarcinogenesis and promoting action of a choline-devoid diet in the rat, *Cancer Res.,* 45, 2834, 1985.

151. **Rushmore, T. H., Ghazarian, D. M., Subrahmanyan, V., Farber, E., and Ghoshal, A. K.,** Probable free radical effects on rat liver nuclei during early hepatocarcinogenesis with a choline-devoid low methionine diet, *Cancer Res.*, 47, 6731, 1987.
152. **Reuber, M. D.,** Development of preneoplastic and neoplastic lesions of the liver in male rats given 0.025 percent N-2-fluorenyldiacetamide, *J. Natl. Cancer Inst.*, 34, 697, 1965.
153. **Teebor, G. W. and Becker, F. F.,** Regression and persistence of hyperplastic hepatic nodules induced by N-2-fluorenylacetamide and their relationship to hepatocarcinogenesis, *Cancer Res.*, 31, 1, 1971.

Chapter 2

CLONAL ANALYSIS OF HEPATIC NEOPLASMS BY MOSAIC PATTERN

W. C. Weinberg, Y. K. Ng, and P. M. Iannaccone

TABLE OF CONTENTS

ISBN 0-8493-4746-7
© 1992 by CRC Press, Inc.

I. INTRODUCTION

An understanding of the causes and controls of cancer development requires clarification of the number of target cells contributing to the tumor. A unicellular origin of cancer was suggested as early as 1914 by Boveri,[1] whose microscopic observations of atypical mitotic activity led him to propose that abnormal chromosome content in occasional cells might result in unlimited growth potential. Bauer[2] encompassed and extended this view with the proposal that a rare gene mutation might be responsible for cancerous growths. His theory of somatic mutation has been supported by the discovery of single base changes in cellular genes capable of imparting a transformed phenotype.[3] At variance with the concept of clonal expansion of rare cells was the field theory proposed by Willis,[4] to whom the gradual transition observed histologically between a cancerous growth and its normal surrounding tissue was more suggestive of recruitment of normal cells in combination with cell multiplication. Such cooperation might be achieved by mechanisms including lateral transmission of a cancer-causing virus, intercellular transmission of information, e.g., paracrine signaling, and/or physiological defects in control mechanisms.

II. CLONAL ANALYSIS OF HUMAN TUMORS

Cellular mosaicism within an individual has been utilized to determine the clonal origins of several human tumors. The composition of tumors in a target organ with at least two genetically distinct cell populations can be evaluated with respect to the proportion of the different cell populations present in a tumor. A tumor multiclonal in origin would be expected to contain each cell type represented in the normal organ. Likewise, a monoclonal tumor should have only one cell type present. Much evidence regarding tumor clonality has accumulated from examining the composition of tumors in individuals displaying cellular mosaicism.

Substantial evidence that human spontaneous tumors arise from a single cell is derived principally from four cell lineage marker systems: immunoglobulin (Ig) expression; karyotype analysis; isozyme determination; and analysis of restriction fragment length polymorphisms (RFLP). The first system employed to this end was that of Ig mosaicism. As B lymphocytes become committed to differentiate, rearrangement of the Ig genes occurs, so that Ig from individual B lymphocytes can be distinguished on the basis of light-chain class (kappa or lambda), heavy-chain class, and idiotypic specificity. The uniform increased expression of specific Ig subclasses in patients with Burkitt lymphoma and multiple myeloma provides strong evidence that clonal proliferation of cells expressing a specific rearrangement of Ig genes has occurred.[5-7]

The most commonly cited karyotypic evidence of tumor clonality is based on the observation of Nowell and Hungerford[8] in 1960 that most chronic myelogenous leukemia (CML) cells, while containing a normal chromosome content, had an unusually small chromosome. This was later identified as a reciprocal translocation between chromosomes 9 and 22, involving the c-abl oncogene.[9,10] This "Philadelphia chromosome" has been identified in 85% of CML cases. Other translocations seen consistently, in which oncogenes have been mapped to chromosomal breakpoints, include the 8;21 and 8;14 translocations involving c-mos and c-myc, in acute myloblastic leukemia and Burkitt lymphoma, respectively.[11,12]

The most widely applied system for determining cell lineage in both experimental models and human spontaneous tumors has been isoenzyme analysis. In eutherian females, X chromosome inactivation occurs in the embryo proper at the time of blastocyst formation, so that only one X chromosome will be active in a given somatic cell. X chromosome inactivation occurs randomly and heritably; therefore, in a female heterozygous for an X-linked gene product, approximately 50% of her cell population will express the product of the paternally derived X and 50% will express that of the maternally derived X chromosome.[13] The distinct cell populations are distributed according to the position of cells at the time of X chromosome inactivation and cell migration patterns. Leiomyomas were examined in five females heterozygous for the X-linked enzyme glucose-6-phosphate dehydrogenase (G6PD).[14] While 85/86 normal tissue samples assayed from these individuals were shown by starch gel electrophoresis to contain cells of both the fast migrating isozyme, A, and the slow migrating variant, B, all 27 tumor samples studied expressed only a single isozyme variant. Since tumors of each variant were present within a given patient, both cell types seem equally capable of proliferation, and there does not appear to have been selection for one cell type over another within an individual. A number of other human tumors found to be clonal using G6PD analysis include Burkitt lymphoma,[7] carcinoma of the cervix,[15] and several pediatric embryonal tumors.[16]

Another method for distinguishing the presence of maternal vs. paternal X chromosomes in X-linked mosaics employs endonuclease digestion of DNA and analysis of resulting fragments (RFLP analysis). RFLP patterns have demonstrated monoclonal origins of several human tumors, including acute myelogenous leukemia, leiomyoma, and Wilms' tumor.[17]

III. EXPERIMENTAL MOSAICS AS MODELS FOR THE STUDY OF TUMOR CLONALITY

Several experimental systems utilizing mosaic pattern analysis have been applied to the study of cellular origins of chemically induced tumors, as well as normal development. Mice mosaic for isozymes of the X-linked enzymes phosphoglycerate kinase (PGK) and ornithine carbamoyl transferase (OCT) provide animal models similar to the human X-linked heterozygote.

Experimental mosaicism can be produced by aggregation of preimplantation embryos derived from two genetically distinct cell populations. The term chimera refers to an individual composed of cells derived from more than one fertilized egg. Experimental primary chimeras were first described in mammals in the early 1960s by Mintz[18] and Tarkowski,[19] working independently. Morula-stage embryos from different strains of mice were aggregated and transferred into the uteri of recipient pseudopregnant females and carried to term. The resulting offspring had cells of each parental strain represented in each tissue. With the proper marker system, organ formation within these animals can be analyzed with respect to cell migration and clonal proliferations. Isozymes of glucosephosphate isomerase (GPI) have been defined which are electrophoretically distinguishable. Chimeras produced between homozygous mouse strains differing in such isozyme expression serve as an experimental model similar to the X-linked PGK mosaic. Electrophoretic analysis of normal tissue samples from these animals demonstrates the presence of both variants of the enzyme.[20]

In a study of the clonality of carcinogen-induced skin tumors, 93 primary tumors were examined from 21 mouse chimeras mosaic for GPI isoenzyme expression.[20] Of 96 tumor samples analyzed, only 7 showed both variants present. The tumors included 78 benign papillomas, 7 squamous cell carcinomas, 1 basal cell carcinoma, and 7 fibrosarcomas. None of the fibrosarcomas were mixed.[20] Similar studies of fibrosarcomas, using heterozygous mice mosaic for PGK isozyme expression, yielded conflicting results. In two studies,[21,22] a total of 12 fibrosarcomas were induced and each was found to contain both isozyme variants. In contrast, the majority of tumors induced, rhabdomyosarcomas, displayed a single isozyme in 80% of cases. A further study of fibrosarcomas from this laboratory explored the possibility that the mixed genotype was due to contaminating cells within the samples.[23] In this study, 82 samples of 10 tumors were examined; 7 of the tumors contained mixed populations upon direct examination, but after early passaging 6 were seen to be exclusively composed of cells containing the electrophoretically slow migrating variant (PGK-1B), 3 contained only the fast migrating variant (PGK-1A), and 1 was mixed. The tumor with both isozymes present was noted before analysis to be shaped like an hourglass, and sampling patterns suggest it represented a coalescence of two tumors. The cell composition of several tumors was then examined histologically.[24] Mouse chimeras were produced containing cells of *Mus musculus* and *M. caroli,* two species of mouse with differences in heterochromatin recognizable with a cDNA probe, and in PGK expression. While electrophoretic isozyme analysis suggested mixed phenotype tumors, *in situ* hybridization demonstrated that the second genotype was due to normal infiltrating cells.

Thus the disadvantage inherent in most of the isozyme studies is the tissue homogenization required for sample analysis requiring loss of architectural integrity, and underlines the importance of careful histological examination of the tissues under study. More recent systems employing histological markers are discussed below.

IV. LIVER EMBRYOLOGY AND REGENERATION: INSIGHTS FROM CHIMERIC PATCH ANALYSIS*

Clonal development refers to a period of growth following the stage of cell commitment when proliferation predominates over cell migration and daughter cells remain in contact. Mosaics have been used to identify periods in organogenesis during which cell migration or clonal growth occurs. Following clonal growth, cell distribution with regard to lineage within a chimera will no longer be random. Clusters of cells of shared genotype in a mosaic system are called patches; patch size is related to the proportion of components within a tissue. The juxtaposition of clones of like genotype, and the dispersion of clones into subclones, complicate the analysis of cell migratory patterns. The relationship between patch size in mosaics and that observed in randomly generated patterns by computer modeling has been used to determine clone size and the generation of clonal growths.[25-27] Based on observations of cuticular mosaicism in Drosophila it has been argued that if the patch size observed in mosaics is larger, clonal growth has occurred. The extent of the size differences observed could reflect the number of generations passed since movement ceased. Factors to be considered in such analyses include preferential sorting of cells of like genotype and the assumption that all cells divide at a constant rate.[28-30] Such studies require a marker system appropriate for histological analysis.

In studies in this laboratory, congenic strains of PVG rats differing in the MHC were used as embryo donors. Morulae-stage embryos were collected from oviducts of PVG and PVG-$RT1^a$ females. The embryos were treated with acidified Tyrode's solution to remove the surrounding zonae pellucidae, and pairs of embryos of the two strains were juxtaposed along the flat surfaces (Figure 1). Following 18 to 24 h of incubation to allow the cells to intermingle, the resulting embryos were transferred into the oviducts of pseudopregnant recipient Sprague-Dawley females. Pups born approximately 21 d later were identified as chimeric based on FACS analysis of blood with rat monoclonal antibodies directed to the class I antigen products $RT1A^a$ and $RT1A^c$. These antibodies were conjugated to ^{125}I or to bacterial glucose oxidase, and binding of the antibody within frozen tissue sections was visualized by autoradiography or by histochemistry. Frozen sections of the parental animals demonstrated clear distinctions in reactivity with antibody to the $RT1A^a$ determinant.[27,31-33] As well as displaying blood chimerism, histological evalu-

* Portions of the following section are derived from a discussion which appears in greater detail in Y.K. Ng and P.M. Iannaccone, *Curr. Top. Dev. Biol.*, 27, 235, 1992.

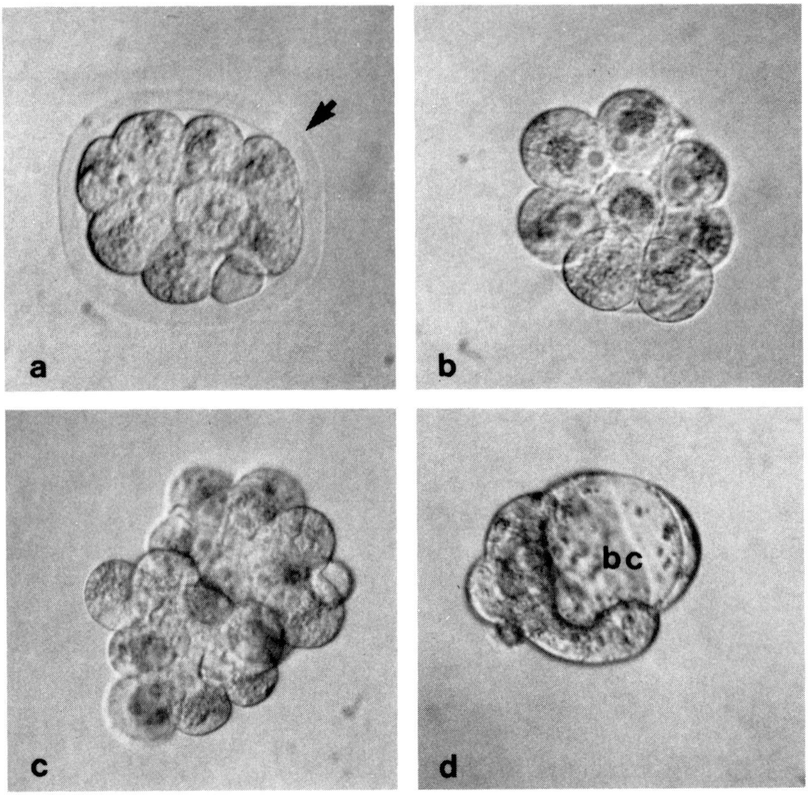

FIGURE 1. Photomicrographs of living preimplantation rat embryos. (a) Eight cell conceptus surrounded by zona pellucida (arrow); (b) the zonae are removed by brief exposure to acidified Tyrode's solution; (c) one embryo of each haplotype (PVG-$RT1^a$ and PVG-$RT1^c$) are pushed together to form the chimera; (d) following 18 h of culture, the embryo develops into a blastocyst with a well-formed blastocyst cavity (bc). Hoffman contrast modulation optics; the embryos are approximately 100 μm in diameter.

ation of multiple organs demonstrated that each of these animals was an amalgam of the two cell lineages present initially.

Considerable effort has gone into the analysis of patches in chimeras over the years.[34-37] The most common approach to determining the size of patches in tissue on the basis of biochemical markers was to establish the binomial distribution of proportion of isozyme activity. Although limited by sampling error and other factors, this approach has been used by us to establish a similarity in epidermal patch size between X chromosome-linked mosaic animals and chimeric animals.[38] The importance of *in situ* methodology in mosaic patch analysis has grown in recent years (see reviews in References 26, 36). The obvious advantage of *in situ* analysis is that it allows direct examination of patterns within tissues of visceral organs. As these patterns are the remnants of the primary embryologic processes (cell movement, cell division, and cell death)[39] which formed them, they provide an independent marker of these primary processes. The principal disadvantage is that mosaic pattern analysis is retrospective.

Mosaic pattern comprises patches which are formed in a dynamical, oscillatory manner. Cell division will place daughter cells of primordia in adjacent positions; cell division among neighboring cells, some from different marker lineages, will cause placement of these cells within a patch of the other marker lineage. This causes fragmentation of the patch. Further division can cause cells previously split away from a patch to rejoin it. Thus, the observation of a patch at any given point in time is the result of clonal expansion, fragmentation, and reaggregation. The concept that finely variegated patch patterns imply extensive cell movement is the basis of numerous studies in which patch size is related either to the number of founder cells or to the time of immobilization of cells. Computer simulation of the process of patch formation indicates the pattern may be independent of both primordial cell number and cell movement.[27] On the other hand, patch pattern is dependent on proportion of cell types in the mosaic and on cell division rules.

Extensive iteration of cell division and placement can result in a highly complex mosaic pattern. These iterations produce statistically self-similar objects (those in which a portion of the object is similar, but not identical to the entire object). Such statistically self-similar objects are often fractal. We have recently proven that liver patches in chimeric rats are fractal[33] (Figure 2). A fractal object is one in which scale-dependent behavior with respect to parameters such as perimeter length occurs. That is, the closer one looks, the more perimeter length one finds even if a correction for the scale of observation is made. This unsettling property is most intuitively understood for coastlines, where the length when determined, for example, from a satellite cannot predict the much greater length and detail found when the same coastline is viewed from an airplane. Neither predicts the much greater length or detail observed by walking along the same coastline. A mathematical consequence of this is that morphometric parameters such as perimeter length, area, or volume of a fractal object cannot be known absolutely, as they are dependent on the scale of observation.[33,40]

The mathematic definition of a fractal requires that for any parameter, measurement of the topologic dimension of the parameter is less than the fractal dimension of the object. The topologic dimension for a line is 1, for a plane it is 2, and for a shape it is 3. The fractal dimension has been proven to be identical to the capacity dimension and to the Hausdorf-Besicovitch dimension.[41] The value of the fractal dimension can be estimated using the equation

$$[D_F = \frac{\log N}{\log 1/m}\ ;\ \text{since } D_C \equiv D_H = \lim_{\epsilon \to 0} \frac{\log N(\epsilon)}{1/\epsilon}]$$

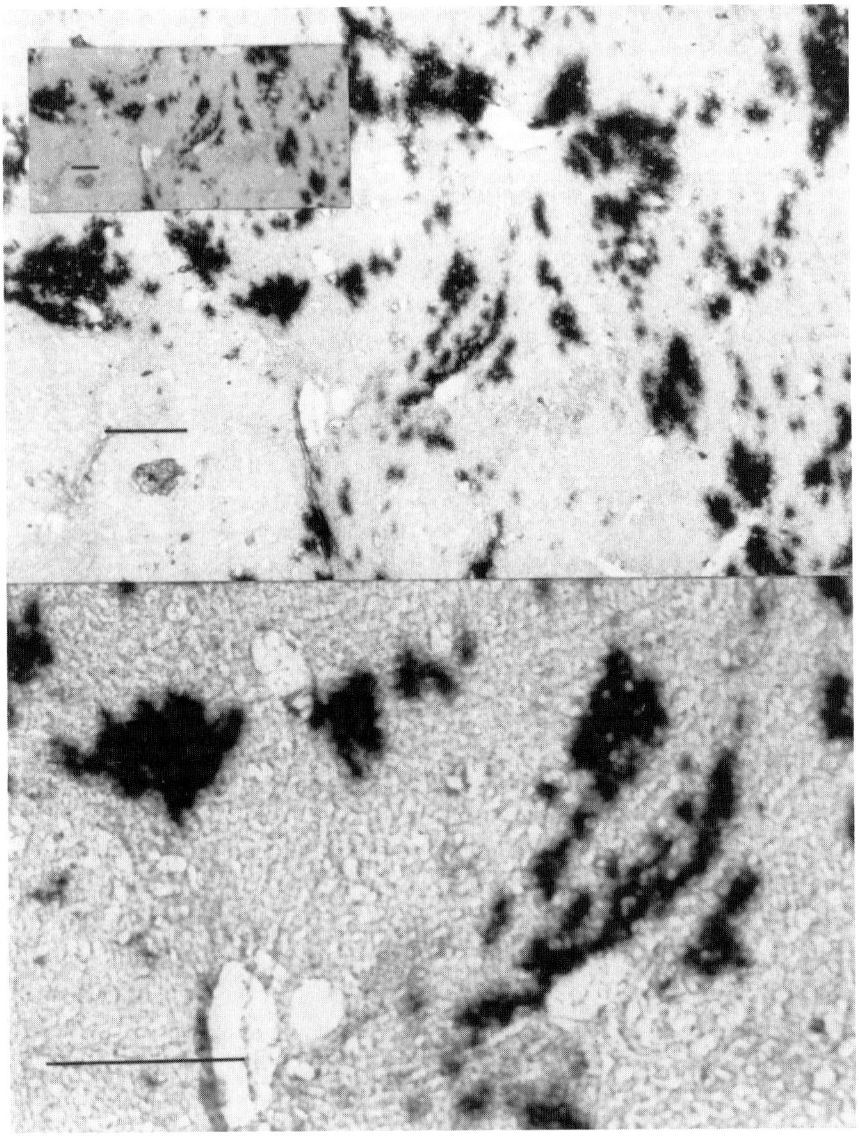

FIGURE 2. Photomicrographs of patches from normal chimeric rat liver taken at different magnifications. The perimeter or area of the black patches (cells of PVG-*RT1*[a] lineage) was used to establish fractal dimension as described in the text. When one cell type (in this case PVG-*RT1*[c]) is more prevalent, then the patches appear as islands in a sea of the other cell type. The bar represents 0.25 mm. (Reprinted from P. M. Iannaccone, *FASEB J.*, 4, 1508, 1990. With permission.)

where D_F is the fractal dimension; N is the uncorrected length of a curve, area of a surface, or volume of a shape determined at various scales of observation (m); D_C is the capacity dimension; D_H is the Hausdorf-Besicovitch dimension; ϵ is the size of circles covering a surface; and $N(\epsilon)$ is the number of these circles required to cover the surface.

The degree of complexity of a fractal, that is, the extent to which detail is nested within detail, can be estimated by examining the objective at several scales of observation. The fractal dimension is estimated from the slope of the log of perimeter length (or some other topologic parameter) plotted against the log of the reciprocal of the scale of observation.[33,40] The greater the fractal dimension the greater the amount of information obtained by higher-scale observation. Further, geometric complexity of liver patches might increase or decrease as a result of cell division patterns which produce these patches. This would result in an increase or a decrease in the fractal dimension. In a cell division program in which the geometric complexity of the patch remained the same, the fractal dimension would not change. This was the result obtained when the fractal dimensions of patches in regenerating and nonregenerating liver were determined. The fractal dimension of patches did not change during compensatory growth. The result is consistent with probabilistic self-similar division rules causing the increase in tissue mass.[27,33,42]

The oscillatory manner by which patches form can be observed in the regenerating liver where patch size increases until the end of the first wave of mitosis when it achieves a steady state.[42] The fractal dimension of patches, however, is unchanged. This behavior is consistent with oscillatory growth by clonal expansion, patch fragmentation, and patch reaggregation. Liver regeneration is an ideal model in which to study the process of patch generation in normal growth, since it has been clearly shown to proceed as uniform and self-limited growth.[43-45] Our results suggest that the repetitive application of simple division rules (e.g., in the most simple case cells divide in a randomly chosen order and the daughter cell of each division is placed in a randomly chosen adjacent position) can produce a complex and realistic mosaic patch pattern. Whether or not such iterating self-similar division rules operate in normal organogenesis of the liver could be established by following marked clones in the embryonic liver (self-similar division rules are those which create objects in which any part is similar, but not identical to the whole). This could be accomplished with transgenic rats expressing a β-galactosidase marker. Chimeras between strains that express the *lac Z* gene and those that do not will allow the kinetic analysis of patch generation within the developing fetus and allow a direct test of the hypothesis that simple iterating division rules are sufficient for the formation of parenchymal mass in this organ. Establishing the validity of such division rules is important as it suggests a way in which complex positional information can be stored more efficiently than using individual genes to establish the correct position at the correct time of all cells in all organs.[46,47]

The presence of cells from both lineages in chimeric organs, even when the animal is unbalanced with respect to the proportion of cells from the two lineages, indicates that extensive cell movement occurs early in embryogenesis and that the allocation of cells to organ primordia is probabilistic.[25] Computer simulation models of cell division, following allocation to organ primordia, in a population of cells comprising two distinguishable lineages suggest that the iteration of simple cell division programs is sufficient to generate complex mosaic patterns.[27,37] These programs do not invoke cell movement, and the results from analysis of the simulations predict that once organ primordia are allocated, extensive cell movement is not required for organ development. The simulation results suggest that the growth of tissue would result in an increase in patch number without dramatic increases in patch area.

We have created and analyzed computer simulations of the formation of patches utilizing both mainframe computers for two-dimensional analysis[25,27] and in three dimensions using supercomputers.[48] The results of these studies suggest that, at least to a first approximation, iterating division rules produce a realistic mosaic pattern. Since the findings were similar based on several simple assumptions not involving cell migration, we conclude that the cell positioning observed in chimeric rat liver can be achieved by random decisions regarding daughter cell positioning and no cell migration.[25,27,37]

The generation of mosaic pattern during organ formation proceeds in three-dimensional space and in time. Although the complexity of this process seems overwhelming, the simplicity with which fractal geometry can be generated suggests an analytical approach. It is therefore necessary to establish that the patches generated are fractal in three dimensions. To do this one must perform extensive three-dimensional reconstructions and to determine the fractal dimensions of these objects based on volume measurements. In simulated growth, however, we have found that mosaic patches are fractal in three dimensions.[48]

V. CLONALITY OF TUMORS

The cellular origins of tumors have been studied in mosaic tissues by many investigators over the past 2 decades. The general conclusion of these studies is that tumors arise by clonal expansion of rare, genetically altered cells. Tumorigenesis is a multistep process, and if any of the steps are rare events then tumors appear to be clonal growths. There are many reviews describing the theories of clonal expansion of genetically altered cell populations in neoplasia.[30,49]

Premalignant lesions have been analyzed as well. *In situ* carcinomas were revealed as clonal[50] in experiments with chimeric mice designed to study colonic epithelial dysplasia. Papillomas induced in chimeric mice were shown to be clonal using GPI as a marker of mosaicism.[20] We obtained similar results utilizing X-linked mosaic animals[30,51] and interspecific chimeras.[24] While other observations suggested that methylcholanthrene- and benzpyrene-induced skin papillomas were mixed tumors, indicating they had arisen from large numbers of cells,[52] the experiments were performed under conditions that resulted in a mixture of neoplastic and nonneoplastic tissues. Fibroconnective core tissue is the result of a host response to the overgrowth of epithelial tissue and will therefore have the same mosaic proportion as other normal tissues in the animal. Separating these components demonstrated that the epithelial neoplastic portion of these tumors was clonal, composed of a single population, either PGK-1A or PGK-1B, while the fibroconnective core of the tumor was composed of both.[51] Another problem was to establish whether or not inflammatory cells could cause the tumors to be mistakenly concluded to be mixed when in fact they were clonal. We achieved this by inducing fibrosarcomas in interspecific chimeras produced between *M. musculus* and *M. caroli*. These species vary in highly repetitive satellite DNA. cDNA probes to *M. musculus* satellite sequences were used to perform *in situ* hybridization. This allowed us to unequivocally determine the marker lineage of both neoplastic fibrocytes and host inflammatory cells. The results were very clear — host inflammatory cells came from both marker lineages within the same tumor, while the neoplastic cells of a given tumor were derived from only a single lineage.[24]

More recently, Winton et al.[53] described studies with an *in situ* marker of mosaicism and examined chemically induced skin tumors in chimeric mice. The authors reported numerous mixed tumors, implying a nonclonal origin; however, the lesions studied were not papillomas, but interepithelial hyperplasias. No evidence from well-formed papillomas was presented. Indeed it has been previously established that hyperplasias are nonclonal. The best-studied system is parathyroid hyperplasia, shown to be nonclonal in chimeras, whereas parathyroid adenomas (capable of autonomous growth) are clonal.[54,55]

VI. CAUTIONS ON THE APPLICATION OF MOSAIC PATTERN ANALYSIS AS A SYSTEM FOR EVALUATING THE CLONAL ORIGIN OF TUMOR FORMATION

Several issues must be considered in applying experimental chimeras to the study of tumor clonality. (1) The degree of variegation observed in target tissues; the lesion under study must be large relative to the observed patch size. This depends in part on a lack of differential sorting of the cellular components of the tissue early in embryonic development. Generally, the proportion of the two contributing components in a chimera is not as uniform within and between animals as in X-linked mosaics, and it has been argued that the patch size is larger in chimeras.[56] This would increase the likelihood of only one component being identified within a tumor even if it arose from multiple cells.[57] In addition, an unbalanced chimera may mask tumors derived from many cells unless the tumor contains the minor cell type. A study from this laboratory examined patch size in small samples of isolated epidermis of X-linked PGK mosaics and has found it to be at least as large as previously reported in chimeras.[38,58] (2) The presence of contaminating populations of cells in samples analyzed for genotype, particularly in isozyme analysis requiring dissection and homogenization of samples; the application of a histological marker capable of distinguishing between parental genotypes allows direct analysis of infiltrating cells and simplifies the interpretation. (3) Differential susceptibility of the parental components to tumor induction and/or differential selection of a particular component by increased cell proliferation; the use of congenic strains differing at a single gene locus offers decreased likelihood of preferential cell sorting during development and resulting larger patch size, as well as diminishing the likelihood of differential susceptibility to carcinogenesis. In our study in PVG rats, parental strains were demonstrated to respond similarly to diethylnitrosamine (DEN) treatment, and foci within chimeras appeared at a frequency paralleling the representation of parental components within the liver. This suggests equal susceptibility of the cellular components to the carcinogenic protocol. (4) Stability of the marker system through the carcinogenic process; marker stability was analyzed in γ-glutamyl transpeptidase (GGT) lesions of PVG parental strains, and multiple cancers involving three organ systems. There was no evidence of significant modulation of the lineage marker in response to the carcinogenesis protocol.

VII. CLONALITY OF EXPERIMENTAL LIVER TUMOR FORMATION

As described previously, many clinically overt tumors are seen to be monoclonal. In view of the multistep nature of the carcinogenic process, attempts have been made to identify steps in the sequence leading to neoplasia. Rat liver offers an ideal model for the study of sequential alterations leading to carcinoma formation.

Following treatment of rat with hepatocarcinogens, but prior to the onset of hepatocellular carcinoma, a variety of alterations are seen in the liver.[59-62] Islands of hepatocytes displaying aberrant expression of several enzymes are believed to represent expanding colonies of initiated cells, some of which remodel, others fated to progress to carcinomas. The earliest of these lesions, identifiable only by special stains, include foci of cells with increased expression of GGT and decreased expression of glucose-6-phosphatase (G-6-Pase) and adenosine triphosphatase (ATPase), as well as decreased glycogen mobilization. Larger lesions (e.g., basophilic foci and hyperplastic nodules) are evident with routine histology. It is believed that these changes represent a continuum in the neoplastic process; while the majority appear to remodel and disappear, occasional foci may progress to form cancers.[62,63] In the context of a multistep process of carcinogenesis, it is necessary to determine the earliest recognizable step at which clonal selection occurs.

The clonal origin of liver tumors was studied by Howell et al.[64] using mice mosaic for the X-linked sparse-fur (Spf) gene locus. This gene codes for variants of the urea cycle enzyme OCT, which can be distinguished histochemically. In Spf mosaics, 327 tumors, examined 12 months after induction by DEN and ranging in size from microscopic to 20 mm in diameter, were analyzed; no evidence of more than one genotype present in a given tumor was seen. Small tumors also showed evidence of only one component cell population, suggesting the uniformity of larger tumors was not due to the loss of the second component as the tumor increased in size. Rabes et al.[65] looked at earlier lesions in X chromosome-linked PGK mosaic mice induced by long-term feeding of 2-acetylaminofluorene. ATPase-deficient foci were located on stained tissue sections; samples corresponding to these foci were punched out from adjacent unstained 40-μm-thick sections. Larger preneoplastic nodules (1 mm wide) and three tumors larger than 3 mm were visualized directly. A sensitive microfluorometric assay was developed and applied following gel electrophoretic separation of the samples, so that samples as small as 3 μg, averaging 200 hepatocytes, could be assayed. In this way, 39/40 preneoplastic nodules were determined to be pure populations of cells expressing either PGK-1A or PGK-1B variants. The study is well designed for evaluation of larger preneoplastic nodules; however, difficulties arise due to the small size of ATPase-deficient foci and the indirect sampling method.

We undertook a study to determine the clonal origin of early lesions induced by carcinogen treatment, using direct histological analysis of chimeras between congenic PVG rat strains, as previously described. Lesions were induced in chimeric rats following the protocol of Pitot et al.,[60] resulting in large numbers of enzyme-altered foci and enhanced production of foci with multiple phenotypic changes. Chimeric and control rats, 10 weeks old, were subjected to partial hepatectomy and liver lesions were induced by a single initiating injection (20 mg/kg) of DEN after 24 h followed by long-term promotion with phenobarbital (0.05% in a high-casein diet).

The animals were fasted overnight before sacrifice at 13 to 14.5 months following initiation of the protocol, so periodic acid-Schiff (PAS) staining would reveal areas of decreased glycogen mobilization. Serial sections were stained for H&E, deficiencies in ATPase and G-6-Pase activity, increases in GGT activity, and decreased glycogen mobilization as determined by staining reaction. These sections were stacked against each other and one was reacted with antibody to determine cell distribution of genotype. A range of chimerism was observed in the rats in this study. Reactivity of erythrocytes from these animals with antibody to $RT1A^a$ ranged from 7.5 to 70%. Liver chimerism was established by patch analysis of the hepatectomy specimen. The distribution of genotype components in the liver closely reflected the proportions seen in other organs as well as in peripheral blood. Patch analysis[66,67] of the animals in the hepatectomy specimen demonstrated patches ranging in size from 100 to 5500 μm.[2]

Differential responsiveness of the two parental strains to DEN was ruled out by following weights of animals injected up to three times with 150 to 200 mg/kg i.p.; the strains displayed similar response in terms of weight loss. This similarity in responsiveness to DEN-induced carcinogenesis was confirmed by the observation that the number of clonal lesions induced in chimeras of each genotype was proportional to the distribution of each genotype to the liver. Stability of the histological marker was examined in parental strain rats treated i.p. with high doses of DEN. In three organ systems of nine animals, 14 cancers appeared, including lung, kidney, and liver, and were screened with expression of the class I antigen determinant used as a genotype marker. The expression of $RT1A^{a}$ antigen in these cancerous tissues was also appropriate to the strain. In addition, enzyme-altered hepatocellular foci were induced in control rats and liver samples were examined for induction of enzyme-altered foci. In these control animals, 615 foci of GGT-positive cells were examined for marker expression, 366 in $RT1^{a}$, and 249 in PVG strain. The lesions were distributed across several size categories, some reaching greater than 2.0 mm in diameter. Of 615 foci, only a single GGT-positive focus in a PVG-$RT1^{a}$ rat demonstrated a diminished, but not absent, expression of the antigen. This represents possible modulation of the marker in <0.2% of lesions.

Serial sections of liver were stained to demonstrate discrete areas of altered enzyme expression and decreased glycogen mobilization and the specific genotype marker. Photomicrographs were projected onto a digitizing tablet through a first surface mirror[25] and the patches and lesions were traced with a tablet cursor. The number of genotype patches/unit area, the percentage of each parental genotype, and the average area of each patch or lesion was computed.[25,66,68] Lesions were evident in all size categories, ranging from less than 0.05 mm to greater than 2.0 mm in diameter. The cellular lineages of these foci were determined by aligning sections stained for these foci with serial sections reacted with antibody, recognizing the genotype distribution within the chimera (Figure 3). A total of 1101 enzyme-altered foci were examined for genotypic composition. Only a small proportion of foci expressing each alteration were of mixed genotype; 181/187 PAS-positive, 714/744 GGT-positive, 101/110 ATPase-negative, and 58/59 G-6-Pase negative foci were composed of a single genotype (Figures 4 and 5). The majority of those lesions of mixed lineage were as small or smaller than the average patch size, and thus at the limit of resolution of this system. Foci with multiple alterations in enzyme expression are also seen (Figures 3 through 6). Foci were scored as consisting of up to the four phenotypic alterations studied. Even foci expressing all four alterations in enzyme expression appeared to

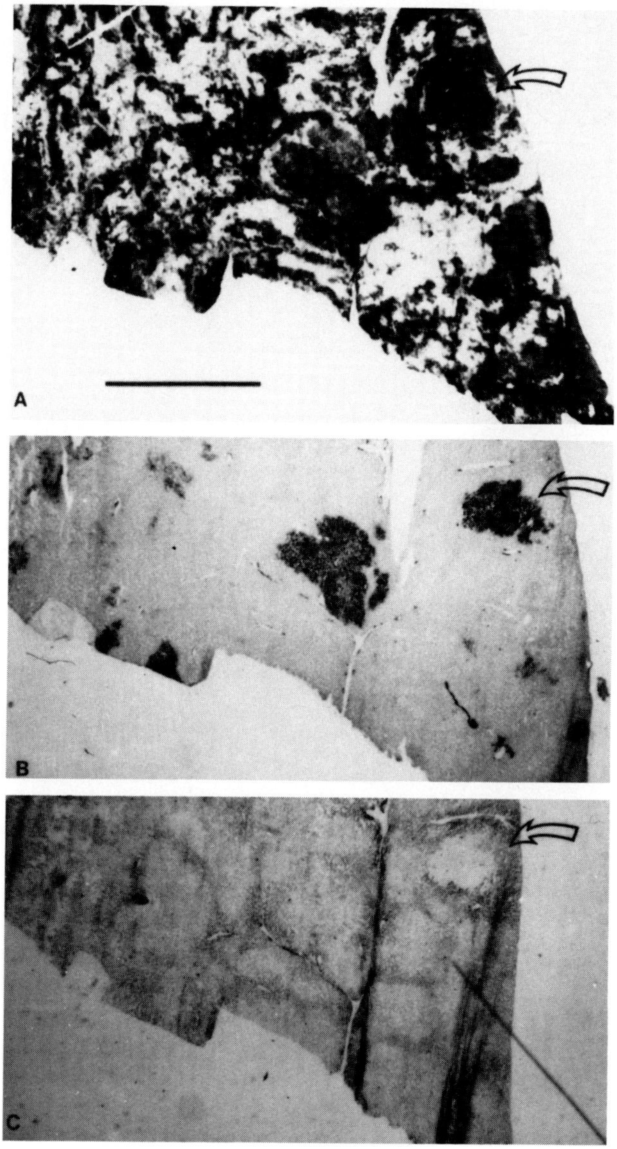

FIGURE 3. Photomicrographs of autoradiograms of chimeric rat liver treated with DEN and partial hepatectomy as described in the text. Sections A through C are from chimera 0906 liver section 17b; (A) stained for the cell lineage marker with an iodinated monoclonal antibody (R2/15S) which recognizes cells of the PVG-*RT1*^{a} lineage; (B) adjacent serial section stained to elucidate γ-glutamyl transpeptidase; (C) adjacent serial section stained for ATPase. Curved arrow demonstrates a lesion which comprises a single cell lineage, indicating its clonal nature. The bar represents 2 mm. (Reprinted from Weinberg and Iannaccone, *J. Cell Sci.*, 89, 423, 1988. With permission.)

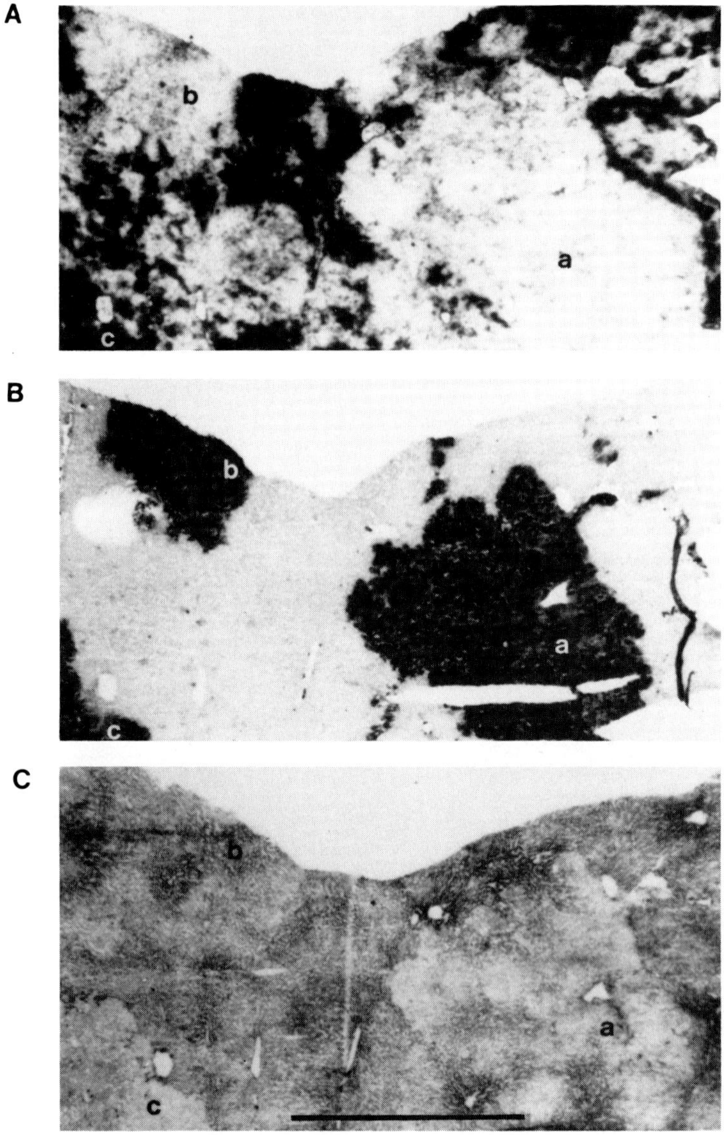

FIGURE 4. Photomicrographs of autoradiograms of chimeric rat liver treated with DEN and partial hepatectomy as described in the text. Sections A through C are from chimera 0906 liver section 23a; (A) stained for the cell lineage marker with an iodinated monoclonal antibody (R2/15S) which recognizes cells of the PVG-*RT1*[a] lineage; (B) adjacent serial section stained to elucidate γ-glutamyl transpeptidase; (C) adjacent serial section stained for ATPase. a, b, and c represent corresponding lesions in adjacent serial sections. The bar represents 2 mm. (Reprinted from Iannaccone et al., *Int. J. Cancer,* 39, 778, 1987. With permission.)

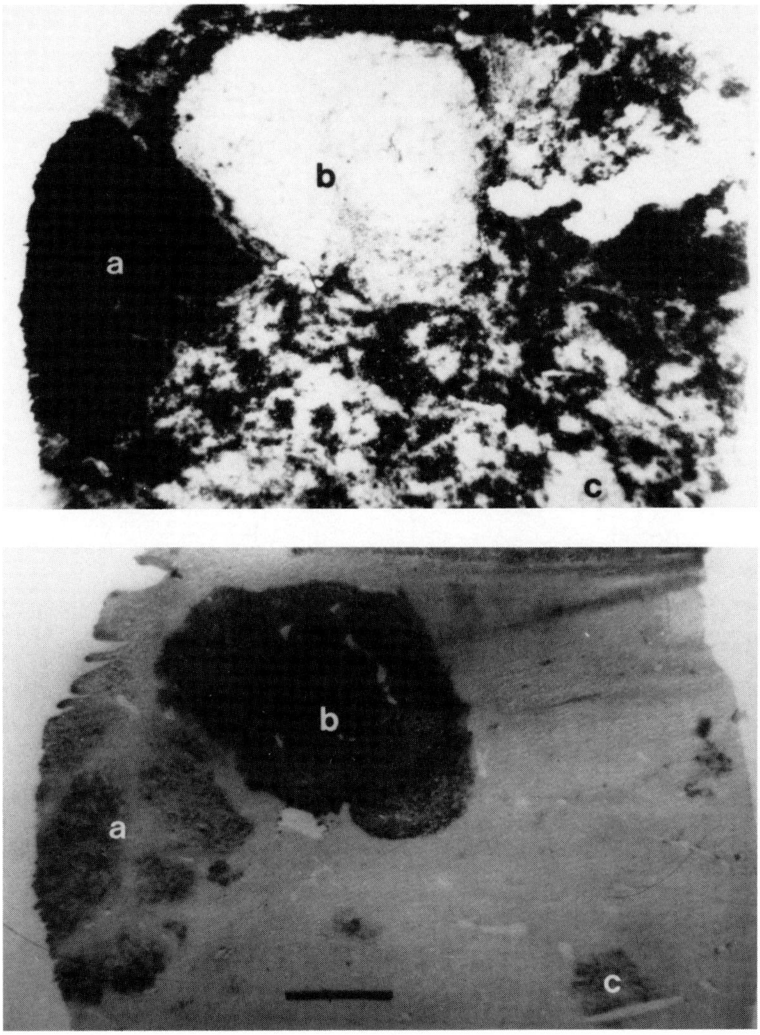

FIGURE 5. Photomicrographs of autoradiograms of chimeric rat liver treated with DEN and partial hepatectomy as described in the text. Adjacent serial sections from chimera 0906 liver section 30a; (A) stained for the cell lineage marker with an iodinated monoclonal antibody (R2/15S) which recognizes cells of the PVG-*RT1*[a] lineage; (B) adjacent serial section stained to elucidate γ-glutamyl transpeptidase; a, b, and c demonstrate lesions which comprise a single cell lineage, indicating their clonal nature. The bar represents 1 mm.

be of one genotypic lineage only, indicating that the phenotypic heterogeneity is more likely due to microenvironmental influences or progression than a multiclonal origin (Figure 7). Given the complex patch shape and distribution in normal chimeric liver, it is unlikely that areas of enzyme alterations and

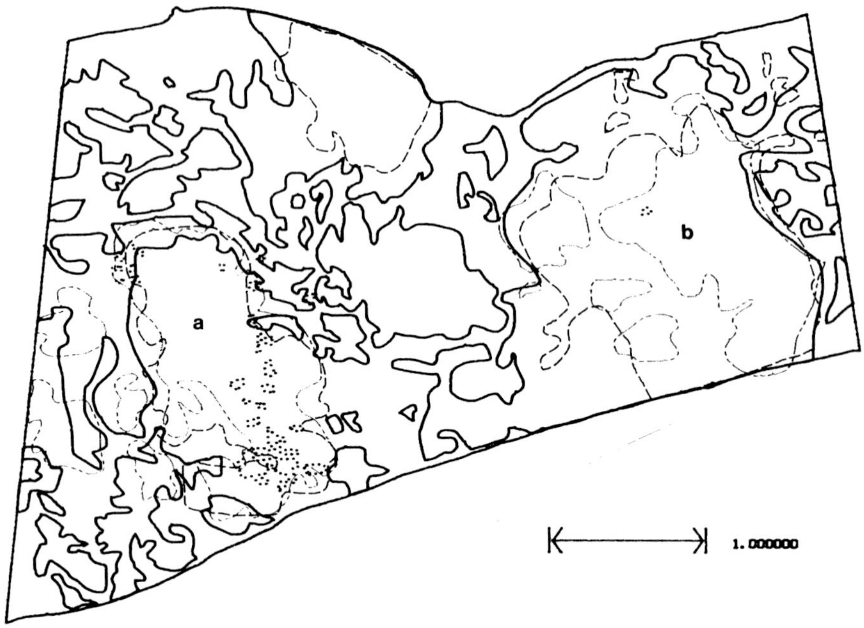

FIGURE 6. Computer-generated plot of composite serial digitized microscopic sections of autoradiograms from chimeric rat liver treated with DEN and partial hepatectomy as described in the text. Liver section 23a from chimera 0906 is presented. The solid black lines represent the limits of PVG-*RT1*^{a} patches. The ···· lines represent areas of PAS staining. The - - - - lines represent areas of γ-glutamyl transpeptidase positivity. The - - ·· - - lines represent areas of ATPase deficiency. a and b refer to clonal lesions demonstrating phenotypic heterogeneity. The bar represents 1 mm. (Reprinted from Weinberg and Iannaccone, *J. Cell Sci.*, 89, 423, 1988. With permission.)

genotype marker would be aligned by chance alone. Numerous examples were seen of regions heterogeneous for enzyme expression. When the outline of all overlapping foci were taken as one large lesion, the entire lesion was found to conform to the patch pattern and lay directly over a single patch. In addition, in several cases there was distortion of the patch pattern surrounding the lesions, indicating the expanding nature of the lesions. In aggregate, these data suggest that phenotypic heterogeneity is the result of the early acquisition of the ability to survive genetic instability in a clonally-expanding population.

A later study by Tsuji et al.[69] evaluating the clonality of early GGT-positive foci utilized heterozygous Spf mice mosaic for OCT activity. Early GGT-positive foci, ranging from as small as three cells to 3 mm in diameter, were analyzed for genotype by double staining of a single section for GGT activity and OCT activity. In this way, 482 lesions were analyzed; 52.1 and 47.9% were entirely positive and negative for OCT activity, respectively.

FIGURE 7. Computer-generated plot of composite serial digitized microscopic sections of autoradiograms from chimeric rat liver treated with DEN and partial hepatectomy as described in the text. Liver section 15c from chimera 0906 is presented. The solid black lines represent the limits of PVG-*RT1*[a] patches. The ···· lines represent areas of PAS staining. The - - - - lines represent areas of γ-glutamyl transpeptidase positivity. The - - ·· - - lines represent areas of ATPase deficiency. The - - · - - lines represent areas of G-6-Pase deficiency. The bar represents 1 mm. (Reprinted from Weinberg and Iannaccone, *J. Cell Sci.*, 89, 423, 1988. With permission.)

This study confirms previous findings in both mouse and rat that indicate early enzyme-altered foci represent clonal growths, and that clonal selection of neoplastic growths occurs at early stages of tumor formation.

VIII. HUMAN HEPATOCELLULAR CARCINOMA

Given that both hepatic carcinoma and the preneoplastic lesions associated with them are clonal in mouse and rat, it should perhaps be evident that human hepatic cancer would arise similarly. This conclusion cannot be arrived at *a priori,* however. Aside from the obvious species difference the animal studies were performed with chemically induced tumors involving partial hepatectomy. Human hepatocellular carcinoma (HCC) is most commonly associated with hepatitis B virus (HBV). The possibility that viral induction can recruit transformed cells by lateral infection, the presence of diffuse forms of HCC and the absence of well-documented mechanisms of viral induction of HCC leave an answer to the question of cellular origins of human HCC an experimental one. There are several places in the world where HBV-associated HCC is very prevalent, among them Taiwan. However, these populations are not known for common X chromosome-linked polymorphisms that would allow mosaic analysis of tumors in the manner described above. Several groups have taken a different approach.[70] Reasoning that viral integration events would be multiple and unique within a population of infected hepatocytes, these investigators have characterized the integration sites of HBV in HCC and in nonmalignant hepatocytes by restriction analysis. Digestion of genomic DNA with various endonucleases, followed by Southern hybridization with HBV-based probes, should reveal unique band patterns in clonally related populations, while many integrants in a nonclonal population from a multicellular process would result in hybridization to a continuous size range of DNA fragments isolated from the tissue. Consequently, autoradiograms should demonstrate a smear of radioactivity. Moreover, the banding pattern of a clonal process can be used as a ''fingerprint'' of the clonal lesion, which in principle should allow one to establish the genetic relationship between separate lesions (e.g., a primary tumor, local extensions of the tumor, and metastatic lesions). A similar approach to relating primary and metastatic HCC has been taken by analyzing DNA ploidy with Feulgen cytophotometry.[71]

The same cautionary notes which have been discussed for mosaic analysis apply to the integration site analysis as well. Notably, it is important to know that the marker used to establish the clonal origin of a tumor is independent of and not affected by the carcinogenic process. This is particularly important if the marker is potentially modulated by tumor formation. Obviously, this independence is not the case for HBV.[72] Since little is known about what relation the integration events play in tumorigenesis (or, indeed, what role nonintegrated episomal forms play), it seems difficult at this time to state unequivocally that integration pattern is useful in determining the clonality of HCC. Nevertheless, since HCC DNA and nontumorous hepatocyte DNA can be compared, it is useful to attempt to use HBV itself as a marker for its associated lesions. Despite the potential difficulties, the results seem quite clear: unifocal and multifocal (both diffuse and multicentric) tumors show various unique integration patterns consistent with clonal expansion of transformed hepatocyte populations. In those situations where nontumorous hepatocytes have integrated HBV, random integration patterns are seen (hybridization to continuously sized fragments evident as a smear in autoradiograms), indicating nonclonal relationships between cells in these populations.[71,73,74]

IX. SUMMARY AND CONCLUSIONS

Experimental mosaics have been utilized to study the clonal origin of tumors and preneoplastic lesions in carcinogen-treated livers of rats and mice. The use of X-linked heterozygotes and experimental chimeras allows the tracing of cell lineage within a given lesion by a variety of techniques. The cell composition of animals mosaic for an X-linked enzyme polymorphism has been assessed by recognizing the differential mobility of the isozymes by gel electrophoresis. Samples can be dissected from tumor samples or thick sections and analyzed in this way. Other studies have utilized differential staining of one of the components within a histological section; this includes recognition of a specific isozyme by enzymatic staining reactions as well as specific antibody reactivity to class I antigens unique to one component of a chimera. Results suggest that liver tumors as well as the vast majority of enzyme-altered foci which occur early following exposure to carcinogens are composed of one genotype and are likely to be clonal in origin. Thus evidence from several systems supports the notion that clonal selection occurs in liver tumor development at the earliest recognizable step (altered enzyme foci) of the multistep process, and that heterogeneity of enzyme expression within lesions reflects genetic instability during progression or microenvironmental differences rather than a multiclonal origin of the lesion. Mosaic pattern analysis has shown that probabilistic cell division rules are sufficient to create a complex fractal pattern. Regenerating liver reconstitutes variegated patch patterns by brief increases in patch size without changing fractal dimension. These observations support the view that highly iterated simple cell division rules may produce a complex pattern. Since it is conceivable that such rules can be genetically encoded, they provide a way in which information necessary for complex morphogenesis could be stored in a small amount of DNA.

ACKNOWLEDGMENTS

This work was supported in part by USPHS grants HD28992 and ES03498 from DHHS and by the Lucille P. Markey Charitable Trust.

REFERENCES

1. **Boveri, T.,** Zür Frage der Entstehung maligner Tumoren, Gustav Fisher, Jena (1914), translated by Boveri, M., *On the Problem of the Origin of Malignant Tumors,* Williams & Wilkins, Baltimore, 1929.
2. **Bauer, K. H.,** *Mutationstheorie der Geschwulst Entstehung,* Springer-Verlag, Berlin, 1928.
3. **Barbacid, M.,** ras Genes, *Annu. Rev. Biochem.,* 56, 779, 1987.
4. **Willis, R. A.,** *Pathology of Tumors,* Appleton-Century-Crofts, New York, 1967, 105.
5. **Martensson, L.,** On "a key point of modern biochemical genetics", *Lancet,* 1, 946, 1963.
6. **Mellstedt, H., Hammarstrom, S., and Holm, G.,** Monoclonal lymphocyte population in human plasma cell myeloma, *Clin. Exp. Immunol.,* 17, 371, 1974.
7. **Fialkow, P. J., Klein, E., Klein, G., Clifford, P., and Singh, S.,** Immunoglobulin and glucose-6-phosphate dehydrogenase as markers of cellular origin in Burkitt lymphoma, *J. Exp. Med.,* 138, 89, 1973.
8. **Nowell, P. C. and Hungerford, D. A.,** A minute chromosome in human chronic granulocytic leukemia, *Science,* 132, 1497, 1960.
9. **DeKlein, A., Van Kessel, A. G., Grosveld, G., Bartram, C. R., Hagemeijer, A., Bootsma, D., Spurr, N. K., Heisterkamp, N., Groffen, J., and Stephenson, J. R.,** A cellular oncogene is translocated to the Philadelphia chromosome in chronic myelocytic leukaemia, *Nature,* 300, 765, 1982.
10. **Rowley, J. D.,** A new consistent chromosomal abnormality in chronic myelogenous leukaemia identified by quinacrine fluorescence and giemsa staining, *Nature,* 243, 290, 1973.
11. **Rowley, J. D.,** Identification of a translocation with quinacrine fluorescence in a patient with acute leukemia, *Annu. Rev. Genet.,* 16, 109, 1973.
12. **Zech, L., Haglung, U., Nilsson, K., and Klein, G.,** Characteristic chromosomal abnormalities in biopsies and lymphoid-cell lines from patients with Burkitt and non-Burkitt lymphomas, *Int. J. Cancer,* 17, 47, 1976.
13. **Lyon, M. F.,** Gene action in the X-chromosome of the mouse (*Mus musculus* L.), *Nature,* 190, 372, 1961.
14. **Linder, D. and Gartler, S. M.,** Glucose-6-phosphate dehydrogenase mosaicism: utilization as a cell marker in the study of leiomyomas, *Science,* 150, 67, 1965.
15. **Smith, J. W., Townsend, D. E., and Sparkes, R. S.,** Genetic variants of glucose-6-phosphatase dehydrogenase in the study of carcinoma of the cervix, *Cancer,* 28, 529, 1971.
16. **Dow, L. W., Jenkins, J. J., Wilimas, J., Hayes, F. A., and Fialkow, P. J.,** Evidence for clonal development of embryonal pediatric tumors, *Proc. Am. Assoc. Cancer Res.,* 27, 152, 1986.
17. **Vogelstein, B., Fearon, E. R., Hamilton, S. R., and Feinberg, A. P.,** Use of restriction fragment length polymorphisms to determine the clonal origin of human tumors, *Science,* 227, 642, 1985.
18. **Mintz, B.,** Formation of genotypically mosaic mouse embryos, *Am. Zool.,* 2, 432, 1962.
19. **Tarkowski, A. K.,** Mouse chimaeras developed from fused eggs, *Nature,* 190, 857, 1961.
20. **Iannaccone, P. M., Gardner, R. L., and Harris, H.,** The cellular origin of chemically induced tumours, *J. Cell Sci.,* 29, 249, 1978.
21. **Reddy, A. L. and Fialkow, P. J.,** Multicellular origin of fibrosarcomas in mice induced by the chemical carcinogen 3-methylcholanthrene, *J. Exp. Med.,* 150, 878, 1979.
22. **Reddy, A. L. and Fialkow, P. J.,** Effect of solvents on methyl-cholanthrene-induced carcinogenesis in mice, *Int. J. Cancer,* 27, 501, 1981.

23. **Deamant, F. D. and Iannaccone, P. M.,** Evidence concerning the clonal nature of chemically induced tumors: phosphoglycerate kinase-1 isozyme patterns in chemically induced fibrosarcomas, *J. Natl. Cancer Inst.*, 74, 145, 1985.
24. **Deamant, F. D., Vijh, M., Rossant, J., and Iannaccone, P. M.,** In situ identification of host derived infiltrating cells in chemically induced fibrosarcomas of interspecific chimeric mice, *Int. J. Cancer,* 37, 283, 1986.
25. **Iannaccone, P. M., Weinberg, W. C., and Berkwits, L.,** A probabilistic model of mosaicism based on the histological analysis of chimaeric rat liver, *Development,* 99, 187, 1987.
26. **Iannaccone, P. M., Howard, J. C., and Berkwits, L.,** Mosaic pattern and lineage analysis in chimeras, *Cell Diff. Dev.*, 25, 77, 1988.
27. **Iannaccone, P. M., Berwits, L., Joglar, J., Lindsay, J., and Lunde, A.,** Probabilistic division systems modeling the generation of mosaic fields, *J. Theor. Biol.*, 141, 363, 1989.
28. **West, J. D.,** Analysis of clonal growth using chimaeras and mosaics, in *Development in Mammals,* Johnson, M. H., Ed., North-Holland, New York, 1978, 413.
29. **Iannaccone, P. M.,** The study of mammalian organogenesis by mosaic pattern analysis, *Cell Differ.*, 21, 79, 1987.
30. **Iannaccone, P. M., Weinberg, W. C., and Deamant, F. D.,** On the clonal origin of tumors: a review of experimental models, *Int. J. Cancer,* 39, 778, 1987.
31. **Weinberg, W. C., Deamant, F. D., and Iannaccone, P. M.,** Patterns of expression of class I antigens in the tissues of congenic strains of rat, *Hybridoma,* 4, 27, 1985.
32. **Weinberg, W. C., Howard, J. C., and Iannaccone, P. M.,** Histological demonstration of mosaicism in a series of chimeric rats produced between congenic strains, *Science,* 227, 524, 1985.
33. **Iannaccone, P. M.,** Fractal geometry in mosaic organs: a new interpretation of mosaic pattern, *FASEB J.*, 4, 1508, 1990.
34. **West, J. D.,** A theoretical approach to the relation between patch size and clone size in chimaeric tissue, *J. Theor. Biol.*, 50, 153, 1975.
35. **West, J. D.,** Distortion of patches of retinal degeneration in chimaeric mice, *J. Embryol. Exp. Morphol.*, 36, 145, 1976.
36. **West, J. D.,** Cell markers, in *Chimeras in Developmental Biology,* LeDouarin, N. and McLaren, A., Eds., Academic Press, New York, 1984, 39.
37. **Bodenstein, L.,** A dynamic simulation model of tissue growth and cell patterning, *Cell Differ.*, 19, 19, 1986.
38. **Ng, Y. K., Ohaki, Y., Deamant, F., and Iannaccone, P. M.,** Comparison of patch size in X-chromosome linked mosaic and dizygotic chimeric animals, *Cell Differ. Dev.*, 30, 27, 1990.
39. **Edelman, G. M.,** *Topobiology,* Basic Books, New York, 1988, 1.
40. **Feder, J.,** *Fractals,* 3rd ed., Plenum Press, New York, 1988, 1.
41. **Farmer, J. D., Ott, E., and Yorke, J. A.,** The dimension of chaotic attractors, *Physica Part D,* 7, 153, 1983.
42. **Ng, Y. K. and Iannaccone, P. M.,** Fractal geometry of mosaic pattern demonstrates liver regeneration is a self-similar process, *Dev. Biol.*, 151, 149, 1992.
43. **Fausto, N.,** Messenger RNA in regenerating liver: implications for the understanding of regulated growth, *Mol. Cell. Biochem.*, 59, 131, 1984.
44. **Fausto, N. and Mead, J. E.,** Regulation of liver growth: protooncogenes and transforming growth factors, *Lab. Invest.*, 60, 4, 1989.
45. **Goyette, M., Petropolus, C. J., Shank, P. R., and Fausto, N.,** Expression of a cellular oncogene during liver regeneration, *Science,* 219, 510, 1983.
46. **Barnsley, M. F., Ervin, V., Hardin, D., and Lancaster, J.,** Solution of an inverse problem for fractals and other sets, *Proc. Natl. Acad. Sci. U.S.A.*, 83, 1975, 1986.
47. **West, B. J. and Goldberger, A. L.,** Physiology in fractal dimensions, *Am. Sci.*, 75, 354, 1987.

48. **Iannaccone, P. M., Lindsay, J., Berwits, L., Lescinsky, G., DeFanti, T., and Lunde, A.**, Fractal Geometry in Virtual Organ Growth, in *Biomedical Modeling and Simulation*, Eisenfeld, J., Witten, M., and Levine, D. S., Eds., Elsevier, Amsterdam, in press.
49. **Nowell, P. C.**, Mechanisms of tumor progression, *Cancer Res.*, 46, 2203, 1986.
50. **Ponder, B. A., Schmidt, G. H., and Wilkinson, M. M.**, Immunohistochemistry in the analysis of mouse aggregation chimaeras, *Histochem. J.*, 18, 217, 1986.
51. **Deamant, F. D. and Iannaccone, P. M.**, Clonal origin of chemically induced papillomas: separate analysis of epidermal and dermal components, *J. Cell Sci.*, 88, 305, 1987.
52. **Reddy, A. L. and Fialkow, P. J.**, Papillomas induced by initiation-promotion differ from those induced by carcinogen alone, *Nature*, 304, 69, 1983.
53. **Winton, D. J., Blount, M. A., and Ponder, B. A.**, Polyclonal origin of mouse skin papillomas, *Br. J. Cancer*, 60, 59, 1989.
54. **Arnold, A. and Kim, H. G.**, Clonal loss of one chromosome 11 in a parathyroid adenoma, *J. Clin. Endocrinol. Metab.*, 69, 496, 1989.
55. **Arnold, A., Staunton, C. E., Kim, H. G., Gaz, R. D., and Kronenberg, H. M.**, Monoclonality and abnormal parathyroid hormone genes in parathyroid adenomas, *N. Engl. J. Med.*, 318, 658, 1988.
56. **McLaren, A.**, Mammalian chimeras, in *Developmental and Cell Biology Series*, Abercrombie, M., Newth, D. R., and Torrey, J. G., Eds., University Press, Cambridge, 1976.
57. **Schmidt, G. H. and Mead, R.**, On the clonal origin of tumours — lessons from studies of intestinal epithelium, *Bioessays*, 12, 37, 1990.
58. **Iannaccone, P. M.**, Clone and patch size in chimeric mouse skin, *Math. Biosci.*, 51, 117, 1980.
59. **Peraino, C., Staffeldt, E. F., and Ludeman, V. A.**, Early appearance of histochemically altered hepatocyte foci and liver tumors in female rats treated with carcinogens one day after birth, *Carcinogenesis*, 2, 463, 1981.
60. **Pitot, H. C., Barsness, L., Goldworthy, T., and Kitagawa, T.**, Biochemical characterization of stages of hepatocarcinogenesis after a single dose of diethylnitrosamine, *Nature*, 271, 456, 1978.
61. **Solt, D. E. and Farber, E.**, New principle for the analysis of chemical carcinogenesis, *Nature*, 263, 701, 1976.
62. **Solt, D. B. and Shklar, G.**, Rapid emergence of carcinogen-induced hyperplastic lesions in a new model for the sequential analysis of liver carcinogenesis, *Am. J. Pathol.*, 88, 585, 1977.
63. **Tatematsu, M., Nagamine, Y., and Farber, E.**, Redifferentiation as a basis for remodeling of carcinogen-induced hepatocyte nodules to normal appearing liver, *Cancer Res.*, 43, 5049, 1983.
64. **Howell, S., Wareham, K. A., and Williams, E. D.**, Clonal origin of mouse liver cell tumors, *Am. J. Pathol.*, 121, 426, 1985.
65. **Rabes, H. M., Bucher, T., Hartmann, A., Linke, I., and Dunnwald, M.**, Clonal growth of carcinogen-induced enzymes-deficient preneoplastic cell populations in mouse, *Cancer Res.*, 42, 3220, 1982.
66. **Weinberg, W. C., Berkwits, L., and Iannaccone, P. M.**, The clonal nature of carcinogen-induced altered foci of gamma-glutamyl transpeptidase expression in rat liver, *Carcinogenesis*, 8, 565, 1987.
67. **Weinberg, W. C. and Iannaccone, P. M.**, Clonality of preneoplastic liver lesions: histological analysis in chimeric rats, *J. Cell Sci.*, 89, 423, 1988.
68. **Berkwits, L. and Iannaccone, P. M.**, A microcomputer-based quantitative morphometric analysis system for computing areas of amorphous patches, *Comput. Biomed. Res.*, 18, 278, 1985.
69. **Tsuji, S., Ogawa, K., Takaska, H., Sonoda, T., and Mori, M.**, Clonal origin of gamma-glutamyl transpeptidase-positive hepatic lesions induced by initiation-promotion in ornithine carbamoyltransferase, *Jpn. J. Cancer Res.*, 79, 148, 1988.

70. **Govindarajan, S., Craig, J. R., and Valinluck, B.,** Clonal origin of hepatitis B virus-associated hepatocellular carcinoma, *Hum. Pathol.*, 19, 403, 1988.
71. **Kuo, S. H., Sheu, J. C., Chen, D. S., Sung, J. L., Lin, C. C., and Hsu, H. C.,** DNA clonal heterogeneity of hepatocellular carcinoma demonstrated by Feulgen-DNA analysis, *Liver,* 7, 359, 1987.
72. **Hsu, H. C., Chiou, T. J., Chen, J. Y., Lee, C. S., Lee, P. H., and Peng, S. Y.,** Clonality and clonal evolution of hepatocellular carcinoma with multiple nodules, *Hepatology,* 13, 923, 1991.
73. **Esumi, M., Tanaka, Y., Tozuka, S., and Shikata, T.,** Clonal state of human hepatocellular carcinoma and non-tumorous hepatocytes, *Cancer Chemother. Pharmacol.*, 23 (Suppl. S1-S3), 1, 1989.
74. **Chen, P. J., Chen, D. S., Lai, M. Y., Chang, M. H., Huang, G. T., Yang, P. M., Sheu, J. C., Lee, S. C., Hsu, H. C., and Sung, J. L.,** Clonal origin of recurrent hepatocellular carcinomas, *Gastroenterology,* 96, 527, 1989.

Chapter 3

HISTOGENESIS AND PATHOBIOLOGY OF MOUSE HEPATOCELLULAR NEOPLASMS

S. Goldfarb and T. D. Pugh

TABLE OF CONTENTS

ISBN 0-8493-4746-7
© 1992 by CRC Press, Inc.

I. INTRODUCTION

Studies in mouse hepatocarcinogenesis, in analogy to those in the rat, have utilized three different types of animal models: (1) chronic feeding of a carcinogen; an example commonly used by federal agencies for testing purposes is the 24-month feeding to B6C3 F_1 mice;[1] (2) short exposure to a carcinogen followed by sustained treatment with exogenous tumor promoters such as phenobarbital;[2] and (3) short exposure to carcinogens either in the intact neonate or in the adult following partial hepatectomy.[3] As expected, the pathologic changes induced by these different regimens are quite diverse and present serious problems in developing a single unified nomenclature. In this regard, the chronic exposure to toxins, regardless of their carcinogenic potential, usually induces nodular hyperplasia of hepatocytes and even ductular or oval cell proliferation,[4] while exposure to some agents often induces adaptive centrilobular eosinophilic alterations of hepatocytes[5] whose significance for cancer development may be exaggerated.[6] In the past, the inability to decide whether a lesion is reparative, "putative preneoplastic", or neoplastic has led to a plethora of noncommittal terms for mouse liver nodules, including types A and B nodules, types 1, 2, and 3 nodules (see discussion in Reference 7), and mouse liver tumors types I, II, III, and IV.[8] More recently, the terminology "focus" and "adenoma" have been more generally accepted,[1] but even these terms should be used with more discrimination than is currently practiced (see Section VII).

This indecisiveness and inability to properly correlate morphology with the natural history of induced hepatocellular lesions has caused considerable consternation for environmentalists and regulators; it elicited a hilarious commentary by Dr. H. L. Stewart.[9] In contrast, we believe that in mice exposed in infancy to small doses of carcinogens, even the smallest and earliest appearing hepatocellular lesions are true neoplasms (see below). This is the reason for emphasizing this model in the remainder of this review.

II. THE NEONATAL MOUSE MODEL, GENERAL FEATURES

Although the use of the neonatal mouse liver tumor model dates from the early 1960s, Vesselinovitch and associates[10] in a series of systematic studies drew attention to the very general utility of the model. They found that single or only a few doses of diverse carcinogens when injected into mice that were 1 to 15 d old induced, after 1 year, a very high incidence of large hepatic neoplasms, many of which metastasized to lungs and lymph nodes. Since the mice with metastases always had at least one nodule that showed trabecular or anaplastic appearing histology (Figure 1) in their livers, they concluded that only these lesions should be considered carcinomas. Later,

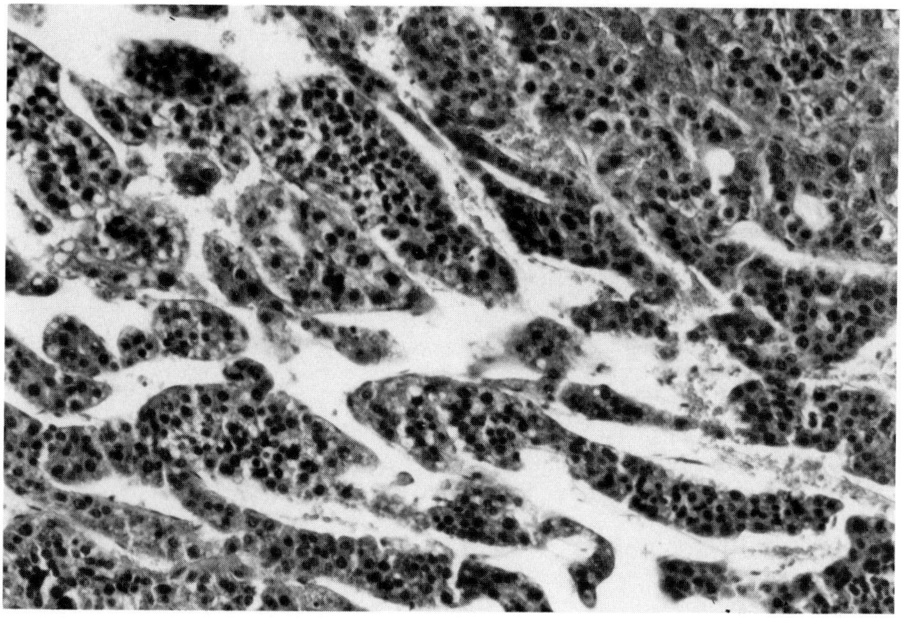

FIGURE 1. Trabecular hepatocellular carcinoma showing papillary features and marked anaplasia from a B6C3 F_1 mouse killed 44 weeks after DEN. (H.E.; magnification × 185.) (Reprinted from Goldfarb et al., *Environ. Health Perspect.*, 50, 149, 1983. With permission.)

we[11-13] and others[14] identified the very earliest microscopic lesions induced by single injections of diethylnitrosamine (DEN). In the protocol (still being used) 15-d-old mice, injected with minimally toxic doses of DEN, developed many microscopic hepatocellular lesions after only 10 weeks. At first, these lesions were merely referred to as ''basophilic foci'' because of the abundance of nucleic acids within the cells (Figures 2 and 3), but we now interpret all the induced proliferative hepatocellular lesions, regardless of their size, as neoplasms and the smallest ones as microneoplasms (Table 1).

III. PATHOLOGY OF THE EARLIEST MICRONEOPLASMS

Hepatocellular lipid accumulation was the earliest alteration appearing in infant B6C3 F_1 mice after injecting them with 5 mg/kg DEN.[12] It appeared at 3 d postinjection, but had entirely regressed by the seventh day. The microneoplasms, although extremely small, were clearly hepatocellular and composed of polygonal basophilic cells. They were generally round, and their constituent cells had high nuclear to cytoplasmic (N/C) ratios which, together with the basophilia, made their recognition unmistakable on well-prepared

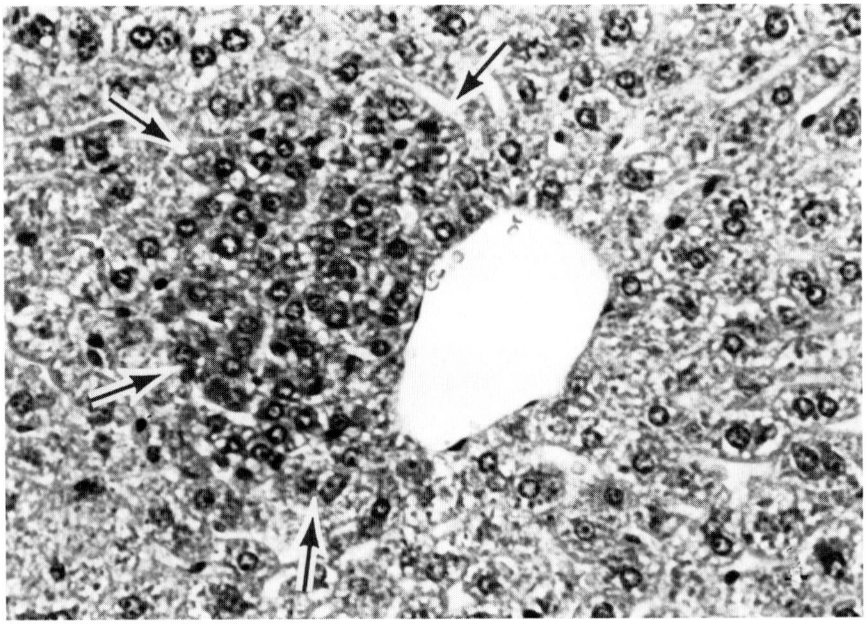

FIGURE 2. Tiny microneoplasm (arrows) adjacent to a terminal hepatic vein. The high N/C ratio results in a crowded appearance. The B6C3 F_1 mouse was killed 10 weeks after injection of DEN. (H.E.; magnification × 380.) (Reprinted from Goldfarb et al., *Environ. Health Perspect.*, 50, 149, 1983. With permission.)

H&E-stained tissue sections (Figure 2). If one considers the nonneoplastic hepatocytes as "normal", then the high N/C ratio took a range of forms from normal-size nuclei in small cells to large nuclei in normal-size cells. Proliferated biliary ductules or oval cells were never seen within or abutting against the microneoplasms at this early time.

IV. HETEROGENEITY AND PROGRESSION IN HEPATOCELLULAR NEOPLASMS

At 20 weeks after DEN and at later times, the enlarging tumors developed an obvious zonal heterogeneity, with more basophilic cells in "shells" at the peripheries of the lesions (Figure 3). However, they were often more clustered in one region at the periphery of the tumor (Figure 4), giving the appearance of one clone arising from another. The basophilic cells usually had larger nuclei than the deeper cells; the latter usually contained hydropic or vasculated appearing cytoplasm. The structural heterogeneity was associated with a functional variation, since cells in the outer 50-μm shell (two-cells thick) of hepatocytes (at 28 weeks post-DEN) had tritiated thymidine ([^{3}H]dThd)

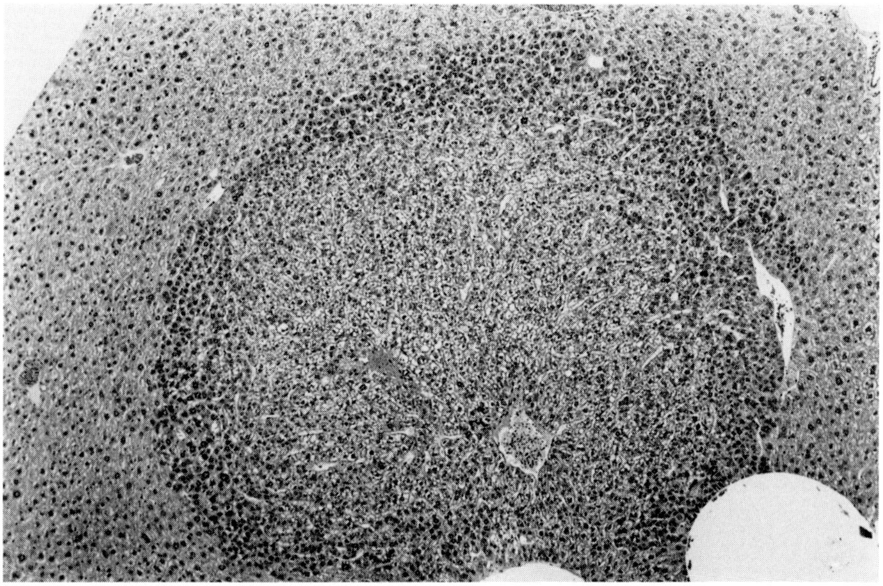

FIGURE 3. Microneoplasm showing outer shell of highly basophilic cells. The B6C3 F_1 mouse was killed 20 weeks after DEN. (H.E.; magnification × 100.)

TABLE 1
Classification of Hepatocellular Neoplasms in the Neonatal Mouse Model

A. Microneoplasms
- Less than 1 mm in diameter
- Always contain at least some basophilic cells, distributed uniformly within the neoplasms or at their peripheries; the cells have high nuclear to cytoplasmic ratios
- Composed of one- or two-cell-thick plates
- When vascular invasion is documented, called invasive microneoplasms
- In the past, often referred to as ''foci''

B. Macroneoplasms, nontrabecular
- More than 1 mm in diameter and therefore easily visible to the naked eye
- Show cytologic features of microneoplasms
- May show extensive areas with two-cell-thick plates and even some three-cell-thick areas; no significant degree of anaplasia
- Invariably develop vascular invasion
- In the past, often referred to as ''hyperplastic nodules'' or ''adenomas''

C. Trabecular carcinomas
- Invariably more than 1 mm in diameter
- Frequent cytologic anaplasia
- Arranged in plates more than three cells in thickness

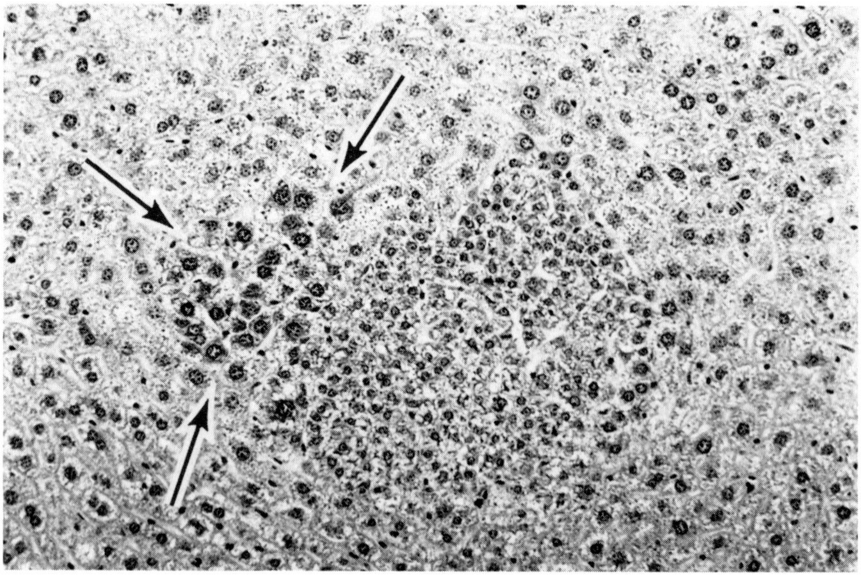

FIGURE 4. Tiny microneoplasm containing two types of cells. While both kinds of cells show high N/C ratios, the more basophilic subclone at the left (arrows) is composed of larger cells. The mouse was killed 20 weeks after DEN. (H.E.; magnification × 195.)

labeling indices that were twice that of the next deeper 50-μm shells of hepatocytes.[15] Although the generally round configuration of the microneoplasms was maintained, the margins of the tumors were often irregular, suggesting infiltration of adjacent nonneoplastic hepatocytes.

Until 28 weeks after DEN the lesions appeared very well differentiated, with one-cell-thick plates of hepatocytes. However, beginning at 28 weeks, the plates took on a "coarsened" appearance with numerous two-cell and occasionally even three-cell-thick plates. This pattern was even more prevalent by 36 weeks, and by 44 weeks there appeared many fully developed and often highly anaplastic trabecular carcinomas.

Infiltration of hepatic vein branches by the microneoplasms was their most remarkable feature (Figure 5). In contrast to human hepatocellular carcinomas, which invade both hepatic and portal vein branches, the mouse neoplasms invaded only branches of the hepatic veins. At times, the resulting tumor thrombi reached very impressive sizes. In fact, in a stereologic study of one neoplasm[12] the thrombus (1.8 mm in length) was 50% longer than the diameter of the tumor! The invasive property, seen in only a few of the lesions at 10 weeks after DEN, was very widespread by 20 weeks. Its prevalence was strikingly size dependent; it was seen in 12% of the tumors measuring 160 to 224 μm in diameter and in 86% of those between 450 to 638 μm in

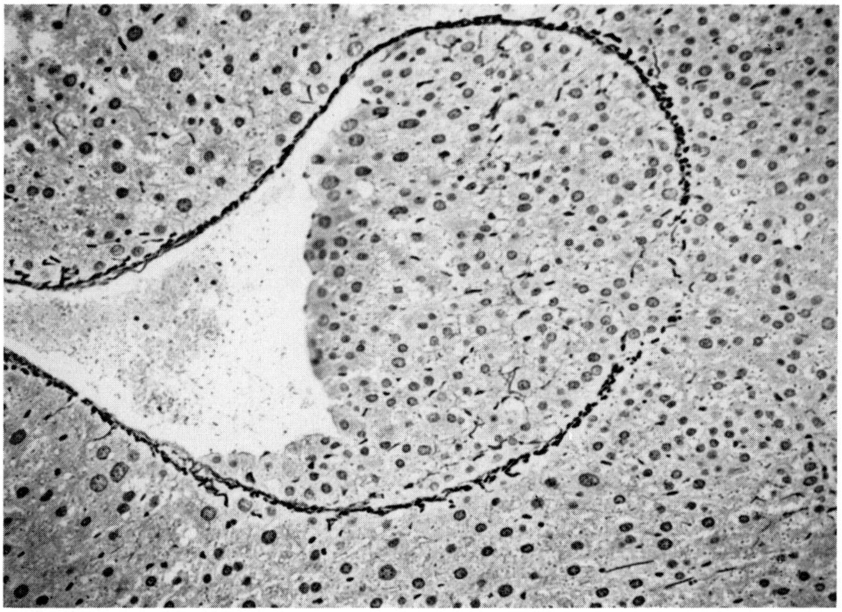

FIGURE 5. Microneoplasm penetrating the wall of a hepatic vein branch and filling much of the lumen as a tumor thrombus. The mouse had been injected with DEN 28 weeks earlier. (Reticulum stain; magnification × 185.) (Reprinted from Goldfarb and Pugh, in *Progress in Liver Diseases,* Popper and Schaffner, Eds., W. B. Saunders Co. With permission.)

diameter. Since the distances between portal triads and terminal hepatic veins is less than 625 μm in more than 95% of such measurements,[16] it appears that the only reason that all microneoplasms do not infiltrate is that they may not be large enough to abut against the veins. Early infiltration was probably also facilitated by the predisposition for the microneoplasms to be localized in centrilobular zones.[16]

Another alteration that is size and time dependent is the focal proliferation of bile ductules within microneoplasms.[12] This was first seen at 20 weeks after DEN in a single 900-μm-diameter neoplasm. Later, at 28 weeks, it was seen in only 20% of lesions measuring between 450 to 638 μm in diameter, but in more than 75% of lesions measuring more than 900 μm in diameter. Stereologic study revealed that the ductules arise from penetrating bile ducts that are accompanied by "feeding" portal vein branches (Figure 6).

V. GROWTH KINETICS OF MICRO- AND MACRONEOPLASMS IN DIFFERENT STRAINS

A remarkable feature of the neonatal mouse model that makes it especially useful for studying the stochastic processes involved in carcinogenesis is the

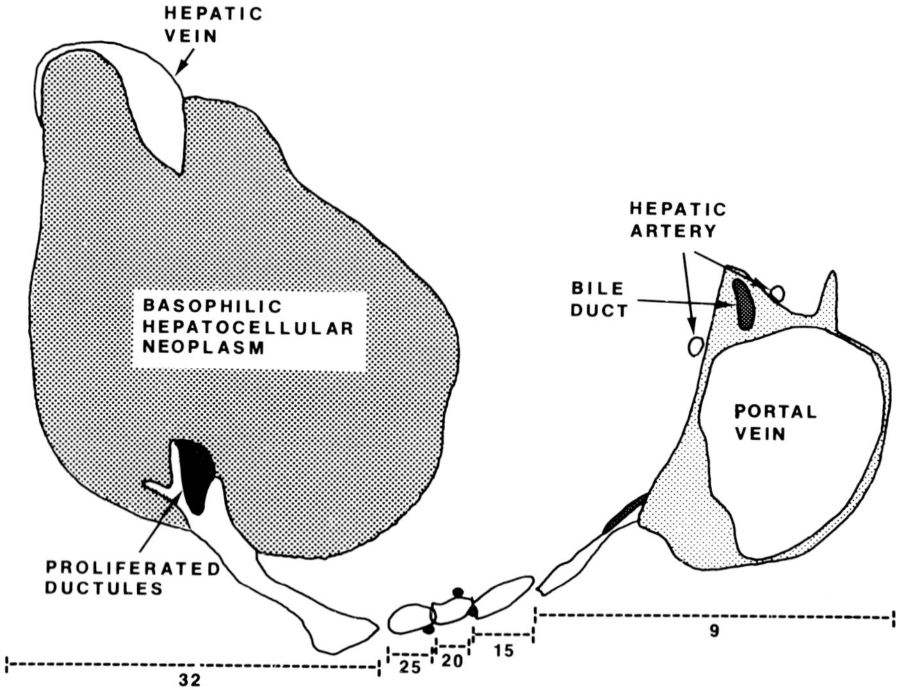

FIGURE 6. Composite drawing of a microneoplasm with its feeding portal vein branch and bile duct. The bile duct and portal vein branch were traced backward through 23 sections (sections 9 to 32) to document continuity with a portal triad. Note the proliferation of ductules at the point of entry and the invasion of a hepatic vein branch on the opposite side of the neoplasm. (H.E.; magnification × 75.) (Reprinted from Koen et al., *Am. J. Pathol.*, 112, 89, 1983. With permission.)

very large number of microneoplasms that are induced. Using mathematical-stereologic methods,[17,18] we have estimated that a single injection of 5 mg of DEN in B6C3 F_1 mice initiated about 1700 neoplasms.[19] However, at 10 weeks after DEN, we estimate that about 300 lesions are large enough (greater than 78 μm in diameter) to be identified. Since the estimate rose to about 900 tumors by 20 weeks and to about 1000 tumors by 28 weeks, the second 10 weeks is the period of most rapid ''emergence'' of the tumors. Of course, the tumors continued to enlarge, and after 36 weeks when they occupied more than 10 to 20% of the liver volume, fusion of lesions was severe and kinetic analysis was no longer feasible.

Recently, several laboratories have reported that the growth rates of autochthonous hepatocellular tumors vary from strain to strain.[20-22] For example, in C57Bl/6 mice which are known to be ''resistant'' to chemical hepatocarcinogenesis, the resistance is primarily a consequence of the slower emergence

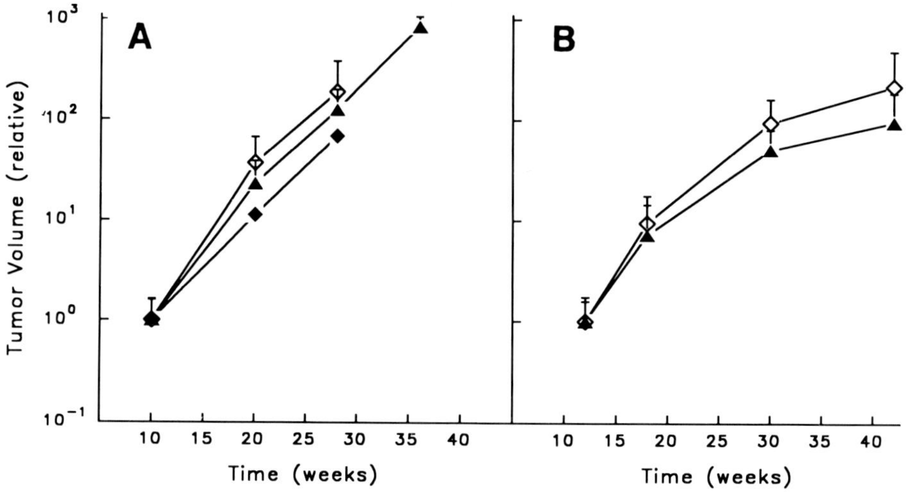

FIGURE 7. Comparison of the change in mean tumor volume for cohorts of the largest 15 tumors (open diamonds) with the aggregate tumor volume (filled triangles) for B6C3 F_1 (A) and C57Bl/6 (B) strains. For the B6C3 F_1 strain, data for a cohort of small tumors (filled diamonds) is also included. For purposes of comparison, the volumes at the earliest sacrifice time were normalized by assigning them the value of 1 (10^0). For each strain, groups of five to ten mice were killed at intervals between 10 to 44 weeks after DEN injection. (Reprinted from Pugh and Goldfarb, *Cancer Res.*, 52, 280, 1992, American Association for Cancer Research. With permission.)

of the neoplasms. Our own studies[15,20] led us to the conclusion that the early tumor growth rates (between 10 to 20 weeks after DEN) were similar for males of the parental C3H and C57Bl/6 strains and for the hybrid B6C3 F_1 mice (tumor doubling times of 2 to 2.5 weeks), but that only the growth of tumors in the C57Bl/6 mice slowed after that time; between 30 to 42 weeks the doubling time was 13 weeks (Figure 7).

One possible explanation for the tumor growth retardation in C57Bl/6 mice came from a study of the individual tumors. We found that within C57Bl/6 microneoplasms, focal collections of cells contained intracytoplasmic inclusions (Figures 8 and 9); these consisted of secretory protein within dilated rough endoplasmic reticulum.[23] These inclusions have been recorded in more advanced hepatocellular tumors in other strains.[24] However, in the C57Bl/6 mice the inclusions appeared very early (10 weeks after DEN) and became increasingly prominent in the ensuring weeks. Most intriguing was the observation that cells within areas containing abundant inclusions had [^{3}H]dThd labeling indices that were about one third that of the uninvolved areas (Figure 10). This led us to speculate that the premature development of the inclusions causes cellular injury, analogous to the hepatocellular injury in hepatocytes of individuals with the genetic defect, α-1-antitrypsin deficiency.[25] We believe

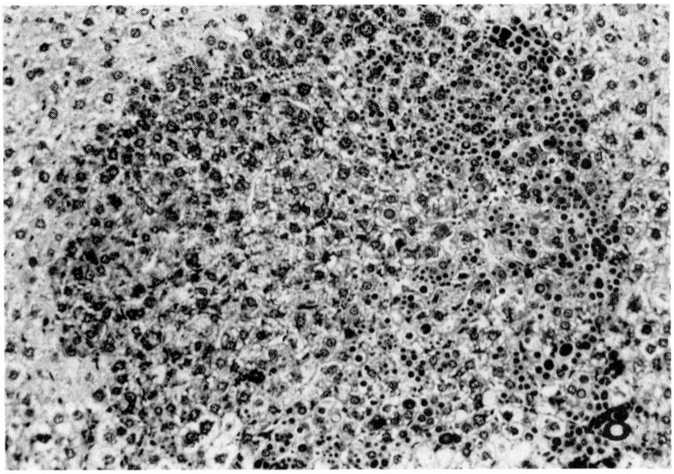

FIGURE 8. Microneoplasm from a C57Bl/6 mouse killed 30 weeks after DEN. Inclusion-rich cells are seen in the cells of the right half of the neoplasm. Overall, the lesion is more basophilic than the surrounding parenchyma, although the staining intensity is greatest in a peripheral zone of tumor cells. (Lendrum stain; magnification × 140.) (Reprinted from Kakizoe et al., *Cancer Res.*, 49, 3985, 1989, American Association for Cancer Research. With permission.)

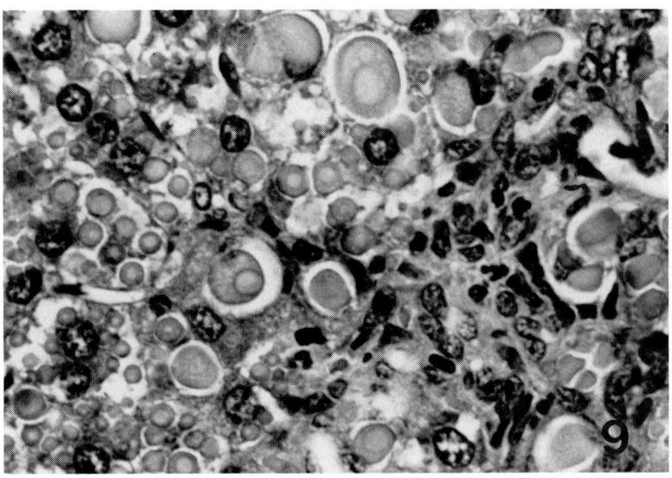

FIGURE 9. Inclusions of varying sizes and complexity within a macroneoplasm from a C57Bl/6 mouse killed 42 weeks after DEN. A cluster of proliferated ductules is noted on the right. (Lendrum stain; magnification × 570.) (Reprinted from Kakizoe et al., *Cancer Res.*, 49, 3985, 1989, American Association for Cancer Research. With permission.)

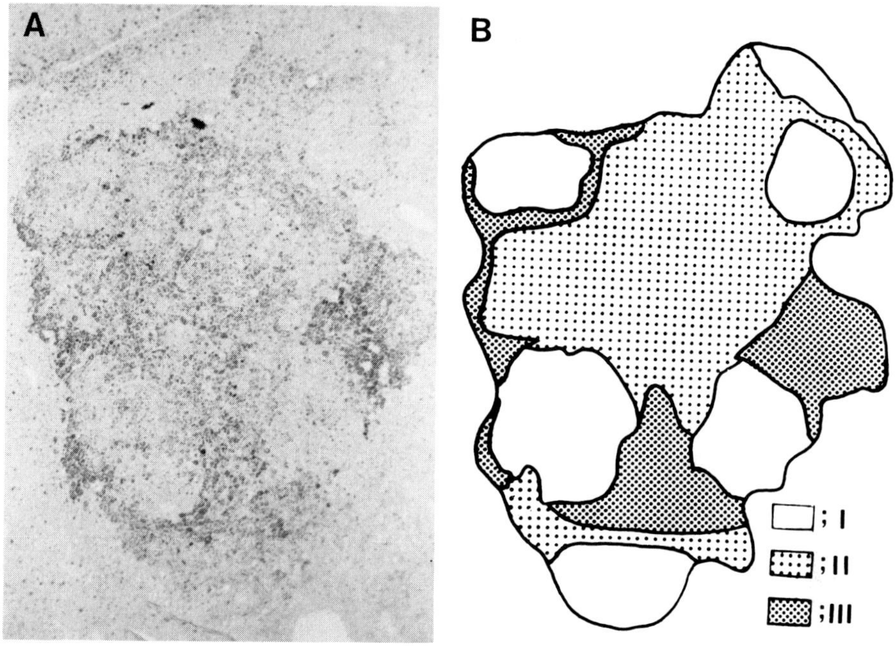

FIGURE 10. Thymidine labeling indices (B) within a complex macroneoplasm that had three types of areas: grade I areas contained no inclusions and had a mean labeling index of 5.1%; grade II areas contained intermediate numbers of inclusions and had a mean labeling index of 1.8%; and grade III areas contained many inclusions and had a mean labeling index of 0.8%. The mean labeling index of ''background'' hepatocytes was 0.17% (11 mice). A section that was cut serial to the autoradiogram was stained (Lendrum stain) for the inclusions (A). Note that areas richest in the inclusions have the lowest labeling indices. The neoplasm was from a C57Bl/6 mouse killed 44 weeks after DEN. (Reprinted from Kakizoe et al., *Cancer Res.*, 49, 3985, 1989, American Association for Cancer Research. With permission.)

that this injury, in turn, resulted in a reduced replication rate that was responsible for the late development of a prolonged doubling time of the tumors.

VI. MODELS OF TUMOR GROWTH

As expected, the mouse liver tumors vary in size; however, the degree of the variation is unusually broad. For example, at 28 weeks the tumors range in size through four orders of magnitude, with most of the tumors in the smaller-size classes.[15] In fact, only 4% of the neoplasms account for more than two thirds of the tumor burden (Figure 11). Two simple mathematical models might explain the very great variation in tumor size; one, referred to as the random rate model, assumes that all tumors start to grow at the same time, but at different growth rates, while the second, referred to as the random

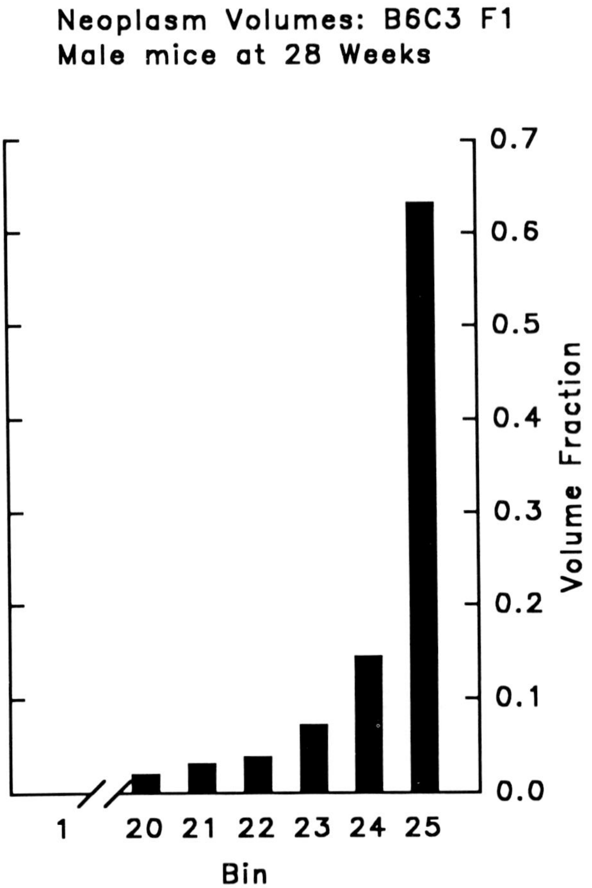

FIGURE 11. Bar graph depicting the contributions of neoplasms to tumor burden in seven B6C3 F_1 mice killed at 28 weeks after DEN. A total of 400 tumors, whose sizes were determined from serial sections, were arranged according to size and distributed in 25 bins (15 tumors per bin). Note that the largest 15 tumors, in bin #25, contributed almost two thirds of the total tumor burden.

delay model, assumes that the neoplasms start their growth at different times, but at the same growth rate (Figure 12). Recently we used several data sets to decide between the two options. Assuming that over time the tumors would maintain their relative rank order, we discovered that between 10 to 20 weeks after DEN the largest 15 neoplasms grew about 50% faster than a similar cohort between the 100th to 115th tumors (Figure 7).[15] More rapid growth of the larger neoplasms is also apparent among tumors induced in the C57Bl/6 and C3H strains. In addition, nontrabecular macroneoplasms in C57Bl/6 mice have [^{3}H]dThd labeling indices that are more than double that of mi-

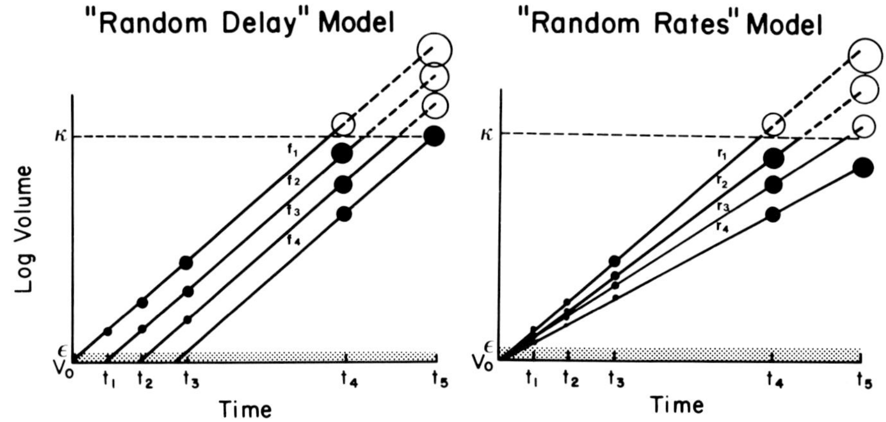

FIGURE 12. Models for growth of tumor populations that might explain variation in tumor size at the end of the study (time t_5). In the "random delay" model, all tumors ($f_{1..n}$) have the same growth rates, but they begin to grow at different times ($t_{0..3}$). In the "random rates" model, tumors have different growth rates ($r_{1..4}$), but begin to grow at the same time. Tumors larger than k are visible to the naked eye, but tumors smaller than ϵ cannot be reliably identified, even with the microscope.

croneoplasms.[23] Thus, the evidence rules out a simple random delay model and favors a random rate model.

VII. SUMMARY OF HISTOGENESIS AND THE NATURAL HISTORY OF MICRONEOPLASMS

Taken together, the data strongly indicate that the hepatocellular lesions induced by DEN are neoplastic from their time of origin. First and most importantly is the capacity of microneoplasms to invade blood vessels, which, to the vast majority of pathologists, signifies malignancy. Second, since they appear many weeks after a very early and transient fatty liver, there is no reason to believe that they reflect a pattern of reparative hyperplasia. Third, a high N/C ratio, generally considered a feature of malignancy or at least premalignancy,[26] is regularly present in the earliest recognizable lesions.[13] Finally, since a cogent argument has been made for considering α-fetoprotein a marker for malignancy in mouse liver tumors,[27] our finding histochemical evidence of the protein in about 15% of microneoplasms,[28] at a minimum, supports the interpretation of neoplasia. Accordingly, we now suggest a rather conservative classification scheme for these lesions (Table 1). Note that the scheme avoids the noncommittal term "focus" and the unsubstantiated and overly restrictive term "adenoma". This latter point is stressed, since on at least one occasion investigators found that DNA from benign "adenomas" had the same transforming efficiencies as DNA from trabecular carcinomas.[29]

We also conclude that hepatocytes are the sole progenitors of hepatocellular microneoplasms induced in the infant mouse model. While many of the tumors do contain ductules, they clearly develop from bile ducts that penetrate the neoplasms subsequent to the development of the tumors. Furthermore, the development of focal heterogeneity within microneoplasms, the gradual thickening of cellular plates, and the late appearance of trabecular and anaplastic carcinomas all support the notion of a continuum of spontaneous neoplastic progression.

REFERENCES

1. **Maronpot, R. R., Haseman, J. K., Boorman, G. A., Eustis, S. E., Rao, G. N., and Huff, J. E.,** Liver lesions in B6C3F1 mice: the National Toxicology Program, experience and position, in mouse liver tumors, *Arch. Toxicol., Suppl.,* 10, 10, 1987.
2. **Ward, J. M., Rice, J. M., Creasia, D., Lynch, P., and Riggs, C.,** Dissimilar patterns of promotion by di(2-ethylhexyl)phthalate and phenobarbital of hepatocellular neoplasia initiated by diethylnitrosamine in B6C3F1 mice, *Carcinogenesis,* 4, 1021, 1983.
3. **Craddock, V. M.,** Cell proliferation and experimental liver cancer, in *Liver Cell Cancer,* Cameron, H. M., Linsell, D. A., and Warwick, G. P., Eds., Elsevier, Amsterdam, 1976, 153.
4. **Lipsky, M. M., Hinton, D. F., Klaunig, J. E., and Trump, B. F.,** Biology of hepatocellular neoplasia in the mouse, *J. Natl. Cancer Inst.,* 67, 365, 1981.
5. **Butler, W. H. and Hempsall, V.,** Histochemical observations on nodules induced in the mouse liver by phenobarbitone, *J. Pathol.,* 125, 155, 1978.
6. **Reuber, M. D.,** Histogenesis of hyperplasia and carcinomas of the liver arising around central veins in mice ingesting chlorinated hydrocarbons, *Pathol. Microbiol.,* 43, 287, 1975.
7. **Goldfarb, S., Pugh, T. D., and Cripps, D. J.,** Increased alkaline phosphatase activity — a positive histochemical marker for griseofulvin-induced mouse hepatocellular nodules, *J. Natl. Cancer Inst.,* 64, 1427, 1980.
8. **Becker, F. F.,** Morphological classification of mouse liver tumors based on biological characteristics, *Cancer Res.,* 42, 3918, 1982.
9. **Stewart, H. L.,** Comparative aspects of certain cancers, in *Cancer, A Comprehensive Treatise,* Vol. 4, Becker, F. F., Ed., Plenum Press, New York, 1975, 328.
10. **Vesselinovitch, S. D., Rao, K. V. N., and Mihailovich, N.,** Neoplastic response of mouse tissues during perinatal age periods and its significance in chemical carcinogenesis, *Natl. Cancer Inst. Monogr.,* 11, 239, 1979.
11. **Goldfarb, S., Vesselvinovitch, S. D., Pugh, T. D., Mihailovich, N., Koen, H., and He, Y.,** Remarkable uniformity of premalignant and malignant hepatocellular foci after single dose injection of diethylnitrosamine (DEN) in infant mice, *Gastroenterology,* 79, 1021, 1980.
12. **Koen, H., Pugh, T. D., and Goldfarb, S.,** Hepatocarcinogenesis in the mouse: combined morphologic-stereologic studies, *Am. J. Pathol.,* 112, 89, 1983.
13. **Goldfarb, S., Pugh, T. D., Koen, H., and He, Y.,** Preneoplastic and neoplastic progression during hepatocarcinogenesis in mice injected with diethylnitrosamine in infancy, *Environ. Health Perspect.,* 50, 149, 1983.

14. **Moore, M. R., Drinkwater, N. R., Miller, E. C., Miller, J. A., and Pitot, H. C.,** Quantitative analysis of the time-dependent development of glucose-6-phosphatase deficient foci in the livers of mice treated neonatally with diethylnitrosamine, *Cancer Res.*, 41, 1585, 1981.
15. **Pugh, T. D. and Goldfarb, S.,** Growth kinetics of microscopic hepatocellular neoplasms in carcinogen-resistant and carcinogen-responsive strains of mice, *Cancer Res.*, 52, 280, 1992.
16. **Koen, H., Pugh, T. D., and Goldfarb, S.,** Centrilobular distribution of diethylnitrosamine-induced hepatocellular foci in the mouse, *Lab. Invest.*, 49, 78, 1983.
17. **Pugh, T. D., King, J. H., Koen, H., Nychka, D., Chover, J., Wahba, G., He, Y., and Goldfarb, S.,** Reliable stereological method for estimating the number of microscopic hepatocellular foci from their transections, *Cancer Res.*, 43, 1261, 1983.
18. **Nychka, D. N., Pugh, T. D., King, J. H., Koen, H., Wahba, G., Chover, J., and Goldfarb, S.,** Optimal use of sampled tissue sections for estimating the number of hepatocellular foci, *Cancer Res.*, 44, 178, 1984.
19. **Goldfarb, S. and Pugh, T. D.,** Multistage rodent hepatocarcinogenesis, in *Progress in Liver Diseases,* Popper, H. and Schaffner, F., Eds., Grune & Stratton, New York, 1986, 597.
20. **Goldfarb, S., Pugh, T. D., and Chover, J.,** Genetic resistance to hepatocarcinogenesis in C57Bl/6 mice: post initiation slowing of tumor growth correlates with distinctive morphology, *Proc. Am. Assoc. Cancer Res.*, 28, 166, 1987.
21. **Hanigan, M. H., Kemp, C. J., Ginsler, J. J., and Drinkwater, N. R.,** Rapid growth of preneoplastic lesions in hepatocarcinogen-sensitive C3H/HeJ male mice relative to C57Bl/6J male mice, *Carcinogenesis,* 9, 885, 1988.
22. **Dragani, T. A., Manetti, G., and Porta, G. D.,** Genetic susceptibility to murine hepatocarcinogenesis is associated with high growth rate of NDEA-initiated hepatocytes, *J. Cancer Res. Clin. Oncol.*, 113, 223, 1987.
23. **Kakizoe, S., Goldfarb, S., and Pugh, T. D.,** Focal impairment of growth in hepatocellular neoplasms of C57BL/6 mice: a possible explanation for the strain's resistance to hepatocarcinogenesis, *Cancer Res.*, 49, 3985, 1989.
24. **Liebelt, A. G., Liebelt, R. A., and Dmochowski, L.,** Cytoplasmic inclusion bodies in primary and transplanted hepatomas of mice of different strains, *J. Natl. Cancer Inst.*, 47, 413, 1971.
25. **Perlmutter, D. H.,** The cellular basis for liver injury in alpha-1-antitrypsin deficiency, *Hepatology,* 13, 172, 1991.
26. **Koss, L. G.,** *Diagnostic Cytology and its Histopathologic Bases,* Vol. 1, Lippincott, Philadelphia, 1979, 76.
27. **Abelev, G. I., Perova, S. C., Khramova, N. I., Postnikova, Z. A., and Irlin, I. S.,** Production of embryonal alpha-globulin by transplantable mouse hepatomas, *Transplantation,* 1, 174, 1963.
28. **Koen, H., Pugh, T. D., Nychka, D., and Goldfarb, S.,** Presence of alphafetoprotein-positive cells in hepatocellular foci and microcarcinomas induced by single injections of diethylnitrosamine in infant mice, *Cancer Res.*, 43, 702, 1983.
29. **Reynolds, S. H., Stowers, S. J., Patterson, R. M., Maronpot, R. R., Aaronson, S. A., and Anderson, M. W.,** Activated oncogenes in B6C3F1 mouse liver tumors: implications for risk assessment, *Science,* 237, 1309, 1987.

Chapter 4

ORIGIN AND PATHOLOGY OF HEPATOBLASTOMA IN MICE*

B. A. Diwan, J. M. Ward, and J. M. Rice

TABLE OF CONTENTS

* This chapter is a U.S. Government work and as such is not subject to copyright.

I. INTRODUCTION, DEFINITION, AND BACKGROUND

Hepatoblastomas are malignant "embryonal" liver tumors. In humans, hepatoblastomas are rare tumors that primarily occur in children under 3 years of age.[1-3] Like other embryonal neoplasms, e.g., nephroblastomas, neuroblastomas, and medulloblastomas, hepatoblastomas are composed of poorly differentiated cells. Typically, these are arranged in the form of rosettes, sheets, or solid cords, separated by vascular channels. The tumors occasionally display osseous, chondrous, or squamous metaplasia, reflecting the capacity for heterologous and divergent differentiation characteristic of this tumor.[4-6] Hepatoblastomas are thus morphologically and clinically different from hepatocellular carcinomas, which may also occur in children, but are more commonly encountered in adults.

Hepatoblastomas are also known to occur spontaneously, although rarely, in some strains of mice.[7-10] Unlike the human tumor, however, naturally occurring murine hepatoblastomas are always seen in aged animals. There is much confusion in the early literature between these lesions and poorly differentiated cholangiomas and cholangiocarcinomas.[11,12] Cholangiocellular tumors that in some ways resemble hepatoblastomas have been shown to arise from periportal areas in mice carrying the viable yellow allele (A^{vy}) at the agouti locus.[11] A few tumors resembling human hepatoblastomas were described by one of us[13] in C3H mice exposed transplacentally to *N*-nitrosoethylurea. Turusov et al.[14] found similar tumors in untreated and dichlorodiphenyltrichloroethane (DDT)-treated CF1 mice and were the first to tentatively call them hepatoblastomas. Hepatoblastomas have been recently reported[15,16] to occur in BALB/c and B6C3F1 mice after chronic exposure to 2-acetylaminofluorene. The incidence of spontaneous or chemically induced tumors reported by the investigators cited above, however, never exceeded 6 to 7% even in lifetime studies.

II. EXPOSURE TO PROMOTERS OF HEPATOCELLULAR CARCINOGENESIS AS A CAUSAL FACTOR IN CHEMICALLY INDUCED HEPATOBLASTOMAS

While attempting to clarify the genetics of susceptibility to promotion of hepatocellular carcinogenesis in susceptible DBA/2NCr (DBA) and resistant C57BL/6NCr (C57) mice, we found that phenobarbital (PB) fed in diet or in drinking water consistently led to development of malignant hepatoblastomas, as well as promoted hepatocellular carcinogenesis, in *N*-nitrosodiethylamine (NDEA)-initiated D2B6F1 males. This barbiturate did not elicit comparable frequency of hepatoblastoma development, however, in reciprocal hybrid

TABLE 1
Incidence of Neoplastic Hepatic Lesions in D2B6F1 and Reciprocal F_1 Hybrid Offspring Given NDEA by i.p. Injection Followed by PB in Drinking Water

Mice	Treatment groups (10 mice/group)	No. of mice with hepatocellular tumors	Hepatocellular tumors per mouse with tumors: Adenomas (mean ± S.E.)	Hepatocellular tumors per mouse with tumors: Carcinomas (mean ± S.E.)	No. of mice with hepatoblastomas
B6D2F1	1 (NDEA)	6	2.2 ± 0.41	0	0
	2 (NDEA,PB)	10	15.0 ± 1.36[a]	2.7 ± 0.41	1
D2B6F1	1 (NDEA)	7	2.7 ± 0.22	1.0 ± 0.0	0
	2 (NDEA,PB)	10	12.5 ± 2.55[a]	2.6 ± 0.3	8

[a] Statistically significant compared with corresponding group 1 ($p < 0.01$).

From Diwan et al., *Carcinogenesis,* 10, 1345, 1989. With permission.

B6D2F1 males that were genetically identical to D2B6F1 males except for the reverse origin of their X and Y chromosomes (Table 1),[17] although the B6D2F1 hybrid is known to develop such tumors naturally.[10] No hepatoblastomas were found in mice in this study given NDEA alone, whereas 80% of similarly initiated D2B6F1 males developed hepatoblastomas within a short latency period averaging 42 weeks after postinitiation treatment with PB. Histologically, hepatoblastomas in D2B6F1 mice were consistent with published descriptions of similar tumors occurring in other studies.[7,14-16] The early appearance (42 weeks of age) of hepatoblastomas in these mice was especially striking since these rare tumors characteristically occur in the mouse later in life.[7,14]

III. STRUCTURE-ACTIVITY RELATIONSHIPS IN PROMOTION OF HEPATOBLASTOMAS IN D2B6F1 MICE

In an earlier study[18] we observed an increase in the incidence of hepatoblastomas in another F1 hybrid with a C57 parent, B6C3F1, when male mice were initiated with NDEA and promoted by feeding di(2-ethylhexyl)phthalate. The observations of Turusov et al.[14] also involved yet another liver promoter, DDT. Thus liver tumor promoters of chemical classes structurally unrelated to PB also had significant ability to enhance development of hepatoblastomas in susceptible strains of mice. To investigate such a relationship systematically, we have examined the tumor-promoting effects of two structurally similar barbiturates, pentobarbital (PTB) and monoethylbarbituric acid (MEB); two hydantoin derivatives, ethylphenylhydantoin (EPH) and diethylhydantoin (DEH); and two polychlorinated hydrocarbons and well-established rodent liver tumor promoters, Aroclor®-1254 (Ar-1254) and DDT.[19-21] In addition,

1,4-bis[2-(3,5-dichloropyridyloxy)]benzene (TCPOBOP), like PB a potent inducer of biotransformation enzymes and also an established liver tumor promoter in mice,[22-24] was also tested for its ability to promote hepatoblastomas in D2B6F1 mice. Male mice of this hybrid were initiated with a single i.p. dose of 90 mg NDEA per kilogram body weight. They were given 2 weeks later a diet containing different test compounds at doses equimolar to 500 ppm of PB. Interestingly, as shown in Figure 1A, PTB and EPH, which are strong promoters of hepatocellular carcinogenesis in the rat,[25,26] were significantly weaker than PB in mice. Both incidence and multiplicity (Table 2) of hepatocellular carcinomas in NDEA-EPH or NDEA-PTB groups were not different than those in mice given NDEA alone. In contrast, many hepatocellular adenomas and carcinomas appeared in mice given NDEA-PB, NDEA-Ar-1254, NDEA-DDT, or NDEA-TCPOBOP. No significant differences were observed among these groups in the incidences of hepatocellular tumors (either adenomas or carcinomas). These groups, however, exhibited pronounced differences in the incidences of hepatoblastomas in this study (Figure 1A). Thus, the incidence of hepatoblastomas in the NDEA-PB group was 73%, while it was 33% in low-dose and 9.0% in high-dose AR-1254 groups, respectively. DDT promoted hepatoblastomas in only 27% of NDEA-initiated mice ($p < 0.002$ as compared to the NDEA-PB group). The weak promoters of hepatocellular carcinogenesis, EPH and PTB, failed significantly to increase the incidence of hepatoblastomas while MEB and DEH, non-promoters of hepatocellular carcinogenesis in both rats and mice, were totally ineffective in promoting hepatoblastomas. Interestingly, TCPOBOP given alone without an initiating dose of NDEA induced a high incidence of hepatoblastomas in these mice (Figure 1B).

Ten animals in the NDEA-TCPOBOP group and four animals in the TCPOBOP only group had hepatoblastoma metastatic to the lungs. Six animals in the NDEA-PB group also had lung metastasis (Figures 2 and 3). Hepatoblastomas found in Ar-1254- and DDT-treated animals were small lesions (Figure 4), None of these lesions in Ar-1254-promoted animals metastasized to the lungs, while only one hepatoblastoma in DDT-treated animals did so.

Among the barbiturate and hydantoin derivatives the ability to promote hepatoblastomas appeared to be related to their potency to enhance hepatocellular carcinogenesis in these mice. This relationship, however, does not appear to hold true for other compounds. For example, when three halogenated compounds, DDT, Ar-1254, and TCPOBOP were compared for their ability to promote hepatoblastomas, only TCPOBOP was found to be as effective as PB, while DDT and Ar-1254 were much weaker. DDT and Ar-1254 were at least as strong as PB in promoting hepatocellular carcinogenesis in these mice. Thus, the efficacy of a given agent to promote development of hepatoblastomas is related to its ability to promote hepatocellular carcinogenesis, but potency for the first effect does not necessarily parallel the latter. A corollary to this observation is that hepatoblastoma development appears directly related to exposure to the promoting compound, and is not simply a

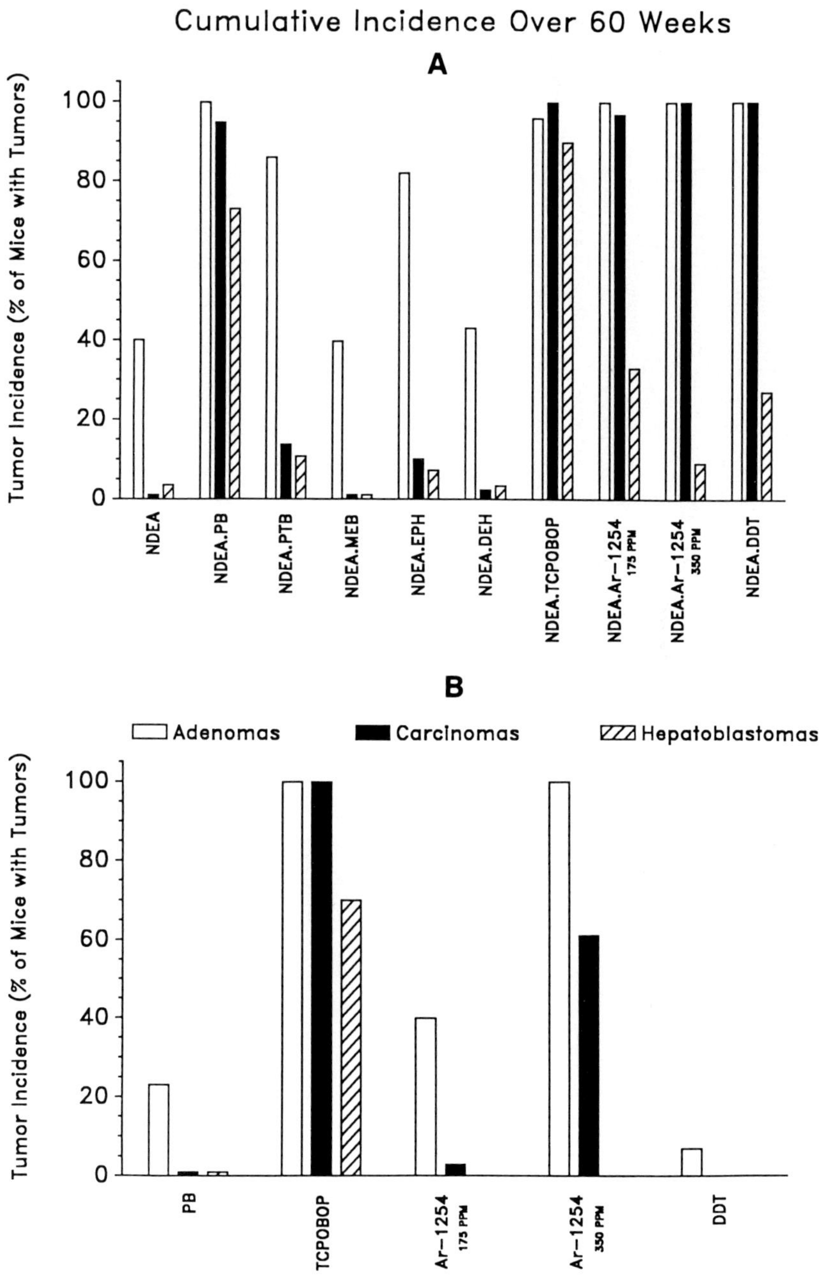

FIGURE 1. Incidence of hepatocellular adenomas, carcinomas, and hepatoblastomas in NDEA-initiated (A) or uninitiated (B) D2B6F1 mice subsequently exposed to various promoting compounds in the diet.

TABLE 2
Cumulative Incidence and Multiplicity of Hepatic Tumors in Various Groups of D2B6F1 Male Mice

Treatment groups	Number of mice with tumors/total number at risk	Incidence (%) of hepatocellular adenoma (mean number/tumor-bearing animal)	Incidence (%) of hepatocellular carcinoma (mean number/tumor-bearing animal)	Incidence (%) of lung metastasis	Incidence (%) of hepatoblastomas (mean number/tumor-bearing animal)	Incidence (%) of lung metastasis
1 NDEA	12/30	40 (3.4)	—	—	3.3 (1.0)	0
2 NDEA.PB	29/29[a]	100 (16.1)[a]	95 (7.9)	3.5	73 (2.7)	29.0
3 NDEA.Ar-1254 (175 ppm)	24/24[a]	100 (10.8)[a]	100 (15.4)[b]	29.2	33 (1.5)	—
4 NDEA.Ar-1254 (350 ppm)	23/23[a]	100 (17.3)[a]	100 (8.2)	21.8	9 (1.0)	—
5 NDEA.DDT	22/22[a]	100 (21.5)[a]	82 (5.1)	—	27.3 (1.0)	16.6
6 NDEA.PTB	24/28[a]	86 (4.8)	13 (1.5)	—	10.7 (1.0)	0
7 NDEA.MEB	12/30	40 (1.9)	0	—	0	0
8 NDEA.EPH	23/29[a]	82 (5.6)	10 (1.3)	—	7.2 (1.0)	0
9 NDEA.DEH	13/30	43 (3.5)	2.3 (3.0)	—	3.3 (1.0)	0
10 NDEA.TCPOBOP	30/30[a]	100 (6.2)[a]	100 (11.3)	40.0	90 (3.2)	37.0
11 Vehicle.TCPOBOP	30/30[a]	100 (6.8)[a]	100 (3.8)	13.3	70 (2.1)	19.0

[a] Significantly different ($p < 0.05$) than group 1 (NDEA only).
[b] Significantly different ($p < 0.05$) than group 2 (NDEA.PB).

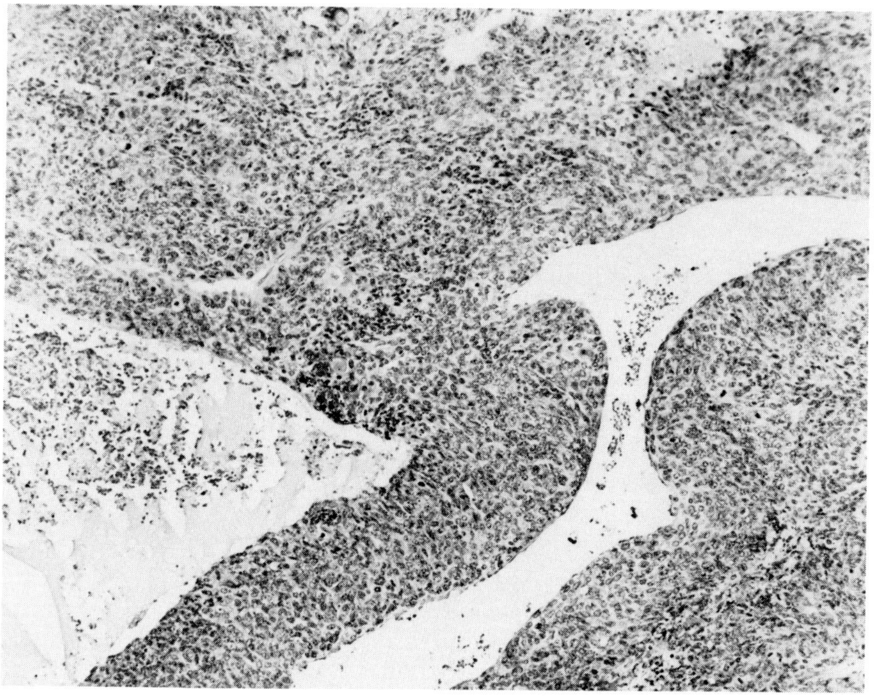

FIGURE 2. Low magnification of hepatoblastoma initiated by NDEA and promoted by PB, illustrating cords of tumor cells separated by large vascular spaces. (H.E.; magnification × 100.)

stochastic event in the natural evolution of hepatocellular tumors in these mice.

IV. HISTOLOGICAL FEATURES OF HEPATOBLASTOMAS IN MICE

Although mouse hepatoblastomas share many histological characteristics with the corresponding neoplasm in humans, some important differences exist between hepatoblastomas in the two species. Histologically, the murine hepatoblastomas were consistent with published illustrations[7,14] and descriptions.[15,16] In PB- and TCPOBOP-promoted animals, hepatoblastomas were large lesions and in some cases occupied an entire liver lobe. Hepatoblastomas were mostly found within or adjacent to hepatocellular carcinomas or adenomas and were often multifocal. In contrast to the eosinophilic hepatocellular neoplasms in H&E preparations, hepatoblastomas were intensely basophilic, containing small, poorly differentiated cells with scant cytoplasm and large

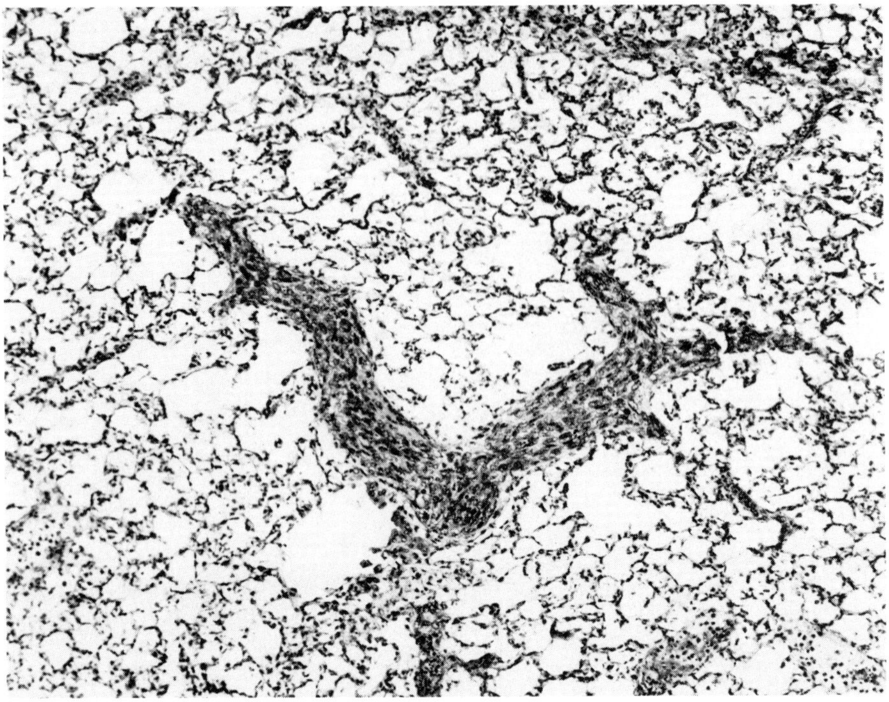

FIGURE 3. Pulmonary metastases of hepatoblastoma in Figure 2. (H.E.; magnification × 100.)

pleomorphic, hyperbasophilic nuclei (Figure 5). Between these small basophilic neoplastic cells were scattered small nests or rows of elongated (spindle-shaped) neoplastic cells with hyperchromatic nuclei and eosinophilic cytoplasm. These cells were often arranged in the form of rosettes, ribbons, or solid cords of cells. In some animals, there were multicentric masses of hepatoblastomas, many of which were hemorrhagic and some cystic. Some hepatoblastomas consisted of multilobular structures, occasionally with distinct lumens. The space between these structures was filled with blood elements. Vascular lakes, large cysts, and necrotic debris common occurred in these lesions. α-Fetoprotein (AFP) was consistently absent in hepatoblastomas, while focal areas of many large hepatoblastomas and the surrounding normal bile duct epithelial cells were positive for keratin by ABC immunohistochemistry (Figure 6). Focal squamous metaplasia was found in a few cases (Figure 7), an apparent progression of this tendency to form keratin, but we did not see osteoid or cartilage formation in hepatoblastomas in our study. Although the hepatoblastomas were well demarcated from surrounding neoplastic hepatocytes in most cases, in some tumors we saw infiltrative growth of hepatoblastoma cells into the surrounding parenchyma (Figure 8). Some early lesions were also found without distinct demarcation from the

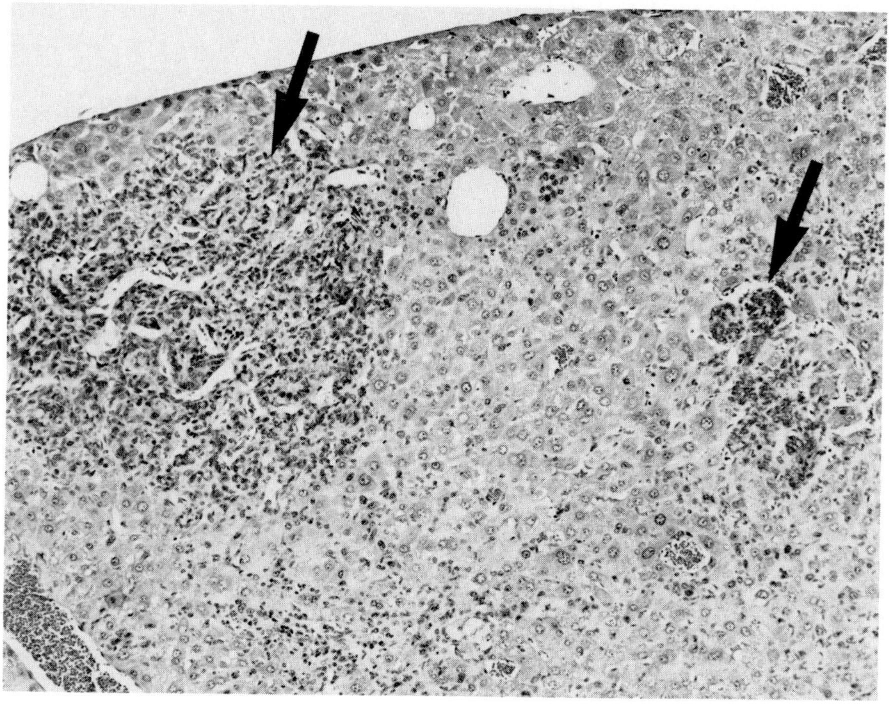

FIGURE 4. Multiple primary hepatoblastomas (arrows) within a liver after NDEA initiation and DDT promotion. (H.E.; magnification × 100.)

surrounding hepatocellular lesions. Mitoses were numerous in most of these tumors.

Based on the proportion of metaplastic mesenchymal components in the lesion, human hepatoblastomas are classified as epithelial or mixed.[2,5,27] A third type, the rhabdomyoblastic mixed tumor, has been described by Willis.[1] Ishak and Glunz[2] identified two types of cells, the fetal type and the embryonal type, within epithelial components of human hepatoblastomas. The fetal type cells are large, round cells with clear cytoplasm and small, round darkly stained nuclei, while the embryonal type cells are less well differentiated and form rows or rosettes. This ''small cell'' variant of human hepatoblastoma closely resembles the predominant component of murine hepatoblastomas. It is also important to note here that the embryonal or ''small cell'' subtype in human hepatoblastoma also lacks immunohistochemically demonstrable AFP.[6] Thus, as suggested by Turusov et al.,[28] the cells of murine hepatoblastoma appear to be less differentiated than most such tumors in humans. Recently, Nonoyama et al.[15] described some ultrastructural differences between human and mouse hepatoblastomas. Their finding indicated that mouse hepatoblastoma cells, unlike those in human lesions, lacked crystalloid inclusions in

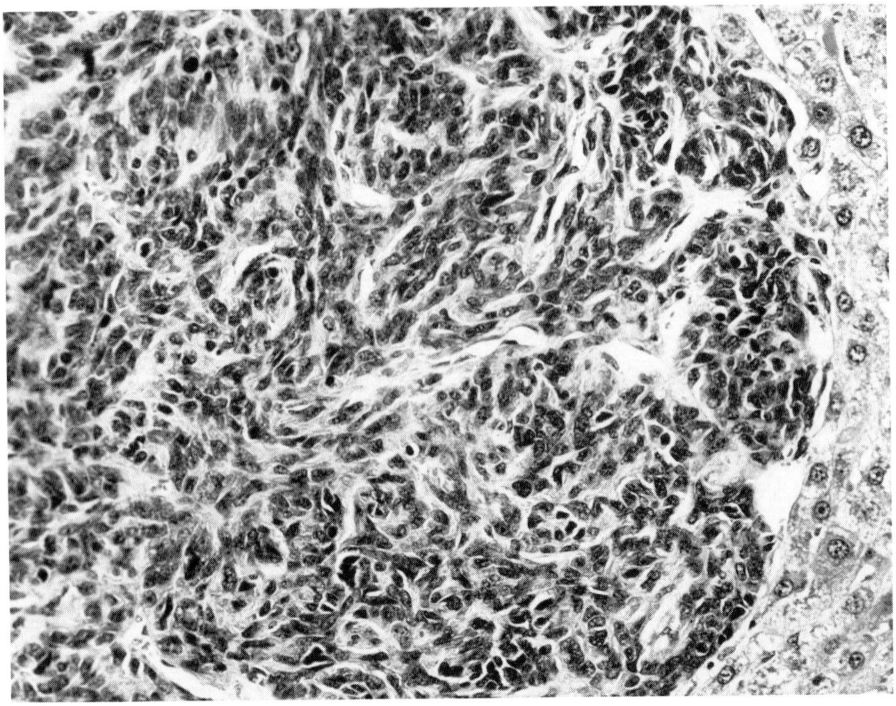

FIGURE 5. Hepatoblastoma initiated by NDEA and promoted by TCPOBOP, showing intensely basophilic cells with scant cytoplasm. (H.E.; magnification × 250.)

mitochondria and were characterized by scarcity of smooth endoplasmic reticulum, Golgi apparatus, and glycogen granules.

V. HISTOGENESIS OF HEPATOBLASTOMAS IN MICE

The histogenesis of hepatoblastomas in mice remains to be fully characterized. Cholangiocarcinomas, hepatocarcinomas, mesenchymal tumors, and hepatoblastomas all developed in neonatal F344 rats injected with cell clones derived from a chemically transformed rat liver epithelial cell line which has properties virtually identical to bile ductular cells and oval cells.[29] It is thus possible to formulate at least two pathways for the evolution of hepatoblastomas in mice subjected to tumor promoters:

1. Evolution from a nonhepatocyte precursor cell, possibly oval cells or bile duct lining cells. Consistent with this are the observations that hepatoblastomas do not express AFP, but do express keratin, an antigen present in normal bile ducts.[16,17] Kharkovskaya et al.[30] recently found that some of the cells of canals of Hering were consistently negative for A6 antigen (found in epithelial cells lining bile ducts and ductules,

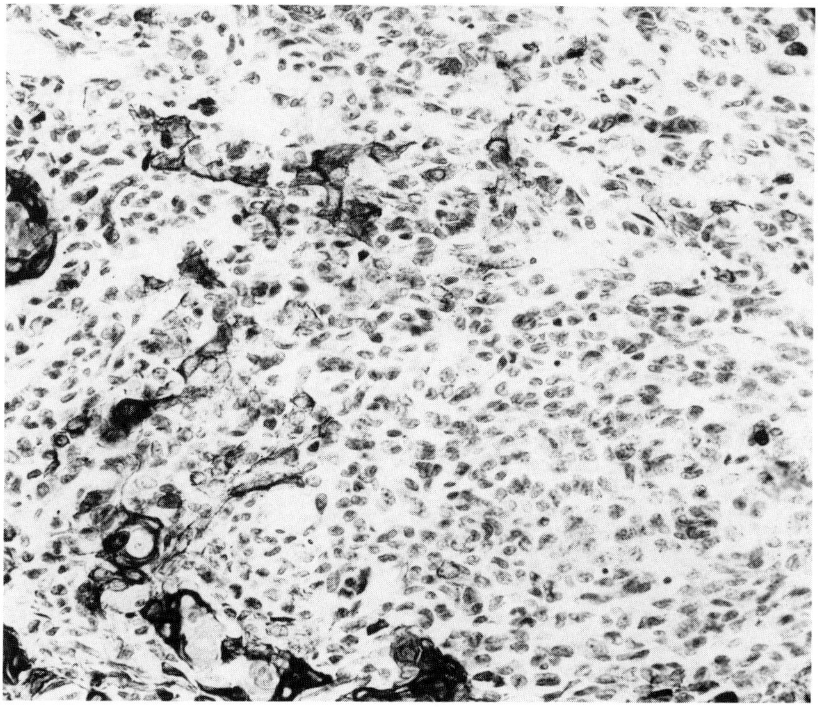

FIGURE 6. Hepatoblastoma cells focally express keratin, probably a precursor to squamous metaplasia. The animal received NDEA initiation and PB promotion. The DAKO rabbit anti-human keratin antibody A575 was used. This antibody immunoreacts exclusively with normal bile duct epithelium (not hepatocytes) in normal mouse liver. (ABC immunohistochemistry for keratin; magnification × 250.)

including the terminal Hering canal) under normal conditions. Since transplanted hepatoblastomas were also negative for this antigen, the authors suggested that A6-negative cells might be the cells of hepatoblastoma origin.

2. Progression of the hepatocyte lineage, known to be susceptible to promotion, from foci to adenomas to hepatocellular carcinomas and further as a result of loss of differentiation (or dedifferentiation) to hepatoblastomas. Consistent with this pathway are the following observations: (a) almost all hepatoblastomas were found within or adjacent to hepatocellular lesions, including hepatocellular foci (Figure 9); (b) hepatoblastomas are invariably found in eosinophilic hepatocellular neoplasms *promoted* by PB or PB-type promoters and not in basophilic lesions occurring spontaneously or induced by a carcinogen; (c) in several cases, dark basophilic, poorly differentiated hepatoblastoma cells appeared to evolve from hepatocellular carcinoma cells in trabeculae (Figures 10 and 11). Such lesions, as well as small and presumably early foci of

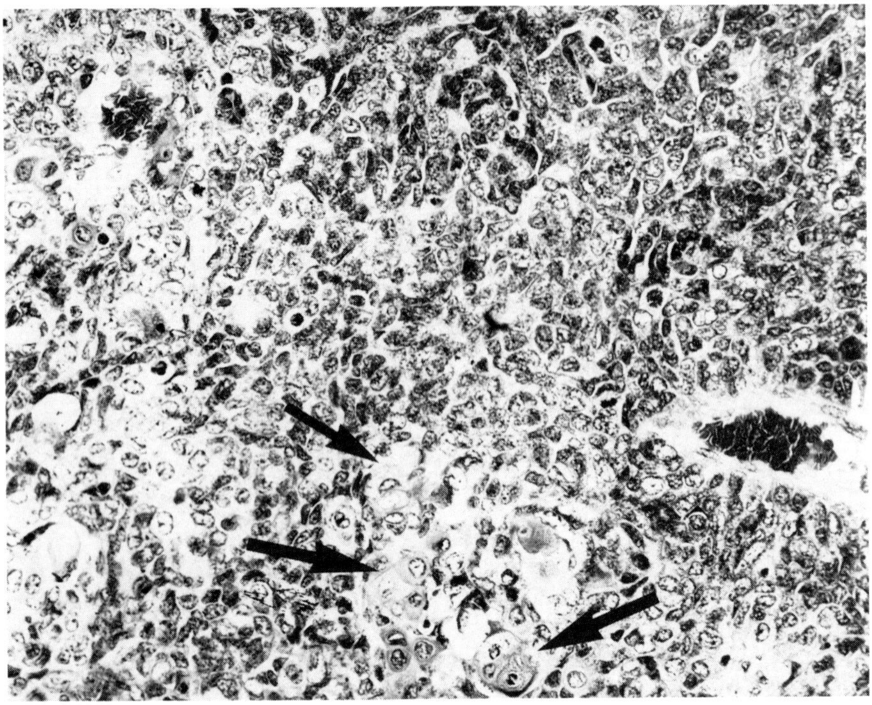

FIGURE 7. Squamous metaplasia (arrows) within an area of a hepatoblastoma initiated by NDEA and promoted by TCPOBOP. (H.E.; magnification × 250.)

hepatoblastoma, consisting of basophilic cells forming ductlike structures, were consistently negative for keratin and appeared to be different from bile ducts; (d) no proliferation of bile ducts, oval cells, or other candidate cells of origin other than neoplastic hepatocytes was apparent in NDEA-initiated, promoter-treated mice.

While the immunohistochemical characteristics of murine hepatoblastoma cells resemble those of cholangiolar cells more closely than those of hepatocytes, particularly in regard to the absence of AFP in hepatoblastoma cells, this discrepancy does not preclude an origin from neoplastic hepatocytes. Based on our observations on a large number of hepatoblastomas in various stages of development, we propose that hepatoblastomas in D2B6F1 mice fed PB following a carcinogen treatment are stimulated to evolve from preexisting preneoplastic or neoplastic hepatocytes and that this process is one of tumor progression. We do not, however, rule out the possibility that some mouse hepatoblastomas may evolve from a nonhepatocyte precursor cell, possibly bile duct lining cells, oval cells, or cells lining the Hering canals.[30] Clearly, more histochemical, ultrastructural, and *in situ* hybridization studies

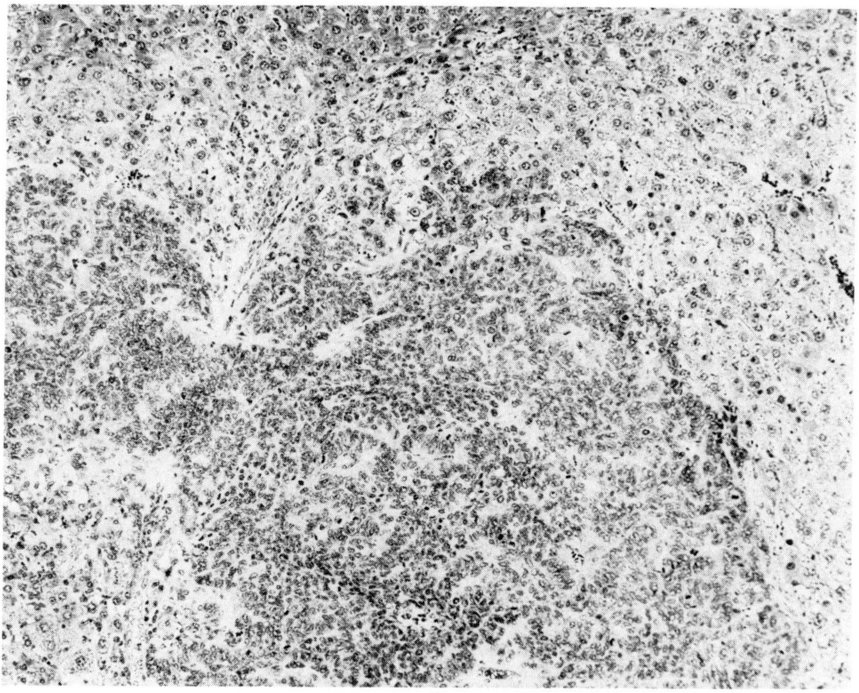

FIGURE 8. Infiltrative growth of hepatoblastoma cells into surrounding parenchyma after NDEA initiation and TCPOBOP promotion. (H.E.; magnification × 100.)

should be carried out to characterize the sequence of molecular events that accompany, and may cause, evolution of this malignant embryonal tumor.

ACKNOWLEDGMENTS

This project has been funded at least in part with federal funds from the Department of Health and Human Services under contract number NO1-CO-74012 with PRI/DynCorp, Inc. The content of this publication does not necessarily reflect the views or policies of the Department of Health and Human Services, nor does mention of trade names, commercial products, or organizations imply endorsement by the U.S. Government.

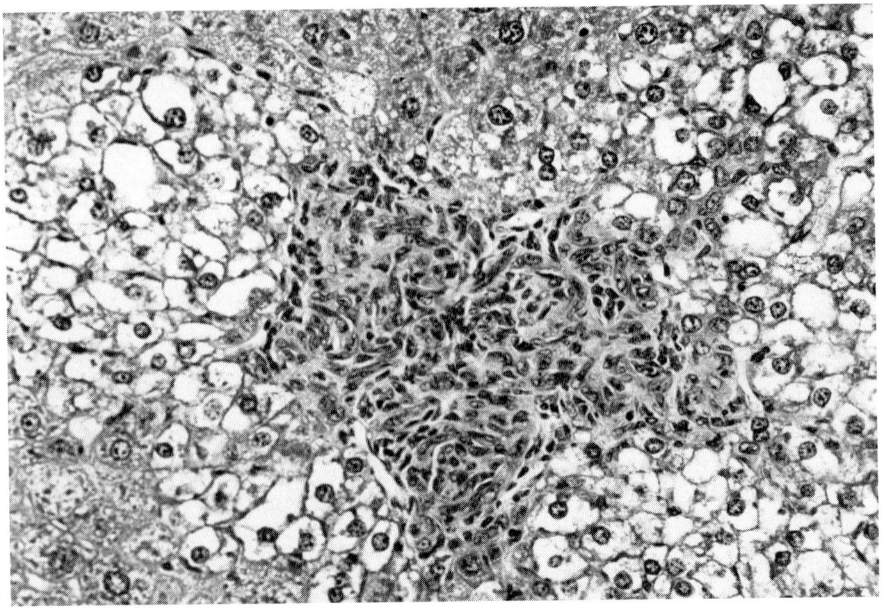

FIGURE 9. Early hepatoblastoma arising within a hepatocellular clear cell focus after NDEA initiation and PB promotion. (H.E.; magnification × 250.)

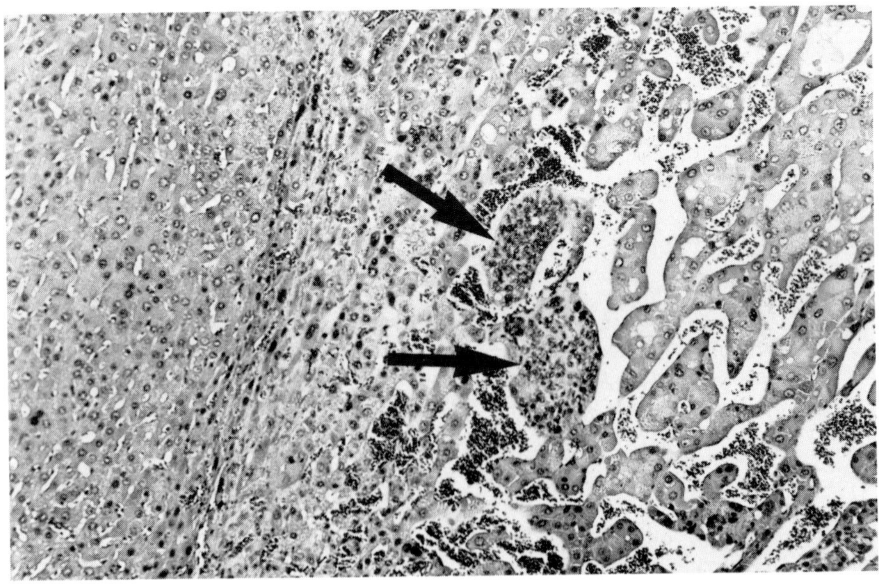

FIGURE 10. Hepatoblastoma (arrows) arising within a hepatocellular carcinoma from neoplastic hepatocytic trabeculae after NDEA initiation and Aroclor® (350 ppm) promotion. (H.E.; magnification × 250.)

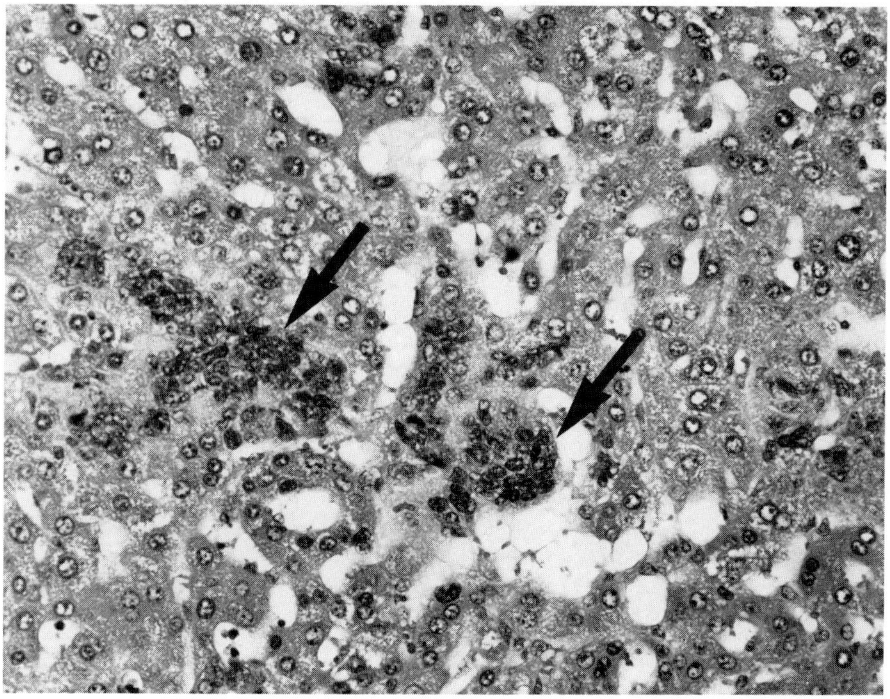

FIGURE 11. Hepatoblastoma (arrows) arising within a hepatocellular carcinoma after NDEA initiation and Aroclor® (175 ppm) promotion. (H.E.; magnification × 100.)

REFERENCES

1. **Willis, R. A.,** *The Pathology of the Tumors of Children,* Charles C Thomas, Springfield, IL, 1962, 57.
2. **Ishak, K. G. and Glunz, R. R.,** Hepatoblastoma and hepatocarcinoma in infancy and childhood: report of 47 cases, *Cancer,* 20, 396, 1967.
3. **Misugi, K., Okajima, H., and Misugi, N.,** Classification of primary tumors of liver in infancy and childhood, *Cancer,* 20, 1760, 1967.
4. **Dehner, L. P.,** Hepatic tumors in the pediatric age group: a distinctive clinicopathologic spectrum, *Perspect. Pediatr. Pathol.,* 4, 217, 1978.
5. **Pollice, L.,** Primary pediatric tumors of the liver, in *Progress in Surgical Pathology,* Vol. III, Penoglio, C. M. and Wolff, M., Eds., Masson, New York, 1981, 195.
6. **Adenova, P., Manivel, J. C., Wick, M. R., Hagen, K., and Dehner, L. P.,** Hepatoblastoma: an immunohistochemical and ultrastructural study, *Hum. Pathol.,* 18, 1025, 1987.

7. **Turusov, V. S. and Takayama, S.,** Tumors of the liver, in *Pathology of Tumors in Laboratory Animals,* Vol. II, Turusov, V. S., Ed., IARC Scientific Publications No. 23, International Agency for Research on Cancer, Lyon, 1979, 193.
8. **Maronpot, R. R., Haseman, J. K., Boorman, G. A., Eustis, S. E., Rao, G. N., and Huff, J. E.,** Liver lesions in B6C3F1 mice: the National Toxicology Program, experience and position, *Arch. Toxicol.,* Suppl. 10, 10, 1987.
9. **Jones, G. and Butler, W. H.,** Morphology of spontaneous and induced neoplasia, in *Mice Hepatic Neoplasia,* Butler, W. H., Ed., Elsevier, Amsterdam, 1975, 21.
10. **Yamate, J., Tajima, M., Kudow, S., and Sannai, S.,** Background pathology in BDF_1 mice allowed to live out their life-span, *Lab. Anim.,* 24, 332, 1990.
11. **Vlahakis, G. and Heston, W. E.,** Spontaneous cholangiomas in strain C3H-A^{vy}*f*B mice and their hybrids, *J. Natl. Cancer Inst.,* 16, 677, 1971.
12. **Reuber, M. D.,** Poorly differentiated cholangiocarcinomas occurring "spontaneously" in C3H and C3H × Y hybrid mice, *J. Natl. Cancer Inst.,* 38, 901, 1967.
13. **Rice, J. M.,** The biological behavior of transplacentally induced tumors in mice, in *Transplacental Carcinogenesis,* Tomatis, L. and Mohr, U., Eds., IARC Scientific Publication No. 4, International Agency for Research on Cancer, Lyon, 1973, 71.
14. **Turusov, V. S., Day, H. E., Tomatis, L., Gati, E., and Charles, R. T.,** Tumors in CF-1 mice exposed for six consecutive generations to DDT, *J. Natl. Cancer Inst.,* 51, 983, 1973.
15. **Nonoyama, T., Reznik, G., Bucci, T. J., and Fullerton, F.,** Hepatoblastoma with squamous differentiation in a B6C3F1 mouse, *Vet. Pathol.,* 23, 619, 1986.
16. **Nonoyama, T., Fullerton, F., Reznik, G., Bucci, T. J., and Ward, J. M.,** Mouse hepatoblastomas: a histologic, ultrastructural, and immunohistochemical study, *Vet. Pathol.,* 25, 286, 1988.
17. **Diwan, B. A., Ward, J. M., and Rice, J. M.,** Promotion of malignant "embryonal" liver tumors by phenobarbital: increased incidence and shortened latency of hepatoblastomas in (DBA/2 × C57BL/6)F1 mice initiated with N-nitrosodiethylamine, *Carcinogenesis,* 10, 1345, 1989.
18. **Ward, J. M., Rice, J. M., Creasia, D., Lynch, P., and Riggs, C.,** Dissimilar patterns of promotion by di(2-ethylhexyl)phthalate and phenobarbital of hepatocellular neoplasia initiated by diethylnitrosamine in B6C3F1 mice, *Carcinogenesis,* 4, 1021, 1983.
19. **Ito, N., Nagasaki, H., Arai, M., Makisura, S., Sugihara, S., and Hirao, K.,** Histopathological studies on liver tumorigenesis by technical polychlorinated biphenyls and its promoting effect on liver tumors induced by benzene hexachloride, *J. Natl. Cancer Inst.,* 51, 1037, 1973.
20. **Preston, B. D., Van Miller, J. P., Moore, R. W., and Allen, J. R.,** Promoting effects of polychlorinated biphenyls (Aroclor 1254) and polychlorinated dibenzofuran-free Aroclor 1254 on diethylnitrosamine-induced tumorigenesis in the rat, *J. Natl. Cancer Inst.,* 66, 509, 1981.
21. **Williams, G. M. and Numoto, S.,** Promotion of mouse liver neoplasms by the organochlorine pesticides chlordane and heptachlor in comparison to dichlorodiphenyltrichloroethane, *Carcinogenesis,* 5, 1089, 1984.
22. **Poland, A., Mak, I., and Glover, E.,** Species differences in responsiveness to 1,4,bis[2-(3,5-dichloropyridyloxy)]benzene, a potent phenobarbital-like inducer of microsomal monooxygenase activity, *Mol. Pharmacol.,* 20, 442, 1981.
23. **Dragani, T. A., Manenti, F., Gallani, G., and Della Porta, G.,** Promoting effects of 1,4-bis[2-(3,5-dichloropyridyloxy)]benzene in mouse hepatocarcinogenesis, *Carcinogenesis,* 6, 225, 1985.
24. **Diwan, B. A., Lubet, R. A., Ward, J. M., and Rice, J. M.,** Liver tumor promoting and/or hepatocarcinogenic effects of 1,4-bis[2-(3,5-dichloropyridyloxy)]benzene in C57BL/6NCr mice and in F344 rats, *Toxicologist,* 10, 232, 1990.

25. **Diwan, B. A., Rice, J. M., Hu, H., and Ward, J. M.,** Barbiturate structure/tumor promotion relationship: tumor promoting effects of two long-acting hypnotic barbiturates, 5,5-diallylbarbituric acid and 5-allyl-5-isopropylbarbituric acid in the rat liver, *Proc. Am. Assoc. Cancer Res.*, 27, 141, 1986.
26. **Diwan, B. A., Rice, J. M., Nims, R. W., Lubet, R. A., Hu, H., and Ward, J. M.,** P450 enzyme induction by 5-ethyl-5-phenylhydantoin and 5,5-diethylhydantoin, analogs of barbiturate tumor promoters phenobarbital and barbital, and promotion of liver and thyroid carcinogenesis initiated by N-nitrosodiethylamine in rats, *Cancer Res.*, 48, 2492, 1988.
27. **Gibson, J. B.,** Histological typing of tumors of the liver, biliary tract and pancreas. 5. Hepatoblastoma, in *International Histological Classification of Tumors,* No. 20, World Health Organization, Geneva, 1978, 24.
28. **Turusov, V. S., Deringer, M. K., Dunn, T. B., and Stewart, H. L.,** Malignant mouse liver tumors resembling human hepatoblastomas, *J. Natl. Cancer Inst.*, 51, 1689, 1973.
29. **Tsao, J. S. and Grisham, J. W.,** Hepatocarcinomas, cholangiocarcinomas, and hepatoblastomas produced by chemically transformed cultured rat liver epithelial cells. A light- and electron-microscopic analysis, *Am. J. Pathol.*, 127, 168, 1987.
30. **Kharkovskaya, N. A., Svinolupova, S. I., Khrustalev, S. A., Engelhardt, N. V., Kondalenko, V. F., Poltoranina, V. S., and Turusov, V. S.,** Transplantable mouse hepatoblastoma: histologic, ultrastructural and immunohistochemical study, *Exp. Pathol.*, 40, 283, 1990.

Chapter 5

OVAL CELLS IN LIVER CARCINOGENESIS: CELL LINEAGES IN HEPATIC DEVELOPMENT AND THE IDENTIFICATION OF FACULTATIVE STEM CELLS IN NORMAL LIVER

N. Fausto, J. M. Lemire, and N. Shiojiri

TABLE OF CONTENTS

ISBN 0-8493-4746-7
© 1992 by CRC Press, Inc.

I. IDENTIFICATION, ISOLATION, AND CULTURE OF OVAL CELLS

The recognition that liver nonparenchymal epithelial cells may proliferate during hepatocarcinogenesis dates from the earliest experiments that showed that chemical agents can cause hepatic tumors.[1-3] An important landmark in the analysis of these cells occurred more than 20 years after Sasaki and Yoshida's[1] original observations with the systematic description by Farber[4] of the histologic changes occurring during hepatocarcinogenesis caused by three chemicals with different structures: ethionine, 2-acetylaminofluorene (AAF) and 3′-methyl-4-dimethylaminoazobenzene (Me-DAB). Farber[4] described the similarities between the histologic alterations induced by these chemicals and pointed out that the proliferation of small epithelial cells (which were referred to as "oval cells") was a common feature of the three experimental models. Some aspects of the early stages of liver carcinogenesis described in this paper are particularly relevant for the present discussion because the observations served as a basis for future investigations of the role of oval cells in hepatic neoplasia:

1. Oval cells appeared to be bile ductular cells that radiated from the portal spaces into the parenchyma.
2. Their proliferation occurred very early after the start of the carcinogenic regimens and preceded the formation of hyperplastic nodules.
3. Oval cell proliferation by itself was considered unlikely to be a premalignant change, but was seen as a condition that could set the stage for neoplastic development.
4. Although no transitions between oval cells and hepatocytes were observed in ethionine and AAF-induced hepatocarcinogenesis, apparent "merging" of oval cells and hepatocytes in regenerating nodules occurred in livers of animals fed Me-DAB.[4]

The study of oval cell/hepatocyte transitions during azo dye carcinogenesis was actively pursued in Japan.[5-8] Other important accomplishments of the period from 1960 to the early 1980s included the development of carcinogenic diets (choline-deficient diet containing 0.1% ethionine; CDE) that proved to be particularly suitable for oval cell analysis[9] and the observation that oval cells, but not hepatocytes are the cells that produce α-fetoprotein (AFP) at the earlier stages of liver carcinogenesis induced by many different chemicals.[7,8,10-12] A major conceptual advance was the formulation by Grisham[13,14] that liver nonparenchymal cells are clonogenic and can function as facultative stem cells.

A new stage in the work on oval cells was launched with the development by several laboratories of procedures that permitted the isolation of these cells in relatively high yields from livers of carcinogen-treated rats.[15-18] Our own intent in developing these techniques was to obtain oval cells in sufficient numbers to perform biochemical analysis and to establish cell lines.[16,17,19] As these cells, isolated from livers of rats fed the CDE diet, were characterized and maintained in culture, it became clear that they had very similar properties to liver epithelial cells (LEC) obtained from normal adult or fetal rat livers.[20,21] In particular, the LE2 and LE6 oval cell lines developed in our laboratory[22,23] are very similar to the WB-F344 line derived from normal adult rat liver by Tsao and co-workers.[20,21,24,25] Work with oval cells and LEC lines proceeded in parallel and led to major advances in the field, opening up new perspectives on the developmental capacity of some types of liver nonparenchymal cells. The major conclusions derived from the first phase of our own work involving oval cells include:

1. Oval cells isolated from livers of carcinogen-treated rats are heterogeneous cell populations that contain cells with different developmental potential. Because of this heterogeneity, we have used the term "oval cell compartment" to refer to the nonparenchymal epithelial cells that proliferate at the early stages of hepatocarcinogenesis.[19]
2. About half of the cells in this compartment contain AFP mRNA (the proportion of cells in which the protein is detected by immunohistochemistry is somewhat smaller) and 20 to 40% of the cells in the compartment have albumin mRNA.[16,17] These proportions vary with the type of carcinogen used and the time elapsed from the start of the carcinogenic regimen.
3. A variable proportion of cells in this compartment contain pyruvate kinase, aldolase, and lactic dehydrogenase isozymes found in fetal liver and duct cells.[17] At least some cells produce both fetal and adult isozymes (pyruvate kinase K and L or aldolase A and B expression in the same cell). In addition, more than 90% of cells in the compartment are γ-glutamyl transpeptidase (GGT)-positive and a high proportion express the placental form of alkaline phosphatase.[17,26]
4. Both *in vivo* and in culture, cells of the oval cell compartment have glucose-6-phosphatase activity.[27] In culture, enzyme activity is detected in confluent cultures, but not in exponentially growing cells. Cells in confluent cultures contain peroxisomes and the number of these organelles can be increased by the addition to the cultures of clofibrate, an agent that stimulates peroxisome proliferation.[27]
5. Oval cells transplanted into fat pads of syngeneic rats can form duct structures resembling bile ducts 2 to 3 weeks following transplantation. No hepatocytes were apparent in the transplants.[27a]
6. The established oval cell lines LE2 and LE6 are diploid or quasidiploid (mostly lacking chromosome 1), do not grow in soft agar, and are not

tumorigenic when transplanted into syngeneic rats or nude mice.[22,23] After approximately 50 passages in culture the cells can form colonies in soft agar if grown in medium containing a high concentration of EGF, but cells from these colonies do not form tumors in nude mice.

7. Oval cell lines can be transformed in culture by oncogene transfection or by infrequent passage.[22,23,28] With either technique the treated cells give rise to hepatocellular carcinomas when transplanted subcutaneously into nude mice or inoculated into the liver or spleen of syngeneic rats. The tumors formed are mostly trabecular hepatocellular carcinomas with a moderate to high degree of differentiation. The histology of tumors produced in nude mice by approximately one dozen clonal isolates of the LE6 line that were transfected with the EJ-*ras* gene varied from poorly differentiated to well-differentiated hepatocellular carcinomas with frequent generation of tumors with a predominant glandular component. However, in separate transfection experiments the same clones could produce hepatocellular carcinomas with different degrees of differentiation.
8. The expression of hepatocyte and oval cell antigens can be modulated in cultured cells obtained from the tumors generated from oval cells.[28]
9. Oval cells in culture express high levels of c-*myc*, p53, TGFβ1, and IGF-II mRNA among other genes.[23,29-31] Cell cycle association of c-*myc* expression is maintained in nontransformed and ''partially transformed'', but nontumorigenic cells, but the acquisition of tumorigenicity is associated with constitutive c-*myc* expression.[23] During the process of transformation in culture, oval cell progressively lose sensitivity to TGFβ1.[30] DNA synthesis in fully transformed cells is not inhibited by TGFβ1 and may even be stimulated by it.

II. OVAL CELLS AS PROGENITOR CELLS

The results obtained with isolated oval cells and oval cell lines and the *in vivo* studies of rat hepatocarcinogenesis[32-36] support the notion that at least some cells in the oval cell compartment are dual-lineage progenitors that have the capacity to differentiate into hepatocytes and may function as progenitors for hepatocellular carcinomas. In addition, oval cells are rich sources of growth factors which may contribute to liver carcinogenesis *in vivo*.[30,37] The lineage-generation property of oval cells might depend on the capacity of the cells to produce or respond to growth factors, but a linkage between these properties has not been firmly established.

As an oval cell compartment is not evident in the liver during regeneration after partial hepatectomy (tissue replenishment results from hepatocyte replication),[38,39] it is logical to assume that oval cells are induced to proliferate by types of injury that cause massive hepatocyte death and inhibition of hepatocyte proliferation.[13,40] The degree and extent of oval cell proliferation may thus depend on the type and extent of injury produced by a carcinogenic agent. Further, the presence of an oval cell compartment at the early stages of carcinogenesis does not necessarily indicate that elements of the compartment are generating hepatocytes. Lineage generation may also be variable and depends on the type of injury induced by a carcinogen. Since it has been established that hepatocellular carcinomas can originate from oval cells,[22,23,28] perhaps the most important problem regarding the role of the oval cell compartment in hepatocarcinogenesis is to determine the precise conditions that may make these cells function as tumor progenitors *in vivo* and the mechanisms that regulate their differentiation into normal and transformed hepatocytes.

The analysis of oval cell proliferation during liver carcinogenesis and of oval cell growth and transformation in culture led to the formulation of new questions regarding the origin, lineage commitment, and differentiation of these cells. In particular, we explored three basic problems that are central to these questions:

1. The origin of ''oval cells'' during liver development
2. The localization of ''oval cell'' precursors in normal adult rat liver
3. The proliferation of ''oval cells'' in noncarcinogenic liver injury

Similar questions apply to LEC obtained from normal livers. These cells can also be transformed in culture by chemicals or oncogene transfection and are capable of generating a spectrum of tumors which include hepatoblastomas and hepatocellular carcinomas.[40,41]

III. THE ORIGIN OF OVAL CELLS: ANALYSIS OF LIVER DEVELOPMENT

We analyzed liver development with the goal of determining whether or not there are cells in the fetal liver which may correspond to the lineage-generating cells in the oval cell compartment in hepatic carcinogenesis. To establish markers for the earliest committed hepatic cells we determined the time at which AFP and albumin mRNA as well as the corresponding proteins are first expressed during rat liver embryonic development. As oval cells are likely to originate from bile ductular cells (or from cells located near bile ductules in portal areas),[13,14,19,32,42-44] we were particularly interested in the formation of intrahepatic biliary ducts in fetal livers, a process that has been studied in detail by various investigators.[45-49]

The endodermal component of the liver primordium arises from cells of the ventral foregut.[50,51] AFP mRNA was detected in 10.5-d fetal liver[52] in the thickened area of the ventral foregut which is in direct contact with the septum transversus mesenchyma. Thus, AFP mRNA transcription occurs in the cells of the ventral foregut from which the liver originates, before hepatic morphogenesis is initiated. Both AFP and albumin mRNA as well as AFP are expressed[52] 1 d later at a time when hepatic cord formation is in progress; albumin is detected by immunohistochemical methods at 12.5 d. This sequence of events is very similar in rats and mice[47] (in mice all these steps occur one gestational day earlier than in rats) and demonstrates that all epithelial cells that constitute the adult liver (hepatocytes, intrahepatic, and extrahepatic bile ducts) originate from AFP and albumin-positive hepatoblasts.[45,47,52,53]

One of the identifying characteristics of oval cell populations is their expression of both hepatocyte and bile duct lineage markers.[19,32,43] Starting at about the 15th day of rat liver development, AFP and albumin-positive hepatoblasts located near large vascular spaces give rise to primitive intrahepatic bile ducts.[45,47,54] When first formed, cells of these primitive bile ducts express both hepatocyte (AFP, albumin, and the surface antigen HES6) and duct lineage markers (cytokeratins 7 and 19, calf muzzle keratin, and the surface antigen BDS7).[52] The simultaneous expression of dual-lineage markers and the location of these structures in the hepatic tissue led us to suggest that the primitive intrahepatic bile ducts formed during liver development are similar to cells in the oval cell compartment in hepatocarcinogenesis.[52,53]

The suggested linkage between oval cells in adult liver and primitive intrahepatic ducts in hepatic development (Figure 1) carries several important implications. It provides support for the notion that oval cells derive from some portion of the biliary tree and makes it likely that segments of the biliary tree of adult liver may correspond to the primitive intrahepatic ducts of developing liver. Alternatively, it is conceivable that primitive intrahepatic ducts that may remain in adult hepatic tissue could form separate structures which are adjacent, but not directly connected to the biliary duct system. Yamamoto and Phillips[55] have described an anastomosing bile ductular plexus in rat liver, but this system is interconnected with the rest of the biliary tree.

The primitive intrahepatic ducts of the developing liver are transitional structures between hepatoblasts and mature bile ducts. They appear capable of generating cells of both the hepatocyte and duct lineages as has been shown for rat liver hepatoblasts at the 12th day of development.[52,56] Interactions with extracellular matrix components such as laminin,[49,57] growth factor gradients, position of the cells in the lobule, etc., are the types of stimuli that may determine lineage commitment of primitive intrahepatic ducts in fetal development and of oval cells during hepatocarcinogenesis (Figure 1).

It is interesting to note that in rats, intrahepatic bile ducts continue to be formed until the first or second postnatal week.[45,52] However, ducts formed postnatally do not originate from AFP-expressing hepatoblasts, but instead are formed from hepatocytes that are almost fully mature,[52,53] although they are still replicating cells. This suggests that it should be possible for mature or nearly mature hepatocytes in adult hepatic tissue to give rise to bile duct

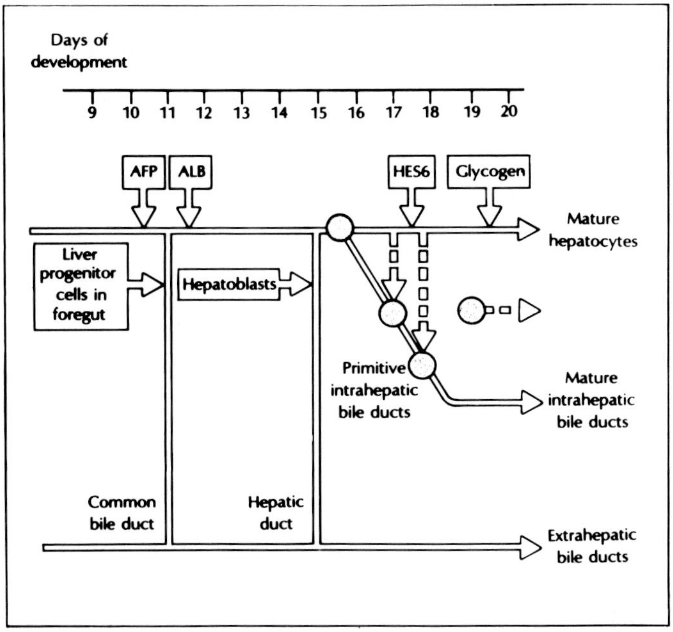

FIGURE 1. Cell lineages in rat liver development and the developmental origin of biopotential progenitor cells in adult livers. The figure indicates: (a) days of development; (b) hepatocyte maturation and the emergence of various markers: AFP, albumin, the cell surface marker HES6, and glycogen; (c) the origin of extrahepatic bile ducts (common and hepatic ducts) from hepatoblasts; and (d) the origin of intrahepatic bile duct from hepatoblasts. Note that intrahepatic bile ducts first appear at 15.5 d of development, but continue to be formed at later stages. The lineage commitment and the degree of differentiation of the cells from which intrahepatic bile ducts develop differs depending on the time during development when these ducts are formed. Hepatoblasts forming primitive intrahepatic bile ducts during development (gray circles) are probably the cells which form a facultative stem cell compartment in adult rat liver. (Reprinted from N. Fausto, *Curr. Opinion Cell Biol.*, 2, 1036, 1990. With permission.)

structures. This has indeed been observed by Desmet,[58] who has described the transdifferentiation or bile duct metaplasia of limiting plate hepatocytes in cases of chronic incomplete obstruction of intra- or extrahepatic bile ducts in humans. If bile duct differentiation from hepatocytes takes place during rat liver carcinogenesis, the duct structures formed from this process become incorporated into the oval cell compartment. However, these ducts as well as those formed by the replication of existing fully differentiated ducts would more likely resemble the seemingly mature intrahepatic ducts formed in rat liver during the first 10 d of life, and should not express AFP or be capable of generating hepatocytes. The presence of bile duct components that differ in their degree of maturation, development potential, and marker expression could account for the cellular heterogeneity of the oval cell compartment.

IV. THE ORIGIN OF OVAL CELLS: AFP EXPRESSION IN NORMAL ADULT LIVER CELLS

One of the strategies that can be used to search for oval cell precursors in normal liver is to look for AFP-expressing cells located in or near the bile duct system in the portal spaces. The rationale for this choice is that as discussed above, AFP mRNA is one of the earliest markers expressed during the genesis of the liver primordium,[52,53] is expressed together with other hepatocyte and duct markers in primitive bile ducts in fetal liver,[52,53] and is highly expressed in oval cells at the early stages of liver carcinogenesis and in most hepatic tumors.[19,32-34,59,60]

Although it is generally accepted that AFP transcription is shut off shortly after birth,[61] analysis of AFP RNA expression in the rat revealed the existence in adult normal liver of two AFP RNA variants of smaller size[62-64] than the 2.1-kb molecule found in fetal livers, oval cells, and hepatocellular carcinomas. It has now been established that variant AFP RNAs of 1.4 and 1.0 kb, which are truncated forms corresponding to the 3′ portion of the 2.1-kb full length AFP mRNA,[63,64] are expressed by both parenchymal and nonparenchymal cells in normal adult rat liver.[64] The 1.4-kb species includes an alternate exon which is not part of the 2.1-kb molecule and encodes translation products which are smaller than the 68- to 70-kd complete AFP molecule, but can be precipitated by anti-AFP antibodies.[64] The physiological significance, if any, of these smaller AFP polypeptides is not known at this time, but it is possible that they retain the capacity present in the complete protein of binding hormones and other molecules.

The expression of the 2.1-kb AFP mRNA and the smaller variants in rat liver is developmentally regulated. The 2.1-kb mRNA is highly expressed in fetal liver (together with a newly described 1.7-kb variant present in much lower amounts), but is barely detectable after the third week of life.[62-64] In contrast, the 1.4- and 1.0-kb variants are not expressed in fetal liver, but can be detected in normal adult livers. The amounts of the 1.4- and 1.0-kb RNA in normal adult rat liver are about 100-fold lower than the level of 2.1-kb RNA present in fetal liver.[64]

It needs to be emphasized that because variant AFP RNAs are present in normal liver, only those cells that might express 2.1-kb AFP mRNA are of interest in the search for oval cell precursors. Detection of these cells requires the use of probes that hybridize with this RNA form, but not with the variants. Using the pBAF 700 ribo probe (it hybridizes only to the 5′ portion of the full-length molecule corresponding to exons Z to E) we detected the 2.1-kb AFP mRNA by *in situ* hybridization in a small proportion of intrahepatic bile ductular cells in portal spaces and in a few hepatocytes (one per 20,000 hepatocyte nuclei).[64] The proportion of AFP-positive ductules was difficult to estimate because very few of these structures may be present in a histological section. Further, we do not know whether or not AFP mRNA expression is confined to very short terminal ductular segments (canals of Hering?). In positive ductules, 2.1-kb AFP mRNA was detected in four to six serial sections in cells that occupied about half of the diameter of the ductule.[64] In other cases AFP-positive cells did not appear to line a duct structure, but a lumen became evident with further sectioning. AFP mRNA-positive hepatocytes in the parenchyma had no preferential lobular localization and no special morphologic features. Surprisingly, in these cells, AFP mRNA expression did not seem to correlate with cell proliferation as shown by the combined analysis of AFP mRNA by *in situ* hybridization and cell replication by bromodeoxyuridine incorporation.[64]

It is tempting to speculate that the ductular cells that express 2.1-kb AFP mRNA in normal adult liver are the source of oval cells or the progenitors of some components of the oval cell compartment. The significance of the apparently nonproliferating, scattered hepatocytes that contain AFP mRNA in adult liver is not clear at this time, but it would be interesting to investigate their response to carcinogenic agents. Nevertheless, it appears that in some ductular cells and scattered hepatocytes the expression of the 2.1-kb AFP mRNA is not shut off at the neonatal stage and persists into the adult. Such cells may be the site of synthesis of the small amount of AFP present in the serum of normal adult rats.

The identification of bile ductular cells that express fetal AFP mRNA as a probable source for the oval cell compartment in hepatocarcinogenesis implies that some ductular cells may have ''stemlike cell'' properties. Although there is not yet direct proof that these normal liver ductular cells can directly generate cells of both the hepatocyte and the duct lineage, it is reasonable to assume that the ductular cells give origin to components of the oval cell compartment that have dual-lineage capabilities. The term stem cell, although simple, is not, however, an accurate descriptor for these cells which are more properly defined as ''committed and partially differentiated progenitor cells that can further differentiate in alternate pathways''.[65] In this context the often discussed functional dichotomy that is supposed to exist between ductal cells and stem cells is not meaningful. More specifically, one cannot discard the possibility that oval cells can serve as progenitor cells by simply stating that they are nothing more than bile ductular cells. Instead, the work reviewed above suggests that various segments of the biliary system differ in their degree of differentiation[32] and that ductular cells may function both as components of the biliary tree and as facultative progenitors capable of lineage generation. The proliferation of ductular cells and the expansion of the compartment appears to be triggered by carcinogens that induce hepatocyte death and at the same time inhibit a regenerative response of surviving hepatocytes. On the other hand, procedures such as bile duct ligation and the action of some chemical agents lead to hyperplasia of mature bile ducts and ductules with little involvement of terminal ductular components.[32,66-70] The mature bile ducts that proliferate under these conditions do not express 2.1-kb AFP mRNA and have no duct/hepatocyte transitional cells.

V. THE ORIGIN OF OVAL CELLS: PROLIFERATION OF AFP-EXPRESSING DUCTULAR CELLS IN NONCARCINOGENIC INJURY

If hyperplasia of normal liver ductular cells that express 2.1-kb AFP mRNA can be triggered by carcinogenic agents that cause hepatocyte cell death and inhibit hepatocyte regeneration, it is logical to expect that such cells might also be induced to proliferate by noncarcinogenic agents that produce similar types of injury. It is noteworthy that oval cell proliferation has never been detected in liver regeneration after partial hepatectomy and that the expression of 2.1-kb AFP mRNA[71] as well as that of the smaller AFP RNA variants changes little during this process.[64] In CCl_4 injury there is an increase in 2.1-kb mRNA expression[64] and the proliferation of some epithelial cells with oval cell characteristics has been noted.[72] However, it seems unlikely that in CCl_4 injury oval cells would play a major role in replenishing the parenchymal cells, as hepatocyte proliferation takes place rapidly and reaches a maximum (in rats) 48 h after CCl_4 administration. On the other hand, galactosamine injury differs from CCl_4 in significant ways in that 24 h after administration of the agent there is rapid proliferation of bile ducts initially confined to portal areas.[73-75] In subsequent days, duct hyperplasia continues and ductular structures and individual ductlike cells invade the parenchyma. During the first 2 to 3 d after galactosamine administration hepatocyte damage is widespread, but hepatocyte proliferation does not start until the third day and reaches a maximum on day 5.[75] Given this sequence of events, we reasoned that in this type of injury, ductular cells with precursor properties might proliferate and have a significant role in generating new hepatocytes.

Kuhlmann and Wurster[76] have detected AFP by immunohistochemical methods in ''canalicular cells'', portal bile ducts, and some hepatocytes in galactosamine-injured rat liver. We found by *in situ* hybridization (using the probe specific for the 2.1-kb AFP mRNA)[77] that in galactosamine-treated rats this mRNA was localized in some bile ducts within portal spaces, in ducts radiating into the parenchyma, in individual ductulelike cells that invaded the parenchyma and in some hepatocytes in proximity to invading ductular cells. A very similar, although not completely identical distribution of AFP RNA in livers injured by galactosamine has been described by Tournier et al.[78] By serial section analysis it could be demonstrated that ductular cells that produce AFP mRNA also expressed bile duct markers (cytokeratins 7, 8, 18, 19, GGT activity, and binding of peanut agglutinin).[77] A major conclusion of this work is that the ductular cells that proliferate during galactosamine injury appear similar if not phenotypically identical to some cells of the oval cell compartment in livers of carcinogen-treated rats.[77]

As stated, one of the important properties of this compartment during hepatocarcinogenesis is their potential capacity to generate hepatocytes by cell differentiation. We asked whether or not in the noncarcinogenic injury caused by galactosamine, ductlike cells could give rise to hepatocytes during the first 5 d after administration of the chemical. Histological analysis of galactosamine-injured livers revealed the presence of duct/hepatocyte transitional forms in a pattern reminiscent of that described in Me-DAB carcinogenesis. In addition, 3 to 5 d after galactosamine injection the parenchyma contains a high proportion of small hepatocytes (cells of less than 16-μm diameter, otherwise identical to normal hepatocytes) and many ductular structures, containing cells expressing AFP and the 2.1-kb AFP mRNA. Such structures are similar to Hering canals of normal liver.[77]

The proportion of the various LEC populations changes during galactosamine injury. In particular, 5 d after galactosamine injection there was a major increase in the number of individual ductular cells invading the parenchyma (because they express AFP, they will be referred to as oval cells) and a fivefold increase in the proportion of small hepatocytes. Analyses of cell type transitions *in vivo* are feasible in this experimental model because duct cells proliferate 2 d before the start of hepatocyte replication. ^{3}H thymidine injection given 30 h after galactosamine administration labeled 70% of the duct cells, but only 0.58% of hepatocytes, making it possible to analyze the fate of originally labeled duct cells. Using this experimental protocol we found that small hepatocytes constituted an increasing proportion of the labeled epithelial cells, while the proportion of labeled normal (''normal-sized'') hepatocytes remained relatively constant. A detailed analysis of labeling index and grain counts for each cell type through a 4-d period after label injection led us to conclude that the small hepatocytes most likely originated from duct cells. The data obtained in the labeling experiments were not consistent with the possibility that small hepatocytes arise from the replication of normal hepatocytes.[77]

The interpretation of labeling and cell fate analysis in the liver *in vivo* is complex and often distorted by problems of label reutilization, differential proliferative rates and extent of cell death among various cell types, etc. Nevertheless, keeping in mind the limitations of the method, it is possible to conclude that ductular cells similar to oval cells proliferate in noncarcinogenic injury and that in galactosamine injury these cells are likely to be progenitors of small hepatocytes (Figure 2) needed to replenish the parenchyma.[77]

A seemingly more dramatic production of hepatocytes by oval cell differentiation has been reported to occur in mice injected with the alkylating agent dipin and subjected to partial hepatectomy.[79,80] Although this treatment eventually leads to liver tumorigenesis, it initially causes almost complete elimination of preexisting hepatocytes, prolonged oval cell hyperplasia, and

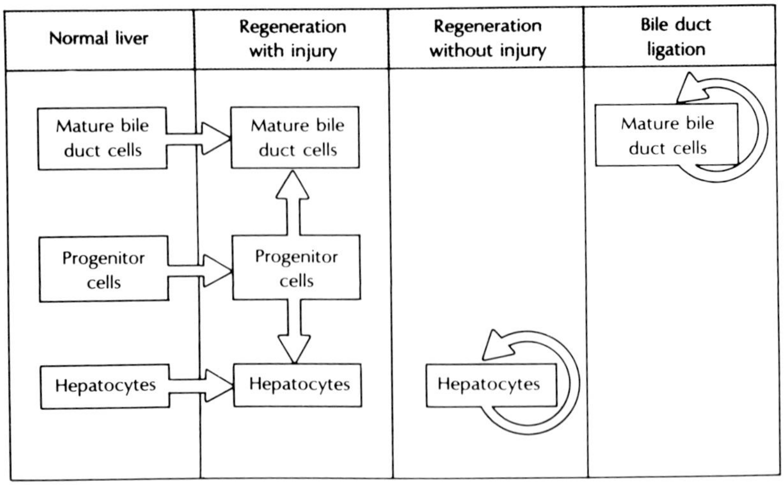

FIGURE 2. Proliferation of progenitor cells in noncarcinogenic liver injury. Bipotential progenitor cells proliferate in noncarcinogenic liver injury in which there is widespread hepatocyte death coupled with inhibition of the replication of surviving hepatocytes. The galactosamine-induced liver injury, ductal cells differentiate into small hepatocytes and contribute to the restoration of the normal parenchyma. In the absence of parenchymal injury (partial hepatectomy, bile duct ligation) hepatocytes and mature bile ducts replicate and the progenitor cell compartment is not activated. (Reprinted from N. Fausto, *Curr. Opinion Cell Biol.*, 2, 1036, 1990. With permission.)

the generation of new hepatocytes by precursor cell differentiation. Oval cell proliferation has also been reported to occur in livers of rats injected with thioacetamide.[81] An unusual aspect of this injury is that oval cell proliferation apparently takes place in central rather than peripheral lobular areas.

VI. SUMMARY AND FUTURE PERSPECTIVES

Nonparenchymal epithelial cells that proliferate during carcinogenesis and form the oval cell compartment are heterogeneous cell populations (Figure 3). Some cells in this compartment are capable of generating both hepatocytes and duct cells, and the lineage-generating cells or other cell types in the compartment are a major source of growth factors during hepatocarcinogenesis. Oval cells have been established in culture as nontumorigenic lines that can be transformed by oncogene transfection or infrequent passage and give rise to hepatocellular carcinomas upon injection into nude mice or syngeneic rats.

The work reviewed here addressed the questions of oval cell origins in the normal rat liver, oval cell counterparts in developing liver, and oval cell

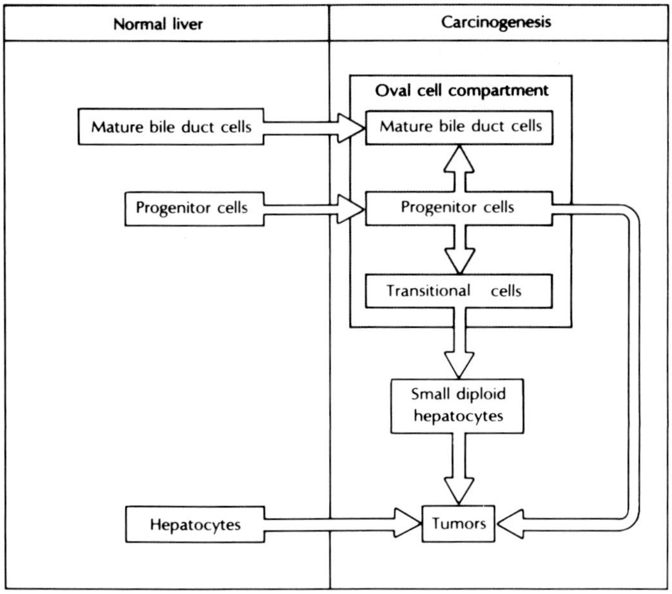

FIGURE 3. Proliferation of progenitor cells in liver carcinogenesis. During hepatocarcinogenesis induced by most chemicals, a cell compartment formed by different cell populations proliferates (oval cell compartment). The compartment contains mature bile duct cells as well as undifferentiated bipotential progenitor cells. These cells can: (a) differentiate into small hepatocytes going through intermediate stages (transitional cells) or (b) function directly as tumor progenitor cells. (Reprinted from N. Fausto, *Curr. Opinion Cell Biol.*, 2, 1036, 1990. With permission.)

proliferation in noncarcinogenic injury. The picture that emerges from this work can be summarized as follows:

1. Cells of the primitive intrahepatic bile ducts in liver development are equivalent to oval cells in adult rat liver (Figure 1).
2. In adult rat liver, a small portion of the bile duct tree, probably corresponding to the ductular segment, has similarities with primitive intrahepatic ducts; these ductular cells express the fetal form of AFP mRNA and may be the facultative stem cells described by Grisham[13,14] and Lombardi.[42]
3. Ductular cells with bipotential lineage-generating capacity proliferate in both carcinogenic and noncarcinogenic injury in which there is parenchymal cell loss coupled with inhibition of hepatocyte replication (Figures 2 and 3).
4. Differentiation of oval (ductular) cells into hepatocytes occurs *in vivo* in carcinogenic and noncarcinogenic injury, but it may not be an obligatory feature associated with oval cell hyperplasia.

Underlying these propositions is the general question of the individuality of the duct and hepatocyte lineages.[32,43,53,82] While for mature forms these lineages appear as clearly distinct and nonoverlapping differentiation paths, in practice such distinctions may turn out to be too rigid. Indeed, the examination of the patterns of cytokeratin expression in human alcoholic liver disease and hepatocellular carcinomas shows that "normal" and transformed hepatocytes can express bile duct-type cytokeratin (see Reference 83–85 and Chapter 11). This is not a surprising finding if one considers the sequence of events in liver development[45-47,52,53,56] as well as the observation that in early devleoping human liver, immature hepatocytes express not only cytokeratin 8, but also cytokeratin 19, a bile duct-type keratin.[86] One suspects that transitions between hepatocytes and ducts occur rather frequently even in the adult liver and that the direction of the change, i.e., duct → hepatocyte or vice-versa may vary under different conditions. The important point, however, is to establish that such transitions are possible and perhaps common and that cells capable of undergoing these shifts exist in normal livers. We suggest that ductular cells in normal livers are themselves capable of undergoing these transitions and that they can give rise to cells which may differentiate into hepatocytes. In the parenchyma, limiting plate hepatocytes can undergo the hepatocyte → duct transdifferentiation and form mature bile duct structures.[58] The notion that bile ductular cells are capable of differentiating into hepatocytes in carcinogenic and noncarcinogenic injury that cause severe parenchymal injury is not new.[3,87,88] This idea was first proposed by the demonstration of Wilson and Leduc[88] that bile ducts give rise to hepatocytes in mice subjected to severe nutritional injury.

Transitional cell types presumably capable of lineage generation are found not only in rodents, but also in human livers (see Reference 89 and Chapter 10) as shown by the expression of common antigenic markers among seemingly different cell types.[90,91] However, it is now clear that the lineage-generating capabilities of some liver nonparenchymal cells go beyond the origination of liver cell progeny and include the generation of intestinal glandular elements and goblet cells[92] (see Chapter 9). Conversely, the capacity to generate hepatocytes is maintained in pancreatic duct cells[93] and quite likely also in other cells of the intestinal system. All of these observations indicate that there may be a good deal of plasticity between hepatic cell lineages and that cells capable of lineage transition exist not only in the liver, but also in other segments of the lower (distal) digestive tract. Placed into this context, the existence of liver "stem cells" becomes a logical concept that should not be regarded as an "exotic" idea.

Progress in the understanding of the role of nonparenchymal epithelial cells in liver development, regeneration, and injury has been significant. As increased emphasis is placed on these cells, it is important to emphasize some issues about these cells that often create misunderstandings. In the first place, the fact that transformed oval cells give rise to hepatocellular carcinomas does not in any way imply that these cells are the exclusive progenitors of these tumors. On the contrary, there are probably many diverse molecular mechanisms and different progenitor cells (hepatocytes and nonparenchymal cells) involved in the generation of liver tumors, depending on the nature of the carcinogen and the type of injury it produces.[94] Secondly, the often debated issue of whether oval cells are bile ductular cells or stem cells is inappropriately formulated as an either/or proposition. Grisham and Hartroft[95] had already recognized the similarities between cells of the oval cell compartment and normal ductular cells, and the evidence gathered so far indicates that certain ductular cells can function both as structural components of the biliary tree as well as lineage-generating cells. Thirdly, although the term "stem cell" is routinely used to describe lineage-generating cells in the liver, the term is not completely accurate and may give the misleading impression that "stem cell" concepts as applied to the liver do not differ from those established in continuously renewing tissues. Finally, it should be recognized as pointed out by Sirica et al.[32] that very little is known about the mechanisms that trigger oval cell proliferation. It is not known whether the proliferative response is a consequence of a direct stimulatory effect by certain agents, an indirect consequence of hepatocyte damage, if it is caused by the release of specific growth factors, by alterations in the extracellular matrix, or by any other of a number of possible mechanisms. What is required for further progress is not only the identification of the mechanisms that regulate the proliferation of nonparenchymal epithelial cells in carcinogenic and noncarcinogenic injury, but also an understanding of the lineage commitment process and the conditions under which progenitor cells differentiate into hepatocytes.

ACKNOWLEDGMENTS

We thank Molly Bartlett, Ann Baxter, and Carol White for their help in preparing the manuscript. The work of the authors was supported in part by grant CA 35249 from the National Cancer Institute.

REFERENCES

1. **Sasaki, T. and Yoshida, T.,** Experimentelle Erzeugung des Lebercarcinoms durch Fütterung mit o-amidoazotoluol, *Virchows Arch. Pathol. Anat. Physiol.,* 295, 175, 1935.
2. **Price, J. M., Harman, J. W., Miller, E. C., and Miller, J. A.,** Progressive microscopic alterations in the livers of rats fed the hepatic carcinogens 3′-methyl-4-dimethylaminoazobenzene and 4′-fluoro-4-dimethylaminoazobenzene, *Cancer Res.,* 12, 192, 1952.
3. **Firminger, H. I.,** Histopathology of carcinogenesis and tumors of the liver in rats, *J. Natl. Cancer Inst.,* 15 (Suppl.), 1427, 1955.

4. **Farber, E.,** Similarities in the sequence of early histological changes induced in the liver of the rat by ethionine, 2-acetylaminofluorene, and 3′-methyl-4-dimethylaminoazobenzene, *Cancer Res.,* 16, 142, 1956.
5. **Inaoka, Y.,** Significance of the so-called oval cell proliferation during azo-dye carcinogenesis, *Gann,* 58, 355, 1967.
6. **Ogawa, K., Minase, T., and Onoé, T.,** Demonstration of glucose-6-phosphatase activity in the oval cells of rat liver and the significance of the oval cells in azo dye carcinogenesis, *Cancer Res.,* 34, 3379, 1974.
7. **Dempo, K., Chisaka, N., Yoshida, Y., Kaneko, A., and Onoé, T.,** Immunofluorescent study on a-fetoprotein-producing cells in the early stage of 3′-methyl-4-dimethylaminoazobenzene carcinogenesis, *Cancer Res.,* 35, 1282, 1975.
8. **Onda, H.,** Immunohistological studies on alpha-1-fetoprotein and alpha-1-acid glycoprotein during azo dye hepatocarcinogenesis in rats, *Gann,* 67, 253, 1976.
9. **Shinozuka, H., Lombardi, B., Sell, S., and Iammarino, R. M.,** Early histological and functional alterations of ethionine liver carcinogenesis in rats fed a choline-deficient diet, *Cancer Res.,* 38, 1092, 1978.
10. **Kuhlmann, W. D.,** Localization of alpha-fetoprotein and DNA synthesis in liver cell populations during experimental hepatocarcinogenesis in rats, *Int. J. Cancer,* 21, 368, 1978.
11. **Sell, S., Osborn, K., and Leffert, H. L.,** Autoradiography of "oval cells" appearing rapidly in the livers of rats fed N-2-fluorenylacetamide in a choline devoid diet, *Carcinogenesis,* 2, 7, 1981.
12. **Tchipysheva, T. A., Guelstein, V. I., and Bannikov, G. A.,** Alpha-feto-protein-containing cells in the early stages of liver carcinogenesis induced by 3′-methyl-4-dimethylaminoazobenzene and N-2-acetyl-aminofluorene, *Int. J. Cancer,* 20, 388, 1977.
13. **Grisham, J. W.,** Cell types in long-term propagable cultures of rat liver, *Ann. N.Y. Acad. Sci.,* 349, 128, 1980.
14. **Grisham, J. W.,** Cell types in rat liver cultures: their identification and isolation, *Mol. Cell. Biochem.,* 53/54, 23, 1983.
15. **Sells, M. A., Katyal, S. L., Shinozuka, H., Estes, L. W., Sell, S., and Lombardi, B.,** Isolation of oval cells and transitional cells from the livers of rats fed the carcinogen DL-ethionine, *J. Natl. Cancer Inst.,* 66, 355, 1981.
16. **Yaswen, P., Hayner, N. T., and Fausto, N.,** Isolation of oval cells by centrifugal elutriation and comparison with other cell types purified from normal and preneoplastic livers, *Cancer Res.,* 44, 324, 1984.
17. **Hayner, N. T., Braun, L., Yaswen, P., Brooks, M., and Fausto, N.,** Isozyme profiles of oval cells, parenchymal cells, and biliary cells isolated by centrifugal elutriation from normal and preneoplastic livers, *Cancer Res.,* 44, 332, 1984.
18. **Sirica, A. E. and Cihla, H. P.,** Isolation and partial characterizations of oval and hyperplastic bile ductular cell-enriched populations from the livers of carcinogen and noncarcinogen-treated rats, *Cancer Res.,* 44, 3454, 1984.
19. **Fausto, N., Thompson, N. L., and Braun, L.,** Purification and culture of oval cells from rat liver, in *Cell Separation Methods and Selected Applications,* Vol. 4, Pretlow, T. G., II and Pretlow, T. P., Eds., Academic Press, Orlando, FL, 1987, 45.
20. **Tsao, M.-S., Smith, J. D., Nelson, K. G., and Grisham, J. W.,** A diploid epithelial cell line from normal adult rat liver with phenotypic properties of 'oval' cells, *Exp. Cell Res.,* 154, 38, 1984.
21. **Tsao, M.-S., Grisham, J. W., Nelson, K. G., and Smith, J. D.,** Phenotypic and karyotypic changes induced in cultured rat hepatic epithelial cells that express the "oval" cell phenotype by exposure to N-methyl-N′-nitro-N-nitrosoguanidine, *Am. J. Pathol.,* 118, 306, 1985.

22. **Braun, L., Goyette, M., Yaswen, P., Thompson, N. L., and Fausto, N.,** Growth in culture and tumorigenicity after transfection with the *ras* oncogene of liver epithelial cells from carcinogen-treated rats, *Cancer Res.,* 47, 4116, 1987.
23. **Braun, L., Mikumo, R., and Fausto, N.,** Production of hepatocellular carcinoma by oval cells: cell cycle expression of c-*myc* and p53 at different stages of oval cell transformation, *Cancer Res.,* 49, 1554, 1989.
24. **Tsao, M.-S., Earp, H. S., and Grisham, J. W.,** Gradation of carcinogen-induced capacity for anchorage-independent growth in cultured rat liver epithelial cells, *Cancer Res.,* 45, 4428, 1985.
25. **Tsao, M.-S., Grisham, J. W., and Nelson, K. G.,** Clonal analysis of tumorigenicity and paratumorigenic phenotypes in rat liver epithelial cells chemically transformed *in vitro, Cancer Res.,* 45, 5139, 1985.
26. **Yaswen, P., Thompson, N. L., and Fausto, N.,** Oncodevelopmental expression of rat placental alkaline phosphatase: detection in oval cells during liver carcinogenesis, *Am. J. Pathol.,* 121, 505, 1985.
27. **Plenat, F., Braun, L., and Fausto, N.,** Demonstration of glucose-6-phosphatase and peroxisomal catalase activity by ultrastructural cytochemistry in oval cells from livers of carcinogen-treated rats, *Am. J. Pathol.,* 130, 91, 1988.

27a. **Thompson, N. and Fausto, N.,** unpublished observations.

28. **Goyette, M., Faris, R., Braun, L., Hixson, D., and Fausto, N.,** Expression of hepatocyte and oval cell antigens in hepatocellular carcinomas produced by oncogene-transfected liver epithelial cells, *Cancer Res.,* 50, 4809, 1990.
29. **Braun, L., Gruppuso, P., Mikumo, R., and Fausto, N.,** Transforming growth factor b1 in liver carcinogenesis: messenger RNA expression and growth effects, *Cell Growth Differ.,* 1, 103, 1990.
30. **Yaswen, P., Goyette, M., Shank, P. R., and Fausto, N.,** Expression of c-Ki-*ras*, c-Ha-*ras*, and c-*myc* in specific cell types during hepatocarcinogenesis, *Mol. Cell. Biol.,* 5, 780, 1985.
31. **Fausto, N., Mead, J. E., Braun, L., Thompson, N. L., Panzica, M., Goyette, M., Bell, G. I., and Shank, P. R.,** Proto-oncogene expression and growth factors during liver regeneration, *Symp. Fundam. Cancer Res.,* 39, 69, 1987.
32. **Sirica, A. E., Mathis, G. A., Sano, N., and Elmore, L. W.,** Isolation, culture, and transplantation of intrahepatic biliary epithelial cells and oval cells, *Pathobiology,* 58, 44, 1990.
33. **Germain, L., Goyette, R., and Marceau, N.,** Differential cytokeratin and a-fetoprotein expression in morphologically distinct epithelial cells emerging at the early stage of rat hepatocarcinogenesis, *Cancer Res.,* 45, 673, 1985.
34. **Evarts, R. P., Nagy, P., Marsden, E., and Thorgeirsson, S. S.,** *In situ* hybridization studies on expression of albumin and a-fetoprotein during the early stages of neoplastic transformation in rat liver, *Cancer Res.,* 47, 5469, 1987.
35. **Evarts, R. P., Nagy, P., Marsden, E., and Thorgeirsson, S. S.,** A precursor-product relationship exists between oval cells and hepatocytes in the rat liver, *Carcinogenesis,* 8, 1737, 1987.
36. **Evarts, R. P., Nakatsukasa, H., Marsden, E. R., Hsia, C.-C., Dunsford, H. A., and Thorgeirsson, S. S.,** Cellular and molecular changes in the early stages of chemical hepatocarcinogenesis in the rat, *Cancer Res.,* 50, 3439, 1990.
37. **Nagy, P., Evarts, R. P., McMahon, J. B., and Thorgeirsson, S. S.,** Role of TGF-beta in normal differentiation and oncogenesis in rat liver, *Mol. Carcinogen.,* 2, 345, 1989.
38. **Fausto, N. and Mead, J. E.,** Regulation of liver growth: protooncogenes and transforming growth factors, *Lab. Invest.,* 60, 4, 1989.
39. **Fausto, N.,** Hepatic regeneration, in *Hepatology: A Textbook of Liver Disease,* Vol. 1, 2nd ed., Zakim, D. and Boyer, T. D., Eds., W. B. Saunders, Philadelphia, 1990, 49.

40. **Tsao, M.-S. and Grisham, J. W.,** Hepatocarcinomas, cholangiocarcinomas, and hepatoblastomas produced by chemically transformed cultured rat liver epithelial cells: a light- and electronmicroscopic study, *Am. J. Pathol.*, 127, 168, 1987.
41. **Garfield, S., Huber, B. E., Nagy, P., Cordingley, M. G., and Thorgeirsson, S. S.,** Neoplastic transformation and lineage switching of rat liver epithelial cells by retrovirus-associated oncogenes, *Mol. Carcinogen.*, 1, 189, 1988.
42. **Lombardi, B.,** On the nature, properties and significance of oval cells, in *Recent Trends in Chemical Carcinogenesis,* Vol. 1, Pani, P., Feo, F., and Columbano, A., Eds., ESA, Cagliari, Italy, 1982, 36.
43. **Marceau, N.,** Cell lineages and differentiation programs in epidermal, urothelial and hepatic tissues and their neoplasms, *Lab. Invest.*, 63, 4, 1990.
44. **Sell, S.,** Is there a liver stem cell?, *Cancer Res.*, 50, 3811, 1990.
45. **van Eyken, P., Sciot, R., and Desmet, V.,** Intrahepatic bile duct development in the rat: a cytokeratin-immunohistochemical study, *Lab. Invest.*, 59, 52, 1988.
46. **van Eyken, P., Sciot, R., and Callea, F.,** The development of the intrahepatic bile ducts in man: a keratin-immunohistochemical study, *Hepatology,* 8, 1586, 1988.
47. **Shiojiri, N.,** Enzymo- and immunocytochemical analyses of the differentiation of liver cells in the prenatal mouse, *J. Embryol. Exp. Morphol.*, 62, 139, 1981.
48. **Shah, K. D. and Gerber, M. A.,** Development of intrahepatic bile ducts in humans: immunohistochemical study using monoclonal cytokeratin antibodies, *Arch. Pathol. Lab. Med.*, 113, 1135, 1989.
49. **Shah, K. D. and Gerber, M. A.,** Development of intrahepatic bile ducts in humans: possible role of laminin, *Arch. Pathol. Lab. Med.*, 114, 597, 1990.
50. **Du Bois, A. M.,** The embryonic liver, in *The Liver, Morphology, Biochemistry, Physiology,* Vol. 1, Rouiller, C., Ed., Academic Press, New York, 1963, 1.
51. **Wilson, J. W., Groat, C. S., and Leduc, E. H.,** Histogenesis of the liver, *Ann. N.Y. Acad. Sci.*, 111, 8, 1963.
52. **Shiojiri, N., Lemire, J. M., and Fausto, N.,** Cell lineages and oval cell progenitors in rat liver development, *Cancer Res.*, 51, 2611, 1991.
53. **Fausto, N.,** Hepatocyte differentiation and liver progenitor cells, *Curr. Opinion Cell Biol.*, 2, 1036, 1990.
54. **Shiojiri, N.,** The origin of intrahepatic bile duct cells in the mouse, *J. Embryol. Exp. Morphol.*, 79, 25, 1984.
55. **Yamamoto, K. and Phillips, M. J.,** A hitherto unrecognized bile ductular plexus in normal rat liver, *Hepatology,* 4, 381, 1984.
56. **Germain, L., Blouin, M.-J., and Marceau, N.,** Biliary epithelial and hepatocytic cell lineage relationships in embryonic rat liver as determined by the differential expression of cytokeratins, a-fetoprotein, albumin, and cell-surface exposed components, *Cancer Res.*, 48, 4909, 1988.
57. **Shiojiri, N. and Katayama, H.,** Development of Dolichos biflorus agglutinin (DBA) binding sites in the bile duct of the embryonic mouse liver, *Anat. Embryol.*, 178, 15, 1988.
58. **Desmet, V. J.,** Modulation of biliary epithelium, in *Modulation of Liver Cell Expression, Falk Symposium,* Vol. 43, Reutter, W., Popper, H., Arias, I. M., Heinrich, P. C., Keppler, D., and Landmann, L., Eds., MTP Press, Lancaster, England, 1987, 195.
59. **Sell, S. and Dunsford, H. A.,** Evidence for the stem cell origin of hepatocellular carcinoma and cholangiocarcinoma, *Am. J. Pathol.*, 134, 1347, 1989.
60. **Tamaoki, T. and Fausto, N.,** Expression of the a-fetoprotein gene during development, regeneration, and carcinogenesis, in *Recombinant DNA and Cell Proliferation,* Stein, G. S. and Stein, J. L., Eds., Academic Press, Orlando, FL, 1984, 145.
61. **Turcotte, B., Guertin, M., Chevrette, M., and Belanger, L.,** Rat a_1-fetoprotein messenger RNA: 5′-end sequence and glucocorticoid-suppressed liver transcription in an improved nuclear run-off assay, *Nucleic Acids Res.*, 13, 2387, 1985.

62. **Wan, Y.-J. Y., Jimenez-Molina, J. L., and Chou, J. Y.,** Fetal and variant a-fetoproteins are encoded by mRNAs that differ in sequence at the 5′ end, *Biochemistry,* 27, 7269, 1988.
63. **Wan, Y.-J. Y. and Chou, J. Y.,** Expression of the a-fetoprotein gene in adult rat liver, *Arch. Biochem. Biophys.,* 270, 267, 1989.
64. **Lemire, J. M. and Fausto, N.,** Multiple alphafetoprotein RNAs in adult rat liver: cell type specific expression and differential regulation, *Cancer Res.,* 51, 4656, 1991.
65. **Okada, T. S.,** Transdifferentiation in animal cells: fact or artifact, *Dev. Growth Differ.,* 28, 213, 1986.
66. **Tavoloni, N.,** The intrahepatic biliary epithelium: an area of growing interest in hepatology, *Semin. Liver Dis.,* 7, 280, 1987.
67. **Desmet, V. J.,** Intrahepatic bile ducts under the lens, *J. Hepatol.,* 1, 545, 1985.
68. **Mathis, G. A., Walls, S. A., D'Amico, P., Gengo, T. F., and Sirica, A. E.,** Enzyme profile of rat bile ductular epithelial cells in reference to the resistance phenotype in hepatocarcinogenesis, *Hepatology,* 9, 477, 1989.
69. **Sell, S.,** Comparison of oval cells induced in rat liver by feeding *N*-2-fluorenylacetamide in a choline-devoid diet and bile duct cells induced by feeding 4,4′-diaminodiphenylmethane, *Cancer Res.,* 43, 1761, 1983.
70. **Yokoyama, S., Satoh, M., and Lombardi, B.,** Bile ductular cells and the phenotypic heterogeneity of the population of hepatic non-parenchymal epithelial cells induced in rats by chemical carcinogens, *Carcinogenesis,* 7, 1215, 1986.
71. **Petropoulos, C., Andrews, G., Tamaoki, T., and Fausto, N.,** a-Fetoprotein and albumin mRNA levels in liver regeneration and carcinogenesis, *J. Biol. Chem.,* 258, 4901, 1983.
72. **Englehardt, N. V., Baranov, V. N., Lazareva, M. N., and Goussev, A. I.,** Ultrastructural localisation of alpha-fetoprotein (AFP) in regenerating mouse liver poisoned with CCl_4. 1. Reexpression of AFP in differentiated hepatocytes, *Histochemistry,* 80, 401, 1984.
73. **Decker, K. and Keppler, D.,** Galactosamine induced liver injury, in *Progress in Liver Diseases,* Vol. IV, Popper, H. and Schaffner, F., Eds., Grune & Stratton, New York, 1972, 183.
74. **Medline, A., Schaffner, F., and Popper, H.,** Ultrastructural features in galactosamine-induced hepatitis, *Exp. Mol. Pathol.,* 12, 201, 1970.
75. **Lesch, R., Reutter, W., Keppler, D., and Decker, K.,** Liver restitution after acute galactosamine hepatitis: autoradiographic and biochemical studies in rats, *Exp. Mol. Pathol.,* 12, 58, 1970.
76. **Kuhlmann, W. D. and Wurster, K.,** Correlation of histology and alpha$_1$-fetoprotein resurgence in rat liver regeneration after experimental injury by galactosamine, *Virchows Arch. A Pathol. Anat. Histol.,* 387, 47, 1980.
77. **Lemire, J. M., Shiojiri, N., and Fausto, N.,** Oval cell proliferation and the origin of small hepatocytes in liver injury induced by d-galactosamine, *Am. J. Pathol.,* 139, 535, 1991.
78. **Tournier, I., Legres, L., Schoevaert, D., Feldmann, G., and Bernuau, D.,** Cellular analysis of a-fetoprotein gene activation during carbon tetrachloride and D-galactosamine-induced acute liver injury in rats, *Lab. Invest.,* 59, 657, 1988.
79. **Faktor, V. M., Uryvaeva, I. V., Sokolova, A. S., Chernov, V. A., and Brodsky, W. Y.,** Kinetics of cellular proliferation in regenerating mouse liver pretreated with the alkylating drug dipin, *Virchows Arch. B Cell Pathol.,* 33, 187, 1980.
80. **Radaeva, S. A. and Faktor, V. M.,** Time course of development of oval cell population induced in the mouse liver by dipin and partial hepatectomy, *Bull. Exp. Biol. Med.,* (transl.), 109, 514, 1990.
81. **Marleen, P., Hendrik, R., and van Oostveldt, P.,** Early liver cell lesions in rats induced by thioacetamide. An ultrastructural, cytophotometric and autoradiographic study, *Pathol. Res. Pract.,* 184, 69, 1989.

82. **Marceau, N., Blouin, M.-J., Germain, L., and Noël, M.,** Role of different epithelial cell types in liver ontogenesis, regeneration and neoplasia, *In Vitro Cell. Dev. Biol.*, 25, 336, 1989.
83. **van Eyken, P., Sciot, R., and Desmet, V. J.,** A cytokeratin immunohistochemical study of alcoholic liver disease: evidence that hepatocytes can express 'bile duct type' cytokeratins, *Histopathology*, 13, 605, 1988.
84. **van Eyken, P., Sciot, R., Paterson, A., Callea, F., Kew, M. C., and Desmet, V. J.,** Cytokeratin expression in hepatocellular carcinoma: an immunohistochemical study, *Hum. Pathol.*, 19, 562, 1988.
85. **Fischer, H.-P., Doppl, W., Osborn, M., and Altmannsberger, M.,** Evidence for a hepatocellular lineage in a combined hepatocellular-cholangiocarcinoma of transitional type, *Virchows Arch. B Cell Pathol.*, 56, 71, 1988.
86. **Stosiek, P., Kasper, M., and Karsten, U.,** Expression of cytokeratin 19 during human liver organogenesis, *Liver*, 10, 59, 1990.
87. **Fishback, E. C.,** A morphologic study of regeneration of the liver after partial removal, *Arch. Pathol.*, 7, 955, 1929.
88. **Wilson, J. W. and Leduc, E. H.,** Role of cholangioles in restoration of the liver of the mouse after dietary injury, *J. Pathol. Bacteriol.*, 76, 441, 1958.
89. **Gerber, M. A., Thung, S. N., Shen, S., Stromeyer, F. W., and Ishak, K. G.,** Phenotypic characterization of hepatic proliferation: antigenic expression by proliferating epithelial cells in fetal liver, massive hepatic necrosis, and nodular transformation of the liver, *Am. J. Pathol.*, 110, 70, 1983.
90. **Englehardt, N. V., Factor, V. M., Yasova, A. K., Poltoranina, V. S., Baranov, V. N., and Lasareva, M. N.,** Common antigens of mouse oval and biliary epithelial cells. Expression on newly formed hepatocytes, *Differentiation*, 45, 29, 1990.
91. **Faris, R. A., McEntire, K. D., Thompson, N. L., and Hixson, D. C.,** Identification and characterization of a rat hepatic oncofetal membrane glycoprotein, *Cancer Res.*, 50, 4755, 1990.
92. **Tatematsu, M., Kaku, T., Medline, A., and Farber, E.,** Intestinal metaplasia as a common option of oval cells in relation to cholangiofibrosis in liver of rats exposed to 2-acetylaminofluorene, *Lab. Invest.*, 52, 354, 1985.
93. **Rao, M. S., Dwivedi, R. S., Yelandi, A. V., Subbarao, V., Tan, X., Usman, M. I., Thangada, S., Nemali, M. R., Kumar, S., Scarpelli, D. G., and Reddy, J. K.,** Role of periductal and cutular epithelial cells of the adult rat pancreas in pancreatic hepatocyte lineage: a change in differentiation commitment, *Am. J. Pathol.*, 134, 1069, 1989.
94. **Saeter, G. and Seglen, P. O.,** Cell biology of hepatocarcinogenesis, *Crit. Rev. Oncogenesis*, 1, 437, 1990.
95. **Grisham, J. W. and Hartroft, W. S.,** Morphologic identification by electron microscopy of "oval" cells in experimental hepatic degeneration, *Lab. Invest.*, 10, 317, 1961.

Chapter 6

GROWTH AND DIFFERENTIATION OF STEM CELLS IN ADULT RAT LIVER

S. S. Thorgeirsson and R. P. Evarts

TABLE OF CONTENTS

* This chapter is a U.S. Government work and as such is not subject to copyright.

I. INTRODUCTION

It is well established that normal cell renewal in adult rat and human liver proceeds very slowly,[1] and no clear distinction can be made between fully differentiated hepatocytes and those having the capacity to replicate. Indeed, the growth response of the liver following partial hepatectomy appears to involve all the remaining hepatocytes which undergo one or two rounds of replication before returning to the quiescent state.[1] The slow cell turnover and the capacity of the adult differentiated hepatocyte to replicate, and to regenerate the liver following both surgical and chemical partial hepatectomy, has made the need for a stem cell compartment in the liver less critical than, for example, in rapidly self-renewing tissues such as skin and intestinal mucosa. Nevertheless, recent evidence indicates that a stem cell compartment exists in the liver, and that this compartment can become the major source of new hepatocytes in a situation in which the adult differentiated hepatocytes cannot respond to loss of liver mass.[2-4] Because of these characteristics the hepatic progenitor cells have been referred to as ''facultative stem cells''.[5] The precise anatomical location of the stem cell compartment in adult liver has not been defined. However, a considerable body of data indicate that cells located in or near the bile ductules (canal of Hering) constitute the compartment of facultative hepatic stem cells.[3,4]

The developmental potential of the hepatic stem cells as well as the possible role of these cells in hepatocarcinogenesis are important research areas being actively pursued in several laboratories. We have postulated[6] that hepatic stem cells have the capacity to develop into several cell lineages, including hepatocytic, bile and intestinal epithelia, and pancreatic lineages (Figure 1). The evidence supporting this notion is to a large extent derived from research in experimental hepatocarcinogenesis. Extensive research in this area has disclosed that one of the earliest cellular responses to chemical hepatocarcinogens in the rat involves the proliferation of cells located around the terminal bile ductules, followed by the appearance of a rapidly growing population of small distinct epithelial cells characterized by oval nuclei and dense cytoplasm.[4,7] These cells, commonly referred to as ''oval cells'', have been shown to be multipotent cells which can differentiate into hepatocytes as well as other cell lineages.[2,8-11]

Another aspect of the cellular biology of the liver supporting the existence of multipotent progenitor cells is that clonogenic liver-derived diploid or near diploid epithelial cell lines from normal as well as hepatocarcinogen-treated rats have been established by several investigators.[12] Studies of the biochemical and ultrastructural characteristics of newly isolated oval cells, cultured oval cells, bile duct cells, and long-term cultures of rat liver-derived epithelial cells have shown a common pattern of expression of many cellular markers, including cytokeratins.[13-16] The suggestion that oval cells are descendants from putative liver ''stem'' cells[4] raises the possibility that rat liver epithelial

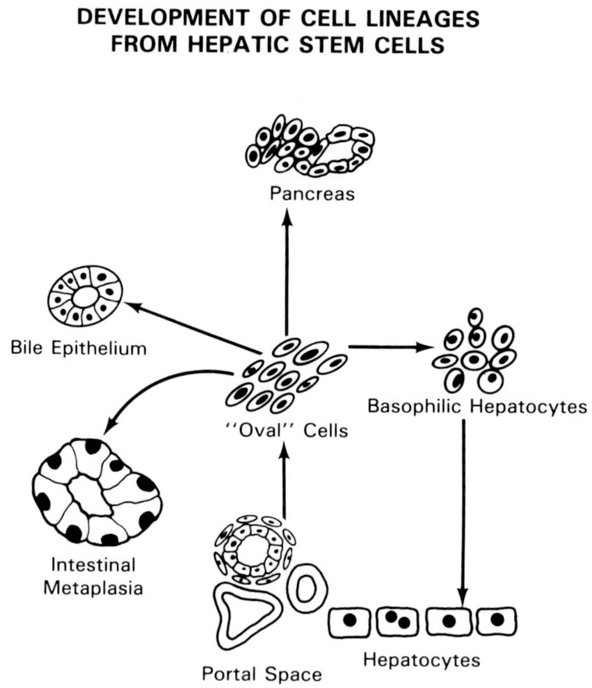

FIGURE 1. Hypothetical scheme for development of cell lineages in rat liver. (Reprinted from *Advances in Applied Biotechnology Series,* Vol. 13, Etiology, Pathology, and Treatment of Hepatocellular Carcinoma in North America, Tabor, E., Di Bisceglie, A. M., and Purcell, R. H., Eds., Gulf Publishing, Houston, 1991, 173. With permission.)

cells, because of their phenotypic similarities to oval cells, are also progeny from such a "stem" cell compartment.

Hepatocytes appear in the pancreas of rats during aging[17] or after feeding a diet deficient in copper.[18,19] The pancreatic hepatocytes in Cu^{2+}-deficient animals seem to arise from periductular or ductular cells in a manner very similar, if not identical, to the oval cell-derived hepatocytes in livers of animals exposed to chemical hepatocarcinogens.[2,7,19] Thus, it has been hypothesized that a stem cell compartment of common embryonic origin is present in both the liver and the pancreas.[4]

Appearance of differentiated hepatocytes in the adult rat pancreas can be experimentally induced.[18] This observation suggests a lineage relationship between cell compartments present in rat liver and pancreas. We have demonstrated that epithelial cell lines with almost identical phenotypes can be established from adult rat liver and pancreas.[20] The established cell lines show similar morphologies as established by light- and electron-microscopic studies and display similar expression patterns of intermediate filament proteins. Vimentin, actin, and β-tubulin are present in all cell lines. In addition, simple epithelial type II cytokeratins 7 and 8 are coexpressed with the type I cytokeratin 14 in several of the cell lines. Neither the type I cytokeratins 18 and 19, which are the normal partners for cytokeratins 8 and 7 in filament formation, or the type II cytokeratin 5 could be detected despite the fact that filaments were formed by both cytokeratins 8 and 14. This novel finding suggests that cytokeratin 14 acts as an indiscriminate type I cytokeratin in filament formation in the established cell lines. The fact that populations of small epithelial cells with common morphological phenotypes emerged after extensive hepatic and pancreatic injury strongly indicates that identical compartments of primitive epithelial cells capable of proliferating *in vitro* are present in these organs.

It has been shown that chemical transformation of rat liver epithelial cell lines results in the generation of a wide range of tumors, including hepatocellular carcinomas, cholangiocarcinomas, and hepatoblastomas, as well as sarcomas and mixed epithelial-mesenchymal tumors.[21,22] Furthermore, we have recently shown that similar tumor types are generated when rat liver-derived epithelial cells are transformed by selective oncogenes.[23] These data indicate that, like the oval cells *in vivo,* the liver-derived epithelial cells possess a blastic multipotent nature and support the notion that hepatic stem cells have the capacity to differentiate along a variety of cell lineage pathways.

We will focus on two aspects of hepatic stem cell biology in this review, namely, activation of facultative stems cells in the adult rat liver and on the factors that may control the growth and differentiation of these cells.

II. ACTIVATION OF FACULTATIVE HEPATIC STEM CELLS

The experimental models most commonly used to activate the hepatic stem cell compartment are the choline-deficient and modified Solt-Farber models.[2,24,25] Our discussion will principally be focused on the use of the Solt-Farber model. The basis for using the modified Solt-Farber model was the observation that short-term dietary administration of 2-acetylaminofluorene (AAF) resulted in virtually complete inhibition of hepatocyte proliferation following partial hepatectomy in the rat.[25] Instead, extensive proliferation of oval cells takes place, and the liver mass is gradually restored without induction of liver tumors.[2,25] In general, the conditions that provide the most powerful stimulus for proliferation of oval cells, and by extension the activation of the hepatic stem cell compartment, involve partial and/or total inhibition of the capacity of adult hepatocytes to respond to either acute (i.e., AAF and partial hepatectomy) or chronic (i.e., choline-deficient diet) loss of hepatic mass. Although the fate of oval cells in both experimental hepatocarcinogenesis and in ‘‘noncarcinogenesis’’ models remains a matter of some controversy, recent data from our laboratory seem to indicate that a precursor-product relationship exists between oval cells and hepatocytes in rat liver.[2,6,11]

A study aimed at examining the cellular distributions of albumin and α-fetoprotein expressions by an *in situ* hybridization technique during early stages of chemical hepatocarcinogenesis in the rat using the complete Solt-Farber model[26] revealed a peculiar distribution of albumin transcripts 2 weeks after the partial hepatectomy.[27] The albumin transcripts were almost exclusively localized in round basophilic areas, some of which were γ-glutamyl transpeptidase (GGT) positive, whereas many were negative for GGT.[27] Thus a severe "inhibition" of albumin expression was observed in the "old" hepatocytes, whereas the "new" basophilic hepatocytes were efficiently expressing the albumin gene. This was an important observation that led to reexamination of the possible precursors for these "foci" of basophilic hepatocytes, with particular emphasis on the possible role of oval cells as progenitors for the basophilic albumin-positive hepatocytes. The experimental approach taken was based on earlier observations[25] (vide supra) that oval cells and not hepatocytes proliferate in the modified Solt-Farber model. It was therefore possible to exclusively label the oval cells with radiolabeled thymidine and to study the possible transfer of thymine from oval cells to hepatocytes.[2] The results from these experiments showed that both oval cells and basophilic hepatocytes prominently expressed albumin, whereas albumin expression in acidophilic preexisting hepatocytes was very low. As expected, the oval cell nuclei were exclusively labeled 1 d after ^{3}H-thymidine administration, and over the next 4 to 5 d it was possible to demonstrate a transfer of radiolabeled thymidine from oval cells to basophilic hepatocytes.[2] These experiments demonstrated that oval cells can be regarded as progenitors, at least in this model, for the basophilic hepatocytes.

What about the other biological options of oval cells outlined in Figure 1? Intestinal metaplasia is frequently seen in rat liver following both the complete and incomplete Solt-Farber protocol.[11,25] Also, Yoshida et al.[28] reported intestinal phenotypic markers in hepatocellular carcinoma induced by 3′-methyl-4-dimethylaminoazobenzene in rats. Formation of intestinal metaplasia from oval cells in rat liver was demonstrated by Tatematsu et al.[29] using the AAF and partial hepatectomy model. Prelabeling oval cells with ^{3}H-thymidine clearly shows that the specialized metaplastic epithelial cells were derived from these cells. Since intestinal metaplasia and/or cylindrical cell metaplasia is frequently seen in areas of cholangiofibrosis induced by several different carcinogens,[30] the observation of Tatematsu et al.[29] provided the basis for the notion that oval cell-derived intestinal metaplasia precedes the formation of cholangiofibrosis.[29]

Development of hepatocytes in rat and hamster pancreas is seen during preneoplasia, neoplasia, and after cell injury.[18] Also, Kimbrough et al.[31] reported pancreatic tissue in close association with cholangiofibrosis in liver of rats fed polychlorinated biphenyls. The presence of exocrine pancreatic tissue similar to that seen in rats has recently been observed in human liver.[32] Although it seems likely that oval cells are the precursors for the pancreatic tissues seen in liver, it remains to be formally demonstrated that this is indeed the case.

III. FACTORS REGULATING GROWTH AND DIFFERENTIATION OF HEPATIC STEM CELLS

Liver regeneration has long been recognized as one of the most dramatic examples of controlled tissue growth. Experimentally, liver regeneration can be induced by either surgical or chemical treatments that remove or destroy a large percentage of the hepatocytes, resulting in rapid restoration of the liver mass from the remaining hepatocytes. Although most of our present knowledge of the molecular mechanism of growth control in the liver stems from *in vivo* and *in vitro* studies on hepatocyte growth, it is known that all the cells of the liver participate in regeneration.[33] This is of considerable interest since recent studies with hepatocytes in culture and gene expression in regenerating liver have identified relatively few growth factors that can be defined as complete mitogens for hepatocytes and therefore implicated in control of liver growth.[33] The complete hepatocyte mitogens are epidermal growth factor (EGF), transforming growth factor-α (TGF-α), heparin-binding growth factor-1 (HBGF-1), and hepatocyte growth factor (HGF). Additionally, there are several co-mitogens that can modulate the effect of the complete mitogens.[33] In light of these observations it is appropriate to inquire if the same complete mitogenic growth factors are involved in the growth regulation of hepatic stem cells.

Since proliferation of the hepatic stem cells is most pronounced in situations in which adult differentiated hepatocytes are incapable of responding to growth signals resulting from acute and/or chronic loss of liver mass, it is necessary to ask if the stem cell compartment is totally unresponsive when normal liver regeneration takes place. The major problem in answering this question is the lack of an unambiguous marker for the hepatic stem cells. Nevertheless, data from several laboratories indicate that expression of α-fetoprotein is a credible marker for early progeny of the hepatic stem cells.[3] This notion is supported by a recent study showing that expression of α-fetoprotein in endodermal cells in the ventral foregut of rat embryo signals an early commitment towards liver lineages.[34]

α-Fetoprotein mRNA constitutes approximately 0.006% of polysomal polyadenylated RNA in normal adult rat liver and is transiently elevated in regeneration induced by either partial hepatectomy (approximately a twofold increase) or CCl_4 injury (approximately a fourfold increase).[35] This increase is associated with the full-length (2.3-kb) α-fetoprotein mRNA species and primarily found in the nonparenchymal liver cells.[36] Increased expression of α-fetoprotein is first seen between 24 and 32 h after partial hepatectomy in the AAF and partial hepatectomy model, and this coincides with prominent thymidine labeling of cells in the periductal space as well as in some of the ductular cells.[11] *In situ* hybridization demonstrated that the α-fetoprotein-positive cells appeared either as individual cells or as pseudoductal formations at the periphery of the periductal space.[11] These α-fetoprotein-positive cells were also positive for an antibody (OV-6) that recognizes oval cells.[11]

The time course for the expression of the full-length α-fetoprotein mRNA after both partial hepatectomy and CCl_4 injury has been documented.[37] The increased expression of α-fetoprotein is first seen 24 h after partial hepatectomy and reaches a maximum 36 h after the operation. The α-fetoprotein expression remains at that level for the next 1.5 d, but is back to "normal" (i.e., barely detectable by Northern analysis) by day 4 after the operation. After CCl_4-induced regeneration the increased expression of α-fetoprotein is first seen at 24 h after the operation and reaches a maximum at 48 h and stays at that level for the next 2 to 3 d, and is back to normal by day 7.

Although EGF, either alone or in combination with insulin, is routinely used to induce DNA synthesis in primary hepatocytes *in vitro,* and infusion of EGF and insulin into rats increases DNA synthesis in the intact normal liver,[38,39] the role of this growth factor in liver regeneration is unclear. Since blood levels of EGF do not change after partial hepatectomy, and the growth factor is not synthesized in the liver,[40] it seems unlikely that EGF is an important player in hepatic regeneration. In contrast to EGF, all the remaining complete liver mitogens show major modulations during hepatic regeneration. The plasma levels of HGF are dramatically increased as early as 2 h after both partial hepatectomy and CCl_4 treatment in the rat, well before the onset of DNA synthesis.[41] Moreover, the level of HGF mRNA in rat liver is also prominently increased from 3 to 6 h posthepatectomy, peaks at 12 h, and gradually returns to undetectable levels between 48 and 72 h.[42] These data suggest that HGF, derived from both extrahepatic and intrahepatic sources, primarily serves to provide the mitogenic signal that initiates the first round of hepatocyte DNA synthesis following partial hepatectomy (and probably also after CCl_4-induced liver injury), and is not involved in activating the hepatic stem cell compartment.

It has been proposed that TGF-α functions as a physiological inducer of hepatocyte DNA synthesis during liver regeneration by means of an autocrine mechanism.[43] This proposal is based on data of TGF-α expression obtained during liver regeneration after partial hepatectomy as well as on the effects of TGF-α on primary hepatocyte cultures.[43] The TGF-α mRNA is increased during liver regeneration after partial hepatectomy, and reaches a maximum (approximately ninefold higher than normal) at the peak of DNA synthesis at 24 h and then gradually declines to undetectable levels at 96 h. Both the message and the peptide are localized in hepatocytes and are found in higher amounts in hepatocytes obtained from regenerating liver.[43] These data support the role of TGF-α in the regenerative response at a time, in contrast to HGF, when hepatocytes have already made a transition from G0 to G1 and entered the cell cycle.[43] However, since the level of α-fetoprotein mRNA reaches a maximum at approximately 36 h after partial hepatectomy[37] it is possible that TGF-α serves to provide the mitogenic signal that activates the stem cell compartment during liver regeneration.

HBGF-1 is an autocrine growth regulator of differentiated hepatoma cells in culture, and the activity is mediated by a receptor of apparently a molecular mass of 100 to 150 kDa.[44] Also, fetal human hepatocytes and adult rat liver membranes exhibit specific binding sites for HBGF-1, and HBGF-1 is capable of stimulating hepatocyte DNA synthesis *in vitro*.[44,45] Following partial hepatectomy the transcripts for HBGF-1 are increased by 4 h to 80% of the peak level at 24 h, remain steady for 4 d, decline significantly on day 5, and are back to normal level by day 7.[45] Both hepatocytes and nonparenchymal cells expressed the HBGF-1 transcripts at 48 h after partial hepatectomy, but information on the precise time course of HBGF-1 expression in these two cell populations is at present lacking.[45] Nevertheless, the data on the temporal HBGF-1 expression in regenerating liver suggest that this growth factor may be capable of activating the hepatic stem cell compartment via paracrine and/or autocrine mechanisms. Taken together, the available data suggest that of the complete hepatic mitogens only two, namely, TGF-α and HBGF-1, are probable candidates to provide the mitogenic signal that initiates the first round of DNA synthesis in hepatic stem cells during liver regeneration.

The precise control that is exhibited during restoration of liver mass following partial hepatectomy has, not surprisingly, led several investigators to search for the existence of both extra- and intrahepatic humoral growth inhibitors.[33] The subject of hepatic growth inhibitors is outside the scope of this chapter and consequently will not be explored. However, we should like to note that at present, TGF-β1 appears to be a contender for a major role as an intrahepatic growth inhibitor, possibly controlling replication of both nonparenchymal and parenchymal cells.[46]

The factors regulating differentiation and lineage commitments of hepatoblasts during liver development and hepatic stem cells when activated in adult liver are to a large extent unknown. However, transcription of liver-specific genes, like other tissue-specific genes, depends on regulatory sequences located close to the mRNA start site as well as those found at more distant locations (i.e., enhancers). Several nuclear DNA-binding proteins that interact with regulatory sequences of liver-specific genes have been identified,[47] and it is to be expected that understanding the regulatory mechanism of these trans-acting factors will provide important insight into the differentiation and lineage commitment of hepatic stem cells.

The cellular biology of the early events after activation of the stem cell compartment in adult rat liver indicates that interaction between different cell types may be a significant factor in determining the fate of the immature cells.[11] The first replicating cell populations that appear after activation of the hepatic stem cell compartment in the AAF partial hepatectomized model are the oval cells and perisinusoidal stellate (Ito) cells.[11] The stellate cell may control to a significant extent the fate of the oval cells by both a paracrine mechanism and the disposition of the extracellular matrix. It is now established that stellate cells are the main source of growth factors such as TGF-α and TGF-β1, and their production of extracellular matrix proteins can lead to fibrosis.[48,49] The stellate cells can therefore create a microenvironment that can either promote or inhibit differentiation of oval cells by isolating the ductular oval cells, by synthesizing the type of extracellular matrix proteins that lead to the formation of adenofibrosis, or by generating matrix proteins (for example, basement membrane proteins) that are important for further differentiation of the oval cells. Since TGF-α and extracellular matrix components are required for kidney tubulogenesis *in vitro,*[50] it seems reasonable to suggest that these factors are involved in stem cell differentiation in both embryonic and adult liver (see also Chapters 5, 8, 9, and 10).

IV. CONCLUSIONS

The existence of hepatic stem cells in adult normal rat liver has been well established. The precise location of these cells in the adult liver remains to be established, but they most likely reside in bile ductules (canals of Hering). The hepatic stem cell compartment is transiently activated in adult liver during regeneration, but only becomes a significant source of new hepatocytes in regeneration when existing adult hepatocytes are incapable of responding to mitogenic signals provoked by reduced liver mass. Substantial evidence exists indicating that hepatic stem cells have the capacity to differentiate into several lineages, including hepatocytic, bile, and intestinal epithelial and pancreatic lineages. Nevertheless, detailed investigations into the fate of the hepatic stem cells *in vivo* are still needed in order to define the developmental potential of these cells as well as to identify both the humoral and the genetic factors regulating the biology of the hepatic stem cells.

REFERENCES

1. **Bucher, N. L. and Malt, R. A.,** *Regeneration of Liver and Kidney,* Little, Brown, Boston, 1971, 1.
2. **Evarts, R. P., Nagy, P., Marsden, E., and Thorgeirsson, S. S.,** A precursor product relationship between oval cells and hepatocytes in rat liver, *Carcinogenesis (London),* 8, 1737, 1987.
3. **Fausto, N.,** Hepatocyte differentiation and liver progenitor cells, *Curr. Opinion Cell Biol.,* 2, 1036, 1990.
4. **Sell, S.,** Is there a liver stem cell?, *Cancer Res.,* 50, 3811, 1990.
5. **Grisham, J. W.,** Cell types in long-term propagable cultures of rat liver, *Ann. N.Y. Acad. Sci.,* 349, 128, 1980.

6. **Thorgeirsson, S. S. and Evarts, R. P.,** Experimental hepatocarcinogenesis: relationship between oval cells and hepatocytes in rat liver, in *Advances in Applied Biotechnology Series,* Vol. 13, Etiology, Pathology, and Treatment of Hepatocellular Carcinoma in North America, Tabor, E., Di Bisceglie, A. M., and Purcell, R. H., Eds., Gulf Publishing, Houston, 1991, 171.
7. **Farber, E.,** The multistep nature of cancer development, *Cancer Res.,* 44, 4217, 1984.
8. **Inaoka, Y.,** Significance of the so-called oval cell proliferation during azo-dye hepatocarcinogenesis, *Gann,* 58, 355, 1967.
9. **Grisham, J. W., Thal, S. B., and Nagel, A.,** Cellular derivation of continuously cultured epithelial cells from normal rat liver, in *Gene Expression and Carcinogenesis,* Gerschenson, L. E. and Thompson, E., Eds., Academic Press, New York, 1974, 1.
10. **Evarts, R. P., Nagy, P., Nakatsukasa, H., Marsden, E., and Thorgeirsson, S. S.,** In vivo differentiation of rat liver oval cells into hepatocytes, *Cancer Res.,* 49, 1541, 1989.
11. **Evarts, R. P., Nakatsukasa, H., Marsden, E., Hsia, C.-C., Dunsford, H. A., and Thorgeirsson, S. S.,** Cellular and molecular changes in the early stages of chemical hepatocarcinogenesis in the rat, *Cancer Res.,* 50, 3439, 1990.
12. **Fausto, N., Thompson, N. L., and Braun, L.,** Purification and culture of oval cells from rat liver, in *Cell Separation: Methods and Selected Applications,* Vol. 4, Pretlow, T. G. and Pretlow, T. P. P., Eds., Academic Press, New York, 1987, 45.
13. **Germain, L., Noël, M., Gourdeau, H., and Marceau, N.,** Promotion of growth and differentiation of ductular oval cells in primary culture, *Cancer Res.,* 48, 368, 1988.
14. **Hayner, N. T., Braun, L., Yaswen, P., Brooks, M., and Fausto, N.,** Isoenzymes of oval cells, parenchymal cells and biliary cells isolated by centrifugal elution from normal and preneoplastic livers, *Cancer Res.,* 44, 332, 1984.
15. **Marceau, N., Germain, L., Goyette, R., Noël, M., and Gourdeau, H.,** Cell of origin of distinct cultured rat liver epithelial cells, as typed by cytokeratin and surface component selective expression, *Biochem. Cell Biol.,* 64, 788, 1986.
16. **Tsao, M. S., Smith, J. D., Nelson, K. G., and Grisham, J. W.,** A diploid epithelial cell line from normal adult liver with phenotypic properties of "oval" cells, *Exp. Cell Res.,* 154, 38, 1984.
17. **Chiu, T.,** Focal eosinophilic hypertrophic cells in the rat pancreas, *Toxicol. Pathol.,* 15, 331, 1987.
18. **Reddy, J. K., Rao, M. S., Qureshai, S. A., Reddy, M. K., Scarpelli, D. G., and Lalwani, N. D.,** Induction and origin of hepatocytes in pancreas, *J. Cell Biol.,* 98, 2082, 1984.
19. **Rao, M. S., Dwivedi, R. S., Yeldandi, A. V., Subbarao, V., Tan, X., Usman, M. I., Thangada, S., Nemali, M. R., Kumar, S., Scarpelli, D. G., and Reddy, J. K.,** Role of periductal and ductular epithelial cells of the adult rat pancreas in pancreatic hepatocyte lineage: a change in the differentiation commitment, *Am. J. Pathol.,* 134, 1069, 1989.
20. **Bisgaard, H. C. and Thorgeirsson, S. S.,** Evidence for a common cell of origin for primitive epithelial cells isolated from rat liver and pancreas, *J. Cell. Physiol.,* 147, 333, 1991.
21. **Tsao, M.-S. and Grisham, J. W.,** Hepatocarcinomas, cholangiocarcinomas and hepatoblastomas produced by chemically transformed cultured rat liver epithelial cells: a light and electron microscopic analysis, *Am. J. Pathol.,* 127, 168, 1987.
22. **Lee, L. W., Tsao, M. S., Grisham, J. W., and Smith, G. J.,** Emergence of neoplastic transformants spontaneously or after exposure to N-methyl-N′-nitro-N-nitrosoguanidine in populations of rat liver epithelial cells cultured under selective or non-selective conditions, *Am. J. Pathol.,* 135, 63, 1989.
23. **Garfield, S., Huber, B. E., Nagy, P., Cordingley, M. G., and Thorgeirsson, S. S.,** Neoplastic transformation and lineage switching of rat liver epithelial cells by retrovirus-associated oncogenes, *Mol. Carcinogen.,* 1, 189, 1988.

24. **Hixson, D. C. and Allison, J. P.,** Monoclonal antibodies recognizing oval cells induced in the liver of rats by N-2-fluorenylacetamide or ethionine in a choline-deficient diet, *Cancer Res.,* 45, 3750, 1985.
25. **Tatematsu, M., Ho, R. H., Kaku, T., Ekem, J. K., and Farber, E.,** Studies on the proliferation and fate of oval cells in the liver of rats treated with 2-acetylaminofluorene and partial hepatectomy, *Am. J. Pathol.,* 114, 418, 1984.
26. **Solt, D. B. and Farber, E.,** New principle for the analysis of chemical carcinogenesis, *Nature,* 263, 701, 1976.
27. **Evarts, R. P., Nagy, P., Marsden, E., and Thorgeirsson, S. S.,** In situ hybridization studies on expression of albumin and α-fetoprotein during the early stage of neoplastic transformation in rat liver, *Cancer Res.,* 47, 5469, 1987.
28. **Yoshida, Y., Kaneoko, A., Chisaka, N., and Onoë, T.,** Appearance of intestinal type of tumor cells in hepatoma tissue induced by 3′-methyl-4-dimethylaminoazobenzene, *Cancer Res.,* 38, 2753, 1978.
29. **Tatematsu, M., Kaku, T., Medline, A., and Farber, E.,** Intestinal metaplasia as a common option in oval cells in relation to cholangiofibrosis in liver of rats exposed to 2-acetylaminofluorene, *Lab. Invest.,* 52, 354, 1985.
30. **Bannasch, P.,** Cellular and subcellular pathology of liver carcinogenesis, in *Primary Liver Tumors,* Remmer, H., Bolt, H. M., Bannasch, P., and Popper, H., Eds., MTP Press, Lancaster, England, 1978, chap. 9.
31. **Kimbrough, R. D., Linder, R. E., and Gaines, T. B.,** Morphological changes in liver of rats fed polychlorinated biphenyls, *Arch. Environ. Health,* 25, 354, 1972.
32. **Wolf, H. K., Burchette, J. L., Garcia, J. A., and Michalopoulos, G.,** Exocrine pancreatic tissue in human liver: a metaplastic process?, *Am. J. Surg. Pathol.,* 14, 590, 1990.
33. **Michalopoulos, G. K.,** Liver regeneration: molecular mechanisms of growth control, *FASEB J.,* 4, 176, 1990.
34. **Shiojiri, N., Lemire, J. M., and Fausto, N.,** Cell lineages and oval cell progenitors in rat liver development, *Cancer Res.,* 51, 2611, 1991.
35. **Petropoulos, C. J., Andrews, G., Tamaoki, T., and Fausto, N.,** α-Fetoprotein and albumin mRNA levels in liver regeneration and carcinogenesis, *J. Biol. Chem.,* 258, 4901, 1983.
36. **Petropoulos, C. J., Yaswen, P., Panzica, M., and Fausto, N.,** Cell lineages in liver carcinogenesis: possible clues from studies of the distribution of α-fetoprotein RNA sequences in cell populations isolated from normal, regenerating, and preneoplastic rat livers, *Cancer Res.,* 45, 5762, 1985.
37. **Nakatsukasa, H., Evarts, R. P., Marsden, E., and Thorgeirsson, S. S.,** Detection of putative liver stem cells and their activation during hepatic regeneration in rats, submitted.
38. **Bucher, N. L. R., Patel, V., and Cohen, S.,** Hormonal factors concerned with liver regeneration, in *Hepatotrophic Factors, Ciba Foundation Symposium No. 55,* Elsevier/Excerpta Medica/North-Holland, New York, 1978, 95.
39. **McGowan, J. A.,** Hepatocyte proliferation in culture, in *Isolated and Cultured Hepatocytes,* Guillouzo, A. and Guguen-Guillouzo, C., Eds., John Libey, London, 1986, 14.
40. **Rall, L. B., Scott, J., Bell, G. I., Crawford, R. J., Penschow, J. D., Niall, H. D., and Coghlan, J.,** Mouse prepro-epidermal growth factor synthesis by the kidney and other tissues, *Nature,* 313, 228, 1985.
41. **Lindroos, P. M., Zarnegar, R., and Michalopoulos, G. K.,** Hepatocyte growth factor (hepatopoietin A) rapidly increases in plasma before DNA synthesis and liver regeneration stimulated by partial hepatectomy and carbon tetrachloride administration, *Hepatology,* 13, 743, 1991.
42. **Zarnegar, R., DeFrances, M. C., Kost, D. P., Lindroos, P., and Michaelopoulos, G. K.,** Expression of hepatocyte growth factor mRNA in regenerating rat liver after partial hepatectomy, *Biochem. Biophys. Res. Commun.,* 177, 559, 1991.

43. **Mead, J. E. and Fausto, N.,** Transforming growth factor α may be a physiological regulator of liver regeneration by means of an autocrine mechanism, *Proc. Natl. Acad. Sci. U.S.A.*, 86, 1558, 1989.
44. **Kan, M., DiSorbo, D., Hou, J., Mansson, P. E., and McKeehan, W. L.,** High and low affinity binding of heparin-binding growth factor to a 130 kDa receptor correlates with stimulation and inhibition of growth of differentiated human hepatoma cells, *J. Biol. Chem.*, 263, 11306, 1988.
45. **Kan, M., Huang, J., Mansson, P.-E., Yasumitsu, H., Carr, B., and McKeehan, W. L.,** Heparin-binding growth factor type 1 (acidic fibroblast growth factor): a potential biphasic autocrine and paracrine regulator of hepatocyte regeneration, *Proc. Natl. Acad. Sci. U.S.A.*, 86, 7432, 1989.
46. **Fausto, N. and Mead, J. E.,** Role of protooncogenes and transforming growth factors in normal and neoplastic liver growth, *Progr. Liver Dis.*, 9, 57, 1990.
47. **Johnson, P. F. and McKnight, S. L.,** Eukaryotic transcriptional regulatory proteins, *Annu. Rev. Biochem.*, 58, 799, 1989.
48. **Nakatsukasa, H., Evarts, R. P., Hsia, C.-C., and Thorgeirsson, S. S.,** Transforming growth factor-α1 and type I procollagen transcripts during regeneration and early fibrosis of rat liver, *Lab. Invest.*, 63, 171, 1990.
49. **Evarts, R. P., Nakatsukasa, H., Marsden, E. R., Hu, Z., and Thorgeirsson, S. S.,** Expression of transforming growth factor alpha (TGF-α) in regenerating liver and during hepatic differentiation, *Mol. Carcinogen.*, 5, 25, 1992.
50. **Taub, M., Wang, Y., Szczesney, T. M., and Kleinman, H. K.,** Epidermal growth factor or transforming growth factor α is required for kidney tubulogenesis in matrigel cultures in serum-free medium, *Proc. Natl. Acad. Sci. U.S.A.*, 87, 4002, 1990.

Chapter 7

THE ROLE OF BIPOTENTIAL PROGENITOR CELLS IN LIVER ONTOGENESIS AND NEOPLASIA

N. Marceau, M.-J. Blouin, M. Noël, N. Török, and A. Loranger

TABLE OF CONTENTS

ISBN 0-8493-4746-7
© 1992 by CRC Press, Inc.

I. INTRODUCTION

The liver of the vertebrate embryo develops as a diverticulum from an endomesodermal rudiment of the foregut. In the rat, the liver emerges at embryonic day 10.5 (E10.5) and shows a clear anatomical contour at E12.[1-4] While the fetal liver is mainly involved in hematopoietic functions,[1,4] prominent changes in the differentiation programs of the epithelial cells (EC) take place during the E12 to E16 period to produce hepatocytes and intrahepatic biliary EC.[1-5] After birth, hepatocytes become the main specialized cell population of the hepatic parenchyma. They exhibit several metabolic activities, with the two most prominent ones being the production of proteins released into the blood circulation via the sinusoids and the formation of bile that is released into the gut via bile canaliculi, biliary ductules, and intra/extrahepatic biliary ducts.[1-4] Through organogenesis, hepatocytes, biliary EC, and the nonparenchymal sinusoidal cells become organized as multilobular units, which in turn form the lobes. As the organ develops, the mesenchymal cells of the septum transversum give rise to the Glisson's capsule, which is comprised of fibroblastic and mesothelial cells.[2,3] This connective tissue capsule defines the final limiting boundaries of the liver.

Although hepatocytes and biliary EC of the adult liver constitute differentiated cells, they retain a capacity to actively proliferate in response to tissue damage.[6-9] This means that in contrast to self-renewal tissues where progenitor (stem) cells continuously proliferate to replace the lost differentiated cells, the need for a progenitor cell compartment in the postnatal liver is not obvious.[5,8,9] However, several studies performed in the adult rat have demonstrated that, in response to a massive necrosis induced by toxic doses of hepatocarcinogens, the remaining tissue undergoes a regenerating process that mimics liver ontogenesis and induces the emergence of atypical EC, designated as "oval cells".[10-19] While the original observations have led to the suggestion that these cells constitute a homogenous progenitor (stem) cell population,[20-22] several lines of evidence accumulated over the last decade suggest that oval cells represent a mixed-cell type compartment of distinct developmental potentials.[5,8,23-25] Therefore, the three questions concerning liver cell lineages and differentiation that have been studied in our and other laboratories over the last few years are whether or not E12 cells are committed stem cells, whether or not hepatocytes and biliary EC are tightly linked to a common progenitor in their lineage, and whether or not a subpopulation of the oval cell compartment corresponds to progenitor (stem) cells remaining in the postnatal liver.

Hepatocytes isolated from postnatal rat liver can be maintained as a functionally active stationary cell monolayer for a few months, or can undergo a few divisions after seeding in the presence of growth-promoting factors, but they cannot be passaged.[26-32] Biliary EC isolated from bile duct-ligated adult rat liver can also be set up in primary culture under conditions that lead to their propagation and retention of their differentiated features for a few passages.[33,34] Moreover, it is possible to obtain enriched preparations of non-parenchymal EC in primary cultures which emerge as well-defined colonies and bear no phenotypic resemblance to authentic hepatocytes and biliary EC.[19,35-41] These liver epithelial cells (LEC) can be established as cell lines and upon long-term culture, some of them can express a few of the hepatocytic metabolic activities.[38] The three main questions about these LEC lines are the identity of their precursors in rat liver *in vivo,* their relationship to progenitors, and their involvement in rat hepatocarcinogenesis.

It is well established that hepatocellular carcinoma and cholangiocarcinoma are the two most frequent types of cancers that can be readily induced in rats fed various chemical carcinogens.[42-56] In these *in vivo* models, the experimental strategy used to identify the precursors is based on detailed cell typing of the tumors with various markers, related to normal liver cell lineages. Attempts to obtain a direct answer about the identity of liver tumor precursors by using an *in vitro* transformation assay with purified populations of liver cells of distinct differentiation status have led so far to very few meaningful observations.[23,32,47,48] In fact, much of the work on rat liver cell transformation *in vitro* has been performed with LEC lines, for which the cell of origin is ill defined.[38,49] Nevertheless, several transformation protocols have been proposed using various chemical carcinogens and/or oncogenes,[50-52] but the significance of the findings with regard to *in vivo* hepatocarcinogenesis and tumor cell lineages remains a major issue (see also Chapter 1).

In recent years, new insights have been provided on EC lineages in normal and neoplastic rat livers, by using a cell-typing strategy which employs a large battery of markers.[16,17,53-59] The work performed in our laboratory is based on the use of three classes of cell markers.[5,24] The first class consists of markers that assess the functional activities of the cells, and includes α-fetoprotein (AFP), albumin, and γ-glutamyl transpeptidase (GGT). The second class is cytokeratins (CK), which constitute a multigene family of intermediate filament (IF) proteins. It is well known that IF proteins can be used as reliable cell lineage markers in tissue differentiation;[60-62] for example, vimentin is largely restricted to mesenchymal cells, whereas CK are usually specific to EC. The third class of markers are cell components detected by monoclonal antibodies (Mab). Some of the most useful Mab are those directed against cell surface components whose expression is dependent on cell lineage and differentiation status. Among these are HES_6 and BDS_7, two surface components found on the surface of hepatocytes and ductular oval cells, respectively.[53,54] OV1, OC.2, and OC.3 correspond to other differentiation-related surface markers.[55,62,63] Few cytoplasmic markers have also been fished out with Mab, and among them are BPC_5 and OV6.[55,63]

This paper reviews recent data on the definition of EC lineages in developing rat liver, the presence of bipotential progenitor cells in regenerating rat liver, the cell lineage relationship between bipotential progenitor cells and

LEC, the relationship of liver carcinoma cells to their putative normal counterparts, and finally, the biological significance of the heterogeneity observed in tumors generated by transformed LEC.

II. EPITHELIAL CELL LINEAGES IN DEVELOPING RAT LIVER

Until recently, much of the work on the organogenesis of rat liver has been focused on the morphological changes of the hepatocytic cells, mainly at the electron microscopy level. From E10.5 to birth, major changes take place in the shape, number, and organization of the cytoplasm of the hepatocytic cells.[1-5,64] Moreover, the time of appearance of liver-specific enzymes indicates that E16 is a critical time for liver cell differentiation; preceding the appearance of immature hepatocytes ($<$E.16), the cells have been named hepatoblasts. At E16, hepatocytic cells clearly exhibit the phenotype of immature hepatocytes, whereas biliary EC form typical ductal structures.[1,4] We have recently examined the differentiation of liver cells *in situ* during the fetal period by determining the differential expression of the three sets of cell markers and have also assessed the *in vitro* differentiation potential of the emerging liver cells.

A. *IN SITU* ANALYSES

1. HES_6- vs. BDS_7-Expressing Cells

Over the last 10 years, our group has used Mab to detect cell lineage-specific markers. HES_6 and BDS_7 were identified by Mab that reacted specifically with adult rat liver hepatocytes and biliary EC.[53,54] In fact, BDS_7 was selected as a specific marker of ductular oval cells isolated from adult rats fed 3′-methyl-4-dimethylaminoazobenzene (Me-DAB).[53] No success has so far been reached in our attempts to identify the antigens recognized by HES_6 and BDS_7 Mab, respectively. Nevertheless, their use in our *in situ* immunohistochemical analyses of embryonic and fetal livers have led to important findings. Indeed, BDS_7-positive cells are already present in the E12 rat liver, whereas HES_6 become suddenly expressed in most E15 to E16 cells. This finding has suggested to us that hepatocytes and biliary EC emerge as distinct cell lineages during the first few days (E10.5 to E12) of liver formation[54] (Table 1). However, data have shown that before E12 (at the time of emergence of the liver), BDS_7 can be detected in non-EC, whereas after E15 it is restricted to biliary EC.[65] The biological significance of this phenomenon is unclear, but nevertheless it is in line with findings reported recently by Sell and Dunsford[43] which suggest that ‘‘mesenchymal cells can express a liver cell phenotype at restricted times during embryogenesis’’. So, it becomes obvious that early embryonic liver cells exhibit a high plasticity in their differentiation-specific features.

TABLE 1
***In Situ* Phenotypic Properties of Progenitors, Hepatocytes, and Biliary EC**

Markers	Progenitors[a]	Hepatocytes[b]	Biliary EC[b]
CK7	−	−	+
CK8	+	+	+
CK14	−	−	−
CKX	+	−	−
CK18	+	+	+
CK19	−	−	+
BPC_5	+	−	−
Alb	+	+	−
AFP	+	+	−
GGT	+	+	+
BDS_7	−	−	+
HES_6	−	+	−

Note: According to the human cytokeratin catalog of Moll et al.,[60] CK55 is equivalent to CK8 and CK39 to CK19. The human equivalent of CK52, as reported in Reference 54, is ill defined. While we proposed before that it might be equivalent to CK14,[5] our recent data clearly establish that the CK detected in progenitors is a new CK, designated for now as CKX.[74] Alb denotes albumin.

[a] Progenitors correspond to EC present in the embryonic liver at the E10.5 to E16 period and specifically expressing BPC_5.
[b] Hepatocytes and biliary EC appearing after E15.

Modified from Germain, L., et al., *Cancer Res.*, 48, 4909, 1988. With permission.

2. AFP/Albumin- vs. GGT-Expressing Cells

While AFP is expressed from E10.5 to postnatal days 10 to 12, albumin is first expressed at E11.5 and continues to be expressed throughout life.[1,66-68] GGT is expressed in all cells during the fetal period and it only becomes restricted to biliary EC after birth.[68] Thus, AFP, albumin, and GGT are not sufficient to provide a clear-cut distinction between the liver cell types and therefore cannot be used as cell lineage markers.

3. Cytokeratin-Expressing Cells

Type II CK8 and type I CK18 are IF components of all EC throughout liver development in the rat.[5] However, while CK8 expression is high during the entire period, CK18 expression is low at E10.5 to E11 and then it gradually increases. CK7 and CK19 are IF components of biliary EC (along with CK8 and CK18), and their expression in fetal rat liver coincides with the emergence of ductular structures of E15 to E16[5,65] (Table 1). However, both CK7 and

CK19 are apparently expressed in mesothelial cells (e.g., Glisson's capsule),[69] which means that they cannot be used by themselves as specific markers of biliary EC. Desmet's team[70] and Shiojiri and collaborators[65] have used several antibodies against CK7 and CK19 to examine the progenitors of intrahepatic biliary EC in rats and concluded that the $CK7^+/CK19^+$ cells emerging at E15 to E16 originate from hepatocytes located proximal to large vascular spaces in the portal area. A similar observation has been reported for humans.[71] It remains, however, that some anti-CK antibodies may react with more than one type of CK and therefore their utility as molecular probes obviously depends on their specificity.

Our group is particularly interested in the biological significance of differential CK gene expression in relation to LEC differentiation.[5] The observation that CK18 expression is low in E10 to E14 liver cells has led us to suspect the presence of another type I CK in fetal rat liver. In this regard, we have reported that a polyvalent antiserum raised in the guineapig against cow hoof keratins is reactive on E12 to E16 cells. Since this antiserum reacted with a CK of 52 kDa present in T51B cells,[49] we have concluded that E12 to E16 cells were expressing CK52[54] (see Section IV). The best candidates for the human CK equivalent for CK52, according to the human CK catalog established by Moll et al.,[60] are CK14 and CK17, which are generally expressed in proliferative squamous epithelia. In a recent study performed in collaboration with Dr. Denis Roop (Baylor College of Medicine, Houston, TX), using an anti-CK14 antibody, very few CK14-positive cells were detected in the liver over the E12 to E18 period.[72,73] Thus, the CK detected by the guinea pig antiserum was not the CK52 present in T51B cells.

We have been examining the expression of a newly detected CK gene (CKX), in both embryonic and adult rat tissues. A cDNA fragment for this gene was subcloned from F9 teratocarcinoma cells by the group of Dr. André Royal (Montreal Cancer Institute, Montreal, Quebec). *In situ* hybridization analyses performed on embryo sections demonstrated the presence of CKX mRNA in cells of the intestinal tube at E10, i.e., before the appearance of the liver.[74] This observation fits with the finding that CKX is present in F9 cells, a cell line that can be induced by retinoic acid to differentiate into a population of primitive endodermal cells. The most significant result is that CKX mRNA are detected in the embryonic liver (Plate 1*, Table 1). Moreover, by using double detection techniques based on *in situ* hybridization of CKX and immunolocalization of AFP, we showed that the two markers largely colocalize in the same cells during the entire fetal period, e.g., E12 and E18 (Plate 1B through 1E). However, in contrast to AFP, CKX expression in the liver is lost early after birth (Plate 1F). The next step for us is to confirm the expression of the CKX gene in E10 to E20 cells at the protein level. Moreover, it is worth noticing that the CKX gene is expressed in the fetal epidermis, but not in the fetal pancreas, despite the fact that this organ, like the liver,

* Plate 1 appears following page 134.

also emerges from the intestinal tube.[74] Cells of the yolk sac express a high level of CKX mRNA as well (Plate 1G). Furthermore, from our *in situ* hybridization analyses performed on adult rat tissues, the data so far demonstrate that the CKX gene is expressed in tissues equivalent to those in humans that express CK16, e.g., expression in outer root sheath cells of hair follicles, but not in liver cells,[74] thus suggesting that CKX is related to CK16. In any event, the main observation emerging from this study is that CKX is a marker of liver cell differentiation status during embryogenesis.

4. BPC_5-Expressing Cells

Because of our success in using Mab to detect new liver cell markers at the postnatal stage, we develop immunization protocols that would allow the detection of a specific marker for emerging rat liver cells. Attempts to use E12 cells as immunogen alone were unsuccessful. Because the number of cells available for hybridoma screening was low, the production of Mab turned out to be tedious and impractical. Since tumor cells exhibit several features of fetal cells, we decided to use hepatoma cells and a combination of several immunization procedures. The Mab obtained against BPC_5 was generated by a highly restrictive and cell-targeting immunization protocol that involves the use of a mixture of Reuber H4 and Morris hepatoma cells incubated with antisera against adult rat hepatocytes and sinusoidal cells prior to the first injection into mice.[63] The subsequent injections were done with E12 liver cells which had also been incubated with the antiserum. The first hybridoma screenings were done with hepatoma cells and the two others with cultured (24-h) E12 cells. It was determined from the cell-typing studies that in the embryo BPC_5 expression is largely restricted to E10 to E16 liver cells[63] (Table 1). In the adult rat, examination of several epithelial tissues revealed no significant staining, except for a few cells lining the small intestine. Our recent work indicates that the BPC_5 Mab preferentially recognizes hyperphosphorylated forms of CK8. In terms of the functional significance of this finding, it seems that the presence of the hyperphosphorylated forms of CK8 is tightly linked to the differentiation status of rat liver cells and that it largely coincides with the transient expression of the CKX at the fetal period. Moreover, the disappearance of the BPC_5 epitope seems to match with the gradual increase of CK18 expression at the protein level and the reorganization of CK IF from a predominant cytoplasmic distribution to a preferential localization close to the surface membrane, parallel to bile canaliculi formation. Our future studies using confocal microscopy should provide data to support this hypothesis. Finally, BPC_5 expression is not linked to cell proliferation, since the postnatal rat hepatocytes remain negative following partial hepatectomy (see Section III).

B. E12 CELL COMMITMENT IN PRIMARY CULTURE

All of the results obtained from cell-typing studies performed *in situ* are supported by our data using cells in primary culture under conditions where cell growth activity was minimal. Freshly isolated E12 cells were plated on a fibronectin-coated substratum and their differentiation potential in response to various combinations of promoting factors was evaluated.[54] At day 1 of culture, the cells expressed BPC_5, CK8, AFP, and albumin, but not HES_6 and BDS_7. After 6 d of culture in the presence of DMSO, more than 60% of the cells which contained CK8, AFP, and albumin also expressed HES_6; the number of BDS_7-positive cells remained extremely low. These cells exhibited typical ultrastructural features of hepatocytes (Figure 1A). With the addition of sodium butyrate to the medium, a large proportion of the cells became BDS_7 positive without losing much of their albumin or AFP. More than 70% of the cells became BDS_7 positive by day 6 and had typical ultrastructural features of biliary EC (Figure 1B).[54] In both conditions (DMSO- and sodium butyrate-supplemented medium) the number of BPC_5-positive cells rapidly decreased with time in culture. We have extended this work by determining the response of E12 cells to additional various physiological growth- and differentiation-promoting factors. Among them are insulin-like growth factor-II and transforming growth factor-β, two factors present in the conditioned medium of BRL EC,[63] which in both cases lead the E12 cells to differentiate along the hepatocytic cell lineage (HES_6^+, BPC_5^-). The addition of tryptose phosphate broth induced the differentiation of the E12 cells toward the biliary ductal cell lineage (BDS_7^+, BPC_5^-).[75] Moreover, nicotinamide, a well-known inhibitor of poly-ADP ribosylation, exerted a permissive effect on growth stimulation by EGF in the presence of insulin and dexamethasone. We also have data indicating that a requirement for the proper growth and differentiation of E12 *in vitro* is the presence of nonparenchymal (mesenchymal) cells in culture.[75] Work is in progress to define the types of interactions that are required between these cells to mimic the sequences of growth and differentiation events occurring *in vivo*. These *in vitro* models confirm the high plasticity of liver differentiation and provide a striking example of environmental influences on liver progenitor cell differentiation (Figure 2).

By using this sort of *in vitro* differentiation assay, it has been possible to evaluate the change in the proportion of the bipotential progenitors during liver development.[54] At E12, more than 70% of the cells exhibit the capacity to express BDS_7, whereas less than 20% of these cells remain at E18; the proportion further drops to less than 5% during the early postnatal period.[54] In fact, the decrease in the number of bipotential progenitors, as assayed *in vitro* matches, to a large extent, the drop in the number of BPC_5-positive cells *in situ*.

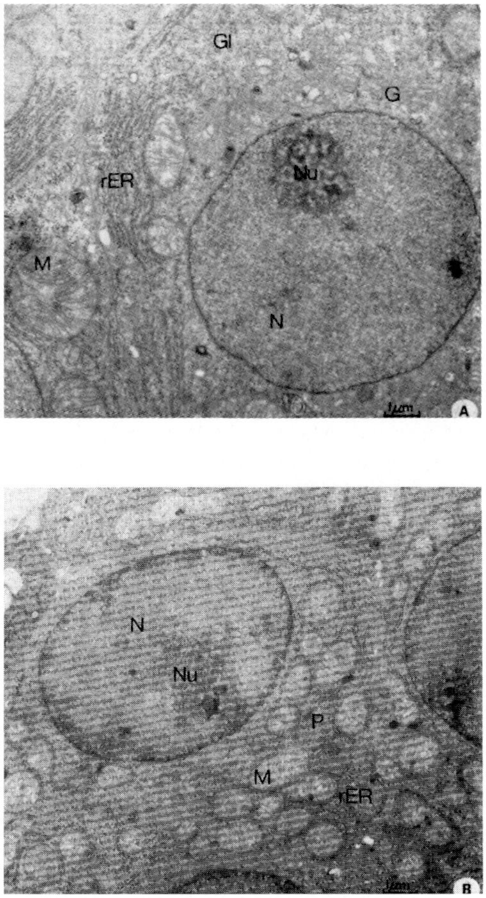

FIGURE 1. Electron microscopy of cells isolated from E12 rat liver and cultured 6 d in medium A* (A) or 4 d in medium B** supplemented with sodium butyrate (B). Note that the cells in (A) have ultrastructural features characteristic of hepatocytes, whereas cells in (B) more resemble bile ductular cells: (a) the rough-surfaced endoplasmic reticulum (rER) is more abundant and is composed of a collection of long cisternae in (A); (b) many free polysomes (P) are present in (B); (c) the mitochondria (M) are bigger and contain longer tansverse cristae in (A); (d) glycogen (Gl) particles are present in (A) only; (e) the nucleus (N) is round and the nucleolus (nu) is large and they contain a distinct nucleolonema in (A); (f) the nucleus (N) of cells in (B) are oval and the nucleolus (nu) is more diffuse. The Golgi apparatus (G) is seen in (A).[54] *Medium A: α-minimal essential medium supplemented with 10% fetal calf serum, 1 μ*M* dexamethasone, 16 n*M* insulin, and 2% DMSO; **medium B: α-minimal essential medium supplemented with 10 ng/ml EGF, 1 μ*M* dexamethasone, 16 n*M* insulin, and 2% DMSO. (Original magnification × 7500.) (Reprinted from Germain, L. et al., *Cancer Res.*, 48, 4909, 1988. With permission.)

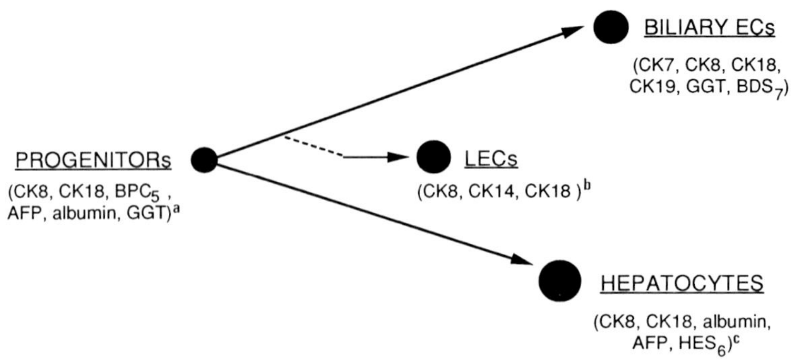

FIGURE 2. Differentiation patterns of rat hepatic EC. Embryonic rat liver EC constitute a population of bipotential progenitor (stem) cells capable of differentiating into hepatocytes or biliary EC. Note the differential expression of CK between the three cell populations. [a]Recent data suggest that progenitors in the embryonic liver express a squamous epithelium-related CK, named CKX.[74] [b]LEC correspond to nonparenchymal EC that emerge in long-term primary cultures of early hepatocytes and exhibit many phenotypic features of progenitor cells that differentiate along a biliary cell sublineage. LEC constitute a distinct EC population, in the sense that they contain a CK (CK14) that is typical of stratified epithelium. However, some LEC variably express CK of biliary EC (CK7 and CK19)[84] and functional activities of hepatocytes (albumin),[91] thus suggesting that they correspond to cells of variable differentiation status. [c]It should be noted that AFP expression is lost as the hepatocytes become polyploidal cells.[66]

III. BIPOTENTIAL PROGENITOR CELLS IN REGENERATING RAT LIVER

A major effort has been made recently to look for the presence of progenitor (stem) cells in regenerating rat liver. Four patterns of ''liver regeneration'' can be induced in the rat, exemplified by: (1) partial hepatectomy, (2) bile duct ligation or selective chronic chemical toxicity, (3) liver damage induced by D-galactosamine, and (4) chemical hepatocarcinogen toxicity.

Following partial hepatectomy, the remaining tissue undergoes a rapid compensatory growth that involves the proliferation of hepatocytes and the other cell types as well.[6,76] Our recent data from the cell-typing analyses on the regenerating livers of rats at the suckling or adult stage supported the notion that no proliferation of the progenitors is taking place, i.e., no BPC_5-positive cell can be detected.[75] Moreover, cultures established with cells isolated from regenerating rat livers yield hepatocyte populations that exhibit no higher proliferative capacity than those isolated from normal rat livers, with no appearance of BPC_5.[75]

Following ligation of the common bile duct or during chronic intoxication with α-naphthyl-isothiocyanate or 4,4′-diaminodiphenylmethane in the rat, a selective proliferation of the intrahepatic biliary ductular cells is observed.[9] The induced proliferative response is maximal at 1 to 2 weeks after bile duct ligation.[33] Interestingly, the use of various hepatocyte vs. biliary ductal cell markers indicated that very few if any of the cells forming the biliary ductal cell structures after bile duct ligation express functional features of hepatocytes (see Chapter 9, this volume). A few techniques have also been developed for the reproducible high-yield isolation and monolayer culture of intrahepatic biliary EC.[9,33] In general, it appears that cultured biliary ductular EC from the liver of bile duct-ligated rats exhibit morphological and enzymatic features of their *in vivo* counterparts. These cells can be induced to proliferate in response to various growth-promoting factors[9] and be maintained in continuous cultures for up to at least ten passages.[34] While they conserve relatively high levels of some enzymatic activities like GGT and leucine aminopeptidase,[34] they rapidly lose alkaline phosphatase activity[34] and CK19.[77] In the light of the selective expression of BPC_5 in progenitors at the early fetal period, it would be of interest to determine whether or not the intact liver tissue and original biliary EC preparations contain few BPC_5-positive cells. In this respect, Sirica and collaborators have recently observed *in situ* that a few intrahepatic hyperplastic ductular cells express phenotypic markers of hepatocytic cells in the liver of rats previously subjected to bile duct ligation followed by CCl_4-induced hepatotoxicity (Chapter 9, this volume). The main question therefore is to determine whether or not these cells correspond to bipotential progenitor cells.

Following liver damage induced by D-galactosamine administration, hepatocytes become severely damaged and a proliferation of biliary EC precedes that of the hepatocytes. This cellular response is similar to that which occurs during the early periods of hepatocarcinogenesis.[78,79] On the basis of AFP, CK7, CK8, CK18, and CK19 expression, Lemire and collaborators[79] have demonstrated that biliary duct cells are mostly responsible for the emergence of the oval cell-like population, as well as provided evidence for the transition of biliary duct cells to hepatocytic cells. However, we have shown that BPC_5 is not expressed during the time of maximum appearance of AFP-positive cells[75] and conclude that bipotential progenitors could hardly be responsible for the induced cellular response to D-galactosamine.

The feeding of rats with Me-DAB leads to a massive necrosis of the hepatic tissue followed by a regeneration that involves the appearance of oval cells.[8,80] A similar response is obtained in the liver of rats fed a choline-deficient diet supplemented with DL-ethionine or 2-acetylaminofluorene (2-AAF).[17,81,82] Recent data obtained from our and other laboratories indicate that the oval cells constitute a mixed-cell type compartment whose composition varies depending on the carcinogen used. We failed to detect in livers of Me-DAB-fed rats BPC_5-positive cells at the time at which the transient AFP production was at maximum. In the livers of rats fed 2-AAF after partial hepatectomy, Evarts and collaborators[21,22] have recently reported cell kinetic analyses using ^3H-thymidine labeling which demonstrated the transfer of the label from AFP-positive oval-like ductal cells to immature AFP-positive hepatocytes. Our analysis on BPC_5 localization in such livers revealed the presence of only a few BPC_5-positive cells at 7 d posthepatectomy (Figure 3). From these cell-typing analyses performed at the early period of hepatocarcinogenesis, we conclude that most of the AFP-positive cells of the oval cell compartment are likely diploid immature hepatocytes, i.e., at a differentiation stage more advanced than those appearing at E16. Still, these cells could be largely derived from bipotential progenitor cells that rapidly progress along the hepatocytic cell lineage.

We have examined the proposal that ductular oval cells might be capable of differentiating into hepatocytes,[53] by using the highly enriched preparations of BDS_7-positive ductular oval cells and evaluating their growth and differentiation capacity in primary culture following the addition of growth- and differentiation-promoting factors. While it was possible to make the cells produce albumin in the absence of growth (Figures 4 and 5), we have been unable to detect the induction of HES_6, the hepatocytic cell-lineage marker that appears at E15 in the embryonic rat liver.[53] Our interpretation is that most ductular oval cells in this system represent a population of immature EC of the ductal cell lineage which can be induced to produce albumin and to express a few other metabolic markers of immature hepatocytes *in vitro,* but not all of the mature hepatocytic features. These observations stress the point that liver cells exhibit a high degree of plasticity in terms of their differentiation potential. Such a conclusion is further supported by the data obtained in fetal rat liver where hepatocytic cells can apparently give rise to biliary EC (as discussed in Section II).

A significant outcome of all these studies performed on regenerating postnatal rat liver is the possibility that hepatocytic and biliary EC can undergo transdifferentiation (Section V). Even if we consider that the two cell types belong to two separate lineages, it could well be that up to a few differentiation steps cells from one lineage can switch to the other. This proposal does not support the presence of pluripotent stem cells per se, but still fits with the view that bipotential progenitors or immature LEC are essentially responsible for the replacement of the EC lost in severely damaged rat liver.

IV. CELL LINEAGE RELATIONSHIP BETWEEN BIPOTENTIAL PROGENITORS AND LIVER EPITHELIAL CELLS

LEC constitute a phenotypically heterogeneous population of nonparenchymal EC. This heterogeneity probably results from differences in the cell of origin *in vivo*.[49] The most prominent feature of LEC is their high growth capacity in culture, which means that they can be readily established as cell lines. Some LEC lines, such as WB-F344, can express a few functional activities (e.g., albumin production) characteristic of hepatocytes.[19] They rarely express

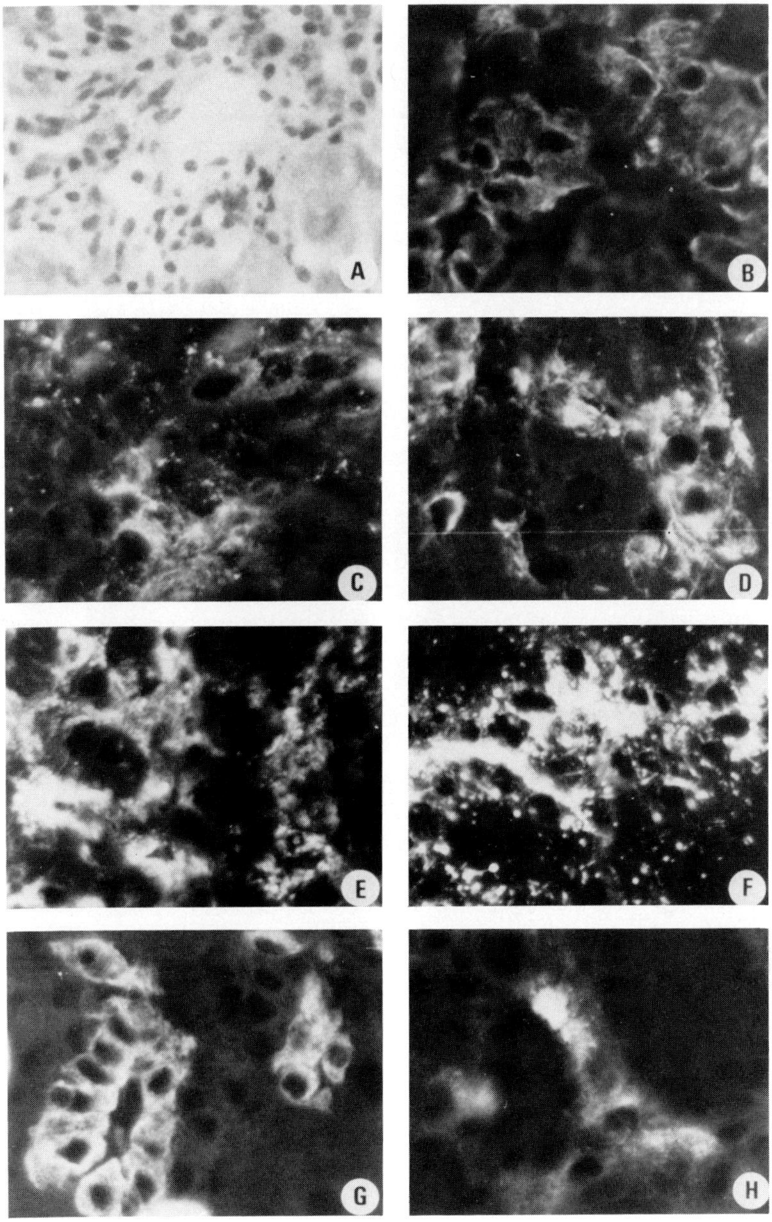

FIGURE 3. Optical microscopy staining with hematoxylin/eosin (A), and immunofluorescence microscopy of rat liver tissue sections using anti-CK8 (B), anti-CK18 (C), antivimentin (D), anti-HES_6 (E), anti-BDS_7 (F), anti-OV-6 (G), and anti-BPC_5 (H) antibodies. Rats were subjected to daily injections of 2-AAF (without diethylnitrosamine) for 5 d before partial hepatectomy (PH) and 4 d after PH.[75] Observations were made from tissue sections obtained 7 d after PH.

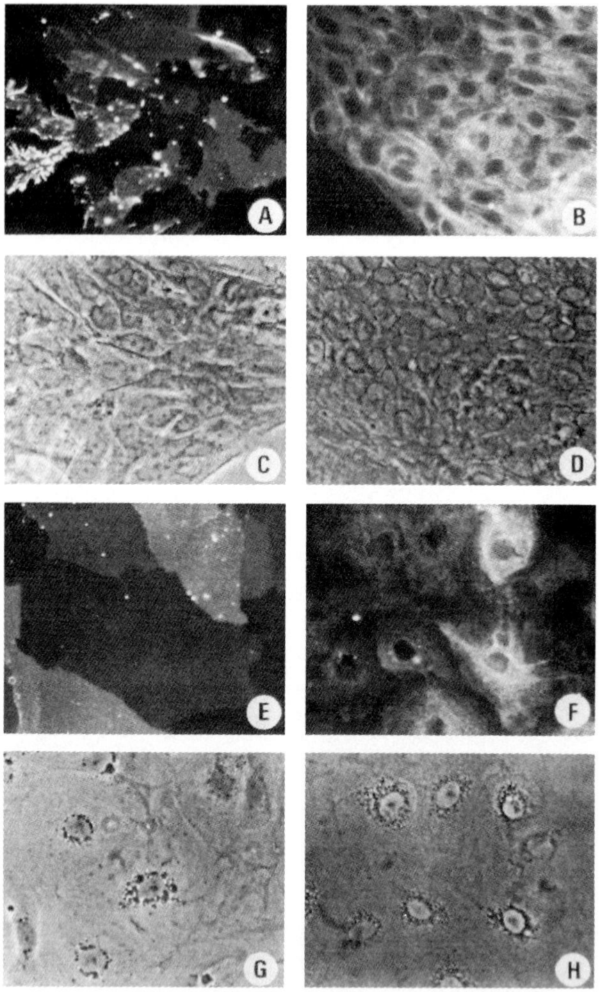

FIGURE 4. Immunofluorescence microscopy of oval cells cultured for 5 d in medium MX* supplemented with DMSO (A to D) or sodium butyrate (E to H) using anti-BDS_7 (A and E) anti-CK52 (B and F) antibodies. (C), (D), (G), and (H) are phase contrast micrographs of (A), (B), (E), and (F), respectively. Note that cells were fixed in either paraformaldehyde (A, C, E, G) or acetone (B, D, F, H). *Medium MX is the following: MEM containing 1 m*M* pyruvate, 0.2 m*M* aspartate, 0.2 m*M* serine, 1 m*M* tyrosine, 1 m*M* proline, 1 m*M* phenylalanine, and supplemented with 20% CLEX, 10 ng/ml oxidized bile acids, 17 μ*M* bilirubin, 10 ng/ml; cholera toxin, 1 μ*M* dexamethasone, 2.5 μg/ml insulin, 50 m*M* β-mercaptoethanol, and 5 μg/ml transferrin. (Original magnification × 400.) (Reprinted from Germain, L. et al., *Cancer Res.*, 48, 368, 1988. With permission.)

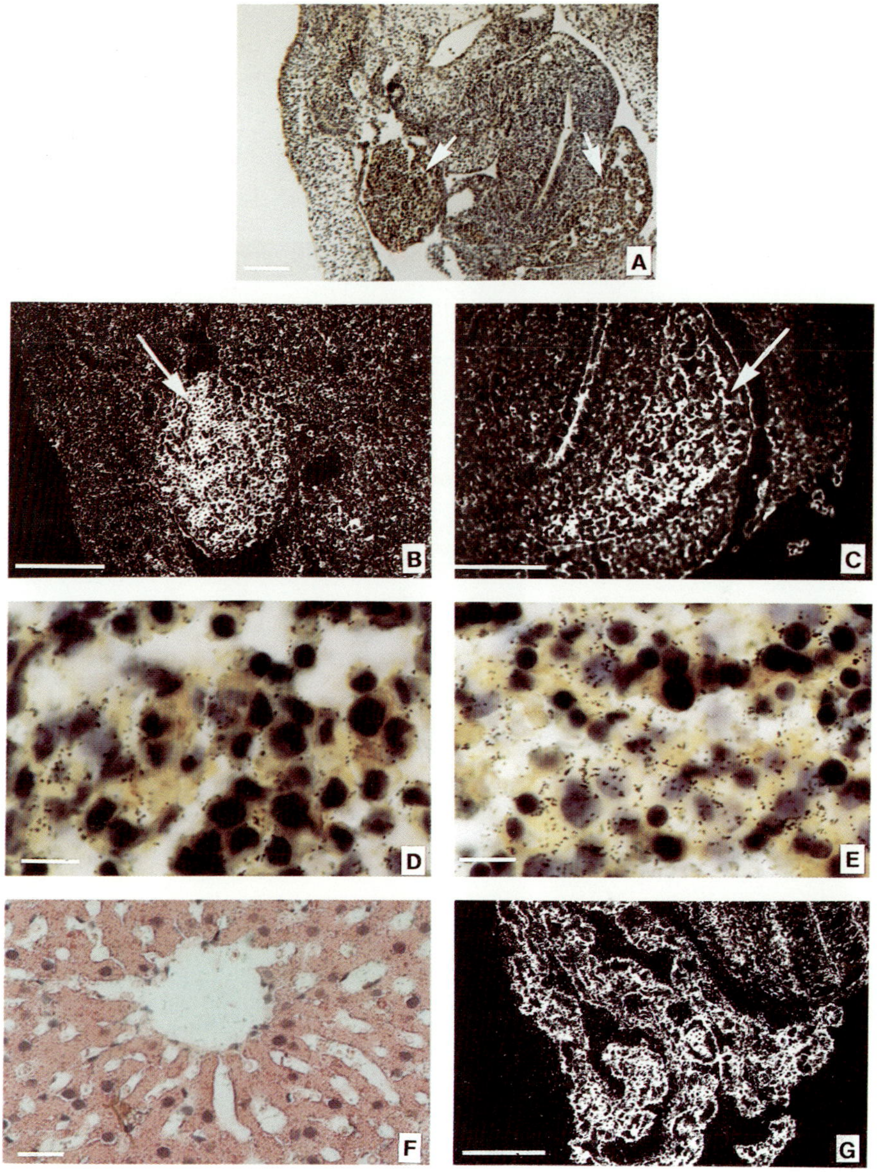

PLATE 1. CKX *in situ* hybridization and AFP immunolocalization in the rat embryo. *In situ* hybridization of CKX was performed with a riboprobe labeled with [^{35}S] and immunolocalization of AFP is demonstrated by AFP immunoperoxidase staining, with a hematoxylin/eosin counter stain. (A) E12 rat embryo transversal section; arrows point to the liver. The brown color of the tissue denotes an AFP-positive reaction (scale bars = 100 μm). (B) and (C) are representative dark-field microscopy results of *in situ* hybridization with CKX in E12 liver tissue. Arrows point out this positive result (scale bars = 100 μm). (D) and (E) show the positive reactivity for both CKX mRNA (black dots) and AFP protein (brown color) in fetal liver; (D) = E12, (E) = E18 (scale bars = 10 μm). (F) is an adult rat liver section; this is illustrative of no reaction hybridization with CKX mRNA and no AFP immunoperoxidase after birth (scale bars = 10 μm). (G) is a dark-field micrograph of CKX expression in E12 yolk sac (scale bars = 100 μm).

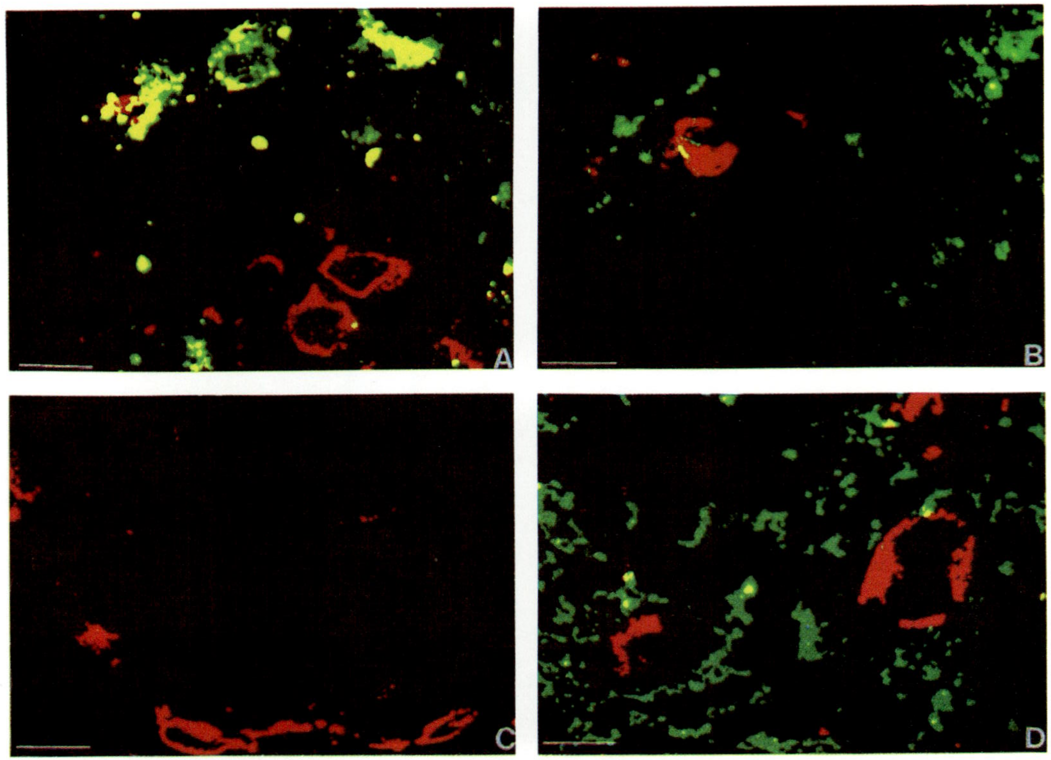

PLATE 2. Confocal immunofluorescence microscopy of frozen sections through the liver of 3-day-old rat, using Texas red anti-CK14 (A, B, C, D), FITC anti-BDS_7 (A and B), and FITC anti-HES_6 (D).[73] The data in (A) and (B) show that few cells have only expressed CK14 or BDS_7, whereas others (in yellow color) are $CK14^+$ and BDS_7^+ (scale bar = 12.3 μm). Notice in (C) the presence of CK14 in the mesothelial cells of the Glisson's capsule (scale bar = 11 μm). In (D), some cells are HES_6^+ while a few other cells are only $CK14^+$ (scale bar = 10 μm).

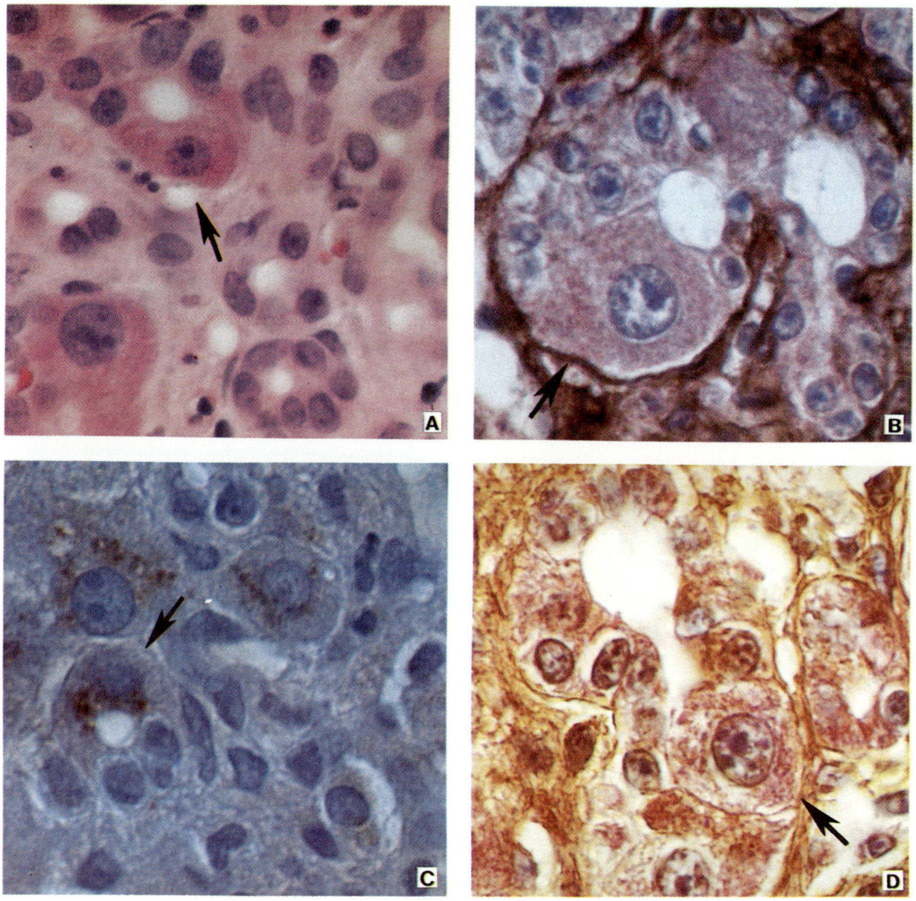

PLATE 3. Light photomicrographs of "ductular hepatocytes" appearing in rat liver after bile duct ligation and subsequent CCl_4 intoxication (A, C) or following severe furan hepatotoxicity (B, D). (A) Newly formed mixed hepatic cell cholangiolarlike structure composed of a single small hepatocyte (arrow) together with biliary epithelial cells (H.E., original magnification × 202); (B) cholangiolarlike structure composed of biliary epithelial cells and a hepatocytic cell (arrow) supported by a common basement membrane immunostained for laminin (H.E. counterstain; original magnification × 320); (C) small "ductular hepatocyte" (arrow) in a newly formed cholangiolarlike structure exhibiting a selective immunohistochemical staining reaction for the rat hepatocyte cytoplasmic antigen H.4. (H. counterstain; original magnification × 320); (D) cholangiolarlike structure composed of biliary epithelial cells and a hepatocytic cell (arrow) on a common basement membrane immunostained for type IV collagen (H.E. counterstain; original magnification × 320).

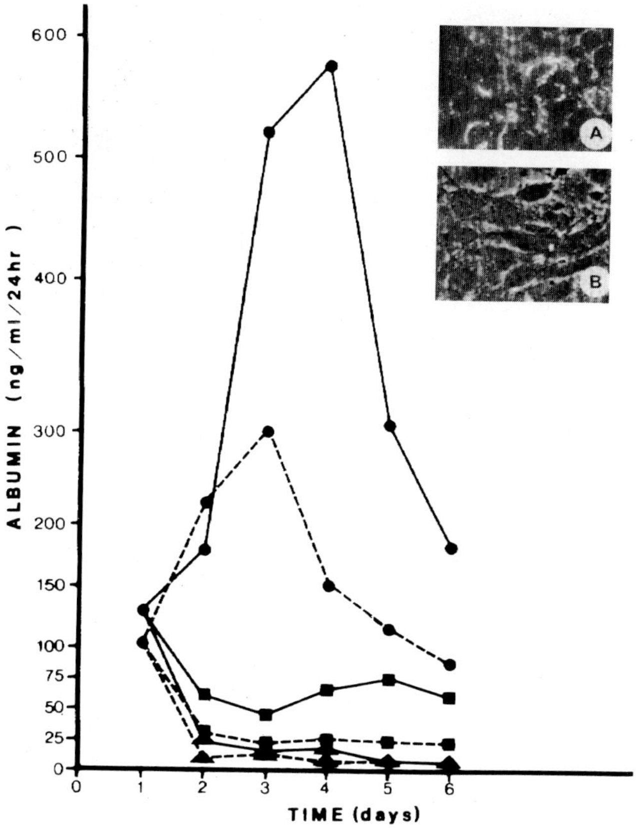

FIGURE 5. Albumin production by oval cells cultured in different media:[53] medium MX without dexamethasone (▲- - -▲); medium MX (▲—▲), medium MX with DMSO without dexamethasone (■- - -■), medium MX with dexamethasone and DMSO (■—■); medium MX with sodium butyrate without dexamethasone (●- - -●); medium MX containing dexamethasone and sodium butyrate (●—●). Inserts: immunofluorescence microscopy using anti-rat albumin monoclonal antibody (A) of oval cells after 3 d in primary culture in presence of medium MX supplemented with sodium butyrate; (B) phase contrast micrograph of (A). (Original magnification × 400.) (Reprinted from Germain, L. et al., *Cancer Res.*, 48, 368, 1988. With permission.)

GGT and are usually BDS_7 and HES_6 negative.[49] The working hypothesis has been that the cell of origin of LEC is located in the terminal biliary ductules (i.e., Hering canals) and that it could constitute a ''facultative hepatocytic stem cell'' population.[37] However, our previous work on the cell typing of LEC isolated by various tissue-dispersion procedures have led to the conclusion that, while some LEC lines may originate from biliary structures, others most likely arise from the Glisson's capsule.[49]

CK are expressed as pairs, and the same epithelial tissue can contain more than one CK pair.[59-61] CK8/CK18 is the most common CK pair of simple epithelium, whereas CK5/CK14 is normally restricted to the basal cell layer of the epidermis.[83] However, we have recently found that T51B cells (an established LEC line derived from adult rat liver) express a mismatched CK pair comprised of CK8 and CK14, i.e., without the corresponding CK18 and CK5, respectively[73] (Figure 6). We have also shown the presence of CK14 in LEC isolated from fetal and neonatal rat liver (Figure 7). All the LEC lines isolated so far from rat liver express vimentin as well. Double immunolocalization analyses on 4-d-old rat liver by confocal microscopy (Plate 2*) using the polyvalent antibody specific to a CK14 peptide and Mab against either BDS_7, HES_6, or BPC_5 have shown that, while some CK14-positive LEC are present in the periportal area, they are not immature hepatocytic (HES_6^+) or bipotential progenitor (BPC_5^+) cells.[73] Interestingly, a few of the CK14-positive LEC also react with BDS_7.[73] Furthermore, we have demonstrated the presence of CK14 in the cells of Glisson's capsule (Plate 2), which support the view that some LEC lines originate from the capsular mesothelium.[49]

In fact, as it turns out, the CK composition of mature rat biliary cells appears to be similar to that of mesothelial cells of Glisson's capsule in that both cell types seem to express CK7, CK8, CK18, and CK19. These observations confirm again that CK provide excellent markers for the various EC lineages vs. those of non-EC present in the liver. The use of other markers would allow a definitive assessment of the EC-differentiation status.[5] Finally, a particularly intriguing finding with regard to CK expression in LEC is the observation that WB-F344, a normal diploid cell line exhibiting oval cell properties, actually lacks CK even at very low passages.[84] The significance of the phenomenon remains to be established (see Section VI).

In general then, LEC seem to constitute a population of (specialized) cells, in the sense that they not only contain CK that are typical of simple epithelium, but they also express a CK that is present in stratified epithelium. Our current view is some LEC represent an early sublineage of progenitors committed along the intrahepatic biliary EC lineage (Figure 2).

V. CELL LINEAGES IN LIVER NEOPLASIA

Many studies have been performed recently in our and other laboratories in an attempt to identify markers other than those based on metabolic activities that would help to define the cellular pathways leading to the emergence of liver cancers, particularly hepatocellular carcinoma.[16,19,53-58,81,82] In most of these studies, the central objective was to identify the carcinogen target cell population. One complication of these experiments is that the hepatocellular response varies depending on the agent used, the mode of administration, and the strain of animals.[85] In general, carcinogen exposure in the rat results in prominent cellular alterations at various successive steps and includes the emergence of numerous foci, the subsequent appearance of fewer remodeling

* Plate 2 appears following page 134.

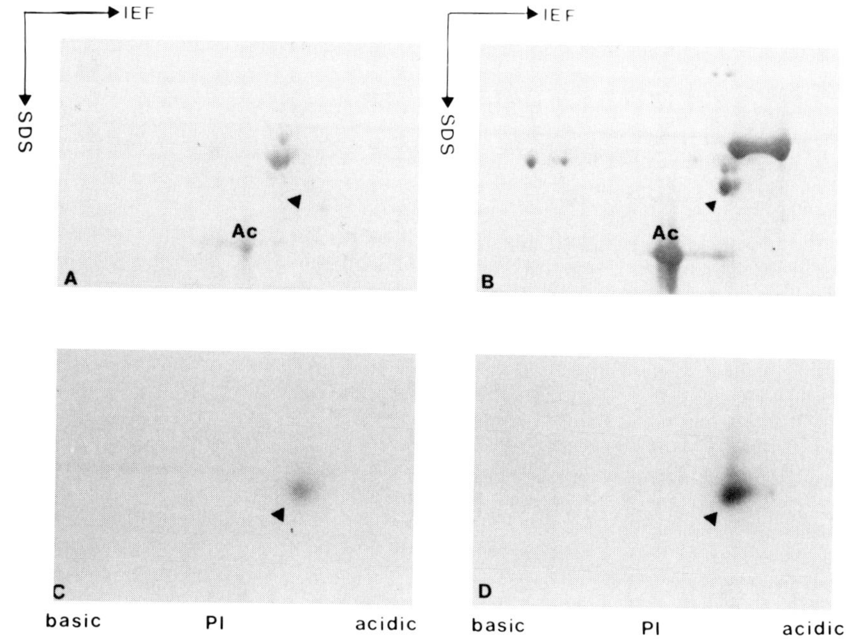

FIGURE 6. Immunoblot detection of CK14 polypeptide in rat skin and T51B cells.[72] Triton®-resistant cytoskeletons from rat skin (A, C) and T51B cells (B, D) were fractionated by two-dimensional gel electrophoresis and stained with Coomassie blue (A, B) or transferred to nitrocellulose membranes for the immunoblot analysis (C, D). Blots were incubated with a polyclonal antibody reacting specifically with CK14. Arrowheads denote the position of CK14 polypeptide. Ac = Actin.

nodules; some of these nodules persist as resistant units, with the result that very few of them become neoplastic.[85,86] It is thus clear that the initiating events occur in very few target cells.

Two opposite views have been proposed about the nature of these infrequent cells. The first one is that they constitute a subpopulation of hepatocytes, exhibiting a resistance to the chemical agent and undergoing a retrodifferentiation to a stage equivalent to that of fetal hepatocytes.[85,86] The evidence for this option is largely based on the use of metabolic markers, such as DT-diaphorase, GGT, and glutathione *S*-transferase-placental, known to play a prominent role in the detoxification of several classes of xenobiotics and their metabolic derivatives. The other view is that liver tumors are derived from progenitor (stem) cells. In this regard, we and others have typed rat cholangiocarcinomas and hepatocellular carcinomas using the new Mab directed against markers present in ductular oval cells, hepatocytes, transplantable hepatomas, and cholangiocarcinomas.[17,55,75,81,82,87,88] Overall, the data demonstrate that cholangiocarcinomas and hepatocellular carcinomas arising within

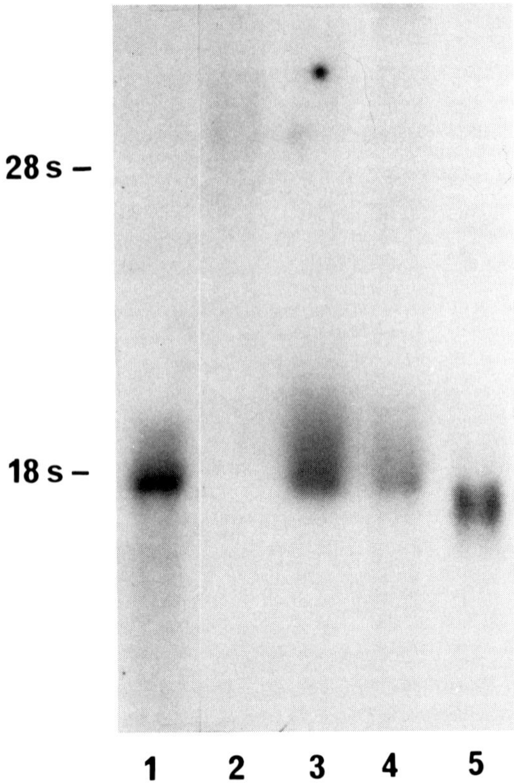

FIGURE 7. Analysis of CK14 mRNA level in LEC.[73] Total RNA (10 μg) isolated from T51B cells (1), rat hepatocytes (2), 12-d fetal rat LEC (3), 5-d newborn rat LEC (4), and ductular oval cells (5) was resolved by formaldehyde-agarose gel electrophoresis, transferred to a nylon membrane by blotting, and hybridized to a ^{32}P-labeled CK14 cDNA probe. The exposure time is 24 h.

nodules are composed of cells that exhibit a mixed phenotype of biliary ductal and hepatocytic cells. For example, the highly phosphorylated form of CK8 detected as BPC_5 in E12 to E14 liver cells is present in both hepatocellular carcinoma and cholangiocarcinoma.[75] Moreover, the cholangiocarcinoma tissue and its derived cell line both express CK7, CK8, CK18, CK19, and BDS_7, which are typical markers of biliary EC (the cholangiocarcinoma-derived cell line acquires vimentin as well). A more complex picture results from the data on hepatocellular carcinoma cell typing.[75] Hepatocellular carcinoma tissue contains CK7, BDS_7, and HES_6 to a relatively low and variable level, and essentially no vimentin. HES_6 and vimentin are rarely expressed, whereas BDS_7 is frequently present. Morris 7777 (and its subclones) and 7795 hepatoma cell lines express large amounts of CK8 and CK18 (i.e., the typical

TABLE 2
Differential Expression of IF Proteins in Cholangiocarcinoma (CC) and Hepatocellular Carcinoma (HCC) Tissues and Derived Cell Lines

	CK7	CK8	CK14	CK18	CK19	Vimentin
Primary tumors						
CC	+	+	−	+	+	−
HCC	(+)[a]	+	−	+	−	(+)[b]
Cell lines						
CC						
H6	+	+	−	+	+	+
HCC						
Reuber						
HF-II-EC3	−	+	−	+	−	−
Morris						
7777	−	+	−	+	−	+
V7.2	−	+	−	+	(+)[c]	−
V7.6	−	+	−	+	(+)[c]	−
7795	−	+	−	+	−	−
8994	+	+	−	+	N.D.[d]	+
Novikoff						
N1-S1	−	+	−	−	+	+

[a] Very low staining reaction.
[b] Selective staining on lining cells.
[c] Very low amount of mRNA was present, but a protein of similar pI and m.w. was detected on the gel.
[d] N.D.: not determined.

hepatocytic CK) with very little or no BDS_7 or HES_6, whereas the Reuber hepatoma cells express not only CK8 and CK18, but to a relatively large extent BDS_7 with no HES_6 [75] (Tables 2 and 3). On these grounds, bipotential progenitors and hepatocellular carcinoma and cholangiocarcinoma cells share several common antigenic markers, the most relevant being BPC_5. Moreover, while cholangiocarcinoma-derived cells seem unable to express the typical markers of hepatocytic cells, hepatocellular carcinoma-derived cells can frequently express some phenotypic features of the biliary EC lineage. This is consistent with our previous finding that the bipotential progenitors present in the E10 to E12 emerging rat liver already express a few features (e.g., AFP) of the perinatal hepatocytic cells, and that the appearance of biliary EC is accompanied by the loss of these ''metabolic'' features and the acquisition of cytoskeletal elements (e.g., CK7, CK19). In this regard, we feel that the expression of these CK seems to be linked to the performance of structural functions (for example, the biliary epithelium form ductules and ducts). In

TABLE 3
Differential Expression of Distinct Cytoplasmic Constituents and Surface-Exposed Components in Cholangiocarcinoma (CC) and Hepatocellular Carcinoma (HCC) Tissues and Derived Cell Lines

	BPC_5	OV6	OV1	BDS_7	HES_6
		Primary tumors			
CC	+	+	+	+	(−)[a]
HCC	+	(−)[b]	−	(−)[b]	(+)[c]
		Cell lines			
CC					
H6	+	+	−	(+)[d]	−
HCC					
Reuber					
H4-II-EC3	+	−	−	+	−
Morris					
7777	+	(+)[c]	−	(+)[d]	−
V7.2	+	(+)[c]	−	−	+
V7.6	−	(+)[c]	−	−	−
7795	+	−	−	(+)[d]	−
8994	+	+	−	−	−
Novikoff					
N1-S1	(+)[d]	−	−	−	−

[a] The tumor area was negative, but the hepatocytes were positive.
[b] Majority of cells were negative, except for those forming a ductular structure.
[c] Intensity varied from cell to cell, but all the cells were positive.
[d] Majority of cells were faintly positive and minority of cells were strongly positive.

short then, these data indicate that hepatocellular carcinoma and cholangiocarcinoma cells can be aligned with bipotential progenitors which are committed to the hepatocytic or biliary EC lineage, but which remain at early intermediate differentiation steps that provide a sort of "differentiation window" (Figure 8). This corresponds to the stage when the cells are most susceptible to transformation. This also fits with the view that hepatocytic cells, biliary EC, and LEC at an early immature stage can undergo transdifferentiation.

VI. DIFFERENTIATION FEATURES OF LEC TRANSFORMED *IN VITRO*

Over the last 2 decades, LEC lines have been used extensively to examine the multistage conversion of normal liver cells into neoplastic cells. In this

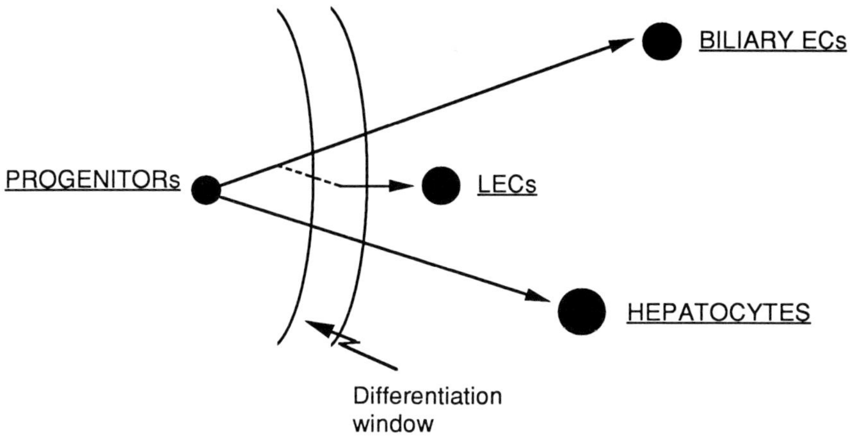

FIGURE 8. Transformation susceptibility of differentiating rat liver cells. Bipotential progenitor cells differentiating along the hepatocytic or biliary EC lineage can be maintained at a "differentiation window" that allows them to undergo transdifferentiation. At that stage, the cells are preferentially susceptible to transformation.

regard, two of the cell lines that have normally been used to study chemical carcinogenesis are WB-F344 and T51B. Several subclones of *N*-methyl-*N*′-nitro-*N*-nitrosoguanidine (MNNG)-transformed WB-F344 have been established and their phenotypes described by Grisham and collaborators.[19,89] Following their transfer to syngeneic animals the transformed cells form a wide variety of tumor types, such as hepatocellular carcinoma, cholangiocarcinoma, sarcomas, mixed epithelial-mesenchymal tumors, and undifferentiated carcinomas.[89] We are currently typing many of these clones in collaboration with Tsao and Grisham's groups. Preliminary data indicate that these cells acquire distinct subsets of CK pairs that include CK8, CK14, CK18, and CK19.[84] We are currently examining the relationship between this differential CK gene expression and the type(s) of tumors produced after transplantation. We have also recently used aflatoxin-transformed T51B cells to assess the type(s) of tumor that they can produce and the changes induced in the differential expression of the various markers.[90] The tumors obtained after transplantation into syngeneic animals were classified as undifferentiated carcinomas. The cells exhibit no significant changes in the expression of CK and vimentin and remained negative for BPC_5, BDS_7, and HES_6.[90]

LEC lines have been exposed to viral or activated cellular oncogenes. Recent work using an LEC clone infected with recombinant retroviruses containing either v-*ras*, v-*myc*, or v-Ha-*ras* or v-*raf*/v-*myc* oncogenes[91,92] indicated that, while the combination v-*raf*/v-*myc* produced differentiated trabecular carcinomas, the tumors derived from v-*raf*- or v-Ha-*ras*-transformed cells exhibited a mixed epithelial-mesenchymal morphology. V-*myc* alone was incapable of transforming these cells.[92] It was suggested that these cell lines are capable of differentiating into both epithelial and mesenchymal lines and that a single transforming oncogene such as v-*raf* or v-Ha-*ras* perturbs the lineage commitment, i.e., induces a switch toward the mesenchymal lineage.[91] However, another interpretation of these observations comes from studies that we performed recently on an LEC line, T51B cells, transfected with c-myc or polyomavirus middle T (MT) oncogene or both and injected into syngeneic animals. While c-myc-transfected T51B cells were not transformed, those expressing MT were highly transformed and produced poorly differentiated carcinomas.[93] Interestingly, while CK gene expression in c-myc-transfected cells was normal, the CK14 gene was inhibited in the MT-transfected T51B cells. All the data provide the first experimental evidence for a possible link between differential CK gene expression and differentiated status of LEC. However, the relative contribution of biliary EC and mesothelial cells to the isolated LEC preparations remains the main question to answer. Once the cell(s) of origin of the various LEC lines will be clearly identified, the data that are coming out of these *in vitro* transformation by oncogenes will turn out to be most significant.

Several studies have been reported by Isom's group[94-97] on the transformation by SV-40 oncogenes of adult rat hepatocytes in primary culture. The transformed cells were selected as very few foci producing large amounts of albumin. Many cell lines have been derived from these foci and their phenotypes examined. Most of them conserved the capacity to produce typical plasma proteins and to express enzyme activities characteristic of immature hepatocytes. In collaboration with this group, we are currently typing the various clones using CK and hepatocyte vs. biliary cell surface markers. The clones are $CK8^+$, CKX^-, $CK18^+$, and $CK19^-$, with very low CK14 gene expression.[98] Only a few cells in the culture are either HES_6 positive or BDS_7 positive, and, in contrast, essentially all of the cells express BPC_5. We propose that the transformed cells are derived from the few bipotential progenitors present at the adult stage. This would fit with our model that at the "differentiation window" cells committed along one lineage can express phenotypic features of the other cell lineages and that the transforming agent simply induces a differentiation arrest at that window (Figure 8).

VII. CONCLUSIONS AND PERSPECTIVES

The emerging rat liver contains bipotential progenitor cells that can differentiate along either the hepatocytic or biliary EC lineage. The latter can also give rise to an LEC sublineage. The evidence provided by an *in vitro* cell-commitment assay and the detection of a selective marker (BPC_5) indicate that the relative proportion of bipotential progenitor cells rapidly decreases during development. Whether or not these cells at E10.5 are derived from a foregut stem cell remains an important question. Moreover, since embryonic pancreas and liver emerge from the foregut, the relationship between hepatic and pancreatic cell lineages is a major issue and, in this regard, it becomes of particular interest to find out whether or not the concept of "differentiation window" can apply to the process of transdifferentiation that can apparently take place.[99,100] The question about the presence of progenitor (stem) cells in postnatal rat liver remains largely unanswered. In contrast to the epidermis and the gut, the liver has a very low rate of cell turnover. In tissues that undergo a rapid cell renewal, the loss of (terminally) differentiated cells, which lack the capacity to proliferate, is counterbalanced by the maturation of progenitor cells usually distinguishable as a separate proliferative cell compartment.[101,102] However, Zajicek and collaborators[103,104] have reported that in normal liver, hepatocytes located near the portal area proliferate continuously while older hepatocytes disappearing near the central vein exhibit a loss in growth potential and eventually disappear. These authors thus hypothesized that hepatocytes are "streaming" from the portal area to the central vein. They have also found that biliary duct cells are "streaming"[105] and proposed that, in both cases, the streams originate in the terminal bile ductules.[106] This would mean that the Potten's model[101,102] of progenitor (stem) cells might apply to the liver. In these terms, the notion of liver "facultative stem cells"[37] should be reexamined within the view that these ductular structures contain few bipotential progenitor cells capable of differentiating along one or the other cell lineage.

While cholangiocarcinoma cells and their derived cell lines exhibit features typical of immature biliary EC, hepatocellular carcinomas and their derived cell lines present a mixed phenotype of hepatocytic and biliary EC. However, bipotential progenitor cells, hepatocellular carcinoma cells, and cholangiocarcinoma cells share several common antigenic markers. Also, the transfer of chemically or oncogene-transformed LEC into syngeneic animals leads to the formation of tumors exhibiting a highly mixed phenotype, including hepatocellular carcinoma, cholangiocarcinoma, and sarcoma, thus suggesting that LEC are at a differentiation status close to that of progenitors. On these grounds, it would seem that the appearance of the commonly cited aberrant, dedifferentiated or oncofetal phenotypic features in tumors are in fact the result of clonal selection of rare bipotential progenitor cells committed along biliary epithelial or hepatocytic cell lineage, but which are maintained at a sort of "differentiation window" that is preferentially (and highly) susceptible to transformation.

ACKNOWLEDGMENTS

We thank Dr. William Waithe for his critical reading of the manuscript and Elisabeth Lemay for her excellent secretarial assistance. M.-J. B. was the recipient of a pre-doctoral fellowship from SRC, N.T. of a post-M.D.

fellowship from the FCAR, and A.L. of a postdoctoral fellowship from the FRSQ. This work was supported by the National Cancer Institute of Canada and the Medical Research Council of Canada.

REFERENCES

1. **Greengard, O., Federman, M., and Knox, W. E.,** Cytomorphometry of developing rat liver and its application to enzymic differentiation, *J. Cell Biol.*, 52, 261, 1972.
2. **Le Douarin, N. M.,** An experimental analysis of liver development, *Med. Biol.*, 53, 427, 1975.
3. **Houssaint, E.,** Differentiation of the mouse hepatic primordium. I. An analysis of tissue interactions in hepatocytic differentiation, *Cell Differ.*, 9, 269, 1981.
4. **Vassy, J., Kraemer, M., Chalumeau, M. T., and Foucrier, J.,** Development of the fetal rat liver: ultrastructural and stereological study of hepatocytes, *Cell Diff.*, 24, 9, 1988.
5. **Marceau, N.,** Cell lineages and differentiation programs in epidermal, urothelial and hepatic tissues and their neoplasms, *Lab. Invest.*, 63, 4, 1990.
6. **Bucher, N. L. R. and McGowan, J. A.,** Regulatory mechanisms in liver regeneration, in *Liver and Biliary Disease: A Pathophysiological Approach,* Wright, R., Albeti, G. G. M. M., Karsan, S., and Millward-Sadler, H., Eds., W. B. Saunders, Philadelphia, 1979, 210.
7. **Popper, H., Kent, G., and Stein, R.,** Ductular cell reaction in the liver in hepatic injury, *J. Mt. Sinai Hosp.*, 24, 551, 1957.
8. **Fausto, N.,** Hepatocyte differentiation and liver progenitor cells, *Curr. Opinions Cell Biol.*, 2, 1036, 1990.
9. **Sirica, A. E., Mathis, G. A., Sano, N., and Elmore, L. W.,** Isolation, culture, and transplantation of intrahepatic biliary epithelial cells and oval cells, *Pathobiology,* 58, 44, 1990.
10. **Farber, E.,** Similarities in the sequence of early histological changes induced in the liver of the rat by ethionine, 2-acetylaminofluorene, and 3′-methyl-4-dimethyl-aminoazobenzene, *Cancer Res.*, 16, 142, 1956.
11. **Grisham, J. W. and Porta, E. A.,** Origin and fate of proliferated ductal cells in the rat: electron microscopic and autoradiographic studies, *Exp. Mol. Pathol.*, 3, 242, 1964.
12. **Inaoka, Y.,** Significance of the so-called oval cell proliferation during azo-dye hepatocarcinogenesis, *Gann,* 58, 355, 1967.
13. **Hixson, D. C., Faris, R. A., and Thompson, N. L.,** An antigenic portrait of the liver during carcinogenesis, *Pathobiology,* 58, 65, 1990.
14. **Sells, M. A., Katyal, S. L., Shinozuka, H., Estes, L. W., Sell, S., and Lombardi, B. L.,** Isolation of oval cells and transitional cells from the livers of rats fed the carcinogen DL-ethionine, *J. Natl. Cancer Inst.*, 66, 355, 1981.
15. **Sell, S., Osborn, K., and Leffert, H. L.,** Autoradiography of "oval cells" appearing rapidly in the livers of rats fed N-2-fluorenylacetamide in a choline devoid diet, *Carcinogenesis (London),* 2, 7, 1981.
16. **Germain, L., Goyette, R., and Marceau, N.,** Differential cytokeratin and α-fetoprotein expression in morphologically distinct epithelial cells emerging at the early stage of rat hepatocarcinogenesis, *Cancer Res.*, 45, 673, 1985.
17. **Hixson, D. C. and Allison, J. P.,** Monoclonal antibodies recognizing oval cells induced in the liver of rats by N-2-fluorenyl acetamide or ethionine in a choline-deficient diet, *Cancer Res.*, 45, 3750, 1985.

18. **Petropoulos, C. J., Yaswen, P., Panzica, M., and Fausto, N.,** Cell lineages in liver carcinogenesis: possible clues from studies of the distribution of alpha-fetoprotein RNA sequences in cell populations isolated from normal, regenerating, and preneoplastic rat livers, *Cancer Res.,* 45, 5762, 1985.
19. **Tsao, M. S., Grisham, J. W., Nelson, K. G., and Smith, J. D.,** Phenotypic and karyotypic changes induced in cultured rat hepatic epithelial cells that express the "oval" cell phenotype by exposure to N-methyl-N′-nitro-N-nitroguanidine, *Am. J. Pathol.,* 118, 306, 1985.
20. **Lombardi, B.,** On the nature, properties and significance of oval cells, in *Recent Trends in Chemical Carcinogenesis,* Vol. 1, Pani, P., Feo, F., and Columbano, A., Eds., ESA, Cagliari, 1982, 36.
21. **Evarts, R. P., Nagy, P., Marsden, E., and Thorgeirsson, S. S.,** A precursor-product relationship exists between oval cells and hepatocytes in rat liver, *Carcinogenesis,* 8, 1737, 1987.
22. **Evarts, R. P., Nagy, P., Nakatsukasa, H., Marsden, E., and Thorgeirsson, S. S.,** In vivo differentiation of rat liver oval cells into hepatocytes, *Cancer Res.,* 49, 1541, 1989.
23. **Fausto, N., Thompson, N. L., and Braun, L.,** Purification and culture of oval cells from rat liver, in *Cell Separation. Methods and Selected Applications,* Pretlow, T. G., II and Pretlow, T. P., Eds., Academic Press, New York, 1987, 45.
24. **Marceau, N., Blouin, M.-J., Germain, L., and Noël, M.,** Role of different epithelial cell types in liver ontogenesis, regeneration and neoplasia, *In Vitro Cell Dev. Biol.,* 25, 336, 1989.
25. **Sell, S.,** Is there a liver stem cell?, *Cancer Res.,* 50, 3811, 1990.
26. **Gugen-Guillouzo, C., Clement, B., Baffet, G., Beaumont, C., Morel-Chany, E., Glaise, D., and Guillouzo, A.,** Maintenance and reversibility of active albumin secretion by adult rat hepatocytes co-cultured with another liver epithelial cell type, *Exp. Cell. Res.,* 143, 47, 1983.
27. **Isom, H. C., Secott, T., Georgoff, I., Woodworth, C., and Mummaw, J.,** Maintenance of differentiated rat hepatocytes in primary culture, *Proc. Natl. Acad. Sci. U.S.A.,* 82, 3252, 1985.
28. **Deschênes, J., Valet, J.-P., and Marceau, N.,** Hepatocytes from newborn and weanling rats in monolayer culture: isolation by perfusion, fibronectin-mediated adhesion, spreading, and functional activities, *In Vitro,* 16, 722, 1980.
29. **Baribault, H., Leroux-Nicollet, I., and Marceau, N.,** Differential responsiveness of cultured suckling and adult rat hepatocytes to growth-promoting factors: entry into S phase and mitosis, *J. Cell. Physiol.,* 122, 105, 1985.
30. **Baribault, H. and Marceau, N.,** Dexamethasone and dimethylsulfoxide as distinct regulators of growth and differentiation of cultured suckling rat hepatocytes, *J. Cell. Physiol.,* 129, 77, 1986.
31. **McGowan, J. A.,** Hepatocyte proliferation in culture, in *Research in Isolated and Cultured Hepatocytes,* Guillouzo, A. and Gugen-Guillouzo, C., Eds., John Libbey Eurotext, London, 1986, 13.
32. **Michaelopoulos, G. K., Strom, S. C., and Jirtle, R. L.,** Use of hepatocytes for studies of mutagenesis and carcinogenesis, in *Research in Isolated and Cultured Hepatocytes,* Guillouzo, A. and Gugen-Guillouzo, C., Eds., John Libbey Eurotext, London, 1986, 333.
33. **Sirica, A. E., Sattler, C. A., and Cihla, H. P.,** Characterization of a primary bile ductular cell culture from the livers of rats during experimental cholestasis, *Am. J. Pathol.,* 120, 67, 1985.
34. **Mathis, G. A., Walls, S. A., and Sirica, A. E.,** Biochemical characteristics of hyperplastic rat bile ductular epithelial cells cultured "on top" and "inside" different extracellular matrix substitutes, *Cancer Res.,* 48, 6145, 1988.

35. **Williams, G. M., Weisburger, E. K., and Weisburger, J. H.,** Isolation and long-term cell culture of epithelial-like cells from rat liver, *Exp. Cell Res.,* 69, 106, 1971.
36. **Chessebeuf, M., Olsson, A., Bournot, P., Desgres, J., Guiguet, M., Maume, G., Maume, B. F., Perissel, B., and Padieu, P.,** Long term cell culture of rat liver epithelial cells retaining some hepatic functions, *Biochimie,* 56, 1365, 1974.
37. **Grisham, J. W.,** Cell types in long term propagable cultures of rat liver, *Ann. N.Y. Acad. Sci.,* 349, 128, 1980.
38. **Grisham, J. W.,** Cell types in rat liver cultures: their identification and isolation, *Mol. Cell. Biochem.,* 53/54, 22, 1983.
39. **Marceau, N., Goyette, R., Deschênes, J., and Valet, J. P.,** Morphological differences between epithelial and fibroblast cells in rat liver cultures, and the roles of cell surface fibronectin and cytoskeletal element organization in cell shape, *Ann. N.Y. Acad. Sci.,* 349, 138, 1980.
40. **Herring, A. S., Raychaudhuri, R., Kelley, S. P., and Iype, T. P.,** Repeated establishment of diploid epithelial cell cultures from normal and partially hepatectomized rats, *In Vitro,* 19, 576, 1983.
41. **Schaeffer, W. I. and Heintz, N. H.,** A diploid rat liver cell culture. IV. Malignant transformation by aflatoxin B_1, *In Vitro,* 14, 418, 1978.
42. **Farber, E.,** Chemical carcinogenesis: a current biological perspective, *Carcinogenesis,* 5, 1, 1984.
43. **Sell, S. and Dunsford, H. A.,** Evidence for the stem cell origin of hepatocellular carcinoma and cholangiocarcinoma, *Am. J. Pathol.,* 134, 1347, 1989.
44. **Bannach, P., Mayer, D., and Hacker, H. J.,** Hepatocellular glycogenesis and hepatocarcinogenesis, *Biochem. Biophys. Acta,* 605, 217, 1980.
45. **Pitot, H. C. and Sirica, A. E.,** The stages of initiation and promotion in hepatocarcinogenesis, *Biochim. Biophys. Acta,* 605, 191, 1980.
46. **Williams, G. M.,** The pathogenesis of rat liver cancer caused by chemical carcinogens, *Biophys. Biochim. Acta,* 605, 167, 1980.
47. **Sato, J.,** Malignant transformation in culture of rat liver cells treated with and without 3′-methyl-4-dimethylaminoazobenzene, in *Cancer Cells in Culture,* Katsuda, H., Ed., University of Tokyo Press, Tokyo, 1968, 335.
48. **Koch, K. S., Lu, X. P., Brenner, D. A., Fey, G. H., Martinez-Conde, A., and Leffert, H. L.,** Mitogens and hepatocyte growth control *in vivo* and *in vitro,* invited review, *In Vitro Cell. Dev. Biol.,* 26, 1011, 1990.
49. **Marceau, N., Germain, L., Goyette, R., Noël, M., and Gourdeau, H.,** Cell of origin of distinct cultured rat liver epithelial cells, as typed by cytokeratin and surface component selective expression, *Biochem. Cell Biol.,* 64, 788, 1986.
50. **Tsao, M. S., Grisham, J. W., Nelson, K. G., and Smith, J. D.,** Phenotypic and karyotypic changes induced in cultured rat hepatic epithelial cells that express the "oval" cell phenotype by exposure to N-methyl-N′-nitro-N-nitroguanidine, *Am. J. Pathol.,* 118, 306, 1985.
51. **Shimada, T., Furukawa, K., Kreiser, D. M., Cawein, A., and Williams, G. M.,** Induction of transformation by six classes of chemical carcinogens in adult rat liver epithelial cells, *Cancer Res.,* 43, 5087, 1983.
52. **Goyette, M., Faris, R., Braun, L., Hixson, D., and Fausto, N.,** Expression of hepatocyte and oval cell antigens in hepatocellular carcinomas produced by oncogene-transfected liver epithelial cells, *Cancer Res.,* 50, 4809, 1990.
53. **Germain, L., Noël, M., Gourdeau, H., and Marceau, N.,** Promotion of growth and differentiation of rat ductular oval cells in primary culture, *Cancer Res.,* 48, 368, 1988.
54. **Germain, L., Blouin, M.-J., and Marceau, N.,** Biliary epithelial and hepatocytic cell lineage relationships in embryonic rat liver as determined by the differential expression of cytokeratins, alpha-fetoprotein, albumin, and cell surface-exposed components, *Cancer Res.,* 48, 4909, 1988.

55. **Dunsford, H. A. and Sell, S.,** Production of monoclonal antibodies to preneoplastic liver cell populations induced by chemical carcinogens in rats and to transplantable Morris hepatomas, *Cancer Res.,* 49, 4887, 1989.
56. **Holmes, C. H., Austin, E. B., Fisk, A., Gunn, B., and Baldwin, R. W.,** Monoclonal antibodies reacting with normal rat liver cells as probes in hepatocarcinogenesis, *Cancer Res.,* 44, 1611, 1984.
57. **Faris, R. A., Monfils, B. A., and Hixson, D. C.,** Antigenic relationship between oval cells, fetal hepatocytes, and myelocytic precursor cells, *Proc. Am. Assoc. Cancer Res.,* 30, 52, 1989.
58. **Faris, R. A., Monfils, B. A., and Hixson, D. C.,** Antigenic phenotypes of ethionine induced primary hepatocellular carcinomas (PHC) and metastatic foci (MF), *Proc. Am. Assoc. Cancer Res.,* 38, 107, 1986.
59. **Gebhardt, R. and Schäfer-Degenhart, I.,** Monoclonal antibodies directed against rat liver epithelial cell lines selectively recognize bile duct epithelium in livers of adult rats, *Cell Biol. Toxicol.,* 4, 379, 1988.
60. **Moll, R., Franke, W., Schiller, D., Geiger, B., and Krepler, R.,** The catalog of human cytokeratins: patterns of expression in normal epithelia, tumors and cultured cells, *Cell,* 31, 11, 1982.
61. **Osborn, M. and Weber, K.,** Intermediate filaments: cell-type specific markers in differentiation and pathology, *Cell,* 31, 303, 1982.
62. **Lazarides, E.,** Intermediate filaments: a chemically heterogenous, developmentally regulated class of proteins, *Annu. Rev. Biochem.,* 51, 219, 1982.
63. **Blouin, M.-J., Germain, L., Noël, M., Dunsford, H., and Marceau, N.,** Common antigenic features between rat liver bipotential progenitors and tumor cells, in preparation.
64. **Vassy, J., Rigaut, J. P., Hill, A.-M., and Foucrier, J.,** Analysis by laser scanning confocal microscopy imaging of the spatial distribution of intermediate filaments in foetal and adult rat liver cells, *J. Microsc.,* 157, 91, 1990.
65. **Shiojiri, N., Lemire, J. M., and Fausto, N.,** Cell lineages and oval cell progenitors in rat liver development, *Cancer Res.,* 51, 2611, 1991.
66. **Bélanger, L., Baril, P., Guertin, M., Gingras, M. C., Gourdeau, H., Anderson, A., and Hamel, D., and Boucher, J. M.,** Oncodevelopmental and hormonal regulation of alpha-fetoprotein gene expression, in *Advances in Enzyme Regulation,* Vol. 21, Weber, G., Ed., Pergamon Press, New York, 1983, 73.
67. **Shiojiri, N.,** Analysis of differentiation of hepatocytes and bile duct cells in developing mouse liver by albumin immunofluorescence, *Dev. Growth Differ.,* 26, 555, 1984.
68. **Daoust, R.,** The histochemical demonstration of γ-glutamyl transpeptidase activity in different populations of rat liver cells during azo dye carcinogenesis, *J. Histochem., Cytochem.,* 30, 312, 1982.
69. **Wu, Y.-J., Parker, L. M., Binder, N. E., Beckett, M. A., Simard, J. H., Griffiths, C. T., and Rheinwald, J. G.,** The mesothelial keratins: a new family of cytoskeletal proteins identified in cultured mesothelial cells and nonkeratinizing epithelia, *Cell,* 31, 393, 1982.
70. **van Eyken, P., Sciot, R., and Desmet, V.,** Intrahepatic bile duct development in the rat: a cytokeratin-immunohistochemical study, *Lab. Invest.,* 59, 52, 1988.
71. **van Eyken, P., Sciot, R., Callea, F., van der Steen, K., Moerman, P., and Desmet, V. J.,** The development of the intrahepatic bile ducts in man: a keratin-immunohistochemical study, *Hepatology,* 8, 1586, 1988.
72. **Blouin, R.,** Expression différentielle des gènes de cytokératine et organisation structurale des filaments dans les cellules de l'épithélium simple, Ph.D. thesis, Laval University, Ste-Foy, Quebec, 1991.
73. **Blouin, R., Blouin, M.-J., Royal, I., Grenier, A., Roop, D. R., and Marceau, N.,** Selective cytokeratin 14 gene expression in rat liver nonparenchymal epithelial cells, submitted.

74. **Török, N., Lampron, C., Royal, A., and Marceau, N.,** unpublished data.
75. **Blouin, M.-J.,** Définition des voies de différenciation des cellules épithéliales hépatiques chez le rat, Ph.D. thesis, Laval University, Ste. Foy, Quebec, 1991.
76. **Fausto, N.,** Hepatic regeneration, in *Hepatology a Textbook of Liver Disease,* Vol. 1, 2nd ed., Zakin, D. and Boyer, T. D., Eds., W. B. Saunders, Philadelphia, 1990, chap. 3.
77. **Sirica, A. E., Elmore, L. W., and Sano, N.,** Characterization of rat hyperplastic bile ductular epithelial cells in culture and in vivo, *Dig. Dis. Sci.,* 36, 494, 1991.
78. **Tournier, I., Legres, L., Schoevaert, D., Feldmann, G., and Bernuau, D.,** Cellular analysis of α-fetoprotein gene activation during carbon tetrachloride and D-galactosamine-induced acute liver injury in rats, *Lab. Invest.,* 59, 657, 1988.
79. **Lemire, J. M., Shiojiri, N., and Fausto, N.,** Oval cell proliferation and the origin of small hepatocytes in liver injury induced by D-galactosamine, *Am. J. Pathol.,* 139, 535, 1991.
80. **Onoë, T., Dempo, K., Kaneko, A., and Watabe, H.,** Significance of alpha-fetoprotein appearance in the early stage of azo-dye carcinogenesis, *Gann Monogr.,* 14, 233, 1974.
81. **Faris, R. A., Monfils, B. A., Dunsford, H. A., and Hixson, D. C.,** Antigenic relationship between oval cells and a subpopulation of hepatic foci, nodules, and carcinomas induced by the "resistant hepatocyte" model system, *Cancer Res.,* 51, 1308, 1991.
82. **Faris, R. A., McEntire, K. D., Thompson, N. L., and Hixson, D. C.,** Identification and characterization of a rat hepatic oncofetal membrane glycoprotein, *Cancer Res.,* 50, 4755, 1990.
83. **Roop, D. R., Krieg, T. M., Mehrel, T., Cheng, C. K., and Yuspa, S. H.,** Transcriptional control of high molecular weight keratin gene expression in multistage mouse skin carcinogenesis, *Cancer Res.,* 48, 3245, 1988.
84. **Marceau, N., Grenier, A., Tsao, M., and Grisham, J.,** unpublished results.
85. **Farber, E. and Sarma, D. S. R.,** Hepatocarcinogenesis: a dynamic cellular perspective, *Lab. Invest.,* 56, 4, 1987.
86. **Farber, E.,** The multistep nature of cancer development, *Cancer Res.,* 44, 4217, 1984.
87. **Dunsford, H. A., Karnasuta, C., Hunt, J. M., and Sell, S.,** Different lineages of chemically induced hepatocellular carcinoma in rats defined by monoclonal antibodies, *Cancer Res.,* 49, 4894, 1989.
88. **Hixson, D. C. and McEntire, K. D.,** Detection using monoclonal antibodies of a structurally altered form of cell CAM 105 on rat hepatocellular carcinomas, in *Molecular Determinants of Animal Form,* Edelman, G. M., Ed., Alan R. Liss, New York, 1985, 253.
89. **Tsao, M. S. and Grisham, J. W.,** Phenotypic modulation during tumorigenesis by clones of transformed rat liver epithelial cells, *Cancer Res.,* 47, 1282, 1987.
90. **Royal, I. and Marceau, N.,** unpublished data.
91. **Garfield, S., Huber, B. E., Nagy, P., Cordingley, M. G., and Thorgeirsson, S. S.,** Neoplastic transformation and lineage switching of rat liver epithelial cells by retrovirus-associated oncogenes, *Mol. Carcinogenesis,* 1, 189, 1988.
92. **Hampton, L. L., Worland, P. J., Yu, B., Thorgeirsson, S. S., and Huggett, A. C.,** Expression of growth-related genes during tumor progression in v-raf-transformed rat liver epithelial cells, *Cancer Res.,* 50, 7460, 1990.
93. **Royal, I., Gourdeau, H., Blouin, R., and Marceau, N.,** Selective inhibition of CK14 gene expression in T51B cells by middle T polyoma virus oncogene, *Cell Growth Differ.,* in press.
94. **Woodworth, C. D. and Isom, H. C.,** Regulation of albumin gene expression in a series of rat hepatocyte cell lines immortalized by simian virus 40 and maintained in chemically defined medium, *Mol. Cell Biol.,* 7, 3740, 1987.
95. **Woodworth, C., Secott, T., and Isom, H. C.,** Transformation of rat hepatocytes by transfection with simian virus 40 DNA to yield proliferating differentiated cells, *Cancer Res.,* 46, 4018, 1986.

96. **Woodworth, C. D. and Isom, H. C.**, Transformation of differentiated rat hepatocytes with adenovirus and adenovirus DNA, *J. Virol.*, 61, 3570, 1987.
97. **Woodworth, C. D., Kreider, J. W., Mengel, L., Miller, T., Meng, Y., and Isom, H. C.**, Tumorigenicity of simian virus 40-hepatocytes cell lines: effect of *in vitro* and *in vivo* passage on expression of liver-specific genes and oncogenes, *Mol. Cell. Biol.*, 8, 4492, 1988.
98. **Marceau, N., Grenier, A., and Isom, H.**, unpublished results.
99. **Rao, M. S., Dwivedi, R. S., Subbarao, V., Usman, M. I., Scarpelli, D. G., Nemali, M. R., Yeldandi, A., Thangada, S., Kumar, S., and Reddy, J. K.**, Almost total conversion of pancreas to liver in the adult rat: a reliable model to study transdifferentiation, *Biochem. Biophys. Res. Commun.*, 156, 131, 1988.
100. **Rao, M. S., Dwivedi, R. S., Yeldandi, A. V., Subbarao, V., Tan, X., Usman, M. I., Thangada, S., Nemali, M. R., Kumar, S., Scarpelli, D. G., and Reddy, J. K.**, Role of periductal and ductular epithelial cells of the adult rat pancreas in pancreatic hepatocyte lineage, *Am. J. Pathol.*, 134, 1069, 1989.
101. **Potten, C. S., Schofield, R., and Lajtha, L. G.**, A comparison of cell replacement in bone marrow, testis and three regions of surface epithelium, *Biochim. Biophys. Acta*, 560, 281, 1979.
102. **Potten, C. S. and Morris, R. J.**, Epithelial stem cells *in vivo*, *J. Cell Sci.*, Suppl. 10, 45, 1988.
103. **Zajicek, G., Oren, R., and Weinreb, M., Jr.**, The streaming liver, *Liver*, 5, 295, 1985.
104. **Zajicek, G., Ariel, I., and Arber, N.**, The streaming liver. III. Littoral cells accompany the streaming hepatocyte, *Liver*, 8, 213, 1988.
105. **Arber, N. and Zajicek, G.**, Streaming liver. IV. Streaming intra-hepatic bile ducts, *Liver*, 10, 205, 1990.
106. **Zajicek, G.**, Hepatocytes and intrahepatic bile duct epithelium originate from a common stem cell, *Gastroenterology*, 100, 582, 1990.

Chapter 8

ANTIGENIC CLUES TO LIVER DEVELOPMENT, RENEWAL, AND CARCINOGENESIS: AN INTEGRATED MODEL

D. C. Hixson, R. A. Faris, L. Yang, and P. Novikoff

TABLE OF CONTENTS

ISBN 0-8493-4746-7
© 1992 by CRC Press, Inc.

I. INTRODUCTION

Determination of the cellular progenitors of liver cancer is important both from the standpoints of understanding the cellular and molecular mechanisms leading to neoplasia as well as for the early detection and intervention of this disease. Using various rat models of chemical carcinogenesis, it has been shown that "carcinogen-initiated" liver epithelial cells proliferate to form foci and nodules of morphologically and biochemically altered cells.[1-8] In several of these model systems, small basophilic epithelial cells with ovoid nuclei, known as oval cells, also proliferate early after carcinogen exposure.[1,3,5,7,8] Despite extensive characterization of these carcinogen-induced cellular alterations, the identity of the rare nodules from which tumors arise and their progenitors remains largely unknown. The difficulties in tracing cell lineages during liver carcinogenesis are in part due to extensive tissue remodeling and phenotypic changes which occur prior to the appearance of hepatocellular carcinomas.

Although current evidence suggests that cancer results from the aberrant proliferation and differentiation of stem cells in a number of model systems, the existence of liver stem cells and their role in the carcinogenic process remain controversial. To address this question, our laboratory and other investigators have produced monoclonal antibodies (Mab) specific for developmental, functional, and cell-associated liver antigens in order to define phenotypic pathways leading to liver cancer.[9] In this chapter, we review the immunocytochemical evidence and propose a working model of lineage relationships in liver development, renewal, and carcinogenesis premised on the existence of liver stem cells. The model serves as a basis for integrating observations, formulating hypotheses, and proposing further experiments to confirm its validity. A brief review of stem cell models of epithelial tissue renewal is also presented to introduce this concept and the characteristics expected of stem cells.

II. ABERRANT TISSUE RENEWAL

Epithelial tissues in adult animals undergo variable rates of turnover. Tissues exposed to mechanical stress such as the intestinal mucosa and skin have relatively high rates of turnover compared to tissues such as liver, which have long-lived cells and correspondingly low rates of renewal.[10-15] For most epithelia, it is generally accepted that replacement of the fully differentiated functional cell population depends on stem cells, multipotent unspecialized cells that can give rise to all the differentiated progeny in a particular tissue.[10] In the small intestine, for example, the stem cell-derived differentiated cells would include absorptive, goblet, and Paneth cells.[12,15] Since stem cells must be maintained, it follows logically that they must be able to undergo divisions that produce both stem cell progeny, which replenish their numbers, and nonstem cell progeny, which ultimately differentiate into functionally mature cells. Present thinking suggests that this is accomplished by a random series of symmetric and asymmetric divisions. The former produces either two new stem cells (stem cell renewal) or two transit cells committed to further differentiation (stem cell loss), while the latter results in both stem and transit cell production.[14] Studies of human skin suggest that as many as 60% of basal cell mitoses may have vertical axes, thereby giving rise to a basal cell and a suprabasal cell — an asymmetric division.[16] Of course, not all basal cells are stem cells. Indeed, only one of the approximately ten basal cells within a given epidermal proliferative unit (EPU) is thought to be a stem cell.[17] It seems likely therefore that most of these vertical divisions represent transit cells migrating from a proliferative to a quiescent compartment rather than asymmetric divisions of stem cells.[16]

Since tissue renewal must continue throughout the entire life of the animal, it also follows that stem cells must be immortal relative to their more differentiated progeny and must have a replication capacity sufficient to meet the lifetime needs of the animal.[10] Furthermore, they must possess efficient repair or protective mechanisms to preserve the integrity of the genome and prevent nonlethal mutations from producing abnormal progeny or subpopulations of genetically altered stem cells. Indeed, evidence gathered to date suggests that stem cells have evolved a number of such mechanisms. In many tissues, stem cells are sequestered in a physically protected area that lowers the probability of insult by physical or chemical agents. Examples would be epidermal stem cells at the bottom of deep rete ridges,[18] intestinal stem cells at the base of the intestinal crypts,[13] and corneal stem cells in the limbus region of the cornea.[19] Kinetic studies in intestine and skin suggest that the presumptive stem cells are often the slowest cycling cells in a renewing tissue. The resulting long cycle times lower the probability of genetic damage and provide increased time for repair.[10,11,14] Further protection is afforded by the production of a multipotent committed population of more rapidly cycling cells that are the major source of new cells.[10,11,14] In the skin, for example, basal cells can be separated by low gravity sedimentation into subgroups composed of either small undifferentiated cells with the long generation times and low RNA content expected of stem cells or intermediate sized cells showing the more rapid rate of proliferation and differentiation characteristic of transit cells.[20] A similar situation exists in the small intestine where the cell cycle time for the stem cell compartment is about twice that of the midcrypt cells which ultimately give rise to all the differentiated cell types in the villus.[12,14] It is estimated that there are usually between two to four amplifying transit cell divisions in epithelial tissues. This is in contrast to as many as 13 transit cell divisions in the erythropoietic lineage in bone marrow.[15] Finally, there is some evidence that stem cells may be able to selectively retain the parental strands of DNA during replication, thus avoiding any errors that might result during DNA replication.[21]

In situations where the renewal capacity of an epithelial tissue has been significantly compromised by treatment with necrogenic doses of chemical agents or high levels of radiation such as during a clonogenic assay in the intestine or skin, regeneration may be mediated in part by transit cells that have retained sufficient proliferative capacity and multipotency to repopulate the tissue. At present, there is no way to distinguish between such clonogenic cells, or what might be termed "facultative" stem cells, and true stem cells. The need for such a distinction, however, is clearly demonstrated by clonogenic assays for hematopoietic stem cells in mice.[22,23] Early studies by Till and McCulloch[24,25] demonstrated that cells present in bone marrow were capable of reconstituting the immune system of lethally irradiated mice. Within a few days after injection of bone marrow cells, colonies containing mixtures of maturing cell types were observed in the spleen. Analysis of chromosomal markers in the donor cells indicated that these colonies were clonally derived.[26] This observation suggested that the donor cell population contained a multipotent stem cell precursor, an idea consistent with the ability of retransplanted colonies to give rise to further colonies.[27] However, subsequent studies indicated that these spleen colony forming cells, designated CFU-S, had a limited capacity for renewal since the number of CFU-S per colony decreased significantly after only a few serial transplantations.[28] Even taking into account the limitations of the assay, it seemed unlikely that the CFU-S had sufficient capacity for renewal to maintain hematopoiesis throughout the life of the mouse. More recent studies by Fleischman et al.[29,30] have provided experimental support for this view and have shown that adult CFU-S from bone marrow have a significantly lower capacity for self-renewal than CFU-S from fetal liver.[31] Recently, Capel et al.[32] have confirmed the existence of short- and long-lived totipotent hematopoietic stem cells in both fetal liver and adult bone marrow by following the fate of individual stem cell clones identified by retroviral integration markers. One interpretation of these results, which is consistent with the large number of amplifying transit cell divisions that occur during hematopoiesis, is that the short-lived totipotent stem cells are a clonogenic cell population derived from early transit cells which are committed to differentiation, but maintain high proliferative capacity, while the long-lived totipotent stem cells represent "true" stem cells.

Although the distinction between clonogenic cells and stem cells may operationally have little significance during normal tissue regeneration, this issue becomes important in epithelial tissues such as liver where significant stem cell involvement is only apparent during severe injury or when the regenerative capacity of existing hepatocytes has been greatly compromised. Moreover, tumors arising from clonogenic vs. stem cell progenitors may manifest significant differences in their behavior and growth characteristics.

This was demonstrated elegantly by Fialkow[33-36] in studies of human leukemias. In normal female tissues, one of the two X chromosome-linked genes for glucose 6-phosphate dehydrogenase (G6PD) is inactivated, producing a random mix of cells expressing the A or B isoenzyme. By analyzing leukemic cells for the isoforms of G6PD, Fialkow and co-workers[34-36] and Dow et al.[37] were able to demonstrate that the overtly leukemic cells were clonally derived either from unipotent transit cells with limited differentiation potential or from multipotent stem cell progenitors capable of giving rise to leukemic progeny in multiple cell lineages.

Since any dividing cell is theoretically vulnerable to genotoxic agents or oncogenic viruses and many of these agents require DNA replication for integration and transformation, it is reasonable to assume that stem cells or long-lived transit cells such as those in bone marrow can be targets of carcinogenic agents and thus give rise to tumors with varying potentials for differentiation. Indeed, Potten and Morris[15] have argued that in rapidly renewing tissues such as the skin or intestinal mucosa, stem cells are the most likely precursor population since they are the only proliferating cells that persist long enough to progress to carcinoma. The more differentiated transit cells are rapidly lost by sloughing as they are carried up the crypt or the EPU by the normal process of migration.[15]

This reasoning is consistent with the view of Pierce and Speers[38] of malignancy as a caricature of normal tissue renewal, their postulate being that carcinomas arise from malignant stem cells that retain the ability to differentiate, but do so infrequently, thus producing an expanding tumor mass composed of constantly proliferating cells. Support for this concept comes from a large number of studies employing multiple tumor models in a variety of animal species. One can go as far back in the phylogenetic tree as the flatworms and find reports of teratological malformations and heterogeneous tumors composed of cells in various stages of differentiation which were derived from mitotically active neoplastic stem cells.[39,40] A similar differentiation phenomenon was observed by Pierce and Wallace[41] within squamous cell carcinomas. They found that within the first few hours following injection of ^{3}H-thymidine into mice bearing these carcinomas, essentially all of the radiolabeled tumor cells were located in the undifferentiated areas of the tumor, but that by 96 hour many of the cells within the well-differentiated squamous pearls had become labeled. Moreover, when the labeled cells within the squamous pearls were transplanted, they no longer gave rise to tumors. On the basis of these results, the authors proposed that the squamous cell tumor was derived from the progeny of malignant stem cells which had retained the capacity to differentiate into postmitotic, well-differentiated squamous cells within the microenvironment of the squamous pearls. Consistent with these earlier studies were more recent findings demonstrating the presence of a subpopulation of doubly labeled basal cells in animals treated with

^{3}H-thymidine and ^{14}C-benzo-(a)-pyrene. The localization, mitotic activation following treatment with 12-O-tetradecanoyl-phorbol-13-acetate (TPA), and turnover of these cells was consistent with a stem cell precursor for carcinoma which became initiated before commitment to terminal differentiation.[15,42]

Perhaps the most striking evidence for the existence of stem cell-derived tumors came from experiments in which mouse blastocysts injected with teratocarcinoma cells gave rise to normal mice chimeric for the genotypes represented in the blastocyst and tumor cells.[43,44] This remarkable finding showed that teratocarcinomas were derived from totipotent stem cells capable of generating, in the appropriate embryonic environment, all somatic tissues of the mouse. Many leukemias of myeloid, B, and T cell origin also appear to be derived from stem cell progenitors and the antigenic and enzymatic phenotypes of these leukemic cells faithfully reproduce stages in the differentiation of their normal counterparts.[45-47]

III. LIVER STEM CELLS

It seems clear from the previous discussion that stem cells are a likely target for carcinogenic agents in various epithelial and lymphoid tissues. What about liver? Do stem cells play a role in hepatocarcinogenesis? For that matter, are there even stem cells in the liver, a question raised by the ability of the liver to regenerate from differentiated hepatocytes, biliary cells, and littoral cells without any apparent input from immature stem cell progenitors.

The question of hepatic stem cells has been a controversial one that has received much attention of late due to recent evidence from a number of investigators suggesting that a significant percentage of primary hepatocellular carcinomas (PHC) generated in rat liver models of carcinogenesis are stem cell derived. The concept of hepatic stem cells is certainly not a new one. As early as 1958, Wilson and Leduc[48] postulated the existence of primitive cells capable of differentiating into parenchymal or bile duct cells during liver restoration following dietary injury. These cells were thought to be located in the cholangioles, the transitional canals that connect the smallest bile ducts with the bile canalicular system.[48-50] The relationship between hepatic stem cells and oval cells, the small basophilic ductule cells that proliferate following treatment with a wide variety of carcinogens, remains unclear.[1,50-52] Although there is now convincing evidence for the ability of oval cells to differentiate into hepatocytes, the rapid and massive expansion seen following treatment with ethionine[7] or azo dyes[8] (i.e., 3′-methyl-4-dimethylaminoazobenzene) makes it likely that most oval cells are committed, partially differentiated transit cells rather than true stem cells. On the basis of electron microscopic autoradiography, Sell and Salman[53] proposed that oval cells may arise from transitional ductule cells in junctional contact with hepatocytes or from even more primitive periductule cells that have not formed tight junctions and do not share canalicular membranes with hepatocytes. If oval cells are stem cell derived, what conditions accelerate their replication and what factors control the end point of their differentiation? Of relevance to this question are the elegant studies of Evarts et al.[54] showing proliferation and comigration of TGFβ-producing Ito cells in close proximity to oval cells. A further question is whether or not oval cells/stem cells play a role in liver maintenance and repair similar to stem cells in the skin, intestine, or mammary gland. Moreover, do cells with oval cell/stem cell phenotypes play a central role in fetal liver development and, more importantly, are these cells capable of differentiation along both hepatocytic and biliary lineages? Studies addressing these questions are reviewed in detail in several of the other chapters of this volume (i.e., see Chapters 5 through 7). In this chapter, we will focus on antigenic evidence for oval cells as stem cells and stem cells as important progenitors of PHC.

IV. ANTIGENIC MARKERS: PAINTING THE STEM CELL PORTRAIT

Hybridoma technology has made it possible to generate highly specific antibodies by immunization with crude antigen preparations, partially purified proteins, whole cells, or tissues. Table 1 lists some of the Mab-defined antigens present on cells in livers of fetal, adult, and carcinogen-treated rats as well as the immunogen employed to generate each hybridoma. A variety of immunization protocols have been used to produce these cell-specific immunological probes, ranging from straightforward immunization with purified or partially purified cell preparations to complex immunization protocols involving immunosuppression or tolerization. We have applied the latter approach successfully to the generation of Mab against oval cell-associated antigens, but have also produced a number of useful Mab by immunization with purified or partially purified proteins. Although many of the antigens listed in Table 1 have yet to be identified, it should be emphasized that this does not detract from their utility as well type-specific or developmentally associated markers. There is no question, however, that the present anonymity of these antigens has significantly hampered analysis of their involvement in the process of malignant transformation.

Several of the Mab listed in Table 1 have been particularly valuable in defining cell lineage relationships during normal liver development and carcinogenesis. BDS_7, a Mab raised by Germain et al.[55,56] against azo dye-induced oval cells, recognizes a bile duct antigen that appears late in fetal development and has been reported to represent commitment to a biliary lineage. OV6, one of several Mab prepared by Dunsford et al.[57,58] and Sell et al.[59] by immunizing mice with partially purified preparations of oval cells or neoplastic nodules induced by AAF, has been widely used and is strongly reactive with virtually all ducts in adult and fetal liver and with both oval and nodular cells in carcinogen-treated liver. Use of OV6 in combination with

TABLE 1
Summary of Mab to Liver Cells

Mab[a]	Antigen[b]	Immunogen[c]	Reactivity[d]	Ref.
270.26*	HBD.1	CD/AAF NPC	AH, FL_{12}, BD	9, 60, 62
258.26*	H.2	CD/AAF NPC	AH, FL_{16}	9, 60, 62
368.7*	H.4	CDE NPC	AH, FL_{16}	9, 60
362.50*	Cell CAM 105	AH_{mem}	AH,	9, 129
HES_6	—	AH	AH, FL_{15}	56, 90
H4	—	GGT^+ Hep	AH, FL_{16}	57
RL16/79	—	AH	AH_{sub}	130
RL23/36	—	AH	AH, FL_{19}	130
21.6*	BD.1	$OC.2^+$ FL_{17}	BD_{sub}, FL_{16}	64
18.11*	BD.2	BDL	O, BD	131
270.38*	OC.2	CD/AAF NPC	O, BD, FL_{12}	9, 60, 62
374.3*	OC.3	CDE NPC	O, BD, FL_{12}	9, 60, 62
OV1	—	CD/AAF NPC	O, BD, FL_{16}	57
OV6	—	S/F HN	O, BD, FL_{18}	57
BDS_7	—	Me-DAB NPC	O, BD, FL_{12}	56, 90
374.8*	H.5	CDE NPC	PN, PHC, AH_{sub}	9, 60
324.5*	TuAg1	THC 1682c and THC 1677	PN_{sub}, PHC, FL_{14}	132
T6	—	Morris hepatoma 7777	PN_{sub}, PHC, FL_{20}	57

[a] Mab that are denoted with an asterisk were developed in our laboratory.
[b] Designated antigen defined by Mab.
[c] Immunogens: CD/AAF NPC, nonparenchymal epithelial cells enriched for oval cells induced by feeding rats a choline-deficient diet containing 2-acetylaminofluorene; CDE NPC, nonparenchymal epithelial cells enriched for oval cells induced by feeding rats a choline-deficient diet containing ethionine; AH, normal adult hepatocytes or plasma membranes (AH_{mem}); GGT^+ hep, γ-glutamyl transpeptidase-positive hepatocytes obtained from rats on the Solt/Farber protocol;[4] $OC.2^+$ FL_{17}, day 17 fetal liver cells expressing OC.2; BDL, biliary epithelial cells obtained from rats 13 weeks following bile duct ligation; S/F HN, hepatic nodular cells isolated from rats on the Solt/Farber protocol,[4] Me-DAB NPC, nonparenchymal epithelial cells enriched for oval cells induced by feeding rats a diet containing 3-methyl-4-dimethylaminoazobenzene; THC 1682c and THC 1677 are ethionine-induced transplantable hepatocellular carcinomas.
[d] Reactivity: AH, adult hepatocytes; FL, fetal liver; BD, bile duct; sub, subpopulation of cells; O, oval cells; PN, preneoplastic hepatocyte nodules; PHC, primary hepatocellular carcinomas.

other Mab has made it possible to dissect these various cell populations into subsets. OV1, for example, another of the Mab described by Dunsford et al.,[57,58] defines a subset of OV6-positive adult and fetal ductal cells and reacts heterogeneously with OV6-positive oval cells.[57,58]

We have also produced and characterized several Mab against carcinogen-induced oval cells.[60,61] The two most thoroughly characterized, designated OC.2 and OC.3, recognize surface antigens which appear very early in fetal liver development and are expressed on oval cells as well as normal adult biliary epithelium.[9] Two other Mab that have proven useful define cytoplasmic hepatocyte-associated antigens designated as HBD.1 and H.2.[60] HBD.1 is one of the earliest markers detected in fetal liver[62] and is one of the few hepatocytic markers that remains invariant in expression during hepatocarcinogenesis.[60,61] Although HBD.1 is also detected on adult biliary epithelial cells, its level of expression on bile ducts is very much lower than that observed on hepatocytes. In contrast, H.2 shows a much more restricted pattern of expression and appears only on hepatocytes in adult and fetal liver.[60] During carcinogenesis H.2 is lost by most preneoplastic nodules produced by the Solt/Farber protocol,[61] but is expressed at normal levels on nodules induced by ethionine in a choline-deficient diet (CDE).[63]

Most recently, we have defined a new bile duct-associated antigen, designated BD.1, which shows developmentally related patterns of expression in fetal liver and heterogeneous reactivity with normal bile ducts in adult liver.[64] Unlike many of the Mab reactive with bile duct cells, such as OC2, OC3, BDS_7, OV6, and OV1, BD.1 shows no apparent reactivity with oval cells (Figure 1). Moreover, its reactivity with fetal hepatocytes appears to be limited to those cells with the composite ductal and hepatocytic phenotypes characteristic of transitional cells forming the earliest bile ducts (Figure 2).

Double-label immunofluorescence analysis of up to 45 serial sections has shown that some OV6-positive bile ducts are composed of a mosaic of both BD.1-negative and -positive cells while others contain a homogeneous population of BD.1-positive or BD.1-negative cells. When a third marker, OV1, is considered, additional heterogeneity is introduced with each of the BD.1-negative or -positive subpopulations further divided into OV1-positive and -negative cells.[65] Engelhardt et al.[66] made similar findings in mouse liver using two Mab, designated A6 and G7, raised against an oval cell-enriched fraction from mice treated with the alkylating drug Dipin. Both Mab defined antigens in bile ducts and Hering canals, but A6 showed heterogeneous staining of canals, thus defining two subpopulations of G7-positive cells.

Although the basis for heterogeneity of BD.1 expression is presently unknown, cell cycle- or differentiation-related changes are the two possibilities suggested by our current results. Changes in expression related to differentiation would be consistent with the observation that in fetal and neonatal liver, bile ducts lack the heterogeneity seen in the adult liver and appear uniformly positive for BD.1. On the other hand, the rapid loss of BD.1 on proliferating ducts following bile duct ligation or partial hepatectomy would seem to favor cell cycle-dependent expression.[65] Both possibilities would be consistent with the absence of BD.1 on the antigenically heterogeneous, rapidly proliferating population of relatively undifferentiated oval cells induced by either chemical carcinogenesis with 2-acetylaminofluorene (AAF) or ethionine or injury (galactosamine) protocols. However, the extremely slow recovery of BD.1 following partial hepatectomy, 4 to 5 weeks, and the presence of BD.1-negative and -positive ductule cells labeled with ^{3}H-thymidine in liver sections taken 40 h after injection of galactosamine and 10 h after

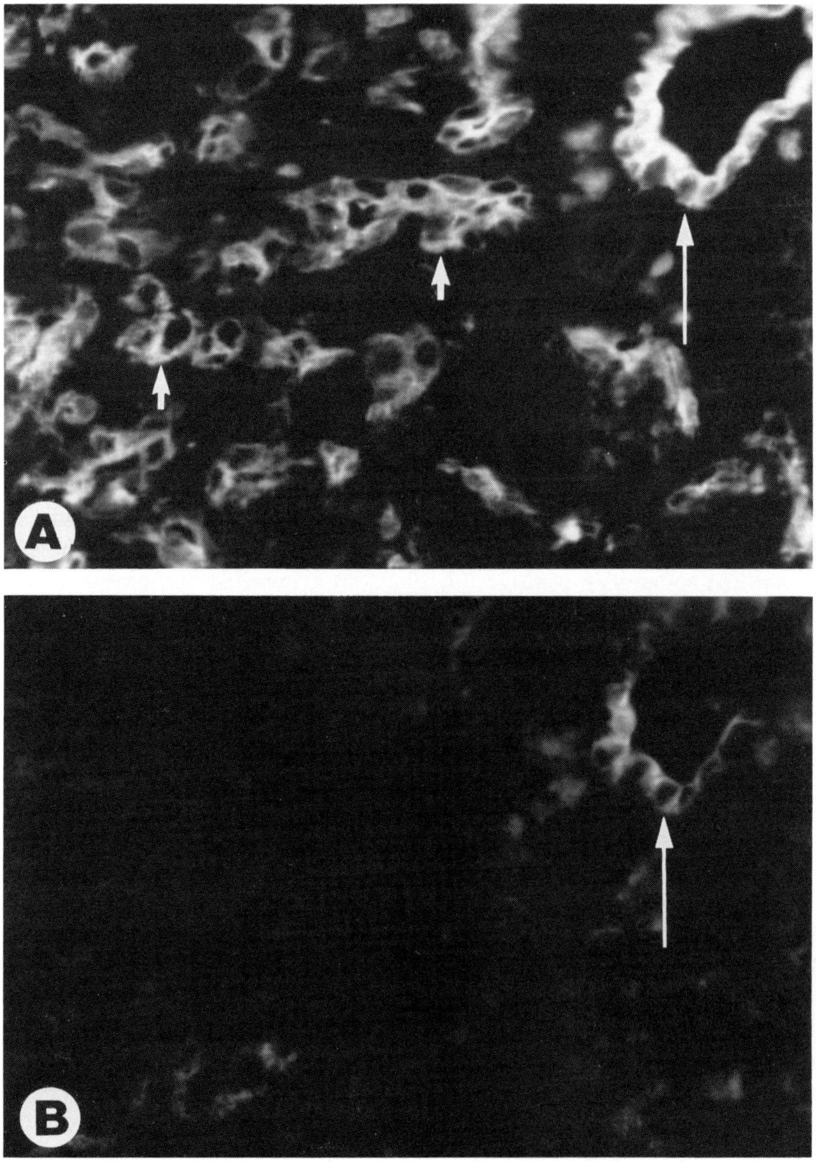

FIGURE 1. Indirect immunofluorescence analysis of ethionine-induced rat liver oval cells using Mab OV6 (panel A) and Mab 21.6 (panel B) demonstrates that OV6 is expressed by oval cells (short arrows) and bile duct (long arrow), whereas the BD.1 antigen defined by Mab 21.6 is heterogeneously expressed by the bile duct, but is not detectable on oval cells.

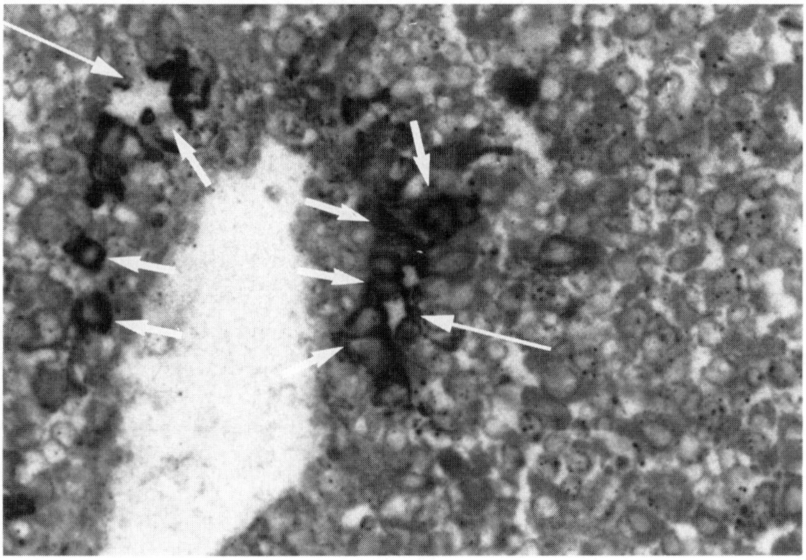

FIGURE 2. Indirect immunoperoxidase analysis of 16-d fetal rat liver using Mab 21.6 demonstrates that the BD.1 antigen is only expressed by newly forming bile duct cells adjacent to a large portal space near the hilus (short arrows). Long arrows denote what appears to be a lumen between primitive ductal cells and neighboring hepatoblasts.

injection with isotope[65] argue against any simple relationship between proliferation and BD.1 expression. In addition, approximately 25% of the ducts remain at least partially positive after partial hepatectomy or bile duct ligation, while conversely, approximately 20% of the bile ducts in adult liver are uniformly negative, a percentage in the latter case much higher than can be accounted for by the low mitotic index (<4%) in resting liver. Our current thinking is that many of the negative ducts represent interlobular branches of the BD.1-negative extrahepatic bile duct, the negative reactivity reflecting the different embryonic origins of the extrahepatic and intralobular ducts.[67] Mosaic ducts are envisioned as representing an intermingling of these two types of ductal cells in regions of transition between the smallest interlobular branches of the extrahepatic duct and the largest branches of the intrahepatic ducts. Regardless of the mechanisms involved in the regulation of its expression, BD.1 is at present one of the few markers that can distinguish between oval cells and intralobular bile ducts. BD.1 also appears to mark the commitment of hepatoblasts to a biliary lineage[65] following contact with the portal mesenchyme.[68] Moreover, BD.1, when used in combination with markers such as OV1, demonstrates the presence of a high degree of phenotypic heterogeneity among biliary ductal cells in adult liver and thus may be of value in the search for the elusive hepatic stem cell.

V. WORKING MODEL FOR LIVER DEVELOPMENT, RENEWAL, AND CARCINOGENESIS

As a means of summarizing and integrating experimental evidence for the involvement of stem cells in liver development, renewal, and carcinogenesis, we have formulated a working model which brings together results and concepts developed in our and other laboratories. Although many aspects of this model are not new and similar schemes have been proposed by other investigators, we have attempted to consolidate and extend previous models while incorporating new concepts, refinements, and questions raised by recent investigations employing monoclonal and polyclonal antibodies specific for developmental, functional, and cell-associated antigens. We have also emphasized in this model the relationship between cellular pathways operative during liver development (Figure 3) and those leading to hepatocellular carcinoma.

In the developmental scheme presented in Figure 3, the earliest epithelial cells, which appear in the liver bud on or about day 10 of gestation, are viewed as multipotential endodermal cells that have been restricted by their interaction with the mesenchyme of the septum transversum to either a ductal or a hepatocytic pathway of differentiation.[50] These cells would be likened to the bipotential hepatoblasts described by Germain et al.[56] and Marceau and associates (see Reference 69 and Chapter 7). It seems likely that these early hepatoblasts are similar, if not identical, to the uncommitted endodermal cells that generate the pancreas and the small intestine during development. This idea is based on the common origin of all three organs as outgrowths of the primitive foregut[75] and on the ability to induce in the adult the development of hepatocytes in the pancreas,[70-72] and intestinal metaplasia[73] or pancreatic acinar cells in the liver.[74] Phenotypically, these earliest hepatoblasts are relatively nondescript null cells with no distinguishing epithelial features. As illustrated in Figure 4, HBD.1 and OC.2 are two of the earliest markers expressed by fetal liver cells. Although at present we are still uncertain which appears first, HBD.1 or OC.2, the most important concept is that the expression of the second of these two antigens represents a commitment to further differentiation. This commitment step is viewed as occurring at random and being transient in nature such that at any given point in time, there are relatively few hepatoblasts doubly positive for HBD.1 and OC.2. A recent report by Shiojiri et al.[76] suggests that the appearance of α-fetoprotein (AFP) also signals commitment to hepatic differentiation.

A key event in determining the subsequent fate of these early cells is the development of the portal veins with their associated mesenchyme, a morphogenic step that first occurs near the hilus. Hepatoblasts that do not come into contact with the portal mesenchyme default into a hepatocytic lineage, a decision heralded by the loss of OC.2, the acquisition of H.2, and around

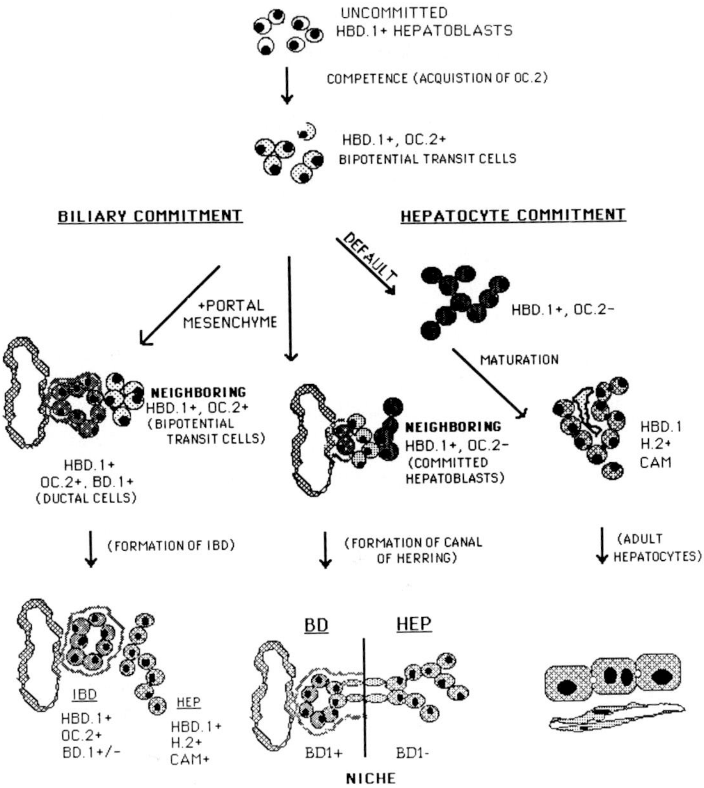

FIGURE 3. Diagramatic model for liver development from bipotential stem cells. Early embryonic hepatoblasts expressing HBD.1 are committed to differentiate along either a hepatocytic or biliary lineage upon acquisition of OC.2. The default pathway is production of mature hepatocytes. However, if these bipotential transit cells interact with the invading portal mesenchyme they form either intralobular bile ducts (IBD) or cholangioles, the so-called canals of Hering. The type of ductal cells produced when these transit cells interact with the mesenchyme and begin to form a basal lamina depends on whether the neighboring cells are bipotential or committed to a hepatocyte lineage. If neighboring cells are already committed to a hepatocyte lineage, there is incomplete formation of the basement membrane, thus creating a stem cell niche between the ductal cells and hepatocytes.

the 15th day of gestation, the appearance of hepatocyte-associated membranc markers such as cell-CAM 105[77] and H.1. In the early stages of liver development, entry into a hepatocytic lineage leads to the formation of cellular dipoles of uncommitted or doubly positive cells and immature hepatocytes. The influence of the latter induces its uncommitted partner towards a hepatocytic lineage, thereby accelerating the formation of the hepatic parenchyma. Alternatively, if a doubly positive hepatoblast contacts the portal mesenchyme, it begins to express BD.1, an event signaling commitment to a biliary lineage.

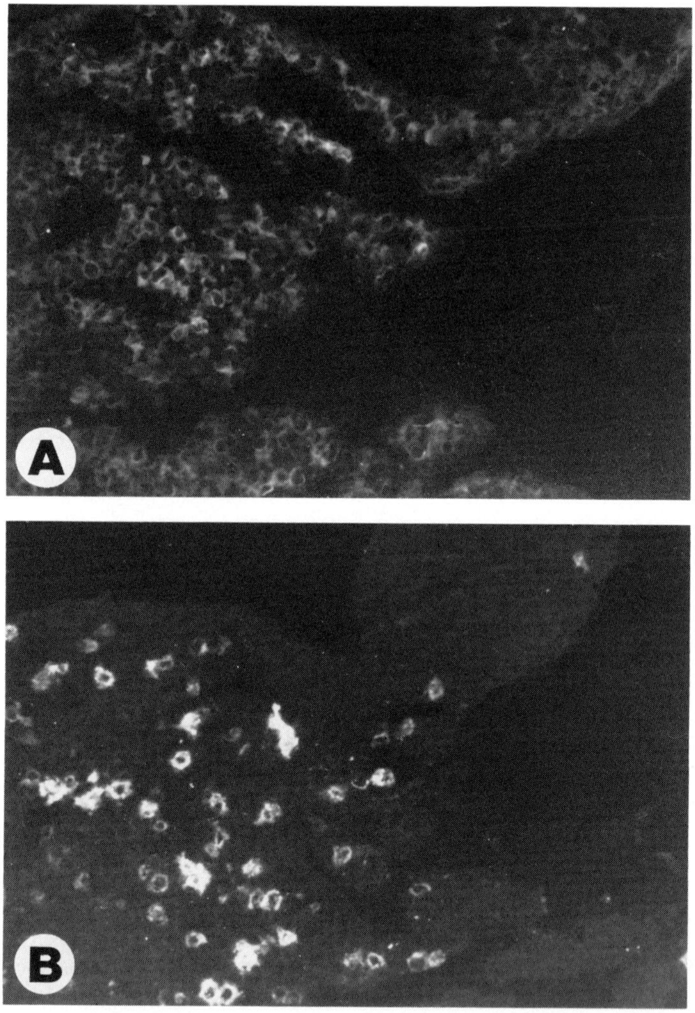

FIGURE 4. Indirect immunofluorescence analysis of 12 fetal rat livers using HBD.1 (panel A) and OC.2 (panel B). HBD.1 is expressed by cords of hepatoblasts, whereas OC.2 is expressed by randomly scattered cells.

Like their hepatocytic counterparts, the BD.1-positive cells lose OC.2, but the loss appears to be transient. Our most recent findings suggest that OC.2 returns with the formation of the laminin-containing basal lamina. Once committed, these first BD.1-positive cells exert an inductive effect[71-73] on neighboring BD.1-negative cells that are not in contact with the portal mesenchyme, causing them to enter a bile duct lineage. This leads to the formation of a ductal plate and ultimately a series of unconnected closed tubes that are positive for BD.1, HBD.1, OC.2, OV6, and cytokeratin 19 (CK19). At this stage, the synthesis of basal lamina by these earliest BD cells and the status of the hepatoblasts on the distal surface of the primitive bile duct become important determinants of subsequent events. If adjacent hepatoblasts are still uncommitted and thus unable to exert a hepatocytic influence, the newly formed ductal cells synthesize a basal lamina that encircles the entire duct, thereby precluding further interaction with adjacent hepatoblasts. If, on the other hand, neighboring hepatoblasts have already made the commitment to a hepatocytic lineage, their influence prevents complete enclosure, creating a region of contact between the two cell types via a transitional zone of BD.1-negative, OC.2-positive, HBD.1-positive cells. Cells within this buffer zone undergo one to three divisions, thereby forming what ultimately will become the cholangioles, the so-called canals of Hering, composed of cells showing a gradient of hepatocytic or bile duct characteristics. Those cells nearest the duct display a weak expression of BD.1 and a lower level of HBD.1 and those in proximity to hepatocytes show the reverse phenotype. Bile duct cells and hepatocytes then begin a maturation process that continues into the adult. Unlike cells in the hepatocytic lineage, which no longer express early fetal antigens, bile duct cells retain fetal characteristics while gaining additional markers such as BD.2. Hepatocytes, in turn, undergo dramatic changes during the fetal to neonatal interface, marked by maturation of the canalicular membranes and increased expression of both membrane and cytoplasmic antigens.

In the cholangioles, under normal resting conditions, further development is prevented by the opposing influences of mature bile duct and hepatocytes, thus creating the hypothetical stem cell niche of Potten.[10] As in other epithelial tissues, such as the skin and the intestine, the uncommitted cells in the midregion of the cholangiole slowly cycle in response to the programmed apoptotic loss of hepatocytes or ductal cells. As the midregion cholangiolar cells divide, they are displaced from the neutral zone and, depending on their direction of migration, become increasingly influenced by the more mature hepatocytes or the bile duct cells at the ends of the canal, ultimately becoming either BD.1 positive or OC.2 negative as they commit to either a biliary or a hepatocytic pathway of differentiation, respectively. It seems likely, based on what occurs in fetal liver, that the early progeny in the hepatocytic lineage will possess characteristics of both ductal cells and hepatocytes while those of the ductal lineage will lack mature ductal markers. Later divisions will lead to mature diploid cell populations with a high renewal capacity. As viewed in the context of the streaming liver model,[78-81] this renewal capacity will gradually diminish as hepatocytes or ductule cells leave their respective proliferative compartments, having completed two to four amplifying divisions, and moved downstream in an increasingly quiescent, nonreplicative state of terminal differentiation ending for hepatocytes in the central lobular zone.

Following cytotoxic or physical insult to the liver, the involvement of hepatic stem cells in repair and regeneration is determined by the proliferative competency of the mature cell populations. If hepatocytes or ductal cells retain their normal capacity for self-renewal, such as after bile duct ligation or partial hepatectomy, stem cell involvement is limited to self-renewal. However, if the ability of hepatocytes or ductal cells has been significantly compromised, by exposure, for example, to azo dyes[8] or AAF,[82] there is a compensatory increase in stem cell proliferation resulting in the appearance of a rapidly dividing transit cell population such as oval cells which, under appropriate conditions, can expand until they compose as much as 80% of the liver.[82] Since these transit cells exit the stem cell niche from both directions, they are heterogeneous in their phenotypic and antigenic makeup.[41,52,55,59,83,84] Their character is further modified by interactions with stromal or cellular elements as they migrate away from the portal mesenchyme along the extracellular matrix within the space of Disse.

During carcinogenesis in constantly renewing epithelia such as skin and intestine, the only cells that usually persist long enough to progress to carcinoma are the long-lived, slowly cycling stem cells.[10,11,14] The life span and/or proliferative capacity of their differentiated progeny is inconsistent with the long latency between carcinogen exposure and the appearance of carcinoma.[15] In contrast, in the liver, stem, transit, and differentiated cell populations are all sufficiently long lived to progress to carcinoma, making it possible to have both differentiated and stem cell progenitors. Which pathway predominates will depend on how a particular carcinogenesis protocol affects the stem cell compartment represented in Figure 3. In the resistant hepatocyte model,[4,5] for example, phenotypic evidence suggests that treatment with diethylnitrosamine will damage both stem cells and their more differentiated progeny,[58,61] the continual slow cycling of stem cells resulting in fixation of DNA damage. The inability of uninitiated hepatocytes to respond to the regenerative stimulus of a partial hepatectomy after exposure to AAF[4,5] results in the recruitment of hepatic stem cells into the renewal process and the production of oval cells, at least some of which are derived from the small percentage of initiated stem cell progenitors. After a long latency period during which additional alterations occur, a few of these initiated stem cells, together with a percentage of their initiated, more differentiated progeny, eventually progress to carcinoma. Agents such as azo dyes or ethionine that produce a more massive proliferation of oval cells than that observed in the resistant hepatocyte model will produce a correspondingly higher percentage of carcinomas with oval cell characteristics.[63] At the other extreme, with protocols such as the Pitot model in which diethylnitrosamine is given after partial hepatectomy,[6] a treatment that produces little or no oval cell proliferation, there will be very little stem cell involvement and most of the resulting tumors will display phenotypes reflecting their derivation from differentiated hepatocytes instead of oval cells.

VI. EXPERIMENTAL BASIS FOR CRITICAL ASPECTS OF THE PROPOSED MODEL

A. EVIDENCE FOR BIPOTENTIAL CELLS IN FETAL AND ADULT LIVER

The origin of intrahepatic bile ducts during liver development has been a question of interest for a large number of investigators over the last 60 years. Recently, the availability of antibodies specific for AFP, albumin, and CK have allowed a more precise identification of the various stages in liver development in both rodent and human liver.[85-87] Results from many of these studies are described in detail in other chapters (5,7,11) of this volume. When taken together, these findings lead to the conclusion put forth by Shiojiri et al.[76] that all endodermally derived cells in the adult liver arise from AFP-positive hepatoblasts that possess, at minimum, a bipotential capacity for differentiation. Intrahepatic bile ducts arise from AFP-positive, albumin-positive hepatoblasts that undergo commitment to a biliary lineage following contact with the portal mesenchyme, an event signaled by the appearance of a single and then a double layer of CK18, γ-glutamyl transpeptidase (GGT)-positive cells. These early ductule structures show a heterogeneous expression of AFP and albumin, but during subsequent maturation lose these markers and begin to express CK7 and CK19 and later acquire the cell-surface marker BDS_7.[9,50,76,85-87] AFP-positive hepatocytes in contrast appear to arise by a gradual maturation process beginning with albumin production. Commitment to a hepatocytic lineage results in an increased expression of the cytoplasmic marker HBD.1, the appearance of H.2, and the gradual acquisition of membrane markers such as HES_6, cell-CAM 105, dipeptidyl peptidase IV, and H.1.[19,62,77,88]

More direct evidence for the bipotentiality of hepatoblasts in fetal liver comes from *in vitro* and *in vivo* demonstrations of their ability to differentiate into cells with phenotypic and morphological characteristics of hepatocytes and biliary cells. Shiojiri[89] showed that both bile ducts and hepatocytes developed within subcutaneous transplants of mouse fetal liver fragments taken at 12.5 to 13.5 gestation days. Since there are as yet no bile ducts at this stage of development, it was concluded that hepatoblasts were capable of differentiating into both types of cells. Germain et al.[56] and Marceau et al.[69,90] reached a similar conclusion based on *in vitro* studies of fetal rat liver cells. These investigators showed that primary cultures of E12 fetal liver cells could be induced to differentiate into cells with antigenic characteristics of hepatocytes or ductal cells, by treatment with dimethyl sulfoxide or sodium butyrate, respectively.

Do bipotential cells exist in adult liver, and if so, what role do they play in the repair of liver damage resulting from physical or toxic insult? Based on an analysis of AFP during acute liver injury induced by carbon tetrachloride and galactosamine, Tournier et al.[91] concluded that expression of high amounts of AFP mRNA and protein during galactosamine intoxication resulted primarily from a minor population of cells composed of proliferating oval cells, bile ductlike structures, and small hepatocytes phenotypically intermediate between oval cells and mature hepatocytes. This was in contrast to carbon tetrachloride-injured livers where oval cells were absent, AFP protein was undetectable, and all hepatocytes exhibited approximately a twofold increase in AFP mRNA, a pattern of expression identical to that observed following partial hepatectomy. Recently, Lemire et al.[92] have examined in greater detail the fate of oval cells generated following intoxication of rat liver with galactosamine. Using a probe specific for the fetal form of AFP in combination with lectins and antibodies specific for bile ducts, these authors concluded that both oval cells and small AFP-positive hepatocytes arose from preexisting ductule cells and not from large hepatocytes, a conclusion substantiated by determination of ^{3}H-thymidine-labeling indices for each cell type.

Bipotentiality has been demonstrated for oval cells generated following treatment with a number of different hepatocarcinogens, AAF,[93-96] ethionine,[84,97,98] hepatitis B virus,[99,100] azo dye,[55] and diethylnitrosamine,[101] to name a few. Using alloantigens as markers for donor cells, we reported the generation of donor-derived colonies of GGT-positive hepatocytes following hepatic transplantation of GGT-positive oval cells induced by AAF.[96] More recently, Evarts et al.[93,95] demonstrated the transfer of radiolabeled thymidine from albumin and AFP-positive oval cells induced by AAF into albumin-positive, transiently AFP-positive basophilic hepatocytes, thereby providing the first direct evidence for a precursor-product relationship between these two cell types.

Results of Tournier et al.[91] discussed above, lend support to the idea that the involvement of bipotential cells in renewal depends on the extent to which the proliferative capacity of hepatocytes has been compromised. This is best exemplified by the ability of the hepatocytes remaining after partial hepatectomy or carbon tetrachloride treatment to completely restore the liver mass.[102] In contrast, a partial hepatectomy performed when rats are exposed to AAF results in significant proliferation of oval cells.[103] In this case the mitoinhibition of normal hepatocyte proliferation by AAF seemingly recruits bipotential cells into the renewal process. Similar considerations may also apply to bile ducts where the increased biliary pressure produced by bile duct ligation causes hyperplasia restricted to existing bile duct cells,[104] but galactosamine intoxication induces damage that recruits both bipotential cells and existing bile ducts into proliferation.[91,92] It is for this reason that bipotential cells in liver have often been viewed as facultative stem cells, the selective nature of their role in renewal distinguishing them from stem cells that bear the entire responsibility for regeneration and renewal in other epithelia. At odds with this view is a recent report by Arber and Zajicek[80] who, on the basis of long-term labeling studies with ^{3}H-thymidine, propose that normal renewal in resting liver is also mediated by stem cells located in the cholangioles. These authors view hepatocytes and bile duct cells as streaming along separate pathways of differentiation, leading from stem cells in the portal tract to either terminally differentiated hepatocytes around the hepatic vein or to mature bile duct cells in the larger intrahepatic ducts with the rate of turnover determining the speed of displacement. Although controversial, the most attractive feature of this model is that it views hepatic renewal as being essentially identical to that in other epithelial tissues, the only major difference being the extremely long life span of hepatocytes and intrahepatic biliary ductule cells.

Fausto[50] has questioned the appropriateness of the term ''stem cells'' to describe bipotential progenitors in the adult liver, his preference being to view them as committed and partially differentiated progenitors until unequivocal evidence can be obtained showing them to be multipotent rather than bipotent cells. Although the point is well taken, there is considerable indirect evidence suggesting that these cells may indeed be multipotent. Recent work by Sirica et al.[105] described in this volume has demonstrated the consistent formation of intestinal epithelium in the livers of rats treated with furan. This finding is consistent with our own studies showing differentiation of a clonal line derived from an azo dye-induced hepatocellular adenocarcinoma into well-differentiated intestinal mucosa when transplanted subcutaneously into juvenile rats.[106] Makino et al.[70] and Rao et al.[71] have reported that pancreatic ducts give rise to hepatocytes during pancreatic regeneration following destruction of acinar cells by treatment with ethionine or chronic copper deficiency. Conversely, Wolf et al.[107] have described the formation of acinar structures histologically and ultrastructurally indistinguishable from normal pancreas in a liver biopsy specimen from a human patient with chronic liver disease. Taken together, these observations suggest the existence in pancreas, liver, and intestine of multipotent stem cells with the capacity to differentiate along multiple cellular lineages given the appropriate microenvironment. Bipotential cells in the liver might thus be regarded as true stem cells whose differentiation capacity has been restricted by the hepatic microenvironment.

B. ROLE OF EXTRACELLULAR MATRIX AND MESENCHYME

More than 60 years ago, Bloom[108] and Horstmann[109] observed a close association between differentiating biliary epithelial cells and the ingrowing portal connective tissue, a relationship which suggested that portal mesenchyme influenced commitment to a biliary lineage. This view was consistent with *in vitro* studies around the same time by Doljanski and Roulet[110] demonstrating the ability of mesenchyme to induce transformation of fetal liver cells into ductlike structures. Further evidence for portal mesenchyme as a determinant of differentiation was provided more recently by Shiojiri,[89] who found that 12.5 to 13.5-d fetal mouse liver fragments proximal or distal to the hilus developed into both hepatocytes and bile ducts if transplanted into the testis of neonatal mice, but more readily formed biliary ducts if transplanted into fat pads or under the ventral skin. Significantly, large hepatocytes were concentrated in the central region of transplants, while bile duct cells were localized in the periphery in proximity to the surrounding connective tissue.

Clues as to which components of the portal mesenchyme are essential elements in the initial development of intrahepatic ducts come from a number of different investigations. Shah and Gerber[68] analyzed the expression of CK and laminin during the development of human liver and observed a close association between laminin deposition and the CK-positive differentiating ductal plate at the epithelial-mesenchymal interface in the portal tracts. Following formation of the second layer of the ductal plate and the appearance of a lumen between the two layers, laminin deposition initiated on the parenchymal side of these early ductal structures. Once these tubular structures became integrated into the center of portal tracts, laminin disappeared from the margin, except in areas of contact with the few scattered ductal plates that persisted into the adult. Laminin continued to be associated with ducts during their ingrowth into the portal mesenchyme and remained after maturation in the form of a laminin-containing basement membrane. Based on these observations, Shah and Gerber[68] proposed that laminin was closely associated with differentiation and ingrowth of ductal cells.

Similar conclusions were made by Ekblom[111] with regard to a role for laminin in the morphogenesis of the kidney. These investigators further showed that the conversion of kidney mesenchyme to epithelium of kidney tubules was closely associated with the appearance of laminin A chain and that initial interaction with laminin at the cell surface triggered the polarization of unpolarized mesenchyme and an adhesion cascade[112,113] starting with the redistribution of uvomorulin and ending with the formation of apical junctional complexes. Whether or not the association with laminin triggers similar events during the formation of intrahepatic bile ducts remains to be determined.

Significantly oval cells can be distinguished from proliferating bile ducts by differences in laminin deposition. This was first reported by Sell,[114] who observed that the AFP-positive, albumin-negative ducts induced by 4,4′-diaminodiphenylmethane were surrounded by a basal lamina that was intensely positive for laminin, but only weakly positive for fibronectin. In contrast, oval cells exhibited almost exactly the opposite phenotype, showing frequent expression of AFP, albumin, and fibronectin, but very weak expression of laminin. These striking differences may be related to the fact that oval cells migrate along the space of Disse away from the influence of the portal mesenchyme and as a consequence may become more hepatocytelike due to their close contact with the hepatic matrix which is rich in fibronectin and low in laminin.[115]

Although the role of other matrix components remains to be established, results from *in vitro* studies by Gall and Bhathal[116] suggest that collagens are probably not directly involved in the transformation of hepatoblasts to biliary cells. Based on previous studies demonstrating the ability of collagen gels to induce differentiation, these investigators examined the development of fetal liver explants embedded in type 1 collagen gels and saw no enhancement of duct morphogenesis relative to explants embedded in agar gels. Treatment with agents known to induce collagen deposition was also ineffective in inducing the development of bile ducts.

C. INDUCTIVE EFFECTS

A key aspect of the proposed model is the ability of mature ductal cells and hepatocytes to influence the differentiation of uncommitted hepatoblasts in fetal liver and stem cells in the canals of Hering in the adult. Evidence for this ability comes largely from the immunocytochemical, ultrastructural, and histochemical analyses by Novikoff et al.[117,118] of hepatocytes and oval cells in rats maintained on a CDE diet. In these studies, oval cells were defined by the positive expression of OC.2, glutathione *S*-transferase, and CK19, markers absent in hepatocytes at all time points examined, and by the absence of the hepatocyte marker HBD.1 and peroxisome-associated catalase activity. In agreement with previous studies, oval cells arranged in ductlike arrays were surrounded by a basement membrane and contained a lumen bordered by apical junctional complexes. Interestingly, even single oval cells were encircled by a basement membrane. The most striking observation was the formation of bile canaliculi bordered by tight and desmosomal junctions along the contiguous membranes of oval cells and hepatocytes (Figure 5). In cases where the attached oval cells were part of a ductulelike array, the apical membrane in common with the lumen defined by the other oval cells remained ATPase negative, while the bile canalicular structures formed on the lateral surface in concert with hepatocytes were strongly ATPase positive. Additionally, the basement membrane which normally encircled the oval cells was lost along the surface in contact with hepatocytes. These results establish that oval cells in contact with hepatocytes are capable of adopting a dual polarity, maintaining oval cell characteristics along the apical membrane held in common with other oval cells while displaying phenotype properties of hepatocytes along the apical membrane shared with hepatocytes.

In vitro studies by Nakamura et al.[119] provide further evidence for the ability of hepatocytes to promote differentiation along a hepatocytic lineage. In this study, differentiation was monitored by immunocytochemical staining for tryptophan 2,3,-dioxygenase (TO), a liver-specific enzyme that is not expressed in the majority of newborn hepatocytes, but gradually increases in expression during the neonatal period until it reaches adult levels 4 weeks after birth. These investigators found that when neonatal (0 day) hepatocytes

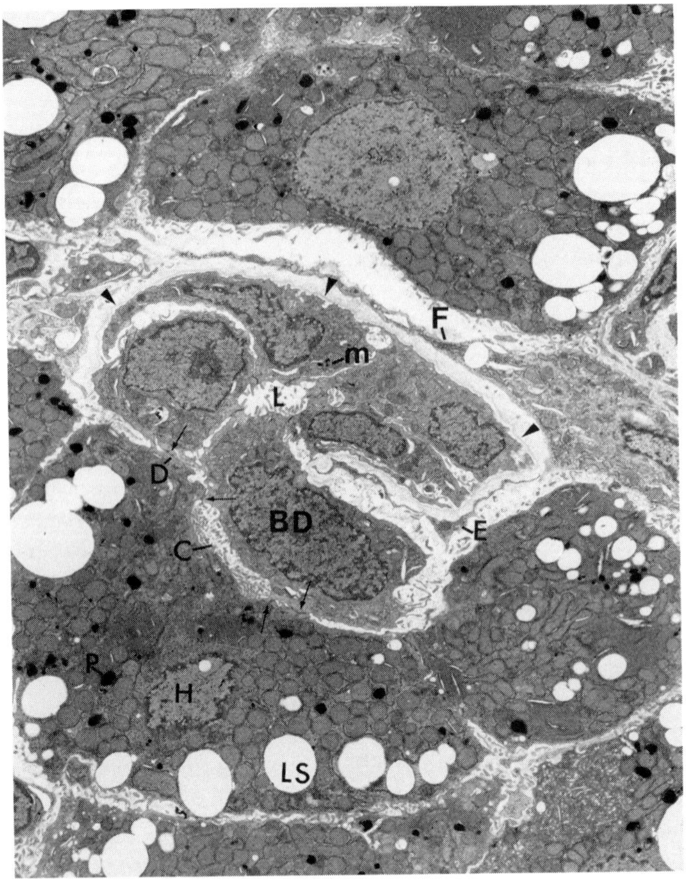

FIGURE 5. Rat liver after 6 weeks of choline-deficient ethionine diet. Electron micrograph of a liver section incubated for catalase activity. Bile canaliculus (C) structure and plasma membrane contact sites (arrows) have formed between a hepatocyte (H) and bile ductulelike cells (BD). At one contact site, desmosomes (D) are evident. BD cells displaying varying morphologic profiles are organized around a central lumen (L) into which microvilli project and have a continuous basal lamina (arrowheads) associated with their basal surfaces. The BD cell in contact with the hepatocyte exhibits two apical surfaces. An extension of a fibroblastlike cell is seen in close association with the basal lamina. Note the absence of basal lamina and endothelial cells (E) where BD cells and a hepatocyte are linked. Also labeled are peroxisomes (P) and lipid spheres (LS) in hepatocytes and microperoxisomes (m) in a BD cell. (Original magnification × 8000.)

were placed in primary culture, they expressed low levels of TO and were able to proliferate in the absence of exogenous growth factors needed to maintain the proliferative status of adult hepatocytes. However, if neonatal hepatocytes were cultured on feeder layers of adult hepatocytes, TO expression was rapidly induced in more than 70% of the cells and proliferation, as measured by ^{3}H-thymidine uptake, decreased sixfold. That this effect was specific to hepatocytes was demonstrated by the inability of feeder layers of nonparenchymal epithelial cells from adult liver or hepatoma cells to induce TO and decrease proliferation. It was further shown that this effect required contact with hepatocytes possessing intact surface membranes and was not produced by extracellular matrix from hepatocyte cultures, hepatocyte-conditioned media, or feeder layers of trypsin- or acid-treated cells. From these results it was suggested that *in vivo,* hepatocytes may differentiate by contact with a component on the surface of mature cells, resulting in a wave of differentiation that spreads from cell to cell.

D. STEM CELLS IN LIVER CARCINOGENESIS

If one accepts the existence of slowly cycling bipotential cells in the adult liver and the concept that oval cells, similar to basal cells in the skin or midcrypt cells in the intestine, represent the undifferentiated progeny of these bipotential cells, it follows logically that at least a percentage of hepatocellular carcinomas will be derived from these bipotential progenitors. In recent years, there has been accumulating experimental evidence to support this contention. In our work, we have focused on determining whether or not the antigenic phenotypes of preneoplastic nodules and hepatocellular carcinomas are consistent with a derivation from oval cell progenitors. Using a battery of Mab specific for antigens associated with bile duct and oval cells and fetal, adult, and neoplastic hepatocytes, we have examined the phenotypic relationships between oval cells, foci, nodules, and hepatocellular carcinomas induced by the resistant hepatocyte model of Solt and Farber[4] and Solt et al.[5] In studies to date, we have detected oval cells, GGT-positive hepatocellular foci, persistent hepatocyte nodules, and primary hepatocellular carcinomas expressing both oval cell and hepatocyte antigens, a finding suggestive of a precursor-product relationship between oval cells and carcinomas.[9,61] Dunsford et al.,[58] using a different panel of Mab raised against oval cells and neoplastic nodules, noted that many of the persistent nodules induced by the resistant hepatocyte protocol were composed of a mixture of OV6-positive oval cells and hepatocytes in close proximity, the number of OV6-positive hepatocytes increasing at later time points. Since most tumors expressed OV6, these authors concluded that primary hepatocellular carcinomas were derived by a progression from oval cells to OV6-positive hepatocytes to OV6-positive carcinomas. Similar lineage relationships have been found in other models of liver carcinogenesis. In animals maintained on a CDE diet, we have detected a significant percentage of nodules and hepatocellular carcinomas expressing both oval cell, hepatocyte, and neoplastic markers.[9,61] Moreover, animals bearing liver tumors expressing both oval cell and hepatocyte markers were also found to have metastatic foci in their lungs with essentially identical phenotypes.[9,61]

More direct evidence that oval cells could progress to hepatocellular carcinoma has come from immunohistochemical analysis of tumors obtained by transplantation of cultured oval cells transformed *in vitro* by treatment with carcinogens such as aflatoxin[120] and *N*-methyl-*N'*-nitrosoguanidine (MNNG),[121,122] or by transfection of *Ha-ras*.[97] In most cases, the resulting tumors displayed a wide range of phenotypes ranging from well-differentiated hepatocellular carcinomas, cholangiomas, and regions of intestinal metaplasia to poorly differentiated or anaplastic tumors. In the studies of Goyette et al.[98] and McMahon et al.,[120] immunohistochemical analysis indicated that the oval cell-derived tumors were composed of an antigenically heterogeneous mixture of cells expressing both oval cell and hepatocyte antigens.

Evidence for oval cell progenitors of primary hepatocellular carcinoma has not been restricted to rodent models of carcinogenesis. Van Eyken et al.[123] have reported that as many as 50% of human hepatocellular carcinomas express CK characteristic of both bile ducts and hepatocytes. Relevant in this regard is a recent report by Vandersteenhoven et al.[100] who, by immunohistochemical analysis, demonstrated the presence of ductal cells with characteristics of both bile duct cells and hepatocytes in the liver of a patient with end-stage cirrhosis resulting from hepatitis B virus infection. Oval cells have also been implicated as progenitors of hepatocellular carcinomas induced by hepatitis B2-type virus infection in woodchucks.[99] In this study, Fu et al.[99] observed extensive oval cell proliferation during the early stages of chronic active hepatitis which was associated with elevated levels of mRNA for insulinlike growth factor II (IGFII). Analysis of the carcinomas revealed an inverse relationship between IGFII mRNA levels and virus replication, high IGFII being associated with low levels of virus. On the basis of these and other results, it was proposed that carcinomas with high IGFII and low virus phenotypes were derived from oval cell progenitors.

As was discussed above the relative percentage of primary hepatocellular carcinomas derived from oval cell progenitors may vary over a wide range, depending on the carcinogenesis protocol and the degree of oval cell involvement in the early stages. Consistent with this view, Dunsford et al.[58] and Sell and Dunsford[101] reported that in animals continuously fed diethylnitrosamine, a treatment which induces little or no oval cell hyperplasia, carcinomas expressing oval cell antigens were only rarely observed; most tumors appeared to arise directly from initiated hepatocytes. It will be of interest to analyze the phenotype of tumors induced by peroxisome proliferators,[124] agents which do not induce oval cell hyperplasia and which produce neoplastic nodules lacking markers commonly expressed by preneoplastic cells in other carcinogenesis models.[125]

VII. CONCLUSIONS AND FUTURE DIRECTIONS

In summary, accumulating findings from immunocytochemical and thymidine-labeling studies strongly suggest that hepatic stem cells exist in adult rat liver and that these cells are one of the progenitors of liver cancer in some model systems.[8,9,50,51,58,59,61,95,96,98,101] Using a panel of Mab that recognize normal hepatocytes, oval cells, and neoplastic liver cells, our laboratory has demonstrated an antigenic relationship between oval cells and a subpopulation of foci, nodules, and carcinomas in the rat,[61,63] an observation consistent with the possibility that oval cells are a precursor of liver tumors in some model systems. Although the precise cellular origin of oval cells remains unknown, there is now strong evidence that these small ductal-like cells are stem cell-derived, bipotential transit cells that can differentiate into both biliary epithelium and hepatocytes.[50,55,56,95,96,126,127] That bipotential liver cells also play a role during liver development is supported by the detection of a subpopulation of cells expressing oval cell antigens as early as embryonic day 12.

Since oval cells morphologically resemble bile duct cells and express many ductal markers, it has been proposed that the true hepatic stem cell may be contained within the bile duct population.[9,48,49-52,59,69,83,128] In support of this view, we have recently identified three populations of antigenically distinguishable bile duct cells in the normal adult rat liver. One subpopulation (30%) lacks BD.1, an antigen expressed by most newly forming bile ducts during fetal liver development, but is heterogeneously expressed by ductal cells in the adult animal. Although the biological significance of BD.1-positive and -negative ductal cells remains to be determined, oval cells also lack detectable levels of BD.1, suggesting that BD.1-negative ductal cells may be precursors of oval cells.

A number of questions await future investigation. What triggers proliferation of oval cells during the neoplastic process? Is it the demand for tissue renewal, impairment of normal hepatocyte function, or aberrant proliferation due to disruption of the niche? What humoral and microenvironmental factors modulate the proliferation and differentiation of oval cells along hepatocytic and biliary lineages? The answer to these questions will provide insight to the mechanisms leading to liver cancer.

Finally, the common embryological origin of liver, pancreas, and intestine raises the possibility that these adult tissues share a pluripotential endodermal stem cell. If this is the case then disruption of the normal mechanisms which control the tissue-specific differentiation of these stem cells would explain the observations by others that oval cells give rise to intestinal epithelium,[73] pancreatic ductal cells become hepatocytes,[70-72] and liver biliary cells form pancreatic acinar cells.[74] In order to address the possibility that pluripotential endodermal stem cells exist, our laboratory is employing Mab to isolate ductal cells from the liver and pancreas and transplanting these cells into fetal, newborn, and adult animals. The fate of these cells is being followed by the expression of dipeptidyl peptidase IV, a model system that allows us to transplant a null cell population and determine if these cells give rise to hepatocytes, bile ducts, pancreatic ducts, or intestinal epithelium.[128] However, it is apparent that before one can successfully identify and isolate true stem cells, we need to develop additional immunological reagents that define cell-surface antigens expressed by rarer subpopulations of ductal cells. For example, probes are needed that distinguish among intralobular ducts, terminal biliary ductules, and cells forming the canal of Hering. The availability of such reagents will ultimately allow one to delineate the *in vitro* and *in vivo* potential of each of these cell populations and will facilitate identification of true hepatic stem cells.

REFERENCES

1. **Farber, E.,** Similarities in the sequence of early histological changes induced in the liver of the rat by ethionine, 2-acetylaminofluorene, and 3′-methyl-4-dimethylaminoazobenzene, *Cancer Res.,* 16, 142, 1956.
2. **Peraino, C., Fry, R. J. M., and Staffeldt, E.,** Reduction and enhancement by phenobarbital of hepatocarcinogenesis induced in the rat by 2-acetylaminofluorene, *Cancer Res.,* 31, 1506, 1971.
3. **Teebor, G. W. and Becker, F. F.,** Regression and persistence of hyperplastic nodules induced by N-2-fluorenylacetamide and their relationship to hepatocarcinogenesis, *Cancer Res.,* 31, 1, 1971.
4. **Solt, D. B. and Farber, E.,** New principle for the analysis of chemical carcinogenesis, *Nature (London),* 263, 701, 1976.
5. **Solt, D. B., Medline, A., and Farber, E.,** Rapid emergence of carcinogen-induced hyperplastic lesions in a new model for the sequential analysis of liver carcinogenesis, *Am. J. Pathol.,* 88, 595, 1977.
6. **Pitot, H. C., Barsness, L., Goldsworthy, T., and Kitagawa, T.,** Biochemical characteristics of stages of hepatocarcinogenesis after a single dose of diethylnitrosamine, *Nature (London),* 271, 456, 1978.
7. **Shinozuka, H., Lombardi, B., Sell, S. A., and Iammarino, R. M.,** Early histological and functional alterations of ethionine liver carcinogenesis in rats fed a choline-deficient diet, *Cancer Res.,* 38, 1092, 1978.
8. **Inaoka, Y.,** Significance of the so-called oval cell proliferation during azo-dye hepatocarcinogenesis, *Gann,* 58, 355, 1967.
9. **Hixson, D. C., Faris, R. A., and Thompson, N. L.,** An antigenic portrait of the liver during carcinogenesis, *Pathobiology,* 58, 65, 1990.
10. **Potten, C. S.,** *Stem Cells: Their Identification and Characterization,* Potten, C. S., Ed., Churchill Livingston, New York, 1983.
11. **Potten, C. S., Schofield, R., and Lajtha, L. G.,** A comparison of cell replacement in bone marrow, testis and three regions of surface epithelium, *Biochim. Biophys. Acta,* 560, 281, 1979.
12. **Potten, C. S.,** Stem cells in gastrointestinal mucosa, *Tumor Viruses and Differentiation,* Scolnick, E. M. and Levine, A. J., Eds., Alan R. Liss, New York, 1983, 381.

13. **LeBlond, C. P.,** The life history of cells in renewing systems, *Am. J. Anat.,* 160, 114, 1981.
14. **Wright, N. and Alison, M.,** *The Biology of Epithelial Cell Populations,* Vol. 1, Clarendon Press, Oxford, 1984, chap. 2.
15. **Potten, C. S. and Morris, R. J.,** Epithelial stem cells in vivo, *J. Cell Sci. Suppl.* 10, 45, 1988.
16. **Wright, N. and Alison, M.,** *The Biology of Epithelial Cell Populations,* Vol. 1, Clarendon Press, Oxford, 1984, chap. 8.
17. **Potten, C. S.,** Stem cells in epidermis from the back of the mouse, in *Stem Cells: Their Identification and Characterization,* Potten, C. S., Ed., Churchill Livingston, New York, 1983, chap. 8.
18. **Lavker, R. M. and Sun, T. T.,** Epidermal stem cells, *J. Invest. Dermatol.,* 81, 121, 1983.
19. **Cotsarelis, G., Cheng, S. Z., Dong, G., Sun, T. T., and Lavker, R. M.,** Existence of slow-cycling limbal epithelial basal cells that can be preferentially stimulated to proliferate: implications on epithelial stem cells, *Cell,* 57, 201, 1989.
20. **Pavlovitch, J. H., Rizk-Rabin, M., Gervaise, M., Metezeau, P., and Grunwald, D.,** Cell subpopulations within proliferative and differentiating compartments of epidermis, *Am. J. Physiol.,* 256, c977, 1989.
21. **Potten, C. S., Hume, W. J., Reid, P., and Cairns, J.,** The segregation of DNA in epithelial stem cells, *Cell,* 15, 899, 1978.
22. **Lord, B. I.,** Haemopoietic stem cells, in *Stem Cells: Their Identification and Characterization,* Potten, C. S., Ed., Churchill Livingston, New York, 1983, chap. 6.
23. **Lord, B. I.,** The relationship between spleen colony production and spleen cellularity, *Cell Tissue Kinet.,* 4, 211, 1971.
24. **Till, J. E. and McCulloch, E. A.,** A direct measurement of the radiation sensitivity of normal mouse bone marrow cells, *Radiat. Res.,* 14, 213, 1961.
25. **Till, J. E., McCulloch, E. A., and Siminovitch, L.,** A stochastic model of stem cell proliferation based on the growth of spleen colony-forming cells, *Proc. Natl. Acad. Sci. U.S.A.,* 51, 29, 1964.
26. **Becker, A. J., McCulloch, E. A., and Till, J. E.,** Cytological demonstration of the clonal nature of spleen colonies derived from mouse marrow cells, *Nature (London),* 197, 452, 1963.
27. **Siminovitch, L., McCulloch, E. A., and Till, J. E.,** Distribution of colony-forming cells among spleen colonies, *J. Cell. Comp. Physiol.,* 62, 327, 1963.
28. **Siminovitch, L., Till, J. E., and McCulloch, E. A.,** Decline in colony-forming ability of marrow cells subjected to serial transplantation into irradiated mice, *J. Cell. Comp. Physiol.,* 64, 23, 1964.
29. **Fleischman, R. A. and Mintz, B.,** Prevention of genetic anemias in mice by microinjection of normal hematopoietic stem cells into the fetal placenta, *Proc. Natl. Acad. Sci. U.S.A.,* 76, 5736, 1979.
30. **Fleischman, R. A., Custer, R. P., and Mintz, B.,** Totipotent hematopoietic stem cells: normal self-renewal and differentiation after transplantation between mouse fetuses, *Cell,* 30, 351, 1982.
31. **Fleischman, R. A. and Mintz, B.,** Development of adult bone marrow stem cells in H-2-compatible and -incompatible mouse fetuses, *J. Exp. Med.,* 159, 731, 1984.
32. **Capel, B., Hawley, R. G., and Mintz, B.,** Long- and short-lived murine hematopoietic stem cell clones individually identified with retorviral integration markers, *Blood,* 75, 2267, 1990.
33. **Fialkow, P. J.,** Clonal evolution of human myeloid leukemias, in *Genes and Cancer,* Bishop, J. M., Rowley, J. D., and Greaves, M., Eds., Alan R. Liss, New York, 1984, 215.

34. **Fialkow, P. J., Denman, A. M., Jacobson, R. J., and Lowenthal, M. N.,** Chronic myelocytic leukemia: origin of some lymphocytes from leukemic stem cells, *J. Clin. Invest.*, 62, 815, 1978.
35. **Fialkow, P. J., Singe, J. M., Adamson, J. W., Berkow, R. L., Friedman, J. M., Jacobson, R. J., and Moohr, J. W.,** Acute nonlymphocytic leukemia: expression in cells restricted to granulocytic and monocytic differentiation, *N. Engl. J. Med.*, 301, 1, 1979.
36. **Fialkow, P. J., Singer, J. W., Adamson, J. W., Vaidya, K., Dow, L. W., Ochs, J., and Moohr, J. W.,** Acute nonlymphocytic leukemia: heterogeneity of stem cell origin, *Blood*, 56, 1068, 1981.
37. **Dow, L. W., Martin, P., Moohr, J., Greenberg, M., Macdougall, L. G., Najifeld, V., and Fialkow, P. J.,** Evidence for clonal development of childhood acute lymphoblastic leukemia, *Blood*, 66, 902, 1985.
38. **Pierce, G. B. and Speers, W. C.,** Tumors as caricatures of the process of tissue renewal: prospects for therapy by directing differentiation, *Cancer Res.*, 48, 1996, 1988.
39. **Hall, F., Morita, M., and Best, J. B.,** Neoplastic transformation in the planarian. I. Cocarcinogenesis and histopathology, *J. Exp. Zool.*, 240, 211, 1986.
40. **Hall, F., Morita, M., and Best, J. B.,** Neoplastic transformation in the planarian. II. Ultrastructure of malignant reticuloma, *J. Exp. Zool.*, 240, 229, 1986.
41. **Pierce, G. B. and Wallace, C.,** Differentiation of malignant to benign cells, *Cancer Res.*, 31, 127, 1971.
42. **Morris, R. J., Fischer, S. M., and Slaga, T. J.,** Evidence that a slowly cycling subpopulation of adult murine epidermal cells retains carcinogen, *Cancer Res.*, 46, 3061, 1986.
43. **Illmensee, K. and Mintz, B.,** Totipotency and normal differentiation of single teratocarcinoma cells cloned by injection into blastocysts, *Proc. Natl. Acad. Sci. U.S.A.*, 73, 549, 1976.
44. **Mintz, B., Cronmiller, C., and Custer, R. P.,** Somatic cell origin of teratocarcinomas, *Proc. Natl. Acad. Sci. U.S.A.*, 75, 2834, 1978.
45. **Greaves, M. F.,** Differentiation-linked leukemogenesis in lymphocytes, *Science*, 234, 697, 1986.
46. **Marks, P. A., Sheffery, M., and Rifkind, R. A.,** Induction of transformed cells to terminal differentiation and the modulation of gene expression, *Cancer Res.*, 47, 659, 1987.
47. **Jimenez, J. J. and Yunis, A. A.,** Tumor cell rejection through terminal cell differentiation, *Science*, 238, 1278, 1987.
48. **Wilson, J. W. and Leduc, E. H.,** Role of cholangioles in restoration of the liver of the mouse after dietary injury, *J. Pathol. Bacteriol.*, 76, 441, 1958.
49. **Grisham, J. W. and Porta, E. A.,** Origin and fate of proliferating hepatic ductal cells in the rat: electron microscopic and autoradiographic studies, *Exp. Mol. Pathol.*, 2, 242, 1964.
50. **Fausto, N.,** Hepatocyte differentiation and liver progenitor cells, *Curr. Opin. Cell Biol.*, 2, 1036, 1990.
51. **Lombardi, B.,** On the nature, properties and significance of oval cells, *Recent Trends Chem. Carcinogen.*, 1, 37, 1982.
52. **Sirica, A. E., Mathis, G. A., Sano, N., and Elmore, L. W.,** Isolation, culture, and transplantation of intrahepatic biliary epithelial cells and oval cells, *Pathobiology*, 58, 44, 1990.
53. **Sell, S. and Salman, J.,** Light and electron-microscopic autoradiographic analysis of proliferating cells during the early stages of chemical hepatocarcinogenesis in the rat induced by feeding N-2-fluorenylacetamide in a choline deficient diet, *Am. J. Pathol.*, 114, 287, 1984.

54. **Evarts, R. P., Nakatsukasa, H., Marsden, E. R., Hsia, C.-C., Dunsford, H. A., and Thorgeirsson, S. S.,** Cellular and molecular changes in the early stages of chemical hepatocarcinogenesis in the rat, *Cancer Res.,* 50, 3439, 1990.
55. **Germain, L., Noël, M., Gourdeau, H., and Marceau, N.,** Promotion of growth and differentiation of rat ductular oval cells in primacy culture, *Cancer Res.,* 48, 368, 1988.
56. **Germain, L., Blouin, M.-J., and Marceau, N.,** Biliary epithelial and hepatocytic cell lineage relationships in embryonic rat liver as determined by the differential expression of cytokeratins, α-fetoprotein, albumin, and cell surface-exposed components, *Cancer Res.,* 48, 4909, 1988.
57. **Dunsford, H. A. and Sell, S.,** Production of monoclonal antibodies to preneoplastic liver cell populations induced by chemical carcinogens in rats and to transplantable Morris hepatomas, *Cancer Res.,* 49, 4887, 1989.
58. **Dunsford, H. A., Karnasuta, C., Hunt, J. M., and Sell, S.,** Different lineages of chemically induced hepatocellular carcinoma in rats defined by monoclonal antibodies, *Cancer Res.,* 49, 4894, 1989.
59. **Sell, S., Hunt, J. M., Knoll, B. J., and Dunsford, H. A.,** Cellular events during hepatocarcinogenesis in rats and the question of premalignancy, in *Adv. Cancer Res.,* 48, 37, 1987.
60. **Hixson, D. C. and Allison, J. P.,** Monoclonal antibodies recognizing oval cells induced in the liver of rats by N-2-fluorenylacetamide or ethionine in a choline-deficient diet, *Cancer Res.,* 45, 3750, 1985.
61. **Faris, R. A., Monfils, B. A., Dunsford, H. A., and Hixson, D. C.,** Antigenic relationship between oval cells and a subpopulation of hepatic foci, nodules, and carcinomas induced by the "resistant hepatocyte" model system, *Cancer Res.,* 51, 1308, 1991.
62. **Faris, R. A., Monfils, B. A., and Hixson, D. C.,** Antigenic relationship between oval cells, fetal hepatocytes, and myelocytic precursor cells, *Proc. Am. Assoc. Cancer Res.,* 30, 52, 1989.
63. **Faris, R. A., Monfils, B. A., and Hixson, D. C.,** Antigenic phenotypes of ethionine induced primary hepatocellular carcinomas (PHC) and metastatic foci (MF), *Proc. Am. Assoc. Cancer Res.,* 38, 107, 1986.
64. **Yang, L., Faris, R. A., Dunsford, H. A., and Hixson, D. C.,** Monoclonal antibody (MAb) defined antigen expressed by a subpopulation of biliary ductular cells (BDC) but absent from oval cells, *Proc. Am. Assoc. Cancer Res.,* 32, 33, 1991.
65. **Yang, L. and Faris, R. A.,** unpublished data.
66. **Engelhardt, N. V., Factor, V. M., Yasova, A. K., Poltoranina, V. S., Baranov, V. N., and Lasareva, M. N.,** Common antigens of mouse oval and biliary epithelial cells. Expression on newly formed hepatocytes, *Differentiation,* 45, 29, 1990.
67. **Shiojiri, N.,** The differentiation of the hepatocytes and the intra- and extrahepatic bile duct cells in mouse embryos, *J. Fac. Sci. Univ. Tokyo Sec. IV Zool.,* 14, 241, 1979.
68. **Shah, K. D. and Gerber, M. A.,** Development of intrahepatic bile ducts in humans, *Arch. Pathol. Lab. Med.,* 113, 1135, 1989.
69. **Marceau, N., Blouin, M.-J., Germain, L., and Noël, M.,** Role of different epithelial cell types in liver ontogenesis, regeneration and neoplasia, *In Vitro Cell. Dev. Biol.,* 25, 336, 1989.
70. **Makino, T., Usuda, N., Rao, S., Reddy, J. K., and Scarpelli, D. G.,** Transdifferentiation of ductular cells into hepatocytes in regenerating hamster pancreas, *Lab. Invest.,* 62, 552, 1990.
71. **Rao, M. S., Dwivedi, R. S., Yeldandi, A. V., Subbarao, V., Tan, X., Usman, M. I., Thangada, S., Nemali, M. R., Kumar, S., Scarpelli, D. G., and Reddy, J. K.,** Role of periductal and ductular epithelial cells of the adult rat pancreas in pancreatic hepatocyte lineage, *Am. J. Pathol.,* 134, 1069, 1989.
72. **Scarpelli, D. G.,** Multipotent developmental capacity of cells in the adult animal, *Lab. Invest.,* 52, 331, 1985.

73. **Tatematsu, M., Kaku, T., Medline, A., and Farber, E.,** Intestinal metaplasia as a common option of oval cells in relation to cholangiofibrosis in liver of rats exposed to 2-acetylaminofluorene, *Lab. Invest.*, 52, 354, 1985.
74. **Rao, M. S., Bendayan, M., Kimbrough, R. D., and Reddy, J. K.,** Characterization of pancreatic-type tissue in the liver of rat induced by polychlorinated biphenyls, *J. Histochem. Cytochem.*, 34, 197, 1986.
75. **Balinsky, B. I.,** *An Introduction to Embryology,* 3rd ed., W. B. Saunders, Philadelphia, 1970.
76. **Shiojiri, N., Lemire, J. M., and Fausto, N.,** Cell lineages and oval cell progenitors in rat liver development, *Cancer Res.*, 51, 2611, 1991.
77. **Petell, J. K., Quaroni, A., Hong, W., Hixson, D. C., Amarri, S., Reif, S., and Bujanover, Y.,** Alteration in the regulation of plasma membrane glycoproteins of the hepatocyte during ontogeny, *Exp. Cell Res.*, 187, 299, 1990.
78. **Zajicek, G., Oren, R., and Weinreb, M., Jr.,** The streaming liver, *Liver,* 5, 295, 1985.
79. **Zajicek, G., Ariel, I., and Arber, N.,** The streaming liver. III. Littoral cells accompany the streaming hepatocyte, *Liver,* 8, 213, 1988.
80. **Arber, N. and Zajicek, G.,** Streaming liver. IV. Streaming intra-hepatic bile ducts, *Liver,* 10, 205, 1990.
81. **Zajicek, G.,** Hepatocytes and intrahepatic bile duct epithelium originate from a common stem cell, *Gastroenterology,* 100, 582, 1990.
82. **Sell, S., Leffert, H. L., Shinozuka, H., Lombardi, B., and Gochman, N.,** Rapid development of large numbers of α-fetoprotein containing "oval" cells in the liver of rats fed N-2-fluorenylacetamide in a choline devoid diet, *Gann,* 72, 479, 1981.
83. **Hayner, N. T., Braun, L., Yaswen, P., Brooks, M., and Fausto, N.,** Isozyme profiles of oval cells, parenchymal cells, and biliary cells isolated by centrifugal elutriation from normal and preneoplastic livers, *Cancer Res.*, 44, 332, 1984.
84. **Plenat, F., Braun, L., and Fausto, N.,** Demonstration of glucose-6-phosphatase and peroxisomal catalase activity by ultrastructural cytochemistry in oval cells from livers of carcinogen-treated rats, *Am. J. Pathol.*, 130, 91, 1988.
85. **Shiojiri, N.,** Enzymo- and immunocytochemical analysis of the differentiation of liver cells in the prenatal mouse, *J. Embryol. Exp. Morphol.*, 62, 139, 1981.
86. **Van Eyken, P., Sciot, R., Callea, F., Van Der Steen, K., Moerman, P., and Desmet, V. J.,** The development of the intrahepatic bile ducts in man: a keratin-immunohistochemical study, *Hepatology,* 8, 1586, 1988.
87. **Marceau, N.,** Cell lineages and differentiation programs in epidermal, urothelial and hepatic tissues and their neoplasms, *Lab. Invest.*, 63, 4, 1990.
88. **Feracci, H., Connolly, T. P., Margolis, R. N., and Hubbard, A. L.,** The establishment of hepatocyte cell surface polarity during fetal liver development, *Dev. Biol.*, 73, 123, 1987.
89. **Shiojiri, N.,** The origin of intrahepatic bile duct cells in the mouse, *J. Embryol. Exp. Morphol.*, 79, 25, 1984.
90. **Marceau, N., Germain, L., Goyette, R., Noel, M., and Gourdeau, H.,** Cell of origin of distinct cultured rat liver epithelial cells, as typed by cytokeratin and surface component selective expression, *Biochem. Cell Biol.*, 64, 788, 1986.
91. **Tournier, I., Legres, L., Schoevaert, D., Feldmann, G., and Bernuau, D.,** Cellular analysis of α-fetoprotein gene activation during carbon tetrachloride and D-galactosamine-induced acute liver injury in rats, *Lab. Invest.*, 59, 657, 1988.
92. **Lemire, J. M., Shiojiri, N., and Fausto, N.,** Oval cell proliferation and the origin of small hepatocytes in liver injury induced by D-galactosamine, *Am. J. Pathol.*, 139, 535, 1991.
93. **Evarts, R. P., Nagy, P., Marsden, E., and Thorgeirsson, S. S.,** A precursor-product relationship exists between oval cells and hepatocytes in rat liver, *Carcinogenesis,* 11, 1737, 1987.

94. **Evarts, R. P., Nagy, P., Marsden, E., and Thorgeirsson, S. S.,** *In situ* hybridization studies on expression of albumin and α-fetoprotein during the early stage of neoplastic transformation in rat liver, *Cancer Res.*, 47, 5469, 1987.
95. **Evarts, R. P., Nagy, P., Nakatsukasa, H., Marsden, E., and Thorgeirsson, S. S.,** *In vivo* differentiation of rat liver oval cells into hepatocytes, *Cancer Res.*, 49, 1541, 1989.
96. **Faris, R. A. and Hixson, D. C.,** Selective proliferation of chemically altered rat liver epithelial cells following hepatic transplantation, *Transplantation,* 48, 87, 1989.
97. **Braun, L., Goyette, M., Yaswen, P., Thompson, N. L., and Fausto, N.,** Growth in culture and tumorigenicity after transfection with the *ras* oncogene of liver epithelial cells from carcinogen-treated rats, *Cancer Res.*, 47, 4116, 1987.
98. **Goyette, M., Faris, R., Braun, L., Hixson, D., and Fausto, N.,** Expression of hepatocyte and oval cell antigens in hepatocellular carcinomas produced by oncogene-transfected liver epithelial cells, *Cancer Res.*, 50, 4809, 1990.
99. **Fu, X.-X., Su, C. Y., Lee, Y., Hintz, R., Biempica, L., Snyder, R., and Rogler, C. E.,** Insulinlike growth factor II expression and oval cell proliferation associated with hepatocarcinogenesis in woodchuck hepatitis virus carriers, *J. Virol.*, 62, 3422, 1988.
100. **Vandersteenhoven, A. M., Burchette, J., and Michalopoulos, G.,** Characterization of ductular hepatocytes in end-stage cirrhosis, *Arch. Pathol. Lab. Med.*, 114, 403, 1990.
101. **Sell, S. and Dunsford, H. A.,** Evidence for the stem cell origin of hepatocellular carcinoma and cholangiocarcinoma, *Am. J. Pathol.*, 134, 1347, 1989.
102. **Fausto, N.,** Hepatic regeneration, in *Hepatology: A Textbook of Liver Disease,* Vol. 1, 2nd ed., Zakim, D. and Boyer, T. D., Eds., W. B. Saunders, Philadelphia, 1990, chap. 3.
103. **Tatematsu, M., Ho, R. H., Kaku, T., Ekem, J. K., and Farber, E.,** Studies on the proliferation and fate of oval cells in the liver of rats treated with 2-acetylaminofluorene and partial hepatectomy, *Am. J. Pathol.*, 114, 418, 1984.
104. **Slott, P. A., Liu, M. H., and Tavoloni, N.,** Origin, pattern and mechanism of bile duct proliferation following biliary obstruction in the rat, *Gastroenterology,* 99, 466, 1990.
105. **Sirica, A. E., Elmore, L. W., Williams, T. W., and Cole, S. L.,** Differentiation potential of hyperplastic bile ductular epithelial cells during cholangiocarcinogenesis and in some other forms of severe hepatic injury, in *The Role of Cell Types in Hepatocarcinogenesis,* Sirica, A. E., Ed., CRC Press, Boca Raton, FL, 1992, chap. 9.
106. **Faris, R. A., Monfils, B. A., Brown, J., Yang, L., Affigne, S., Marceau, N., and Hixson, D. C.,** H5D.7, an azo dye induced transplantable hepatocellular carcinoma cell line that possesses multipotential stem cell properties, *Proc. Am. Assoc. Cancer Res.*, 31, 156, 1990.
107. **Wolf, H. K., Burchette, J. L., Garcia, J. A., and Michalopoulos, G.,** Exocrine pancreatic tissue in human liver; a metaplastic process?, *Am. J. Surg. Pathol.*, 14, 590, 1990.
108. **Bloom, W.,** The embryogenesis of human bile capillaries and ducts, *Am. J. Anat.*, 36, 451, 1926.
109. **Horstmann, E.,** Entwicklung and Entwicklungsbedingungen des intrahepatischen Gallengangsystems, *Wilhelm Roux' Arch. Entwicklungsmech. Org.*, 139, 363, 1939.
110. **Doljanski, L. and Roulet, Fr.,** Ueber die gestaltende Wechselwirkung zwischen dem Epithel und dem Mesenchyme, zugleich ein Beitrag zur Histogenese der sogenannten 'Gallengangswucherungen', *Virchows Arch. Pathol. Anat. Physiol.*, 292, 256, 1934.
111. **Ekblom, P.,** Developmentally regulated conversion of mesenchyme to epithelium, *FASEB J.*, 3, 2141, 1989.
112. **Gumbiner, B., Stevenson, B., and Grimaldi, A.,** The role of the cell adhesion molecule uvomorulin in the formation and maintenance of the epithelial junctional complex, *J. Cell Biol.*, 107, 1575, 1988.

113. **McNeill, H., Ozawa, M., Kemier, R., and Nelson, W. J.,** Novel function of the cell adhesion molecule uvomorulin as an inducer of cell surface polarity, *Cell,* 62, 309, 1990.
114. **Sell, S.,** Comparison of oval cells induced in rat liver by feeding N-2-fluorenylacetamide in a choline-devoid diet and bile duct cells induced by feeding 4,4′-diaminodiphenylmethane, *Cancer Res.,* 43, 1761, 1983.
115. **Reid, L. M.,** Stem cell biology, hormone/matrix synergies and liver differentiation, *Curr. Opinion Cell Biol.,* 2, 121, 1990.
116. **Gall, J. A. M. and Bhathal, P. S.,** Development of intrahepatic bile ducts in rat foetal liver explants in vitro, *J. Exp. Pathol.,* 71, 41, 1990.
117. **Novikoff, P. M., Hixson, D. C., and Yam, A.,** Liver cell interrelations in early stages of rat hepatocellular carcinogenesis, *J. Cell. Biol.,* 109 (part 2), 183a, 1989.
118. **Novikoff, P. M., Ikeda, T., Hixson, D. C., and Yam, A.,** Characterizations of and interactions between bile ductule cells and hepatocytes in early stages of rat hepatocarcinogenesis induced by ethionine, *Am. J. Pathol.,* 139, 1351, 1991.
119. **Nakamura, T., Nagao, M., and Ichihara, A.,** In vitro induction of neonatal rat hepatocytes by direct contact with adult rat hepatocytes, *Exp. Cell. Res.,* 169, 1, 1987.
120. **McMahon, J., Klinke, G., Karnasuta, C., Sell, S., del Campo, A., and Dunsford, H. A.,** Cellular lineage(s) of liver cancer: a clonal analysis, *Proc. Am. Assoc. Cancer Res.,* 28, 122, 1987.
121. **Tsao, M. S., Grisham, J. W., Nelson, K. G., and Smith, J. D.,** Phenotypic and karyotypic changes induced in cultured rat hepatic epithelial cells that express the "oval" cell phenotype by exposure to N-methyl-N′-nitro-N-nitrosoguanidine, *Am. J. Pathol.,* 118, 306, 1985.
122. **Tsao, M. S. and Grisham, J. W.,** Phenotypic modulation during tumorigenesis by clones of transformed rat liver epithelial cells, *Cancer Res.,* 47, 1282, 1987.
123. **Van Eyken, P., Sciot, R., Paterson, A., Callea, F., Kew, M. C., and Desmet, V. J.,** Cytokeratin expression in hepatocellular carcinoma: an immunohistochemical study, *Hum. Pathol.,* 19, 562, 1988.
124. **Rao, M. S. and Reddy, J. K.,** Peroxisome proliferation and hepatocarcinogenesis, *Carcinogenesis,* 8, 631, 1987.
125. **Rao, M. S., Tatematsu, M., Subbarao, V., Ito, N., and Reddy, J. K.,** Analysis of peroxisome proliferator-induced preneoplastic and neoplastic lesions of rat liver for placental form of glutathione-S-transferase and γ-glutamyltranspeptidase, *Cancer Res.,* 46, 5287, 1986.
126. **Dunsford, H. A., Maset, R., Salman, J., and Sell, S.,** Connection of ductlike structures induced by a chemical hepatocarcinogen to portal bile ducts in the rat liver detected by injection of bile ducts with a pigmented barium gelatin medium, *Am. J. Pathol.,* 118, 218, 1985.
127. **Sell, S.,** Is there a liver stem cell?, *Cancer Res.,* 50, 3811, 1990.
128. **Thompson, N. L., Hixson, D. C., Callanan, H., Panzica, M., Flannagan, D., Faris, R. A., Hong, W., Hartel-Schenk, S., and Doyle, D.,** A fischer rat substrain deficient in dipeptidyl peptidase IV activity makes normal steady state RNA levels and an altered protein, *Biochem. J.,* 273, 497, 1990.
129. **Hixson, D. C. and McEntire, K. D.,** Detection using monoclonal antibodies of a structurally altered form of cell CAM 105 on rat hepatocellular carcinomas, in *Molecular Determinants of Animal Form,* Edelman, G. M., Ed., Alan R. Liss, New York, 1985, 253.
130. **Holmes, C. H., Austin, E. B., Fisk, A., Gunn, B., and Baldwin, R. W.,** Monoclonal antibodies reacting with normal rat liver cells as probes in hepatocarcinogenesis, *Cancer Res.,* 44, 1611, 1984.
131. **Faris, R. A.,** unpublished data.
132. **Faris, R. A., McEntire, K. D., Thompson, N. L., and Hixson, D. C.,** Identification and characterization of a rat hepatic oncofetal membrane glycoprotein, *Cancer Res.,* 40, 475, 1990.

Chapter 9

DIFFERENTIATION POTENTIAL OF HYPERPLASTIC BILE DUCTULAR EPITHELIAL CELLS IN RAT MODELS OF HEPATIC INJURY AND CHOLANGIOCARCINOGENESIS*

A. E. Sirica, L. W. Elmore, T. W. Williams, and S. L. Cole

TABLE OF CONTENTS

* This work was supported by USPHS Grant 2RO1 CA 39225 to A.E.S. from the National Cancer Institute. Its contents are solely the responsibility of the authors and do not necessarily represent the official views of the National Cancer Institute.

ISBN 0-8493-4746-7
© 1992 by CRC Press, Inc.

I. INTRODUCTION

Bile ductules are short, fine, epithelial-lined channels which connect the bile canalicular network of the hepatocytes with the interlobular bile ducts within the portal tracts of liver. The transition from the bile canaliculi into bile ductules occurs at the edge of the portal tracts at the level of the limiting plate of hepatocytes. Here, connection of the bile canaliculi and the terminal end of the bile ductules forms what has been described as the "canalicular-ductular junction".[1] These minute connecting structures, commonly known as the canals of Hering or cholangioles, are formed in part by hepatocytes from the limiting plate and in part by simple flattened or low cuboidal biliary epithelial cells.[2,3] The canals of Hering then communicate with the bile ductules located at the periphery of the portal areas. These slightly larger channels are lined solely by simple, low cuboidal biliary epithelial cells, but are not accompanied by branches of the hepatic artery.[2,4] In rat liver, it has also been demonstrated by scanning electron microscopy of biliary tract casts that the periportal bile ductules are organized in the form of a richly anastomosing biliary plexus.[2] The bile ductules join with the larger interlobular bile ducts in the portal tracts. The simple cuboidal epithelial cells which line the interlobular bile ducts are somewhat taller than those of the bile ductules, and as the duct size enlarges so does the height of their lining epithelium. In contrast to the bile ductules, the interlobular bile ducts are accompanied by a branch of the hepatic artery, and these two structures, together with an associated branch of the portal vein, comprise the portal triad.[3,5] The largest branches of the intrahepatic biliary tree are lined by a simple columnar epithelium. Interestingly, the large hilar intrahepatic bile ducts of man have been demonstrated to have many irregular side branches and pouches,[6] and in addition, two types of peribiliary glands have been observed around the intrahepatic biliary ducts near the hilus of normal human liver.[7,8] On the other hand, the hilar biliary plexus of human liver differs from that of other species and the periportal bile ductular plexus of rat liver was not recognized in human biliary tract casts.[6]

Despite its morphological complexity, the intrahepatic biliary tree of normal adult mammalian liver is comprised of epithelial cells (bile ductal and ductular), representing only 3 to 5% of the total hepatic cell population[9,10] and about 0.14 to 0.43% of the liver volume.[11,12] However, intrahepatic biliary epithelial cells, particularly bile ductular cells, are commonly observed to increase in number in adult liver following a variety of clinical and experimental forms of hepatic injury.[13,14] Depending on the species and the nature of the clinical or experimental liver condition, bile ductular epithelial cells may become greatly increased in number by various means, including "typical" bile ductular hyperplasia,[14,15] "atypical" bile ductular hyperplasia,[16] or "oval cell" proliferation.[17-20] An experimental prototype of "typical" bile ductular hyperplasia is the type of proliferative biliary cell reaction which occurs in rat liver following ligation of the common bile duct.[21-23] Clinically in humans, this type of bile ductular hyperplasia predominates in conditions such as extrahepatic biliary obstruction.[14-16] "Typical" bile ductular hyperplasia such as that occurring in the liver of bile duct-ligated rats is characterized by a well-organized proliferation of preexisting biliary epithelial and stromal cells with the resultant formation of well-differentiated hyperplastic bile ductules and small ducts.[15] In contrast, "atypical" bile ductular hyperplasia has been defined as an increase in intrahepatic ductular structures caused by the "metaplastic" or "ductular transformation" of periportal hepatocytes.[16,24] "Atypical" bile ductular hyperplasia has mostly been investigated in human liver under conditions of long-standing incomplete obstructions such as occurring in primary biliary cirrhosis, primary sclerosing cholangitis, focal nodular hyperplasia, and in advanced alcoholic liver disease.[16,24-26] "Oval cell" proliferation, on the other hand, has been mostly studied in rat liver during the early periods of chemical hepatocarcinogenesis[15,17-20,27-29] and in certain forms of chemically induced severe liver injury, like that caused by D-galactosamine.[30] Proliferating oval cells represent a heterogeneous nonparenchymal liver cell population comprised of subpopulations of cells with different developmental capacities, but also possessing characteristics of bile ductular cells.[15,29,31] Although still not fully resolved, it is nevertheless generally held that proliferating oval cells are likely derived from the canals of Hering.[15,32] This view is strengthened by the observations that proliferating oval cells originate at the edge of the portal tracts and can organize into atypical ductularlike structures that are connected to the preexisting intrahepatic biliary tree.[15,32]

"Oval cell" proliferation during hepatocarcinogenesis or in severe hepatic injury is not an exclusive response of rat liver, but has also been reported to occur in the diseased livers of other animal species and possibly in human liver.[33-35] Moreover, in rats, chronic intoxication with either α-naphthyl isothiocyanate or 4,4′-diaminodiphenylmethane has been found to induce a bile ductular reaction that appears to be more closely related to the "typical" bile ductular hyperplasia associated with bile duct ligation than to hepatocarcinogen- or D-galactosamine-induced oval cell proliferation.[22,36-40] However, it has also been proposed that the increase in bile ductules induced by α-naphthyl isothiocyanate in guinea pigs occurred mainly by the tubularization of hepatic laminae and the modulation of hepatocytes towards intrahepatic biliary cells.[41] Thus, diverse pathways exist in various species for increasing the intrahepatic bile ductular cell population in response to carcinogenic and noncarcinogenic insults.

As detailed in other chapters in this book (i.e., Chapters 5 and 6, this volume), there is now compelling evidence to support the existence of a "facultative stem cell" in adult rat liver,[15,28,29,42] which becomes expanded during "oval cell" proliferation and that may serve as an alternative cell of origin for hepatocellular carcinoma and other hepatic neoplasms. In this chapter, our focus will be more on the "typical" hyperplastic bile ductular epithelial cell, particularly in terms of its relevant phenotypic features and growth properties following bile duct ligation in rats. In addition, a significant portion of this chapter will be devoted to demonstrating the ability of hyperplastic bile ductular cells to preferentially differentiate along the small-intestinal lineage during the early stages of furan-induced cholangiocarcinogenesis in rat liver. Finally, we will discuss the appearance of "ductular hepatocytes" observed in association with proliferated cholangiolar structures in certain cases of severe liver injury, including that resulting from bile duct ligation and subsequent CCl_4 intoxication.

II. CHARACTERISTICS OF HYPERPLASTIC BILIARY EPITHELIUM IN RAT LIVER AFTER BILE DUCT LIGATION

A. SELECT PHENOTYPIC FEATURES

Figure 1 presents an immunohistochemical profile of select phenotypic features which specifically distinguish the hyperplastic bile ductular epithelial cells from hepatocytes as well as from sinusoidal and perisinusoidal cell types and accompanying fibrotic stromal cell types in cholestatic rat liver at several weeks after bile duct ligation. As exemplified here and reviewed by us elsewhere,[15,43] the hyperplastic bile ductular epithelial cells present in adult rat liver at 2 to 8 weeks after bile duct ligation uniformly retain a number of the characteristic phenotypic features of nonproliferating intrahepatic biliary epithelial cells of normal adult rat, including strongly positive immunohistochemical staining for the enzymes γ-glutamyl transpeptidase (GGT; Figure 1a) and glutathione *S*-transferase-placental or 7-7 (GST-P; Figure 1b) and for the intermediate filament proteins, cytokeratin 19 (CK19; Figure 1c) and CK7.[44,45] In addition, like normal rat intrahepatic biliary epithelial cells,[14,46] the hyperplastic bile ductular epithelial cells exhibit an intensely positive cytoplasmic staining when reacted immunohistochemically with antibody against S-100 protein (Figure 1d). However, in contrast to the preexisting intrahepatic biliary epithelial cells of normal adult rat liver, the epithelial cells of newly formed hyperplastic bile ductules in the liver of bile duct-ligated rats were demonstrated to become strongly positive for vimentin and leucine aminopeptidase expression.[15,44,47,48] Also, the hyperplastic bile ductular epithelial cells appearing in adult rat liver after bile duct ligation differ from various subpopulations of the proliferating oval cells that arise in rat liver during the early stage of hepatocarcinogenesis by their lack of expression of albumin,[15,22] glucose-6-phosphatase,[15,22] and, perhaps with the exception of an occasional cell, α-fetoprotein.[15,22,49-51] Moreover, the hyperplastic bile ductules induced in rat liver following bile duct ligation are characterized by a well-developed basement membrane[15] exhibiting a strong immunohistochemical staining for laminin[51] (Figure 1e). With respect to this latter phen-

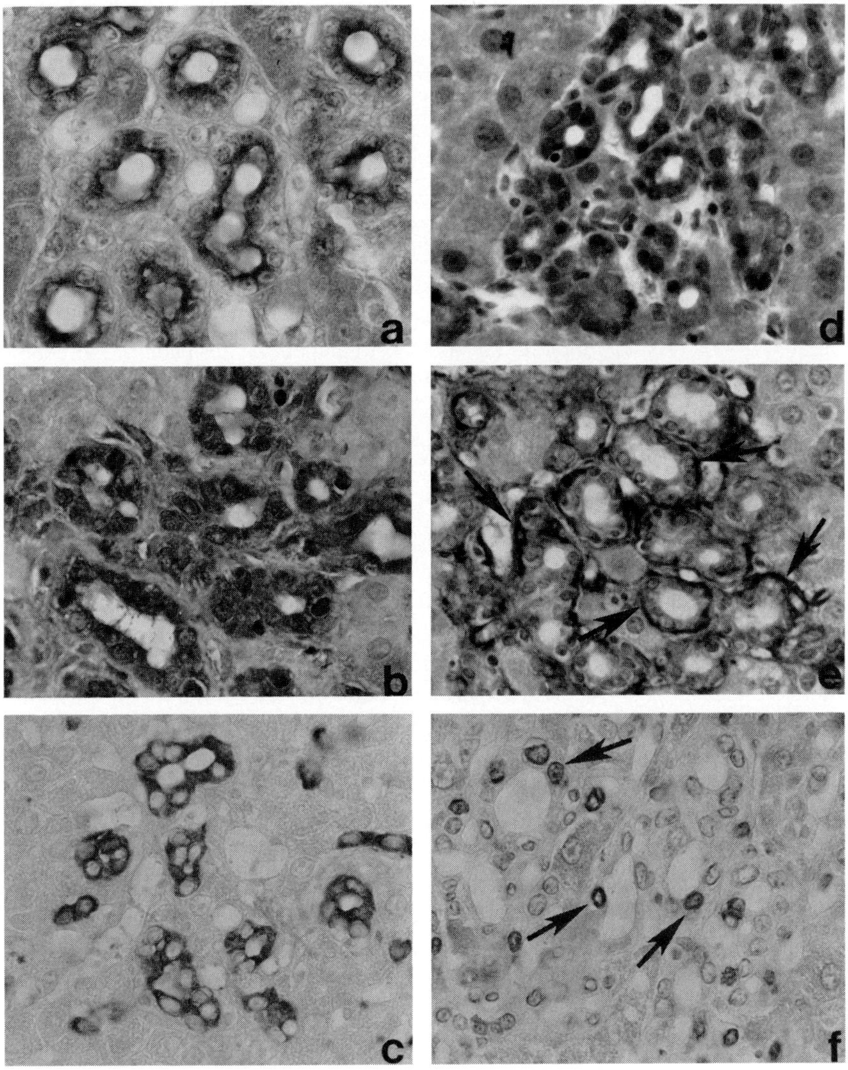

FIGURE 1. Light photomicrographs demonstrating immunohistochemical staining of select phenotypic traits of the epithelial cells of hyperplastic bile ductules in rat liver at several weeks after bile duct ligation. a, Luminal cell surface γ-glutamyl transpeptidase (original magnification × 161); b, cytoplasmic glutathione *S*-transferase-P (original magnification × 161); c, cytoplasmic CK19 (original magnification × 128); d, cytoplasmic immunoreactivity towards anti-S-100 protein (original magnification × 128); e, basement membrane laminin (arrows; original magnification × 128); f, BrdU-labeled dark stained nuclei (arrows; original magnification × 161). a, b, d, e, and f counterstained with hematoxylin, which accounts for the obvious nuclear staining in these respective preparations.

otypic trait, the hyperplastic bile ductules of the bile duct-ligated rat again more closely resemble normal rat intrahepatic bile ducts than they do the atypical ductular structures associated with hepatocarcinogen-induced oval cell proliferation, which were shown to be only faintly delineated by immunohistochemically detectable laminin.[38] "Transitional cells", also characteristically observed in adult rat liver in association with hepatocarcinogen-induced oval cell proliferation and having morphologic features intermediate between those of intrahepatic biliary epithelial cells and hepatocytes, have not been detected in adult rat liver at various times after bile duct ligation.[15,48,51] Furthermore, it is noteworthy that when an essentially pure population of hyperplastic bile ductular epithelial cells isolated from bile duct-ligated donor rats was inoculated into the interscapular fat pads of syngeneic recipient rats, only well-differentiated bile ductular structures supported by a dense connective tissue stroma and without evidence of "transitional cell" differentiation formed at the transplantation site.[51]

B. GROWTH KINETICS *IN VIVO*

Recently, Arber and Zajicek[52] have postulated that the intrahepatic biliary epithelium of normal rat liver is continuously renewed via a "stem cell" located in the canal of Hering, which differentiates along the biliary epithelial cell lineage as it "streams" towards the interlobular bile ducts. Zajicek[53] has further speculated that bile duct ligation in rats may also trigger a common stem cell that can give rise to both hepatocyte and biliary cell progeny. However, carefully conducted kinetic analyses of intrahepatic biliary epithelial cell proliferation in bile duct-ligated rats have provided convincing data to conclude that the origin of the epithelial cells of the hyperplastic bile ductules in adult rat liver after bile duct ligation is essentially from the cell division of preexisting biliary epithelial cells.[13,23,54,55] Nuclear labeling has readily been demonstrated by histoautoradiography following incorporation of ^{3}H-thymidine[13,23,54,55] or by immunohistochemistry after incorporation of 5-bromo-2-deoxyuridine (Figure 1f) into the replicating DNA of the expanding population of biliary ductular epithelial cells forming in rat liver subsequent to bile duct ligation. Moreover, Slott et al.[23] recently reported that after bile duct ligation in rats, replicative DNA synthesis as indicated by ^{3}H-thymidine-labeled nuclei occurred in ductal and ductular structures of all sizes and not just the smallest-sized ductules, leading these authors to conclude that most if not all of the biliary epithelial cells in this model are capable of undergoing cell division. Time-related changes in the ^{3}H-thymidine labeling index was also found to closely parallel increases in the hyperplastic biliary epithelial cell area determined morphometrically in histological sections prepared from adult rat liver at different times during the first week after bile duct ligation.[23] We have further demonstrated, using a histomorphometric procedure, a linear increase in the amounts of total liver tissue section area occupied by newly formed GGT-positive hyperplastic bile ductular tissue during the first 4 weeks after bile duct ligation in young adults rats.[56] Between 5 and 8 weeks after bile duct ligation, a plateau was reached, with approximately 50% of the total area of the liver tissue sections now being occupied by hyperplastic bile ductular tissue. However, at between 10 and 15 weeks after ligation, we noted an apparent decline in the hyperplastic bile ductular tissue area/total liver tissue section area.[56] These findings are consistent with ^{3}H-thymidine labeling indices for proliferating biliary epithelial cells in rat liver at various early and late time points after bile duct ligation.[23,54-56] The highest ^{3}H-thymidine labeling indices have been obtained during the first week of bile duct ligation,[23,54,55] followed by markedly lower labeling indices at between 3 and 12 weeks.[54-56] At more advanced time points after bile duct ligation in rats, there is a noticeable apoptotic cell death occurring within the formed hyperplastic bile ductules[55,57,58] with a concomitant consolidation of their connective tissue stroma.[58] Thus, under the condition of sustained bile duct ligation in rats for periods of up to 12 weeks, it seems quite evident that the increase in biliary epithelial cells in liver is due, for all practical purposes, to the replication of cells of like kind and is not the result of either a liver stem cell that becomes activated along the biliary cell lineage or to a ductular metaplasia of hepatocytes.[12,13,23,51,54,55]

C. THE RESISTANCE PHENOTYPE AND BILE DUCTULAR CELL HYPERPLASIA

At present, there have been only a limited number of studies aimed at defining specific mechanisms and growth factors which may be involved in initiating and regulating the proliferation of intrahepatic biliary epithelial cells during conditions of extrahepatic biliary obstruction. While it is beyond the scope of this chapter to provide a comprehensive treatment of this important area of research (for recent reviews, see References 15 and 43), it is becoming apparent that the intrahepatic biliary epithelium of normal and bile duct-ligated rats exhibit a unique pattern of phase I and phase II metabolizing activities that would favor their resistance to the cytotoxic and mitoinhibitory effects of various hepatotoxic and carcinogenic chemicals. Specifically, normal rat intrahepatic biliary epithelial cells have been found to lack detectable levels of a number of key phase I activities or components, including total spectral cytochrome P-450,[59] cytochrome P-450 B,[60] NADPH cytochrome P-450 reductase,[61] epoxide hydrolase,[62] benzo[a]pyrene hydroxylase,[15] 7-ethoxycoumarin *O*-deethylase,[59,63] and 7-ethoxyresorufin *O*-deethylase.[63] On the other hand, these cells have been demonstrated to possess detectable amounts of various phase II conjugating enzyme activities or isoforms, such as those of GST (isozymes A [3-3] or C [3-4], E [5-5], and P [7-7],[15,64] UDP-glucuronosyltransferase isozymes (phenol UDP-glucuronosyltransferase and 17β-hydroxysteroid UDP-glucuronosyltransferase),[63,65,66] and sulfotransferase.[15,63] Similarly, hyperplastic bile ductular epithelial cells isolated from rat liver to 6 to 15 weeks after bile duct ligation have also been found to lack detectable phase I components (i.e., total spectral cytochrome P-450, cytochrome P-450 P, and amidoypyrene-*N*-demethylase activity),[51,56,67] but to express a number of phase II activities or forms, including the A (or C) and P isozymic forms of GST,[51,56] *p*-nitrophenol UDP-glucuronosyltransferase,[51,56] and a very low sulfotransferase activity.[68] Interestingly, the profile of phase I and II metabolic activities exhibited by both the intrahepatic biliary epithelium of normal rat liver and the hyperplastic bile ductal/ductular epithelial cells of cholestatic rat liver after bile duct ligation resembles the hepatotoxin/hepatocarcinogen resistance phenotype of putative preneoplastic hepatocyte populations induced in rat liver by various hepatocarcinogenic regimens.[56,69-72] Such a pattern would be expected to allow for detoxification reactions by phase II conjugating enzymes while greatly limiting phase I-dependent activations of hepatotoxic and/or hepatocarcinogenic chemicals to their ultimate reactive metabolites. In this context, it is noteworthy that a bile ductular epithelial cell population isolated from bile duct-ligated rats was found to be resistant in aflatoxin B1 cytotoxicity *in vitro*.[73] Thus, as already pointed out in Chapter 1 of this volume, intrahepatic biliary epithelial cells, including those of the bile ductules and canals of Hering (''oval cells''), by virtue of their apparent resistance to various hepatotoxic and hepatocarcinogenic environments, are equipped not only to survive, but may also be permitted to selectively proliferate under conditions that would be mitoinhibitory to normal hepatocytes.[74]

III. PLASTICITY OF HYPERPLASTIC BILE DUCTULAR EPITHELIAL CELLS IN TWO NOVEL RAT MODELS OF SEVERE HEPATIC INJURY

As previously indicated in this chapter, there are at least two fundamental differences which distinguish the typical hyperplastic bile ductular reaction which occurs in rat liver after ligation of the common bile duct from intrahepatic ''oval cell'' proliferation that is induced in rats by various hepatocarcinogenic regimens or following D-galactosamine hepatotoxicity. The first of these relates to the fact that the proliferated ductular structures forming after bile duct ligation more closely resemble in their morphology and differentiated phenotype the preexisting intrahepatic bile ducts/ductules of normal adult rat liver than those developing in association with oval cell proliferation. The second, and perhaps more revealing of these differences, is the fact that the hyperplastic bile ductular cell population that emerges in the liver of the bile duct-ligated rat appears to be capable under these conditions of differentiating only along the intrahepatic biliary cell lineage. In contrast, cells within the oval cell compartment that expands in rat liver under select hepatocarcinogenic/hepatotoxic conditions may potentially serve as nonparenchymal precursors for a number of endoderm-derived cell types, including hepatocytes, biliary epithelial cells, pancreatic acinar cells, and intestinal mucosal cells.[75-81] In order to resolve the apparent differences in differentiation potential between oval cells and those which comprise more typical hyperplastic bile ductules, it is important to determine if injurious conditions to liver can be established which would permit the latter cells to also differentiate along other cell lineages. In this context, we have recently described two novel rat models of severe hepatic injury which strongly suggest that under the appropriate conditions at least some of the cells forming more typical hyperplastic bile ductular structures can selectively differentiate into either small-intestinal mucosal cell types or ''ductular hepatocytes''.[82-84] Both of these models were developed in young adult Fisher 344 males rats and their specific features are described below.

A. HYPERPLASTIC BILE DUCTULAR EPITHELIAL CELLS AS PUTATIVE PRECURSORS OF METAPLASTIC INTESTINAL CELLS IN THE RIGHT AND CAUDATE LOBES OF RAT LIVER DURING THE EARLY STAGES OF FURAN CHOLANGIOCARCINOGENESIS

In a 2-year carcinogenicity bioassay conducted by the National Toxicology Program, furan was found to induce a very high incidence of cholangiocarcinomas in the liver of both male and female Fisher 344 rats.[85] Based on this initial observation, we recently investigated the early cellular changes associated with administering furan by gavage to young adult Fisher 344 male rats at a determined effective dose of 45 mg/kg once daily, 5 d a week for period of up to 4 weeks.[82] This treatment resulted in the rapid development of a striking lobe-specific cholangiofibrosis that was localized essentially to the right and caudate liver lobes[82,83] (Figures 2 and 3). As shown in Figure 2b and c, this lesion is typified by hyperplastic bile ductular structures in close association with metaplastic intestinal-like glands supported by a dense fibrotic stroma. In contrast, histological examination of the median and left lateral liver lobes of these chronic short-term-treated rats generally revealed pathological features that ranged from mild fibrosis (Figure 2a) to a prominent cirrhosis together with sporadic areas of hepatocellular degeneration and necrosis.[82]

Figure 3 presents quantitative data which clearly exemplify the dramatic liver lobe-specific distribution of the furan-induced cholangiofibrosis. As further demonstrated in Figure 3C, the mean numbers of total metaplastic intestinal glands per square centimeter histological liver section area were determined after approximately 4 weeks of furan treatment to be 677 for the right liver lobe, 112 for the caudate liver lobe, 1.9 for the median liver lobe, and 0.51 for the left lateral liver lobe. These numbers correlated directly with actual mean percents of cholangiofibrotic tissue occupying the total tissue section areas of random histological samplings from each of the respective liver lobes (compare Figures 3B and C). In addition, by employing specific

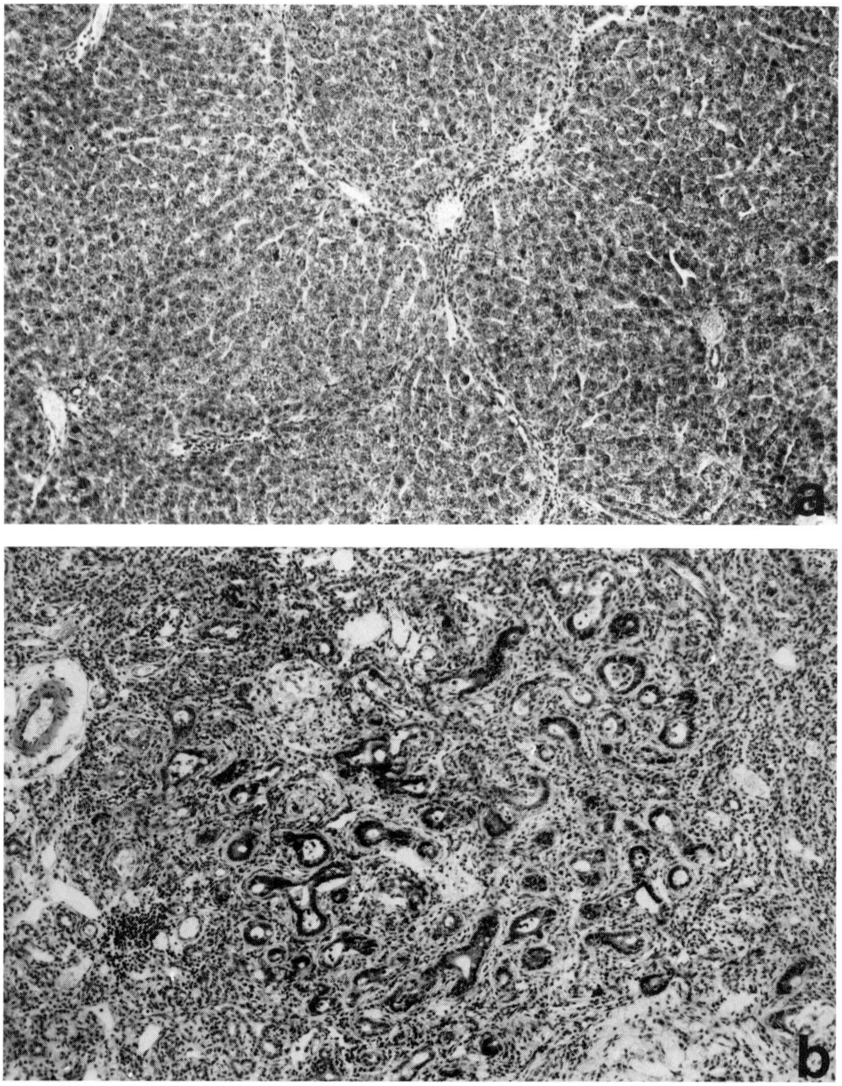

FIGURE 2. Lobe-specific cholangiofibrosis induced in the liver of rats after 3 to 4 weeks of exposure to furan at 45 mg/kg body weight. a, Light photomicrograph of median liver lobe exhibiting only mild fibrosis (H.E.; magnification × 13); b, extensive area of cholangiofibrosis in the right liver lobe (H.E.; magnification × 26); c, higher-power light photomicrograph of cholangiofibrotic tissue formed in the caudate liver lobe, demonstrating hyperplastic bile ductules (left half of photomicrograph) and metaplastic intestinal-like glands (right half of photomicrograph) supported by a dense fibrotic stroma (H.E.; magnification × 128). (The photomicrograph in panel c is reproduced with permission from Elmore and Sirica, *Cancer Res.*, 51, 5752, 1991.)

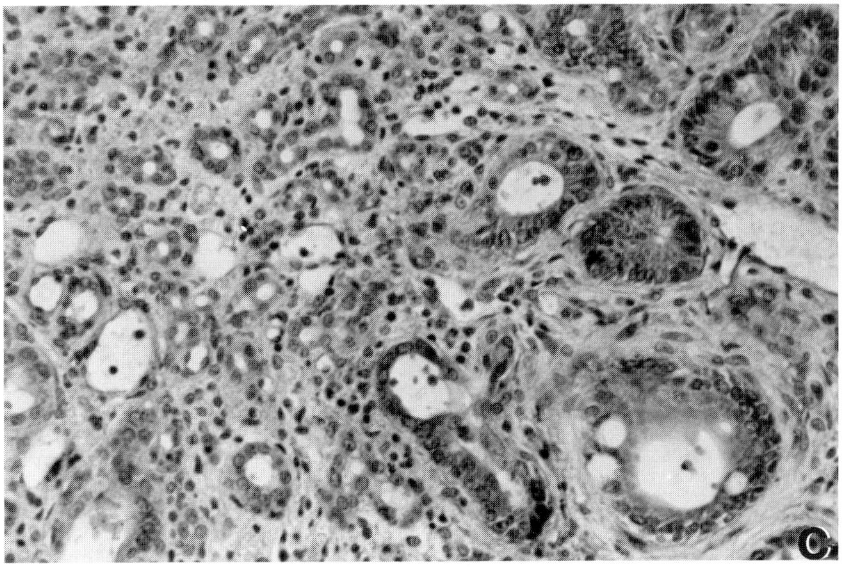

FIGURE 2. *Continued.*

histochemical and immunohistochemical markers and electron microscopy, we have identified four intestinal cell types within these metaplastic glands: tall columnar epithelial cells with well-developed striated borders (absorptive cells), mucin-positive goblet cells, Paneth cells, and serotonin-positive neuroendocrine cells[82,83] (Figure 3D and Figure 4). An analysis of approximately 300 metaplastic intestinal glands (~5000 cells) formed within the cholangiofibrotic tissue areas of the right and caudate liver lobes of the furan-treated rats at the end of the 4-week treatment period further indicated the various cell types comprising these glands to be present in the following percents: 84.3 ± 4.3% absorptive cells, 15.6 ± 4.6% goblet cells, <1.0% Paneth cells, and 0.2 to 0.4% serotoin-positive neuroendocrine cells[83] (Figure 3D). Thus, in terms of both cellular composition and percentage values for the different cell types, the metaplastic intestinal glands appearing in the liver of the furan-treated rats was found by us to closely resemble the crypts of Lieberkühn of normal rat small intestine.[86]

Table 1 demonstrates that both the hyperplastic bile ductular structures and metaplastic small-intestinal-like structures within the furan-induced areas of cholangiofibrosis exhibit a number of similar phenotypic characteristics. Also, in a recently completed time-course study,[86] we have determined in histological sections of right liver lobe obtained at specific time points from Fisher 344 male rats that received furan by gavage at 45 mg/kg body weight once a day, 5 d a week over a 32-d treatment period, the following sequence

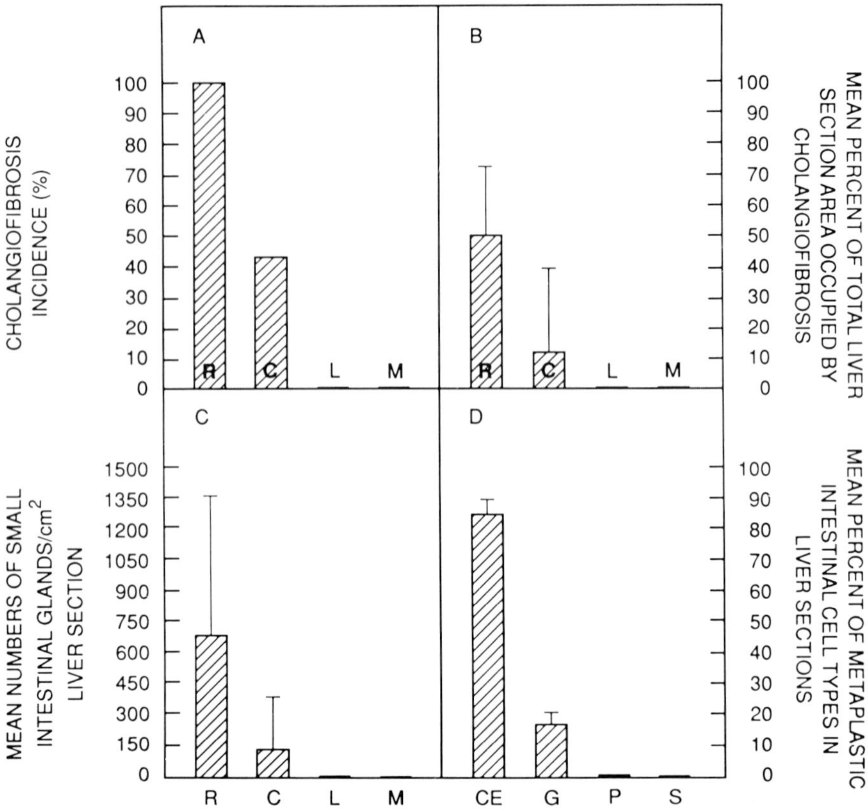

FIGURE 3. Quantitation of cholangiofibrosis with associated intestinal metaplasia in the different lobes of liver of furan-treated rats. Analyses were performed on liver lobes from a group of seven rats that had received furan by gavage at a daily dose of 45 mg/kg body weight, five times per week for a period of four weeks. R, right lobe; C, caudate lobe; L, left lateral lobe; M, median lobe; CE, metaplastic columnar epithelial cell; G, goblet cell; P, Paneth cell; S, serotonin-positive cell.

of changes: at day 1, a prominent zonal hepatonecrosis localized to zone 3, but also extending into zone 2 of the liver acini; at days 3 to 5, the presence of a diffuse inflammatory cell infiltrate within the hepatonecrotic areas, which gradually resolved; at day 7, the distinct appearance of a more typical hyperplastic bile ductular cell reaction; at day 9, a replacement of much of the injured liver parenchyma by hyperplastic bile ductular tissue that also contained occasional metaplastic glands comprised of tall columnar basophilic epithelial cells; at days 12 to 16, an increasing cellular diversity within developing metaplastic glands, with the additional sequential development of goblet cells, Paneth cells, and serotonin-positive neuroendocrine cells; and at day 32, a typical cholangiofibrosis as already described above. Taken together,

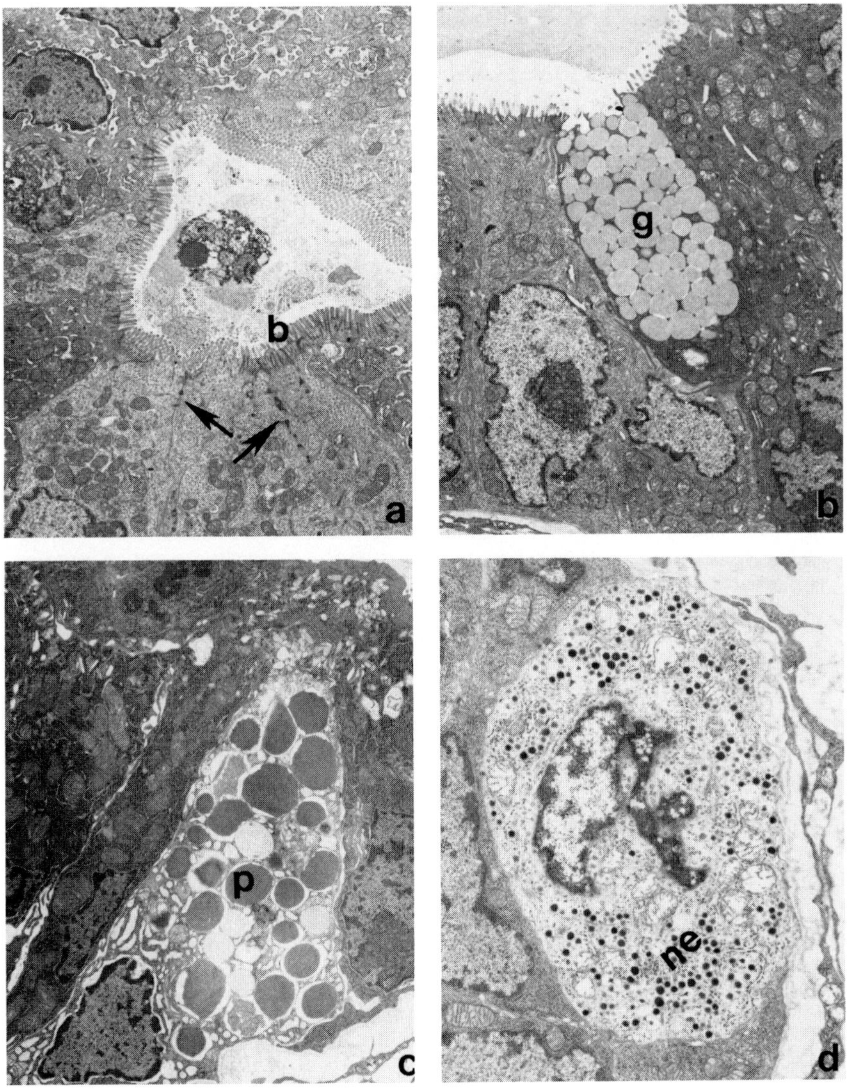

FIGURE 4. Electron photomicrographs of representative cell types comprising furan-induced metaplastic intestinal glands in rat liver. a, Metaplastic columnar epithelial cells exhibiting a well-developed striated-border (b) and juxtaluminal junctional complexes (arrows; original magnification × 5600); b, goblet cell (g; original magnification × 5600); c, Paneth cell (p; original magnification × 7000); d, neuroendocrine cell (ne, original magnification × 9200). (The micrographs in panels a, b, and d are reproduced with permission from Elmore and Sirica, *Cancer Res.*, 51, 5752, 1991.)

TABLE 1
Comparison of the Phenotypes Characterizing Hyperplastic Bile Ductules and Metaplastic Intestinal Glands in Areas of Furan-Induced Cholangiofibrosis in Rats[a]

Phenotypic characteristic	Hyperplastic bile ductules	Metaplastic intestinal glands
Cytokeratin 19	+[b]	+
Basement membrane laminin	+	+
Basement membrane type IV collagen	+	+
BD.2 antigen[c]	+	+
α-Fetoprotein	−	−
γ-Glutamyl transpeptidase	+	±
Glutathione *S*-transferase-P	+	±
"S-100 protein"	+	±
Mucin	−	+[d]
Serotonin	−	+[d]

[a] Summarized from data mostly presented in References 82 and 86.
[b] − Not detected; + moderate to strongly positive immunohistochemical staining; ± heterogeneous staining reaction.
[c] Positive immunohistochemical staining demonstrated with monoclonal antibody 18.11 described in Chapter 8 (Table 1) of this book.
[d] Positive immunohistochemical staining exhibited by select cells only.

the phenotypic data[82] (Table 1) and time-course data[86] strongly suggest a precursor relationship between the hyperplastic bile ductular structures and the metaplastic small-intestinal glands forming in specific lobes of the furan-treated rats, and further indicate that small-intestinal metaplasia of hyperplastic bile ductular epithelial cells is an important intermediate stage in the pathogenesis of cholangiofibrosis in this experimental model.

Intestinal metaplasia has long been observed to occur in association with oval cell proliferation induced early in rat liver as a result of exposures to hepatocarcinogens like 3′-methyl-4-dimethylaminoazobenzene (Me-DAB).[78] More recently, Tatematsu et al.[79] have demonstrated intestinal metaplasia as being a common option of proliferating oval cells in relation to the genesis of cholangiofibrosis in the liver of rats exposed to the hepatocarcinogen 2-acetylaminofluorene (2-AAF) for 2 weeks, with a partial hepatectomy performed midway at 1 week. However, unlike the classical type of oval cell proliferation appearing in rat liver in response to short-term exposures to Me-DAB or 2-AAF, none of the hyperplastic bile ductular epithelial cells associated with the small-intestinal metaplasia in the liver of the furan-treated rats were found to produce immunohistochemically detectable α-fetoprotein[82] (Table 1) or to express α-fetoprotein mRNA.[82] Also, in contrast to the atypical ductular structures forming in association with hepatocarcinogen-induced oval cell proliferation, the hyperplastic bile ductular structures and the metaplastic intestinal glands appearing in the liver of the furan-treated rats were both demonstrated to be characterized by well-developed basement membranes enriched in laminin[82] (Table 1). As further supported by the data in Table 1, it is reasonable to suggest that the hyperplastic bile ductular structures which precede the development of small-intestinal metaplasia in select liver lobes of rats in response to short-term exposures to furan are more related in their phenotype to the proliferated bile ductules and small ducts which are formed in rat liver following bile duct ligation than they are to the atypical proliferated ductular structures which appear in the liver of hepatocarcinogen- or D-galactosamine-treated rats.

It is not yet certain as to whether the early liver lobe-specific small-intestinal metaplasia and associated cholangiofibrosis observed by us in the furan-treated rats simply reflects a reactive change to the furan administration or if this early change is mechanistically related to the subsequent high incidence of intrahepatic cholangiocarcinomas reported in Fisher 344 rats after their long-term exposure to furan.[85] We are also now conducting a long-term carcinogenicity bioassay in Fisher 344 male rats and have very preliminary data that confirm in part the apparent progressive nature of the cholangiofibrosis. In addition, we have very recently identified a cholangioma in the liver of a single rat that was sacrificed at an intermittent time point in this study, which was characterized by the presence of small-intestinal metaplasia. Some intrahepatic cholangiocarcinomas induced by the long-term feeding of Me-DAB to rats have also been previously shown to exhibit intestinal mucosal features.[87,88] However, in the human, intestinal metaplasia occurs in neoplasms of both the intrahepatic and extrahepatic biliary tract[89-91] and typically in the gallbladder under conditions of increased risk for biliary cancer, such as chronic cholecystitis and cholelithiasis.[92,93] Although the role of intestinal metaplasia in the pathogenesis of intra- and extrahepatic cholangiocarcinomas is still debatable, this phenomenon clearly reflects the cell plasticity of the entire biliary tract and not just that of the cells comprising the canals of Hering and bile ductules. However, it is still interesting that at least some of the tumors produced by transplantation of rat liver epithelial cells with oval cell properties which had been chemically transformed *in vitro* were adenocarcinomas showing intestinal cell differentiation.[94]

It has recently been suggested that in the human, both tubular adenocarcinoma and undifferentiated carcinoma of the gallbladder may likely be arising from the neoplastic transformation of metaplastic intestinal epithelium, whereas the papillary carcinomas appear to be originating from the native gallbladder epithelium.[93] Moreover, it was further indicated in this study that gallbladder carcinomas with intestinal metaplasia resulted in poorer prognoses than those of the pure papillary type.[93] In addition, it has recently been reported that predominant neuroendocrine differentiation in carcinomas of the human extrahepatic biliary tract also appears to be associated with a worse prognosis than predominant exocrine-type biliary tract cancers.[95] Thus, it will be interesting to determine if a relationship exists between the biological behavior of intrahepatic cholangiocarcinomas induced in rats by the long-term chronic administration of furan and the frequencies of different metaplastic intestinal mucosal cell types that may be encountered in these tumors as they develop and progress.

Intestinal metaplasia in the human intrahepatic biliary tract occurs in hepatolithiasis and in liver fluke (*Clonorchis sinensis*; *Opisthorchis viverrini*) infection, conditions which are known to predispose to the subsequent development of cholangiocarcinoma in liver.[96-99] Neuroendocrine cells, which have been detected in sparse numbers in the intrahepatic biliary tree of normal adult and infant human livers, were also reported to be significantly increased in number in hepatolithiasis.[100] In addition, rare primary carcinoid tumors of the liver[101] and of the common bile duct[102,103] occur in man. Based on the experimental animal and human data described above, it is tempting to speculate that at least some of the neuroendocrine cells in hepatolithiasis as well as those transformed to give rise to primary ''carcinoid'' tumors of liver and extrahepatic bile duct may have arisen as a result of a neuroendocrine cell metaplasia of biliary cells located in different regions of the biliary tract. In this regard, the recently reported results of Roskams et al. (see Reference 104 and Chapter 11), demonstrating neuroendocrine features of reactive bile ductules in human cholestatic liver disease appear to be particularly relevant.

Before concluding this section, it should be emphasized that in the furan rat model of intrahepatic liver injury and cholangiocarcinogenesis, it is still not known why the initial small-intestinal metaplasia and associated cholangiofibrosis are preferentially induced in the right and caudate liver lobes. We have suggested that this liver lobe-specific effect could be related to ''streaming'' within the portal venous circulation to liver, thereby allowing for higher concentrations of the gavage-administered furan to be selectively routed to right and caudate liver lobes over the left lateral and median lobes.[82] On the other hand, we have not ruled out the possibility of interlobe differences in furan metabolism.[82] Also, it is still not clear if furan is acting through a mutagenic[105,106] or a nonmutagenic mechanism,[107,108] although it has recently been proposed that furan may be a ''cryptic mutagen''.[106] It is noteworthy, however, that the short-term treatment of the rats with furan yielded only very low numbers of putative preneoplastic hepatocellular focal lesions per cubic centimeter of liver in each of the respective liver lobes.[82] Furfural, a derivative of furan, has also been reported not to induce hepatocyte nodules or hepatocellular carcinomas in rats.[109] As described in the next section, it is nevertheless interesting that rats exposed to a severely hepatotoxic (lethal) furan treatment also developed in their liver, along with small-intestinal metaplasia, distinct cholangiolarlike structures composed of both bile ductular epithelial cells and ''ductular hepatocytes'' in apparently different stages of morphological maturation.[82]

B. HYPERPLASTIC BILE DUCTULAR EPITHELIAL CELLS AS PUTATIVE PRECURSORS OF "DUCTULAR HEPATOCYTES" IN RAT LIVER AFTER BILE DUCT LIGATION AND SUBSEQUENT CCl_4 INTOXICATION

One of the most unique examples of epithelial cell plasticity following cell injury is that involving the apparent metaplastic conversion or "transdifferentiation" of ductular cells into hepatocytes in the pancreas of both rats and hamsters under specific conditions of induced acinar cell loss and subsequent pancreatic regeneration.[110-112] Particularly pertinent is the eloquent study of Rao et al.[110] demonstrating that both proliferated ductular and interstitial cells appear to be capable of differentiating into hepatocytes in the pancreas of young adult male Fisher 344 rats during recovery from copper deficiency-induced depletion of greater than 90% of acinar cells. In an effort to demonstrate an analogous response phenomenon in liver, we developed a model of severe noncarcinogenic hepatic injury in young adult Fisher male rats, which combined prior bile ductular cell hyperplasia following bile duct ligation with a subsequent CCl_4-induced zone 3 hepatonecrosis.[84] The most notable consequence of this treatment was the distinct, albeit rare, development in liver of mixed hepatic cell cholangiolarlike structures[84] (Plate 3*). These were observed to be formed only in those areas of preexisting hyperplastic bile ductular tissue in histological liver sections from rats whose common bile duct was ligated for 4 weeks and who then survived for an additional 3 to 5 d after being administered a single hepatonecrogenic dose of CCl_4. In contrast, such canal of Hering-like structures were not detected in liver sections from control rats subjected to either bile duct ligation alone or to CCl_4 treatment without prior bile duct ligation.[84] The newly appearing mixed-hepatic cell, cholangiolarlike structures were found to be composed of cuboidal biliary epithelial cells together with one (Plate 3a) or more[84] hepatocytic cells whose cell and nuclear sizes were intermediate between those of the hyperplastic bile ductular epithelial cell and the nonductular adult rat hepatocyte[84] (i.e., Plate 3a). Phenotypically, the "ductular hepatocytes" of at least some of the mixed-hepatic cell, cholangiolarlike structures were further demonstrated to be distinguished from the biliary epithelial cells of the same structures by virtue of their negative cytoplasmic immunohistochemical staining reaction with monoclonal anti-CK19 antibody[84] and their strongly positive cytoplasmic immunohistochemical staining reacting with monoclonal antibody 368.7 against the hepatocyte-specific antigen H.4 (see Reference 84 and Chapter 8) (Plate 3c). Conversely, the biliary epithelial cells of these cholangiolarlike structures were strongly positive for CK19, but immunohistochemically negative for H.4.[84] On the other hand, both the bile ductular epithelial cells and the "ductular hepatocytes" of the newly formed cholangiolarlike structures in the liver of the bile duct-ligated/CCl_4-treated rats were observed to be supported by a common basement membrane that gave a strong

* Plate 3 appears following page 134. Plate 3A and 3C reprinted with permission from Sirica, A.E. and Williams, T.W., *Am. J. Pathol.*, 140, 129, 1992.

immunohistochemical staining for laminin (data not shown). The mixed-hepatic cell, cholangiolarlike structures, which appeared in association with hyperplastic bile ductules in cholangiofibrotic areas of rat liver only after a lethal hepatotoxic treatment with furan, were also found to be completely surrounded by a basement membrane exhibiting a strong immunohistochemical staining for laminin and type IV collagen[82] (Plate 3b and 3d). In addition, the biliary epithelial cells and "ductular hepatocytes" of these novel structures demonstrated a strongly positive immunohistochemical staining for a common luminal cell-surface GGT, but in agreement to what was observed in the bile duct ligation/CCl_4 rat model, the morphologically obvious "ductular hepatocytes" (i.e., Plate 3b and 3d) seen in rat liver after lethal furan treatment did not immunostain for cytoplasmic CK19.[82]

The similarities between our findings in rat liver after bile duct ligation/CCl_4 treatment[84] and those of Rao et al.[110] in rat pancreas following copper depletion and recovery are particularly noteworthy. In both cases the respective treatments led to a significant loss of parenchymal cells (hepatocytes vs. pancreatic acinar cells). Both models were also characterized by a prominent ductular proliferative response. Finally, in both models, hepatocytes of different apparent stages of morphological development, together with ductular epithelial cells, were actually observed to be lining hyperplastic ductules.[84,110] Based on these similarities, we have suggested that "ductular hepatocytes" appearing in rat liver in close association with hyperplastic bile ductules after bile duct ligation and subsequent CCl_4 treatment arose from some (albeit, very few) proliferated biliary ductular cells that could not be distinguished by their morphology at the light microscopic level from typical hyperplastic biliary epithelium.[84]

This interpretation is further supported by the following observations: (1) in histological sections from the liver of bile duct-ligated/CCl_4-treated rats, the "ductular hepatocytes" were identified only in areas of typical bile ductular hyperplasia and at sites distant from the periportal location of the canals of Hering; (2) as already indicated, "ductular hepatocytes" were identified only within cholangiolarlike structures also lined by biliary epithelial cells; (3) unlike canals of Hering, in which only the bile ductular epithelial cells are supported by basement membrane,[113] both the biliary epithelial and "ductular hepatocytic" cell components of the newly appearing mixed-hepatic cell cholangioles were found to be limited by a common laminin-enriched basement membrane; (4) rare, but distinct "transitional" hepatic cells with morphological features under light microscopy that appeared to be intermediate in various degrees between the hyperplastic bile ductular epithelial cell and the normal rat hepatocyte were also identified in ductular structures within the hyperplastic bile ductular hepatic tissue of the bile duct-ligated/CCl_4-treated rats;[84] (5) the mean maximum cell and nuclear diameters determined for the "ductular hepatocytes" in this model were intermediate in size between those of the hyperplastic bile ductules and the normal-appearing nonductular hepatocytes;[84] (6) in contrast to the findings of Rao et al.[110] in rat pancreas, we found no evidence to indicate that hepatocytic cells were arising from interstitial or periductular cells within the hyperplastic bile ductular tissue in the liver of the bile duct-ligated/CCl_4-treated rats; and (7) to recapitulate, ''ductular hepatocytes'' appeared in the regions of preexisting bile ductular hyperplasia only after the added depletion of zone 3 hepatocytes as a result of the subsequent CCl_4 treatment. With respect to defining the origin of the ''ductular hepatocytes'' in the cholangiofibrotic areas of liver of rats that received the lethal furan treatment, it is also tempting to suggest they too arose from a population of hyperplastic bile ductular cells that were activated to differentiate along the hepatocyte lineage in response to an increased severity of liver injury and parenchymal cell loss. However, additional studies are needed to substantiate this possibility.

Hepatocytes lining rare atypical ductlike structures have also been observed in association with oval cell proliferation in rat liver in response to severe hepatic injury induced by D-galactosamine (see Reference 77 and Chapter 5). In addition, it is important to point out that the appearance of mixed-hepatic cell, cholangiolarlike structures like those described above is not an esoteric finding limited to rodents,[77,82,84,114] but has also been seen to occur in human liver following prominent hepatic necrosis from various causes, including severe forms of viral hepatitis.[34,113,115] In this context, our experimental findings derived from the bile duct ligation/CCl_4 rat model of severe hepatic injury appear to be quite complementary to the recent post-mortem findings of Vandersteenhoven et al.[34] on a 66-year-old man with end-stage cirrhosis resulting from chronic hepatitis B infection and secondary biliary cirrhosis due to biliary obstruction related to a marked cholelithiasis. Here also, unique cholangiolarlike structures lined by biliary epithelium and ''ductular hepatocytes'' were observed to be formed exclusively within areas of intense bile ductular hyperplasia. Interestingly, these authors further demonstrated that the ''ductular hepatocytes'' exhibited immunohistochemically positive staining for hepatitis B surface antigen, which is classically an expression of hepatocytes and not biliary epithelial cells. However, that hepatitis B virus may infect human intrahepatic biliary epithelial cells along with hepatocytes is strongly suggested by the more recent findings of Wang et al.,[116] who detected hepatitis B $\times$ antigen in bile duct cells of chronic hepatitis B virus carriers and in cholangiocarcinoma tissues. This observation raises the possibility that hepatitis B virus may also be playing a role in the pathogenesis of some biliary cell-derived hepatic neoplasms.[116]

IV. SUMMARY AND PERSPECTIVES

In this chapter, we have emphasized two novel adult rat models of severe hepatic injury, which strongly suggest that at least some hyperplastic bile ductular epithelial cells appearing in liver are capable of altering their differentiation commitment in rather specific ways depending upon the nature of the injurious stimulus and the extent and pattern of the resulting liver injury and parenchymal cell loss. In the bile duct ligation/CCl_4 model, the hepatic injury and loss of hepatocytes were extreme, providing an environment in liver that selectively favored the development of rare, mixed hepatic cholangiolarlike structures composed in part of biliary epithelial cells and ductular hepatocytic cells. In comparison, our data from the short-term chronic furan-treatment model indicated liver lobe differences in the severity of the hepatotoxicity and associated reactive responses, leading to an activation of some of the hyperplastic bile ductular epithelial cells in those liver lobes with the greatest injury and hepatocyte loss to preferentially differentiate along the small-intestinal lineage, and with increased furan-induced toxicity to also be associated with newly appearing "ductular hepatocytes". Although these results can be considered compatible with the possibility of "facultative multipotential stem cells" being contained within at least some of the proliferated bile ductules, they can also be interpreted as a metaplastic or "transdifferentiation" process comparable to the "transdifferentiation" of pancreatic ductular epithelial cells into hepatocytes in rats and hamster pancreas and intestinal metaplasia in the human extrahepatic biliary tract described in the previous section. However, regardless of the potential mechanism (facultative biliary stem cell vs. transdifferentiation or metaplastic conversion of a hyperplastic biliary epithelial cell), it seems apparent that cells lining both the intrahepatic and extrahepatic biliary tract can exhibit altered patterns of cellular differentiation under select conditions of persistent noncarcinogenic and carcinogenic injury, and that the plasticity exhibited by these cells may play an important role in the cytogenesis and biological behavior of various histologic types of hepatic and biliary cell cancers, including the metaplastic intestinal type of cholangiocarcinoma and possibly hepatic carcinoid tumor.

REFERENCES

1. **Tanikawa, K.,** *Ultrastructural Aspects of the Liver and Its Disorders,* Igaku-Shoin Ltd., Tokyo, 1979, 33.
2. **Yamamoto, K. and Phillips, M. J.,** A hitherto unrecognized bile ductular plexus in normal rat liver, *Hepatology,* 4, 381, 1984.
3. **Phillips, M. J., Poucell, S., Patterson, J., and Valencia, P.,** *The Liver — An Atlas and Text of Ultrastructural Pathology,* Raven Press, New York, 1987, chap. 1.
4. **Popper, H. and Schaffner, F.,** Pathophysiology of cholestasis, *Hum. Pathol.,* 1, 1, 1970.
5. **Millward-Sadler, G. H. and Jezequel, A. M.,** Normal histology and ultrastructure, in *Liver and Biliary Disease — Pathophysiology, Diagnosis, and Management,* Wright, R., Alberti, K. G. M. M., Karran, S., and Millward-Sadler, G. H., Eds., W. B. Saunders, London, 1979, chap. 2.

6. **Yamamoto, K., Fisher, M. M., and Phillips, M. J.,** Hilar biliary plexus in human liver — a comparative study of the intrahepatic bile ducts in man and animals, *Lab. Invest.,* 52, 103, 1985.
7. **Terada, T., Nakanuma, Y., and Ohta, G.,** Glandular elements around the intrahepatic bile ducts in man; their morphology and distribution in normal livers, *Liver,* 7, 1, 1987.
8. **Ishida, F., Terada, T., and Nakanuma, Y.,** Histological and scanning electron microscopy of intrahepatic peribiliary glands in normal human livers, *Lab. Invest.,* 60, 260, 1989.
9. **Grant, A. G. and Billing, B. H.,** The isolation and characterization of a bile ductule cell population from normal and bile-duct ligated rat livers, *Br. J. Exp. Pathol.,* 58, 301, 1977.
10. **Strazzabosco, M., Mennone, A., and Boyer, J. L.,** Intercellular pH regulation in isolated rat bile duct epithelial cells, *J. Clin. Invest.,* 87, 1503, 1991.
11. **Reddy, J. K. and Rao, M. S.,** Hepatic ultrastructure and adaptation, in *Pathogenesis of Liver Disease,* Farber, E., Phillips, M. J., and Kaufman, N., Eds., Williams & Wilkins, Baltimore, 1987, chap. 2.
12. **Gall, J. A. M. and Bhathal, P. S.,** A quantitative analysis of the liver following ligation of the common bile duct, *Liver,* 10, 116, 1990.
13. **Grisham, J. W. and Porta, E. A.,** Origin and fate of proliferated hepatic ductal cells in the rat: electron microscopic and autoradiographic studies, *Exp. Mol. Pathol.,* 3, 242, 1964.
14. **Tavoloni, N.,** The intrahepatic biliary epithelium: an area of growing interest in hepatology, *Semin. Liver Dis.,* 7, 280, 1987.
15. **Sirica, A. E., Mathis, G. A., Sano, N., and Elmore, L. W.,** Isolation, culture, and transplantation of intrahepatic biliary epithelial cells and oval cells, *Pathobiology,* 58, 44, 1990.
16. **Desmet, V. J.,** Current problems in diagnosis of biliary disease and cholestasis, *Semin. Liver Dis.,* 6, 233, 1986.
17. **Farber, E.,** Similarities in the sequence of early histological changes induced in the liver of the rat by ethionine, 2-acetylaminofluorene, and 3′-methyl-4-dimethylaminoazobenzene, *Cancer Res.,* 16, 142, 1956.
18. **Shinozuka, H., Lombardi, B., Sell, S., and Iammarino, R. M.,** Early histological and functional alterations of ethionine liver carcinogenesis in rats fed a choline-deficient diet, *Cancer Res.,* 38, 1092, 1978.
19. **Sell, S., Osborn, K., and Leffert, H. L.,** Autoradiography of "oval cells" appearing rapidly in the livers of rats fed N-2-fluorenylacetamide in a choline-devoid diet, *Carcinogenesis,* 2, 7, 1981.
20. **Fausto, N., Thompson, N. L., and Braun, L.,** Purification and culture of oval cells from rat liver, in *Cell Separation — Methods and Selected Applications,* Vol. 4, Pretlow, T. G., II and Pretlow, T. P., Eds., Academic Press, Orlando, FL, 1987, 45.
21. **Steiner, J. W., Curruthers, J. S., and Kalifat, S. R.,** The ductular cell reaction of rat liver in extrahepatic cholestasis. I. Proliferated biliary epithelial cells, *Exp. Mol. Pathol.,* 1, 162, 1962.
22. **Sirica, A. E. and Cihla, H. P.,** Isolation and partial characterizations of oval and hyperplastic bile ductular cell-enriched populations from the livers of carcinogen and noncarcinogen-treated rats, *Cancer Res.,* 44, 3454, 1984.
23. **Slott, P. A., Liu, M. H., and Tavoloni, N.,** Origin, pattern, and mechanism of bile duct proliferation following biliary obstruction in the rat, *Gastroenterology,* 99, 466, 1990.
24. **Desmet, V. J.,** Intrahepatic bile ducts under the lens, *J. Hepatol.,* 1, 545, 1985.
25. **Uchida, T. and Peters, R. L.,** The nature and origin of proliferated bile ductules in alcoholic liver disease, *Am. J. Clin. Pathol.,* 79, 326, 1983.

26. **Nakanuma, Y. and Ohta, G.,** Immunohistochemical study on bile ductular proliferation in various hepatobiliary diseases, *Liver,* 6, 205, 1986.
27. **Marceau, N.,** Cell lineages and differentiation programs in epidermal, urothelial and hepatic tissues and their neoplasms, *Lab. Invest.,* 63, 4, 1990.
28. **Sell, S.,** Is there a liver stem cell?, *Cancer Res.,* 50, 3811, 1990.
29. **Fausto, N.,** Hepatocyte differentiation and liver progenitor cells, *Curr. Opinions Cell Biol.,* 2, 1036, 1990.
30. **Tournier, I., Legres, L., Schoevaert, D., Feldmann, G., and Bernuau, D.,** Cellular analysis of α-fetoprotein gene activation during carbon tetrachloride and D-galactosamine-induced acute liver injury in rats, *Lab. Invest.,* 59, 657, 1988.
31. **Germain, L., Blouin, M.-J., and Marceau, N.,** Biliary epithelial cell and hepatocytic cell lineage relationships in embryonic rat liver as determined by the differential expression of cytokeratins, α-fetoprotein, albumin, and cell surface-exposed components, *Cancer Res.,* 48, 4909, 1988.
32. **Makino, Y., Yamamoto, K., and Tsuji, T.,** Three-dimensional arrangement of ductular structures formed by oval cells during hepatocarcinogenesis, *Acta Med. Okayama,* 42, 143, 1988.
33. **Couch, J. A. and Courtney, L. A.,** N-Nitrosodiethylamine-induced hepatocarcinogenesis in estuarine sheepshead minnow (Cyprinodon variegatus): neoplasms and related lesions compared with mammalian lesions, *J. Natl. Cancer Inst.,* 79, 297, 1987.
34. **Vandersteenhoven, A. M., Burchette, J., and Michalopoulos, G.,** Characterization of ductular hepatocytes in end-stage cirrhosis, *Arch. Pathol. Lab. Med.,* 114, 403, 1990.
35. **Fu, X.-X., Su, C. Y., Lee, Y., Hintz, R., Biempica, L., Snyder, R., and Rogler, C. E.,** Insulinlike growth factor II expression and oval cell proliferation associated with hepatocarcinogenesis in woodchuck hepatitis virus carriers, *J. Virol.,* 62, 3422, 1988.
36. **McLean, M. R. and Rees, K. R.,** Hyperplasia of bile ducts induced by alpha-naphthyl-iso-thiocyanate: experimental biliary cirrhosis free from biliary obstruction, *J. Pathol. Bacteriol.,* LXXVI, 175, 1958.
37. **Steiner, J. W. and Carruthers, J. S.,** Electron microscopy of hyperplastic ductular cells in α-naphthyl isothiocyanate-induced cirrhosis, *Lab. Invest.,* 12, 471, 1963.
38. **Fukushima, S., Shibata, M., Hibino, T., Yoshimura, T., Hirose, M., and Ito, N.,** Intrahepatic bile duct proliferation induced by 4,4′-diaminophenylmethane in rats, *Toxicol. Appl. Pharmacol.,* 48, 145, 1979.
39. **Sell, S.,** Comparison of oval cells induced in rat liver by feeding N-2-fluorenylacetamide in a choline-devoid diet and bile duct cells induced by feeding 4,4′-diaminodiphenylmethane, *Cancer Res.,* 43, 1761, 1983.
40. **Dunsford, H. A., Maset, R., Salman, J., and Sell, S.,** Connection of ductlike structures induced by a chemical hepatocarcinogen to portal bile ducts in the rat liver detected by injection of bile ducts with a pigmented barium gelatin medium, *Am. J. Pathol.,* 118, 218, 1985.
41. **Bhathal, P. S. and Christe, G. S.,** A fluorescence microscopic study of bile duct proliferation induced in guinea pigs by α-naphthyl isothiocyanate, *Lab. Invest.,* 20, 480, 1969.
42. **Grisham, J. W.,** Cell types in long-term propagable cultures of rat liver, *Ann. N.Y. Acad. Sci.,* 349, 128, 1980.
43. **Sirica, A. E.,** Biology of biliary epithelial cells, in *Progress in Liver Diseases,* Vol. X, Boyer, J. L. and Ockner, R. K., Eds., W. B. Saunders, Philadelphia, in press.
44. **Milani, S., Herbst, H., Schuppan, D., Niedobitek, G., Kim, K. Y., and Stein, H.,** Vimentin expression of newly formed rat bile duct epithelial cells in secondary biliary fibrosis, *Virchows Arch. A: Pathol. Anat. Histopathol.,* 415, 237, 1989.
45. **Alpini, G., Lenzi, R., Zhai, W.-R., Liu, M. H., Slott, P. A., Paronetto, F., and Tavoloni, N.,** Isolation of a nonparenchymal liver cell fraction enriched in cells with biliary epithelial phenotypes, *Gastroenterology,* 97, 1248, 1989.

46. **Seglen, P. O., Skomedal, H., Saeter, G., Schwarze, P. E., and Nesland, J. M.,** Neuroendocrine dysdifferentiation and bombesin production in carcinogen-induced hepatocellular rat tumours, *Carcinogenesis,* 10, 21, 1989.
47. **Chou, S.-T. and Gibson, J. B.,** A histochemical study of the bile ducts in long-term biliary obstruction in the rat, *J. Pathol.,* 103, 163, 1971.
48. **Sirica, A. E., Sattler, C. A., and Cihla, H. P.,** Characterization of a primary bile ductular cell culture from the livers of rats during extrahepatic cholestasis, *Am. J. Pathol.,* 120, 67, 1985.
49. **Petropoulos, C. J., Yaswen, P., Panzica, M., and Fausto, N.,** Cell lineages in liver carcinogenesis: possible clues from studies of the distribution of α-fetoprotein RNA sequences in cell populations isolated from normal, regenerating, and preneoplastic rat livers, *Cancer Res.,* 45, 5762, 1985.
50. **Alpini, G., Aragona, E., Shafritz, D., and Tavoloni, N.,** Characterization of proliferating bile duct-like cells by northern blotting and in situ hybridization, *Hepatology,* 12, 912, 1990.
51. **Sirica, A. E., Elmore, L. W., and Sano, N.,** Characterization of rat hyperplastic bile ductular epithelial cells in culture and in vivo, *Dig. Dis. Sci.,* 36, 494, 1991.
52. **Arber, N. and Zajicek, G.,** Streaming liver. VI. Streaming intra-hepatic bile ducts, *Liver,* 10, 205, 1990.
53. **Zajicek, G.,** Hepatocytes and intrahepatic bile duct epithelium originate from a common stem cell, *Gastroenterology,* 100, 582, 1991.
54. **Wu, P. C., Ma, L., Gibson, J. B., Hirai, H., and Tsukada, Y.,** Serum alpha-fetoprotein in rats after ligation of the common bile duct: relation to ductular cell (oval cell) proliferation, *J. Pathol.,* 133, 61, 1981.
55. **Gall, J. A. M. and Bhathal, P. S.,** Origin and involution of hyperplastic bile ductules following total biliary obstruction, *Liver,* 10, 106, 1990.
56. **Mathis, G. A., Walls, S. A., D'Amico, P., Gengo, T. F., and Sirica, A. E.,** Enzyme profile of rat bile ductular epithelial cells in reference to the resistance phenotype in hepatocarcinogenesis, *Hepatology,* 9, 477, 1989.
57. **Bhathal, P. S. and Gall, J. A. M.,** Deletion of hyperplastic biliary epithelial cells by apoptosis following removal of the proliferative stimulus, *Liver,* 5, 311, 1985.
58. **Aronson, D. C., De Haan, J., James, J., Bosch, K. S., Ketel, A. G., Houtkooper, J. M., and Heijmans, H. S. A.,** Quantitative aspects of the parenchymal-stroma relationship in experimentally induced cholestasis, *Liver,* 8, 116, 1988.
59. **Parola, M., Cheeseman, K. H., Biocca, M. E., Dianzani, M. U., and Slater, T. F.,** Isolation and characterization of biliary epithelial cells from normal rat liver, *J. Hepatol.,* 6, 175, 1988.
60. **Baron, J., Redick, J. A., and Guengerich, F. P.,** An immunohistochemical study on the localizations and distributions of phenobarbital- and 3-methylcholanthrene-inducible cytochrome P-450 within the livers of untreated rats, *J. Biol. Chem.,* 256, 5931, 1981.
61. **Taira, Y., Redick, J. A., and Baron, J.,** An immunohistochemical study on the localization and distribution of NADPH-cytochrome c (P-450) reductase in rat liver, *Mol. Pharmacol.,* 17, 374, 1980.
62. **Kawabata, T. T., Guengerich, F. P., and Baron, J.,** Effects of phenobarbital, trans-stilbene oxide, and 3-methylcholanthrene on epoxide hydrolase within the centrolobular, midzonal, and periportal regions of rat liver, *J. Biol. Chem.,* 258, 7767, 1983.
63. **Schrenk, D., Eisenmann-Tappe, I., Gebhardt, R., Mayer, D., El Mouelhi, M., Röhrdanz, E., Münzel, P., and Bock, K. W.,** Drug metabolizing enzyme activities in rat liver epithelial cell lines, hepatocytes and bile duct cells, *Biochem. Pharmacol.,* 41, 1751, 1991.
64. **Redick, J. A., Jakoby, W. B., and Baron, J.,** Immunohistochemical localization of glutathione S-transferases in livers of untreated rats, *J. Biol. Chem.,* 257, 15200, 1982.

65. **Knapp, S. A., Green, M. D., Tephly, T. R., and Baron, J.,** Immunohistochemical demonstration of isozyme- and strain-specific differences in the intralobular localizations and distributions of UDP-glucuronosyltransferases in livers of untreated rats, *Mol. Pharmacol.*, 33, 14, 1988.
66. **Franson, K. L., Burchell, B., and Mathis, G. A.,** 17 β-Hydroxysteroid UDP-glucuronosyl transferase is expressed in bile ductular epithelial cells under physiological conditions, *Biochem. Biophys. Res. Commun.*, 173, 1001, 1990.
67. **Mathis, G. A., Wyss, P. A., Schuetz, E. G., Ringer, D. P., Sirica, A. E., and Guzelian, P. S.,** Unique pattern of drug metabolizing enzymes in bile ductular epithelial cells, *Experientia*, 45, A65, 1989.
68. **Ringer, D. P., Norton, T. R., and Howell, B. A.,** 2-Acetylaminofluorene-mediated alteration in the level of liver arylsulfotransferase IV during hepatocarcinogenesis, *Cancer Res.*, 50, 5301, 1990.
69. **Sato, K., Kitahara, A., Yin, Z., Waragai, F., Nishimura, K., Hatayama, I., Ebina, T., Yamazaki, T., Tsuda, H., and Ito, N.,** Induction by butylated hydroxyanisole of specific molecular forms of glutathione S-transferase and UDP-glucuronyltransferase and inhibition of development of γ-glutamyl transpeptidase-positive foci in rat liver, *Carcinogenesis*, 5, 473, 1984.
70. **Roomi, M. W., Ho, R. K., Sarma, D. S. R., and Farber, E.,** A common biochemical pattern in preneoplastic hepatocyte nodules generated in four different models in the rat, *Cancer Res.*, 45, 564, 1985.
71. **Satoh, K., Kitahara, A., Soma, Y., Inaba, Y., Hatayama, I., and Sato, K.,** Purification, induction, and distribution of placental glutathione transferase: a new marker enzyme for preneoplastic cells in the rat chemical hepatocarcinogenesis, *Proc. Natl. Acad. Sci. U.S.A.*, 82, 3964, 1985.
72. **Chowdhury, N. R., Saber, M. A., Lahiri, P., Mackenzie, P. I., Novikoff, P. M., Becker, F. F., and Chowdhury, J. R.,** Expression of specific UDP-glucuronosyltransferase isoforms in carcinogen-induced liver nodules, *Hepatology*, 13, 38, 1991.
73. **Ledda, G. M., Sells, M. A., Yokoyama, S., and Lombardi, B.,** Metabolic properties of isolated rat liver cell preparations enriched in epithelial cells other than hepatocytes, *Int. J. Cancer*, 31, 231, 1983.
74. **Ghoshal, A. K., Mullen, B., Medline, A., and Farber, E.,** Sequential analysis of hepatic carcinogenesis: regeneration of liver after carbon tetrachloride-induced liver necrosis when hepatocyte proliferation is inhibited by 2-acetylaminofluorene, *Lab. Invest.*, 48, 224, 1983.
75. **Evarts, R. P., Nagy, P., Marsden, E., and Thorgeirsson, S. S.,** A precursor-product relationship exists between oval cells and hepatocytes in rat liver, *Carcinogenesis*, 8, 1737, 1987.
76. **Evarts, R. P., Nagy, P., Nakatsukasa, H., Marsden, E., and Thorgeirsson, S. S.,** In vivo differentiation of rat liver oval cells into hepatocytes, *Cancer Res.*, 49, 1541, 1989.
77. **Lemire, J. M., Shiojiri, N., and Fausto, N.,** Oval cell proliferation and the origin of small hepatocytes in liver injury induced by D-galactosamine, *Am. J. Pathol.*, 139, 535, 1991.
78. **Terao, K. and Nakano, M.,** Cholangiofibrosis induced by short-term feeding of 3′-methyl-4-(dimethylamino) azobenzene: an electron microscopic observation, *Gann*, 65, 249, 1974.
79. **Tatematsu, M., Kaku, T., Medline, A., and Farber, E.,** Intestinal metaplasia as a common option of oval cells in relation to cholangiofibrosis in liver of rats exposed to 2-acetylaminofluorene, *Lab. Invest.*, 52, 354, 1985.
80. **Rao, M. S., Bendayan, M., Kimbrough, R. D., and Reddy, J. K.,** Characterization of pancreatic-type tissue in the liver of rat induced by polychlorinated biphenyls, *J. Histochem. Cytochem.*, 34, 197, 1986.

81. **Wolf, H. K., Burchette, J. L., Jr., Garcia, J. A., and Michalopoulos, G.,** Exocrine pancreatic tissue in human liver: a metaplastic process?, *Am. J. Surg. Pathol.,* 14, 590, 1990.
82. **Elmore, L. W. and Sirica, A. E.,** Phenotypic characterization of metaplastic intestinal glands and ductular hepatocytes in cholangiofibrotic lesions rapidly induced in the caudate liver lobe of rats treated with furan, *Cancer Res.,* 51, 5752, 1991.
83. **Sirica, A. E., Elmore, L. W., Williams, T. W., and Cole, S. L.,** Early induction of four intestinal cell types in the right and caudate liver lobes of furan-treated rats, *Hepatology,* 14, 103A, 1991.
84. **Sirica, A. E. and Williams, T. W.,** Appearance of ductular hepatocytes in rat liver after bile duct ligation and subsequent zone 3 necrosis by carbon tetrachloride, *Am. J. Pathol.,* 140, 129, 1992.
85. Toxicology and carcinogenesis studies of furan in F344 rats and B3C3F1 mice, *NTP Technical Report,* 1991.
86. **Elmore, L. W. and Sirica, A. E.,** Sequential appearance of intestinal mucosal cell types in the right and caudate liver lobes of furan-treated rats, *Hepatology,* in press.
87. **Reddy, K. P., Buschmann, R. J., and Chomet, B.,** Cholangiocarcinomas induced by feeding 3′-methyl-4-dimethylaminoazobenzene to rats, *Am. J. Pathol.,* 87, 189, 1977.
88. **Yoshida, Y., Kaneko, A., Chisaka, N., and Onoé, T.,** Appearance of intestinal type of tumor cells in hepatoma tissue induced by 3′-methyl-4-dimethylaminoazobenzene, *Cancer Res.,* 38, 2753, 1978.
89. **Azadeh, B. and Parai, S. K.,** Argentaffin cells, intestinal metaplasia and antral metaplasia in carcinoma of the gall bladder, *Histopathology,* 4, 653, 1980.
90. **Kozuka, S., Kurashina, M., Tsubone, M., Hachisuka, K., and Yasui, A.,** Significance of intestinal metaplasia for the evolution of cancer in the biliary tract, *Cancer,* 54, 2277, 1984.
91. **Koga, A., Momii, S., Eguchi, M., and Makino, T.,** Ultrastructure of well-differentiated adenocarcinoma of the gallbladder, *Ultrastruct. Pathol.,* 15, 41, 1991.
92. **Albores-Saavedra, J., Nadji, M., Henson, D. E., Zeigels-Weissman, J., and Mones, J. M.,** Intestinal metaplasia of the gallbladder: a morphologic and immunocytochemical study, *Hum. Pathol.,* 17, 614, 1986.
93. **Hisatomi, K., Haratake, J., Horie, A., and Ohsato, K.,** Relation of histopathological features to prognosis of gallbladder cancer, *Am. J. Gastroenterol.,* 85, 567, 1990.
94. **Tsao, M.-S. and Grisham, J. W.,** Hepatocarcinomas, cholangiocarcinomas, and hepatoblastomas produced by chemically transformed cultured rat liver epithelial cells — a light- and electron-microscopic analysis, *Am. J. Pathol.,* 127, 168, 1987.
95. **Hsu, W., Deziel, D. J., Gould, V. E., Warren, W. H., Gooch, G. T., and Staren, E. D.,** Neuroendocrine differentiation and prognosis of extrahepatic biliary tract carcinomas, *Surgery,* 110, 604, 1991.
96. **Koga, A., Ichimiya, H., Yamaguchi, K., Miyazaki, K., and Nakayama, F.,** Hepatolithiasis associated with cholangiocarcinoma — possible etiologic significance, *Cancer,* 55, 2826, 1985.
97. **Terada, T. and Nakanuma, Y.,** Morphological examination of intrahepatic bile ducts in hepatolithiasis, *Virchows Arch. A. Pathol. Anat.,* 413, 167, 1988.
98. **Ona, F. V. and Dytoc, J. N.,** Clonorchis-associated cholangiocarcinoma: a report of two cases with unusual manifestations, *Gastroenterology,* 101, 831, 1991.
99. **Pungpak, S., Akai, P. S., Longenecker, B. M., Ho, M., Befus, A. D., and Bunnag, D.,** Tumour markers in the detection of opisthorchiasis-associated cholangiocarcinoma, *Trans. R. Soc. Trop. Med. Hyg.,* 85, 277, 1991.
100. **Kurumaya, H., Ohta, G., and Nakanuma, Y.,** Endocrine cells in the intrahepatic biliary tree in normal livers and hepatolithiasis, *Arch. Pathol. Lab. Med.,* 113, 143, 1989.

101. **Andreola, S., Lombardi, L., Audisio, R. A., Mazzaferro, V., Koukouras, D., Doci, R., Gennari, L., Makowka, L., Starzl, T. E., and Van Thiel, D. H.,** A clinicopathologic study of primary hepatic carcinoid tumors, *Cancer,* 65, 1211, 1990.
102. **Brown, W. M., III, Henderson, J. M., and Kennedy, J. C.,** Carcinoid tumor of the bile duct — a case report and literature review, *Am. Surg.,* 56, 343, 1990.
103. **Angeles-Angeles, A., Quintanilla-Martínez, L., and Larriva-Sahd, J.,** Primary carcinoid of the common bile duct — immunohistochemical characterization of a case and review of the literature, *Am. J. Clin. Pathol.,* 96, 341, 1991.
104. **Roskams, T., van den Oord, J. J., De Vos, R., and Desmet, V. J.,** Neuroendocrine features of reactive bile ductules in cholestatic liver disease, *Am. J. Pathol.,* 137, 1019, 1990.
105. **Reynolds, S. H., Stowers, S. J., Patterson, R. M., Maronpot, R. R., Aaronson, S. A., and Anderson, M. W.,** Activated oncogenes in B6C3F1 mouse liver tumors: implications for risk assessment, *Science,* 237, 1309, 1987.
106. **Rosenkranz, H. S. and Klopman, G.,** 'Cryptic' mutagens and carcinogenicity, *Mutagenesis,* 5, 199, 1990.
107. **Mortelmans, K., Haworth, S., Lawlor, T., Speck, W., Tainer, B., and Zeiger, E.,** Salmonella mutagenicity tests. II. Results from the testing of 270 chemicals, *Environ. Mutagen.,* 8 (Suppl. 7), 1, 1986.
108. **Wilson, D. M., Goldsworthy, T. L., Popp, J. A., and Butterworth, B. E.,** Evaluation of genotoxicity, cytotoxicity, and cell proliferation in hepatocytes from rats and mice treated with furan, *Proc. Am. Assoc. Cancer Res.,* 31, 103, 1990.
109. **Shimizu, A., Nakamura, Y., Harada, M., Ono, T., Sato, K., Inoue, T., and Kanisawa, M.,** Positive foci of glutathione S-transferase placental form in the liver of rats given furfural by oral administration, *Jpn. J. Cancer Res.,* 80, 608, 1989.
110. **Rao, M. S., Dwivedi, R. S., Yeldandi, A. V., Subbarao, V., Tan, X., Usman, M. I., Thangada, S., Nemali, M. R., Kumar, S., Scarpelli, D. G., and Reddy, J. K.,** Role of periductal and ductular epithelial cells of the adult rat pancreas in pancreatic hepatocyte lineage — a change in the differentiation commitment, *Am. J. Pathol.,* 134, 1069, 1989.
111. **Makino, T., Usuda, N., Rao, S., Reddy, J. K., and Scarpelli, D. G.,** Transdifferentiation of ductular cells into hepatocytes in regenerating hamster pancreas, *Lab. Invest.,* 62, 552, 1990.
112. **Konishi, N., Ward, J. M., and Waalkes, M. P.,** Pancreatic hepatocytes in Fisher and Wistar rats induced by repeated injections of cadmium chloride, *Toxicol. Appl. Pharmacol.,* 104, 149, 1990.
113. **Phillips, M. J. and Poucell, S.,** Modern aspects of the morphology of viral hepatitis, *Hum. Pathol.,* 12, 1060, 1981.
114. **Wilson, J. W. and Leduc, E. H.,** Role of cholangioles in restoration of the liver of the mouse after dietary injury, *J. Pathol. Bacteriol.,* LXXVI, 441, 1958.
115. **Gerber, M. A., Thung, S. N., Shen, S., Stromeyer, F. W., and Ishak, K. G.,** Phenotypic characterization of hepatic proliferation — antigenic expression by proliferating epithelial cells in fetal liver, massive hepatic necrosis, and nodular transformation of the liver, *Am. J. Pathol.,* 110, 70, 1983.
116. **Wang, W., London, W. T., and Feitelson, M. A.,** Hepatitis B x antigen in hepatitis B virus carrier patients with liver cancer, *Cancer Res.,* 51, 4971, 1991.

Chapter 10

CELL LINEAGES IN HUMAN LIVER DEVELOPMENT, REGENERATION, AND TRANSFORMATION

M. A. Gerber and S. N. Thung

TABLE OF CONTENTS

ISBN 0-8493-4746-7
© 1992 by CRC Press, Inc.

I. INTRODUCTION

In this chapter, we will review current knowledge about epithelial cell lineages in human liver. Over the last decade, this important area has been investigated extensively in experimental animals. Due to obvious limitations, *in vivo* studies of fetal and adult human livers are not possible. *In vitro* studies of growth and differentiation of human hepatocytes are still in their infancy. For these reasons, current information on human liver cell lineages is indirect and incomplete. It is derived from studies of the development of human liver, of liver regeneration after necrosis of hepatocytes, of benign or malignant liver tumors, and particularly of their precursors. We hope that this information will build a bridge to the many experimental investigations of liver cell types in hepatocarcinogenesis reported elsewhere in this book.

II. DEVELOPMENT

The human liver develops as hepatic diverticulum from the endodermal lining of the foregut during the third and fourth week of gestation.[1-4] This hepatic primordium differentiates cranially into proliferating hepatic cords and caudally into the gallbladder and extrahepatic bile ducts. The anastomosing cords of immature hepatic epithelial cells or hepatoblasts grow into the mesenchyme of the septum transversum. Primitive hepatic sinusoids are formed by ingrowth of capillaries from the vitelline veins at the outer margins of the septum transversum. The hepatic stroma is derived from the mesenchymal cells of the septum transversum. It has been hypothesized that the hepatoblasts in direct contact with the mesenchyme around the portal veins develop into bile duct-type cells. We proposed that the extracellular matrix, particularly the laminin, in contact with the developing epithelial cells provides local instructive stimuli for differentiation and migration of these cells.[5] It appears that the hepatoblasts transform first into a single-layered and then a double-layered cylinder, the so-called ductal plate (Figure 1) as laminin is deposited at the junction of the ductal plate cells and the mesenchyme around the portal veins. At 5 to 6 weeks gestational age, a lumen appears between the two layers of the ductal plate and laminin is formed on the parenchymal side of the tubular structures. At 12 weeks gestational age, the ductal structures begin to grow into the mesenchyme around the portal veins, again accompanied by laminin. As they reach the center of the portal tracts, they differentiate into mature bile ducts surrounded by a laminin-containing basement membrane. The canals of Hering which are lined by hepatocytes and ductular cells connect the bile ducts with the bile canaliculi between hepatocytes. The remaining ductal plates at the margin of portal tracts with the accompanying laminin regress, with the exception of scattered ductal plate remnants that may persist into adult life.

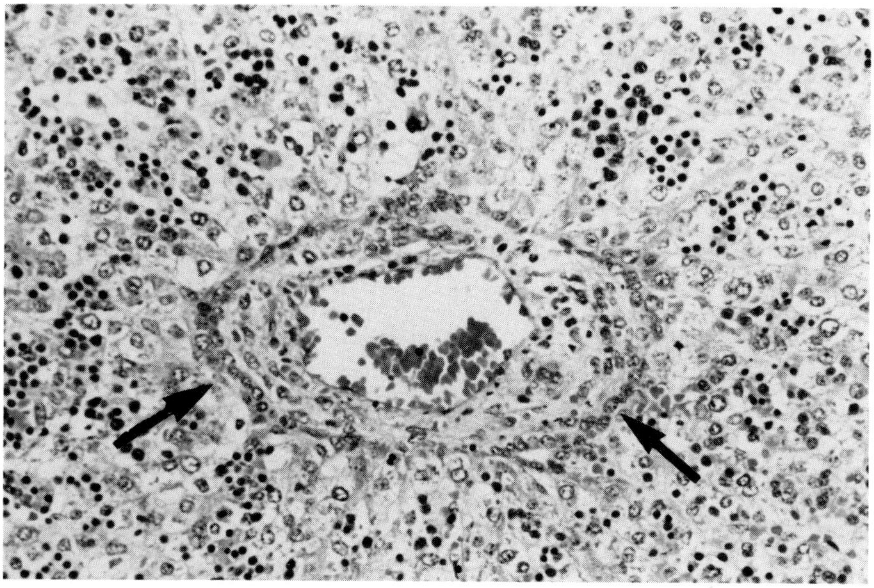

FIGURE 1. Fetal liver at gestational age of 6 weeks. Mesenchymal tissue around a portal vein branch is surrounded by a single layer of small cells with oval nuclei, the so-called ductal plate (arrows). (H.E.; magnification × 100.)

Several investigators attempted to define the development and differentiation of hepatoblasts by immunohistochemical staining for a variety of antigens, particularly cytokeratin (CK) polypeptides.[6-8] At early gestational age, hepatoblasts express CK8, 10, 14/15, 16, 18, and 19.[6] As they differentiate along the hepatocyte lineage, they gradually lose most acidic CK (CK10, 14/15, 16, and 19) at about 10 to 24 weeks of gestational age,[8] such that the main CK polypeptides expressed by mature hepatocytes represent CK8 and 18. In contrast, differentiation of hepatoblasts into ductal plates around the mesenchyme of portal veins resulted in increased expression of CK8, 18, and 19 at 6 weeks of gestational age (Figure 2). In addition, CK7 becomes detectable in ductal plates after 20 weeks of gestation.[7] These CK are retained during development into bile ducts and therefore the main CK polypeptides expressed in mature bile ducts are CK7, 8, 18, and 19. We speculated that undifferentiated hepatoblasts during an early gestational age contain both hepatocyte-type and bile duct-type CK. With cellular differentiation, ductal plate and bile duct cells within and along portal tracts express the bile duct type of CK, while the hepatocytes lose the acidic CK, some of which during later development and in adults are exhibited by bile duct and ductular cells. The expression of bile duct-type CK by hepatoblasts early during gestation suggests the possibility of common bipotential precursor cells for hepatocytes

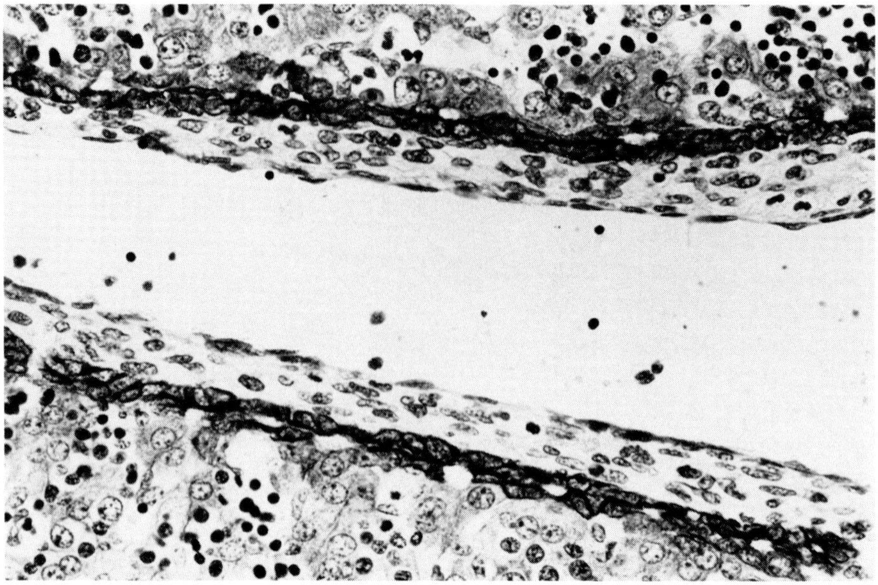

FIGURE 2. Fetal liver at gestational age of 6 weeks stained with monoclonal antibody (AE1) to CK10, 14/15, 16, and 19. There is weak expression of the CK polypeptides in the cytoplasm of hepatoblasts and much stronger expression in the cytoplasm of the ductal plate cells. (Immunoperoxidase staining, counterstained with hematoxylin; magnification × 250.)

and bile duct cells. The periportal ductal plate cells persisting even in adult livers may represent such precursor cells. They may give origin to regeneration of hepatocytes after liver necrosis with formation of ductular hepatocytes as well as to proliferating bile ductules associated with diseases of the biliary tree (see below).

III. REGENERATION AND PUTATIVE LIVER STEM CELLS

Morphologic features of liver regeneration include mitoses, ductular hepatocytes, and hepatocytes in plates which are two or more cells thick.[9] In man, as in experimental animals, the characteristics of liver regeneration appear to depend on the etiopathogenesis and extent of hepatocyte loss. After partial hepatectomy, liver tissue regenerates rapidly and completely, but it takes longer in man (4 or more months) than in the rat (2 to 4 weeks).[10,11] Proliferating hepatocytes may be identified by immunohistochemical staining for DNA polymerase α.[12] Data on the cell types and kinetics of human liver cell replication are scarce, but it appears that adult human hepatocytes are capable of mitotic division. This may be seen in acute viral hepatitis with

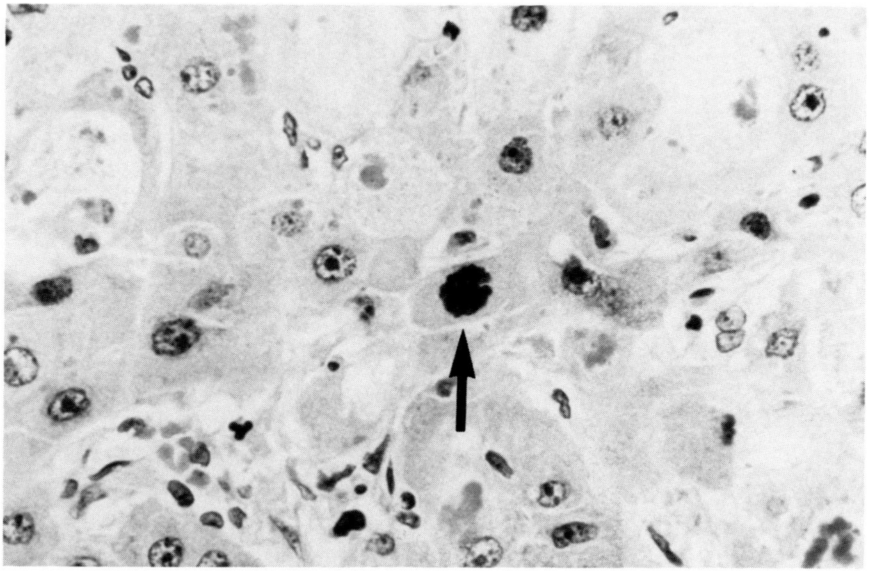

FIGURE 3. Liver of patient with spotty necrotic acute hepatitis. There is regeneration as evidenced by a hepatocyte in mitosis (arrow). (H.E.; magnification × 400.)

spotty necrosis which is associated with mitoses of hepatocytes scattered throughout the parenchyma (Figure 3). There is no morphological or molecular evidence that these hepatocytes "dedifferentiate" before they undergo replication.[13,14]

In contrast, in virus- or toxin-induced massive hepatic necrosis resulting in sudden loss of all hepatocytes, regeneration appears to originate from progenitor cells, perhaps ductal plate remnants or oval cells (see below).[5] In these cases, ductlike structures composed of small cells with oval nuclei, often without a lumen, are seen around the portal tracts and extend into the necrotic lobules.[9,15] The nature and potential of these cells have been debated for many years. While some investigators consider them proliferating biliary epithelium without differentiation into hepatocytes, others suggest that they represent sprouts of interlobular bile ducts with gradual transformation into functional hepatocytes. We demonstrated markers of both hepatocytes (α-1-antitrypsin) and bile ducts (carcinoembryonic antigen [CEA] family of glycoproteins) in these cells.[9] Under the electron microscope, the ductlike structures contain cells that are intermediate in appearance between hepatocytes and bile duct epithelium.[16] Therefore, we termed these cells ductular hepatocytes (Figure 4) and believe that, similar to oval cells in the rat, they can differentiate into hepatocytes or bile ducts.[9] Both differentiation pathways result in two-cell-thick structures. The ductular hepatocytes may be derived

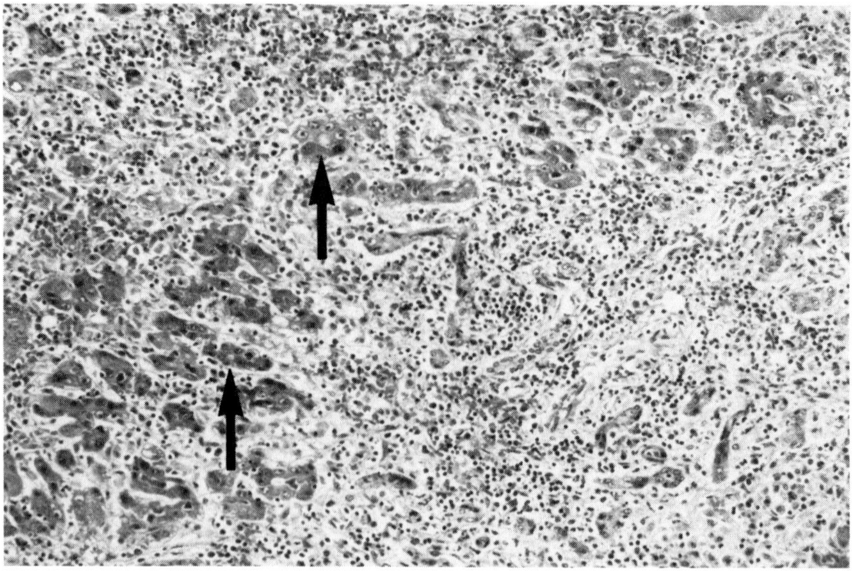

FIGURE 4. Liver showing massive hepatic necrosis. There is extensive regeneration with formation of ductlike structures by immature cells, so-called ductular hepatocytes (arrows). (H.E.; magnification × 100.)

from bipotential progenitor cells or liver stem cells which proliferate when the majority of hepatocytes have been destroyed. It is not clear whether the putative stem cells originate in the canals of Hering, periductular cells, ductal plate remnants, or oval cells, and no definite stem cell population has been identified in adult human liver.[17-23]

Ductular hepatocytes must be distinguished from proliferating bile ductules which are seen in diseases of the biliary tree such as primary biliary cirrhosis, primary sclerosing cholangitis, and mechanical obstruction of bile ducts. In contrast to ductular hepatocytes, bile ductules often exhibit a central lumen which is surrounded by cuboidal cells on a basement membrane. They are located at the border of portal tracts and may represent sprouts of small bile ducts or ductal plate remnants. However, hepatocytes may also undergo ductular metaplasia as suggested by immunohistochemical studies of hepatocytes expressing bile duct-type CK.[24,25] Although based on almost exclusively phenotypic data, we conclude that, depending on the type and extent of liver injury, hepatocytes, bile duct epithelial cells, or bipotential progenitor cells may regenerate. In contrast to the large body of knowledge in experimental liver regeneration, very little is known about the control of growth, growth factors, and gene expression patterns in human liver regeneration.

IV. BENIGN EPITHELIAL TUMORS

In human livers, benign epithelial growths are relatively rare and include tumorlike masses such as focal nodular hyperplasia and nodular regenerative hyperplasia, or true neoplasms such as hepatocellular adenoma and bile duct adenoma. Focal nodular hyperplasia probably represents a local response of the liver to a vascular malformation.[26] It has the appearance of a tumor which is subdivided by fibrous septa into nodules containing normal hepatocytes in plates which are two or more cells thick. Nodular regenerative hyperplasia of the liver may represent a response to obliterative vascular lesions.[16,27-31] It consists of nodules scattered throughout the noncirrhotic liver and is composed of expanding masses of hepatocytes in cell plates which are two or more cells thick. Histochemical, immunohistochemical, and ultrastructural studies confirmed the hepatocellular nature of the hyperplastic cells in the nodules.[9,32,33] Hepatocellular adenomas are true benign neoplasms composed of hepatocytes which are slightly larger than nonneoplastic hepatocytes and arranged in sheets and cords without the normal acinar architecture. Ultrastructural studies demonstrated the similarity of the neoplastic cells to normal hepatocytes. Adenomas are usually related to steroid therapy, such as oral contraceptives or androgens.[34,35] Bile duct adenomas are composed of small, normal-appearing ducts or ductules lined by regular cuboidal cells. The immunophenotype of these cells is similar to that of bile duct epithelial cells. It is not clear whether intrahepatic bile duct adenomas are true neoplasms or a developmental anomaly similar to von Meyenburg complexes.[36,37] Malignant transformation of any of the lesions described above is rare. The occurrence of these benign epithelial tumors in the human liver is consistent with the view that at least two distinct epithelial cell lineages develop in the human liver.

V. PRECURSOR LESIONS AND HEPATOCELLULAR CARCINOMA (HCC)

About 60 to 80% of HCC are associated with cirrhosis, usually macronodular in type.[32] The regenerative nodules in cirrhosis consist of hepatocytes which, in contrast to normal liver, are arranged in two-cell-thick plates and therefore represent hyperplasia (Figure 5). Development of HCC has been documented in these regenerative nodules, particularly macroregenerative nodules (measuring more than 10 mm in diameter) which may show marked steatosis.[38-42] Adenomatous hyperplasia was originally described by Edmondson[43] in 1976 and defined as sizable (up to 6 to 8 cm in diameter) discrete nodules of proliferating hepatocytes without a true capsule. In contrast to adenomas, they may contain small septa, blood vessels, and bile ducts entering from the periphery. These adenomatous hyperplastic nodules may show areas of cytologic atypia and development of hepatocellular carcinoma

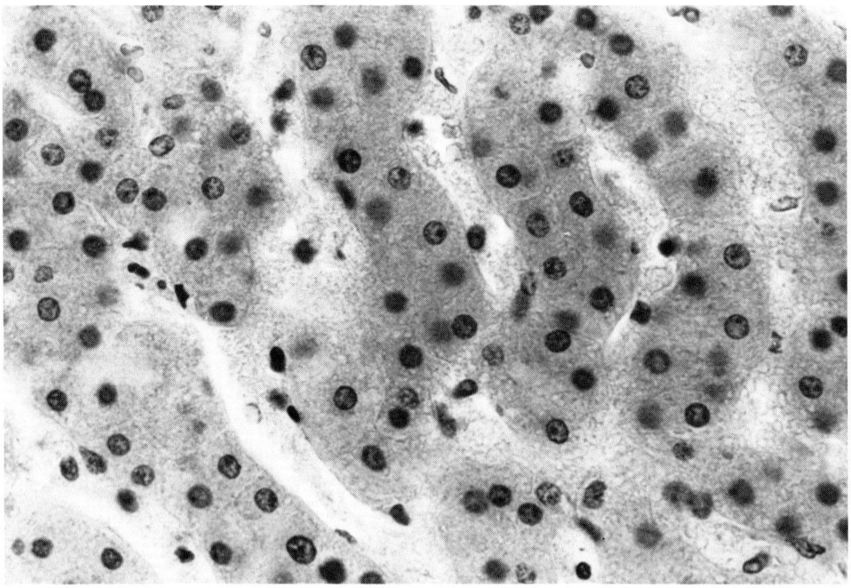

FIGURE 5. Regenerative nodule in liver cirrhosis. Regeneration is evidenced by formation of hepatocytes arranged in two-cell-thick plates. (H.E.; magnification × 200.)

which is characterized by nuclear crowding and increased nuclear/cytoplasmic ratio, increased cytoplasmic basophilia, and microacinar formation.[44-48] These early malignant foci in adenomatous hyperplastic nodules form nodules-in-nodules and represent strong evidence that the adenomatous hyperplastic nodules in cirrhotic livers are preneoplastic.

Liver cell dysplasia was defined by Anthony et al.[49,50] as "cellular enlargement, nuclear pleomorphism and multinucleation of liver cells recurring in groups or occupying whole cirrhotic nodules". Several investigators found a strong relationship between hepatocyte dysplasia and hepatocellular carcinoma, particularly in patients with chronic hepatitis B (HBV) or non-A, non-B virus infection, while others did not.[32,51,52] The incidence of liver cell dysplasia in cases of macroregenerative nodules was significantly higher than in cases without macroregenerative nodules.[53] Attempts to demonstrate HBV antigens or tumor markers in dysplastic cells by immunomorphological or histochemical techniques were generally unsuccessful.[32] Ultrastructural study of dysplastic hepatocytes showed regenerative rather than degenerative features.[54,55] Feulgen-DNA flow cytometry of hepatocyte dysplasia demonstrated discontinuous ploidy levels in contrast to the continuous ploidy levels (from diploid to polyploid) of HCC.[56,57] These observations suggest that dysplastic hepatocytes represent a hyperplastic cell population as a result of repeated cell proliferation.

Characterization of precursor lesions in human liver in association with HCC has been carried out in several laboratories, including our own.[55,58,59] By morphometry, two hepatocyte populations were identified in the precursor lesions: large cells with large nuclei and a normal to moderately high nuclear/cytoplasmic ratio and small cells with a very high nuclear/cytoplasmic ratio, often arranged in two-cell-thick plates. These morphological studies indicate that hypertrophy with a high nuclear/cytoplasmic ratio, hyperplasia with increased thickness of cell plates, and dysplasia are frequently associated with malignant transformation of hepatocytes.

Absence of hemosiderin and excess storage of glycogen have been demonstrated in preneoplastic hepatocytes during experimental hepatocarcinogenesis.[60,61] In cirrhotic human livers, iron deposition was noted in macroregenerative nodules and foci of liver cell dysplasia, but not in borderline lesions or HCC.[62-65] Mallory bodies, a cytoskeletal abnormality, has been detected in an HCC cell line and in hepatocytes of carcinogen (griseofulvin)-treated mice together with γ-glutamyl transpeptidase, an oncofetal enzyme which has been associated with malignant transformation.[66-68] In man, hepatocytes containing Mallory bodies showed features of dysplastic cells and some contained γ-glutamyl transpeptidase and α-fetoprotein (AFP).[69,70] Atypical foci of hepatocytes in adenomatous hyperplastic nodules contained Mallory bodies and Mallory body-positive hepatocellular carcinoma cells showed more DNA hyperploidy or aneuploidy compared to negative cells.[71,72] While the role of hemosiderin, glycogen, or Mallory bodies in malignant transformation has not been defined, their presence in preneoplastic cells suggests that hepatocytes may be transformed while maintaining hepatocellular characteristics.

Silver staining for the nucleolar organizer region-associate proteins (AgNOR) may be useful in identifying hyperplastic, dysplastic, or neoplastic liver lesions. AgNOR counts of HCC and hepatocyte dysplasia were significantly higher than those of cirrhotic areas and normal hepatocytes.[73,74] In another study, there was a progressive increase of AgNOR counts from adenomatous hyperplasia to atypical foci to overt HCC.[75]

Flow cytometric or image analysis of nuclear DNA content may help to define precursor lesions, particularly since these studies can be performed on formalin-fixed, paraffin-embedded tissues after liberation of nuclei by enzymatic treatment following conventional histopathological evaluation of specimens.[76-78] In HCC, a correlation between the DNA ploidy pattern and biological behavior and consequently prognosis, has been demonstrated.[79,80] Separation of distinct subpopulations of preneoplastic cells according to cell size and ploidy by elutriation or by fluorescent-activated cell sorting should be possible and lead to a better definition of the precursor lesions.

Histochemical studies of selected marker enzymes (γ-glutamyl transpeptidase, glucose-6-phosphatase, adenosine triphosphatase, epoxide hydrase activities) in human HCC and its precursor lesions have been carried out in several laboratories, including our own.[32] The enzyme alterations in human HCC were similar to those reported for experimental HCC. However, the liver surrounding the tumors and a hepatocellular adenoma did not show any change of enzyme activity from normal. On the other hand, all hyperplastic nodules studied in nodular regenerative hyperplasia of the liver showed abnormal enzyme patterns supporting the notion that this lesion is preneoplastic.[81,82]

Although numerous receptors are expressed in human liver, they have not been studied in detail in HCC and its precursor lesions. Most HCC contained higher concentrations of androgen receptors in nuclei and cytosol than the surrounding liver parenchyma.[83-85] Low levels of estrogen receptors were found in normal human liver, focal nodular hyperplasia, and some HCC, while progesterone receptors and glucocorticoid receptors were detected in varying quantities.[86-89]

While the expression of AFP in HCC has been studied in detail, little is known about the induction of oncofetal proteins in preneoplastic hepatocytes. We and others detected AFP in histologically normal hepatocytes, in some dysplastic cells, and in putative oval cells in the nontumorous portion of cirrhotic livers of patients with HCC, but there was no preferential expression of AFP in preneoplastic cells.[90-97] Clearly, production of monoclonal antibodies to cell-surface and cytoplasmic antigens of human hepatocellular carcinoma cells and its precursor cells is needed.[98-100] The establishment of libraries of monoclonal antibodies to fetal, adult, regenerating, and transformed hepatocytes will be crucial for exploring lineage relationships between the cell populations involved in development, differentiation, and hepatocarcinogenesis.[101]

A large body of information has been accumulated on changes in gene expression, particularly of protooncogenes and growth factors, based on experimental studies of liver regeneration and transformation.[102] Expression of cellular oncogenes (c-ras, c-myc, p36, p53, and others) has been investigated in human HCC, but not in liver regeneration or transformation.[32] These important genes, as well as growth factors, such as transforming growth factor alpha, insulin-like growth factor II, and hepatocyte growth factor must be studied in order to understand the molecular mechanisms involved in liver development, regeneration, differentiation, and transformation.[103]

VI. PROLIFERATION OF BILE DUCTULES AND PUTATIVE OVAL CELLS

Proliferation of bile ductlike structures occurs in a variety of liver diseases.[24] In some forms of liver injury such as alcoholic liver disease, the proliferating bile ductules grow into the lobular parenchyma. In others, such as primary biliary cirrhosis, they remain confined to the portal tracts or fibrous septa. The origin of the proliferating biliary epithelial cells has not been

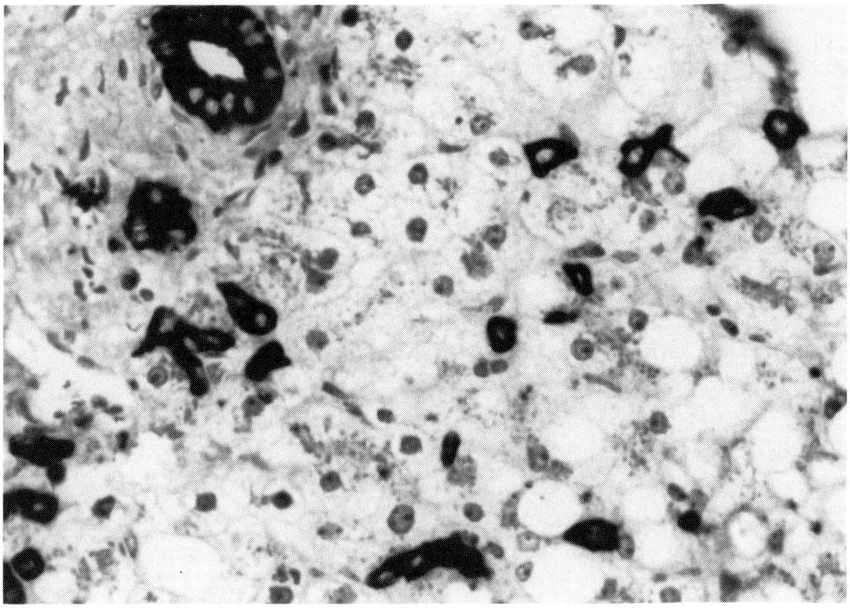

FIGURE 6. Liver of patient with alcoholic liver disease stained with polyclonal antibody to high-molecular weight CK. Small cells with oval nuclei are scattered throughout the hepatic lobule and express the same CK as a mature bile duct. (Immunoperoxidase staining, counterstained with hematoxylin; magnification × 250.)

determined. It is likely that they are derived either from proliferation of existing bile ducts, ductal plate remnants, or from ductular metaplasia of hepatocytes.[24,25,104] In an extensive review of this topic, Desmet[105] concluded that under different conditions either one or both mechanisms may give rise to bile ductules. The CK composition of the bile ductules and bile ducts differs from that of hepatocytes as revealed by immunohistochemistry using monoclonal antibodies to individual CK polypeptides.[106] The immunohistochemical staining pattern may be used to distinguish hepatocytes expressing CK8 and 18 from bile duct epithelial cells expressing CK8, 18, 7, and 19. In several liver diseases, such as alcoholic hepatitis, we found many small cells with oval nuclei scattered throughout the lobules either alone or in groups between hepatocyte plates (Figure 6).[107] Preliminary studies suggest that in contrast to hepatocytes and bile ducts, they express CK14 as well as AFP. Cytologically, these cells resemble oval cells in rats and may be related to pluripotent stem cells or bipotential progenitor cells which can develop into hepatocytes and HCC or into ductular cells and cholangiocarcinoma.[108]

The malignant counterpart of proliferating bile ductules and bile duct adenoma is the cholangiocarcinoma.[109] Precursor lesions of intrahepatic cholangiocarcinomas have not been defined, although occasional cholangiocarcinomas in association with multiple von Meyenburg complexes with atypia have been described.[110,111] Large bile duct carcinomas near the hilus of the liver may be associated with hyperplasia, atypical hyperplasia, and carcinoma *in situ* of bile duct epithelial cells.[112]

VII. CONCLUSIONS

Over the last decade, experimental studies have significantly advanced our knowledge of liver development, regeneration, and malignant transformation. The results of these investigations enable us to interpret the limited observations available on human liver cell lineages and come to the following tentative conclusions: early in gestation, immature, probably bipotential hepatoblasts derived from the endodermal lining of the foregut, terminally differentiate either into hepatocytes or via ductal plates at the margins of portal tracts into bile ducts. Differentiated cells of these two cell lineages, hepatocytes and bile duct epithelial cells, maintain the potential to multiply even during adult life. Thus, mild injury or focal cell loss result in replication of mature hepatocytes or bile ducts. In contrast, massive injury and some carcinogenic stimuli appear to target putative liver stem cells or bipotential precursor cells. Location of these progenitor cells may be related to ductal plate remnants, canals of Hering, or periductal cells. Phenotypically, they resemble oval cells, and after proliferation as ductular hepatocytes, they may differentiate into hepatocytes or biliary epithelial cells. Thus, two pathways appear to be possible for liver regeneration: (1) originating from bipotential precursor cells and (2) from differentiated hepatocytes or bile duct epithelial cells. Similarly, carcinogenesis seems to follow either one of two pathways: proliferation and malignant transformation of bipotential precursor cells resembling oval cells, or proliferation and transformation of differentiated hepatocytes or bile duct epithelial cells. Thus, the liver may present examples for both major theories of the cellular origin of cancer.

REFERENCES

1. **Gerber, M. A. and Thung, S. N.,** Histology of the liver, *Am. J. Surg. Pathol.,* 11, 709, 1987.
2. **Clearfield, H. R.,** Embryology, malformations, and malposition of the liver, in *Bockus Gastroenterology,* 4th ed., Berk, J. E., Haubrich, W. S., Kalser, M. H., Roth, J. L. A., and Schaffner, F., Eds., W. B. Saunders, Philadelphia, 1985, 2659.
3. **MacSween, R. N. M. and Scothorne, R. J.,** Developmental anatomy and normal structure, in *Pathology of the Liver,* 2nd ed., MacSween, R. N. M., Anthony, P. P., and Scheuer, P. J., Eds., Churchill Livingstone, Edinburgh, 1985, 1.

4. **Rappaport, A. M.,** Physioanatomic considerations, in *Diseases of the Liver,* 5th ed., Schiff, L. and Schiff, E. R., Eds., J. B. Lippincott, Philadelphia, 1982, 1.
5. **Shah, K. D. and Gerber, M. A.,** Development of intrahepatic bile ducts in humans: possible role of laminin, *Arch. Pathol. Lab. Med.,* 114, 597, 1990.
6. **Shah, K. D. and Gerber, M. A.,** Development of intrahepatic bile ducts in humans: immunohistochemical study using monoclonal cytokeratin antibodies, *Arch. Pathol. Lab. Med.,* 113, 1135, 1989.
7. **Eyken, P. V., Sciot, R., Callea, F., Van Der Steen, K., Moerman, P., and Desmet, V. J.,** The development of the intrahepatic bile ducts in man: a keratin-immunohistochemical study, *Hepatology,* 8, 1586, 1988.
8. **Stosiek, M. K. and Karsten, U.,** Expression of cytokeratin 19 during human liver organogenesis, *Liver,* 10, 59, 1990.
9. **Gerber, M. A., Thung, S. N., Shen, S., Stromeyer, F. W., and Ishak, K. G.,** Phenotypic characterization of hepatic proliferation, *Am. J. Pathol.,* 110, 70, 1983.
10. **Post, J. and Hoffman, J.,** Hepatic cell replication during growth and regeneration, in *Progress in Liver Diseases,* Vol. 2, Popper, H. and Schaffner, F., Eds., Grune & Stratton, New York, 1965, 155.
11. **Pack, G. T. and Islami, A. H.,** Surgical treatment of hepatic tumors, in *Progress in Liver Diseases,* Vol. 2, Popper, H. and Schaffner, F., Eds., Grune & Stratton, New York, 1965, 499.
12. **Seki, S., Sakaguchi, H., Kawakita, N., Yanai, A., Kim, K., Mizoguchi, Y., and Kobayashi, K.,** Identification and fine structure of proliferating hepatocytes in malignant and nonmalignant liver diseases by use of a monoclonal antibody against DNA polymerase alpha, *Hum. Pathol.,* 21, 1020, 1990.
13. **Grisham, J. W., Tillman, R. L., and Nagel, A. E. H.,** Ultrastructure of the proliferating hepatocyte: sinusoidal surfaces and endoplasmic reticulum, in *Liver Regeneration After Experimental Injury,* Reutter, L. R., Ed., Stratton Medical Book Corp., New York, 1975.
14. **Fausto, N.,** New perspectives on liver regeneration, *Hepatology,* 6, 326, 1986.
15. **Phillips, M. J. and Poucell, S.,** Modern aspects of the morphology of viral hepatitis, *Hum. Pathol.,* 12, 1060, 1981.
16. **Stromeyer, F. W. and Ishak, K. G.,** Nodular transformation (nodular "regenerative" hyperplasia) of the liver, *Hum. Pathol.,* 12, 60, 1981.
17. **Sell, S.,** Is there a liver stem cell?, *Cancer Res.,* 50, 3811, 1990.
18. **Fausto, N.,** Hepatocyte differentiation and liver progenitor cells, *Curr. Opinion Cell Biol.,* 2, 1036, 1990.
19. **Reid, L. M.,** Stem cell biology, hormone/matrix synergies and liver differentiation, *Curr. Opinion Cell Biol.,* 2, 121, 1990.
20. **Sirica, A. E., Mathis, G. A., Sano, N., and Elmore, L. W.,** Isolation, culture, and transplantation of intrahepatic biliary epithelial cells and oval cells, *Pathobiology,* 58, 44, 1990.
21. **Thorgeirsson, S. S. and Evarts, R. P.,** Experimental hepatocarcinogenesis: relationship between oval cells and hepatocytes in rat liver, in *Etiology, Pathology, and Treatment of Hepatocellular Carcinoma in North America,* Vol. 13, Tabor, E., Di Bisceglie, A. M., and Purcell, R. H., Eds., Gulf Publishing, Houston, 1991, 171.
22. **Shiojiri, N., Lemire, J. M. and Fausto, N.,** Cell lineages and oval cell progenitors in rat liver development, *Cancer Res.,* 51, 2611, 1991.
23. **Vandersteenhoven, A. M., Burchette, J., and Michalpoulos, G.,** Characterization of ductular hepatocytes in end-stage cirrhosis, *Arch. Pathol. Lab. Med.,* 114, 403, 1990.
24. **Thung, S. N.,** The development of proliferating ductular structures in liver disease, *Arch. Pathol. Lab. Med.,* 114, 407, 1990.
25. **Van Ekyen, P., Sciot, R., Callea, F., and Desmet, V. J.,** A cytokeratin-immunohistochemical study of focal nodular hyperplasia of the liver: further evidence that ductular metaplasia of hepatocytes contributes to ductular "proliferation", *Liver,* 9, 372, 1989.

26. **Wanless, I. R., Mawdsley, C. and Adams, R.,** On the pathogenesis of focal nodular hyperplasia of the liver, *Hepatology,* 5, 1194, 1985.
27. **Wanless, I. R., Godwin, T. A., Allen, F., and Fedler, A.,** Nodular regenerative hyperplasia of the liver in hematologic disorders: a possible response to obliterative portal venopathy, *Medicine,* 59, 367, 1980.
28. **Solis-Herruzo, J. A., Vidal, J. V., Colina, F., Santalla, F., and Castellano, G.,** Nodular regenerative hyperplasia of the liver associated with the toxic oil syndrome: report of five cases, *Hepatology,* 6, 687, 1986.
29. **Wanless, I. R.,** Micronodular transformation (nodular regenerative hyperplasia) of the liver: a report of 64 cases among 2,500 autopsies and a new classification of benign hepatocellular nodules, *Hepatology,* 11, 787, 1990.
30. **Colina, F., Alberti, N., Solis, J. A., and Martinez-Tello, F. J.,** Diffuse nodular regenerative hyperplasia of the liver (DNRH). A clinicopathologic study of 24 cases, *Liver,* 9, 253, 1989.
31. **Moran, C. A., Mullick, F. G., and Ishak, K. G.,** Nodular regenerative hyperplasia of the liver in children, *Am. J. Surg. Pathol.,* 15, 449, 1991.
32. **Gerber, M. A.,** Recent studies on the developing human hepatocellular carcinoma, *Cancer Surv.,* 5, 741, 1986.
33. **Nakhleh, R. E. and Snover, D. C.,** Use of alpha-1-antitrypsin staining in the diagnosis of nodular regenerative hyperplasia of the liver, *Hum. Pathol.,* 19, 1048, 1988.
34. **Ishak, K. G.,** Hepatic lesions caused by anabolic and contraceptive steroids, *Semin. Liver Dis.,* 4, 113, 1984.
35. **Thung, S. N. and Gerber, M. A.,** Benign and malignant neoplasms of the liver, in *Diseases of the Liver and Biliary Tract,* Gitnick, G. L., Ed., Mosby Year Book Publishers, St. Louis, 1992, 389.
36. **Govindarajan, S. and Peters, R. L.,** The bile duct adenoma: a lesion distinct from von Meyenburg complex, *Arch. Pathol. Lab. Med.,* 108, 922, 1984.
37. **Allaire, G. S., Rabin, L., Ishak, K. G., and Sesterhenn, I. A.,** Bile duct adenoma: a study of 152 cases, *Am. J. Surg. Pathol.,* 12, 708, 1988.
38. **Nakanuma, Y., Ohta, G., Sugiura, H., Watanabe, K., and Doishita, K.,** Incidental solitary hepatocellular carcinomas smaller than 1 cm in size found in autopsy: a morphologic study, *Hepatology,* 6, 631, 1986.
39. **Rapaccini, G. L., Pompili, M., Caturelli, E., Anti, M., Aliotta, A., Cedrone, A., and Amadei, E.,** Focal ultrasound lesions in liver cirrhosis diagnosed as regenerating nodules by fine-needle biopsy, *Dig. Dis. Sci.,* 35, 442, 1990.
40. **Grigioni, W. F., D'Errico, A., Bacci, F., Carella, R., and Mancini, A. M.,** Small liver masses in cirrhotic patients: a pathological clue for the morphogenesis of human hepatocellular carcinoma, *Acta Pathol. Jpn.,* 39, 520, 1989.
41. **Terada, T., Nakanuma, Y., Hoso, M., Saito, K., Sasaki, M., and Nonomura, A.,** Fatty macroregenerative nodule in non-steatotic liver cirrhosis: a morphologic study, *Virchows Arch. A: Pathol. Anat.,* 415, 131, 1989.
42. **Hirooka, N., Nitta, Y., Tsunoda, T., Kitazawa, E., Sato, J., Machii, A., Murakami, Y., Takahashi, M., and Kondo, F.,** Microscopic incipient hepatocellular carcinoma found incidentally in a routine liver biopsy specimen, *Hepatology,* 12, 291, 1990.
43. **Edmondson, H. A.,** Benign epithelial tumors and tumorlike lesions of the liver, in *Hepatocellular Carcinoma,* Okuda, K. and Peters, R. L., Eds., John Wiley & Sons, New York, 1976, 309.
44. **Arakawa, M., Kage, M., Sugihara, S., Nakashima, T., Suenaga, M., and Okuda, K.,** Emergence of malignant lesions within an adenomatous hyperplastic nodule in a cirrhotic liver, *Gastroenterology,* 91, 198, 1986.
45. **Kondo, F., Wada, N., Nagato, Y., Nakajima, T., Kondo, Y., Hirooka, N., Ebara, M., Ohto, M., and Okuda, K.,** Biopsy diagnosis of well-differentiated hepatocellular carcinoma based on new morphologic criteria, *Hepatology,* 9, 751, 1989.

46. **Ohno, Y., Shiga, J., and Machinami, R.,** A histopathological analysis of five cases of adenomatous hyperplasia containing minute hepatocellular carcinoma, *Acta Pathol. Jpn.,* 40, 267, 1990.
47. **Nakanuma, Y., Terada, T., Terasaki, S., Ueda, K., Nonomura, A., Kawahara, E., and Matsui, O.,** 'Atypical adenomatous hyperplasia' in liver cirrhosis: low-grade hepatocellular carcinoma or borderline lesion?, *Histopathology,* 17, 27, 1990.
48. **Sakamoto, M., Hirohashi, S., and Shimosato, Y.,** Early stages of multistep hepatocarcinogenesis: adenomatous hyperplasia and early hepatocellular carcinoma, *Hum. Pathol.,* 22, 172, 1991.
49. **Anthony, P. P., Vogel, C. L., and Barker, L. F.,** Liver cell dysplasia: a premalignant condition, *J. Clin. Pathol.,* 26, 217, 1973.
50. **Anthony, P. P.,** Precursor lesions for liver cancer in humans, *Cancer Res.,* 36, 2579, 1976.
51. **Lefkowitch, J. H. and Apfelbaum, T. F.,** Liver cell dysplasia and hepatocellular carcinoma in non-A, non-B hepatitis, *Arch. Pathol. Lab. Med.,* 111, 170, 1987.
52. **Wada, K., Kondo, F., and Kondo, Y.,** Large regenerative nodules and dysplastic nodules in cirrhotic livers: a histopathologic study, *Hepatology,* 6, 1684, 1988.
53. **Furuya, K. F., Nakamura, M., Yamamoto, Y., Togei, K., and Otsuka, H.,** Macroregenerative nodule of the liver: a clinicopathologic study of 345 autopsy cases of chronic liver disease, *Cancer,* 61, 99, 1988.
54. **Tanikawa, K.,** Regeneration of the hepatocyte after partial hepatectomy, in *Ultrastructural Aspects of the Liver and its Disorders,* Tanikawa, K., Ed., Igaku-Shoin, Tokyo, 1979, 105.
55. **Watanabe, S., Okita, K., Harada, T., Kodama, T., Numa, Y., Takemoto, T., and Takahashi, T.,** Morphologic studies of the liver cell dysplasia, *Cancer,* 51, 2197, 1983.
56. **Kagawa, K., Deguchi, T., Tomimasu, H., Okanoue, T., Okuda, K., Okuno, T., Taknio, T., Fujimiya, T., Kamachi, M., Tsuchibashi, Y., Ashihara, T., and Takeoka, O.,** Feulgen-DNA cytofluorometry of the liver cell dysplasia (LCD) in the liver cirrhosis, *Jpn. J. Gastroenterol.,* 81, 82, 1984.
57. **Koike, Y., Kamiryo, K., Suzuki, Y., Kiyosawa, K., Nagata, A., Furuta, S., and Nagata, T.,** DNA content of hepatocytes in various stages of liver cirrhosis, *Liver,* 5, 156, 1985.
58. **Chen, M.-L., Gerber, M. A., Thung, S. N., Thornton, J. C., and Chung, W. K.,** Morphometric study of hepatocytes containing hepatitis B surface antigen, *Am. J. Pathol.,* 114, 217, 1984.
59. **Roncalli, M., Borzio, M., Tombesi, M. V., Ferrari, A., and Servida, E.,** A morphometric study of liver cell dysplasia, *Hum. Pathol.,* 19, 471, 1988.
60. **Williams, G. M.,** Functional markers and growth behavior of preneoplastic hepatocytes, *Cancer Res.,* 36, 2540, 1976.
61. **Bannasch, P.,** Cytology and cytogenesis of neoplastic (hyperplastic) hepatic nodules, *Cancer Res.,* 36, 2555, 1976.
62. **Terada, T. and Nakanuma, Y.,** Iron-negative foci in siderotic macroregenerative nodules in human cirrhotic liver a marker of incipient neoplastic lesions, *Arch. Pathol. Lab. Med.,* 113, 916, 1989.
63. **Terada, T. and Nakanuma, Y.,** Survey of iron-accumulative macroregenerative nodules in cirrhotic livers, *Hepatology,* 10, 851, 1989.
64. **Terada, T., Kadoya, M., Nakanuma, Y., and Matsui, O.,** Iron-accumulating adenomatous hyperplastic nodule with malignant foci in the cirrhotic liver, *Cancer,* 65, 1994, 1990.
65. **Henmi, A., Uchida, T., and Shikata, T.,** Karyometric analysis of liver cell dysplasia and hepatocellular carcinoma, *Cancer,* 55, 5594, 1985.

66. **Borenfreund, E., Higgins, P. J., and Bendich, A.,** In vivo-in vitro rat liver carcinogenesis: modification in protein synthesis and ultrastructure, *Ann. N.Y. Acad. Sci.*, 349, 357, 1980.
67. **Denk, H., Gschnait, F., and Wolff, K.,** Hepatocellular hyalin (Mallory bodies) in long term griseofulvin-treated mice: a new experimental model for the study of hyalin formation, *Lab. Invest.*, 32, 773, 1975.
68. **French, S. W.,** The Mallory body: structure, composition and pathogenesis, *Hepatology,* 1, 76, 1981.
69. **Nakanuma, Y. and Ohta, G.,** Is Mallory body formation a preneoplastic change?, *Cancer,* 55, 2400, 1985.
70. **Nakanuma, Y. and Ohta, G.,** Expression of Mallory bodies in hepatocellular carcinoma in man and its significance, *Cancer,* 57, 81, 1986.
71. **Terada, T., Hoso, M., and Nakanuma, Y.,** Mallory body clustering in adenomatous hyperplasia in human cirrhotic livers, *Hum. Pathol.*, 20, 886, 1989.
72. **Hoso, M. and Nakanuma, Y.,** Cytophotometric DNA analysis of hepatocellular carcinoma with Mallory bodies, *Virchows Arch. A: Pathol. Anat.*, 416, 51, 1989.
73. **Crocker, J. and McGovern, J.,** Nucleolar organiser regions in normal, cirrhotic, and carcinomatous livers, *J. Clin. Pathol.*, 41, 1044, 1988.
74. **Nonomura, A., Mizukami, Y., Fujitsugu, M., and Nakanuma, Y.,** Identification of nucleolar organizer regions in non-neoplastic and neoplastic hepatocytes by the silver-staining technique, *Liver,* 10, 229, 1990.
75. **Terasaki, S., Terada, T., Nakanuma, Y., Nonomura, A., Unoura, M., and Kobayashi, K.,** Argyrophilic nucleolar organizer regions and alpha-fetoprotein in adenomatous hyperplasia in human cirrhotic livers, *Am. J. Clin. Pathol.*, 95, 850, 1991.
76. **Schwarze, E., Pettersen, E. O., Shoaib, M. C., and Seglen, O.,** Emergence of a population of small, diploid hepatocytes during hepatocarcinogenesis, *Carcinogenesis,* 5, 1267, 1984.
77. **Trask, B., van den Engh, G., Landegent, J., Jansen, N., and van der Ploeg, M.,** Detection of DNA sequences in nuclei in suspension by in situ hybridization and dual beam flow cytometry, *Science,* 230, 1401, 1985.
78. **Coon, J. S., Landay, A. L., and Weinstein, R. S.,** Flow cytometric analysis of paraffin-embedded tumors: implications for diagnostic pathology, *Hum. Pathol.*, 17, 435, 1986.
79. **Ishizu, H.,** Flow cytometric analysis of the nuclear DNA content of hepatocellular carcinoma, *Jpn. Surg.*, 19, 662, 1989.
80. **Wenming, C. and Mengchao, W.,** The biopathologic characteristics of DNA content of hepatocellular carcinomas, *Cancer,* 66, 498, 1990.
81. **Thung, S. N. and Gerber, M. A.,** Enzyme pattern and marker antigens in nodular 'regenerative' hyperplasia of the liver, *Cancer,* 47, 1796, 1981.
82. **Uchida, T., Miyata, H., and Shikata, T.,** Human hepatocellular carcinoma and putative precancerous disorders. Their enzyme histochemical study, *Arch. Pathol. Lab. Med.*, 105, 180, 1981.
83. **Wilkinson, M. L., Iqbal, M. J., and Williams, R.,** Characterization of high affinity binding sites of androgens in primary hepatocellular carcinoma, *Clin. Chim. Acta,* 152, 105, 1985.
84. **Nagasue, N., Yukaya, H., Chang, Y.-C., Ogawa, Y., Kohno, H., and Ito, A.,** Active uptake of testosterone by androgen receptors of hepatocellular carcinoma in humans, *Cancer,* 57, 2162, 1986.
85. **Ohnishi, S., Murakami, T., Moriyama, T., Mitamura, K., and Imawari, M.,** Androgen and estrogen receptors in hepatocellular carcinoma and in the surrounding noncancerous liver tissue, *Hepatology,* 6, 440, 1986.
86. **Nagasue, N., Ito, A., Yukaya, H., and Ogawa, Y.,** Estrogen receptors in hepatocellular carcinoma, *Cancer,* 57, 87, 1986.

87. **Bojar, H., Petzinna, D., Brolsch, Ch., and Staib, W.,** Steroid receptor status of focal nodular hyperplasia of the human liver, *Klin. Wochensch.,* 10, 446, 1984.
88. **Nagasue, N., Kohno, H., Chang, Y., Yamanoi, A., Nakamura, T., Yukaya, H., and Hayashi, T.,** Clinicopathologic comparisons between estrogen receptor-positive and -negative hepatocellular carcinomas, *Ann. Surg.,* 212, 150, 1990.
89. **Nagasue, N., Kohno, H., Yamanoi, A., Kimoto, T., Chang, Y., and Nakamura, T.,** Progesterone receptor in hepatocellular carcinoma correlation with androgen and estrogen receptors, *Cancer,* 67, 2501, 1991.
90. **Okita, K., Kodama, T., Harada, T., Noda, K., Fukumoto, Y., Takenami, T., Shigeta, K., Mizuta, M., and Takemoto, T.,** Early lesions and development of primary hepatocellular carcinoma in man-association with hepatitis B viral infection, *Gastroenterol. Jpn.,* 12, 51, 1977.
91. **Peyrol, S., Grimaud, J. A., Chayvialle, J. A., Veyre, B., Paliard, P., and Lambert, R.,** Tissular immunoenzymatic detection of hepatic alphafetoprotein in human hepatomas, *Digestion,* 18, 351, 1978.
92. **Sakamoto, S. I., Kawaharada, M., Yachi, A., and Wata, T.,** Precancerous changes in cirrhotic nodules of human livers, in *Carcino-Embryonic Proteins: Chemistry, Biology, Clinical Applications,* Vol. 2, Lehmann, F. G., Ed., Elsevier/North-Holland Biomedical Press, New York, 932, 1979.
93. **Roncalli, M., Borzio, M., DeBiagi, G., Servida, E., Cantaboni, A., Sironi, M., and Taccagni, G. L.,** Liver cell dysplasia and hepatocellular carcinoma: a histological and immunohistochemical study, *Histopathology,* 9, 209, 1985.
94. **Roncalli, M., Borzio, M., DeBiagi, G., Ferrari, A. R., Macchi, R., Tombesi, V. M., and Servida, E.,** Liver cell dysplasia in cirrhosis, *Cancer,* 57, 1515, 1986.
95. **Gerber, M. A., Thung, S. N., and Popper, H.,** Localization of fetal antigens in relation to HBsAg in hepatocytes, in *Virus and the Liver,* Bianchi, L., Gerok, W., Sickinger, K., and Stalder, G. A., Eds., MTP Press, Lancaster, England, 1980, 205.
96. **Sakamoto, S., Yachi, A., Anzai, T., and Wada, T.,** AFP-producing cells in hepatitis and in liver cirrhosis, *Ann. N.Y. Acad. Sci.,* 259, 253, 1975.
97. **Otsuru, A., Nagataki, S., Koji, T., and Tamaoki, T.,** Analysis of alpha-fetoprotein gene expression in hepatocellular carcinoma and liver cirrhosis by in situ hybridization, *Cancer,* 62, 1105, 1988.
98. **Takahashi, H., Ozturk, M., Wilson, B., Maki, A., Ozawa, K., Koizumi, M., Endo, K., Strauss, W., Shouval, D., and Wands, J.,** In vivo expression of two novel tumor-associated antigens and their use in immunolocalization of human hepatocellular carcinoma, *Hepatology,* 9, 625, 1989.
99. **Carlson, R. I., Ben-Porath, E., Shouval, D., Strauss, W., Isselbacher, K. J., and Wands, J. R.,** Antigenic characterization of human hepatocellular carcinoma, *J. Clin. Invest.,* 76, 40, 1985.
100. **Shouval, D., Eilat, D., Carlson, R. I., Adler, R., Livni, N., and Wands, J. R.,** Human hepatoma-associated cell surface antigen: identification and characterization by means of monoclonal antibodies, *Hepatology,* 5, 347, 1985.
101. **Holmes, C. H., Austin, E. B., Fisk, A., Gunn, B., and Baldwin, R. W.,** Monoclonal antibodies reacting with normal rat liver cells as probes in hepatocarcinogenesis, *Cancer Res.,* 44, 1611, 1984.
102. **Fausto, N. and Mead, J. E.,** Role of protooncogenes and transforming growth factors in normal and neoplastic liver growth, *Progress in Liver Disease,* Vol. 2, Popper, H. and Schaffner, F., Eds., Grune & Stratton, New York, 1989, 57.
103. **Castilla, A., Prieto, J., and Fausto, N.,** Expression of transforming growth factors $\beta 1$ and α in chronic liver disease, *N. Engl. J. Med.,* 324, 933, 1991.
104. **Gall, J. A. M. and Bhathal, P. S.,** Origin and involution of hyperplastic bile ductules following total biliary obstruction, *Liver,* 10, 106, 1990.

105. **Desmet, V. J.,** Modulation of biliary epithelium, in *Modulation of Liver Cell Expression,* Reutter, W., Popper, H., Arias, I. M., Heinrich, P. C., Keppler, D., and Landmann, L., Eds., MTP Press, Lancaster, England, 1987, 195.
106. **Cooper, D., Schermer, A., and Sun, T.-T.,** Biology of disease: classification of human epithelia and their neoplasms using monoclonal antibodies to keratins: strategies, applications, and limitations, *Lab. Invest.,* 52, 243, 1985.
107. **Lai, Y.-S., Thung, S. N., Gerber, M. A., Chen, M.-L., and Schaffner, F.,** Expression of cytokeratins in normal and diseased livers and in primary liver carcinomas, *Arch. Pathol. Lab. Med.,* 113, 134, 1989.
108. **Sell, S. and Dunsford, H. A.,** Evidence for the stem cell origin of hepatocellular carcinoma and cholangiocarcinoma, *Am. J. Pathol.,* 134, 1347, 1989.
109. **Nakajima, T., Kondo, Y., Miyazaki, M., and Okui, K.,** A histopathologic study of 102 cases of intrahepatic cholangiocarcinoma: histologic classification and modes of spreading, *Hum. Pathol.,* 19, 1228, 1988.
110. **Dekker, A., Fiebo, J. W., Kate, T., and Terpstra, O. T.,** Cholangiocarcinoma associated with multiple bile-duct hamartomas of the liver, *Dig. Dis. Sci.,* 34, 952, 1989.
111. **Burns, C. D., Kuhns, J. J., and Wieman, T. J.,** Cholangiocarcinoma in association with multiple biliary microhamartomas, *Arch. Pathol. Lab. Med.,* 114, 1287, 1990.
112. **Kurashina, M., Kozuka, S., Nakasima, N., Hirabayasi, N., and Ito, M.,** Relationship of intrahepatic bile duct hyperplasia to cholangiocellular carcinoma, *Cancer,* 61, 2469, 1988.

Chapter 11

DEVELOPMENT OF INTRAHEPATIC BILE DUCTS, DUCTULAR METAPLASIA OF HEPATOCYTES, AND CYTOKERATIN PATTERNS IN VARIOUS TYPES OF HUMAN HEPATIC NEOPLASMS

P. Van Eyken and V. J. Desmet

TABLE OF CONTENTS

ISBN 0-8493-4746-7
© 1992 by CRC Press, Inc.

I. INTRODUCTION

This chapter starts with a brief discussion of cytokeratins and their usefulness as markers of directions of epithelial differentiation. A following section deals with current insights into the development of the intrahepatic bile ducts. The transformation of hepatocytes into bile duct cells in the postnatal human liver, a process known as "ductular metaplasia of hepatocytes", is discussed in a separate section. A final section is devoted to the cytokeratin patterns of various benign and malignant human hepatic neoplasms.

II. CYTOKERATINS AS MARKERS OF EPITHELIAL DIFFERENTIATION

A. CYTOKERATINS AND THEIR USE IN PATHOLOGY

The cytoskeleton of cells is made up of microtubules, microfilaments, and intermediate filaments.[1] Cytokeratins constitute the most complicated group within the latter family and are characteristic of epithelial cells.[2-4] Using two-dimensional gel electrophoresis, Moll and co-workers[2] have identified and catalogued 19 different cytokeratin polypeptides in human epithelia with molecular weights ranging from 40 to 68 kDa. Human cytokeratins are usually referred to by their number in this "catalog of Moll".[2] The 19 different cytokeratins are encoded by different genes. According to isoelectric point, immunoreactivity, messenger RNA hybridization, and peptide mapping, the 19 cytokeratins can be divided into an acidic ($pI < 5.5$) family comprising cytokeratins 9 to 19 and a basic ($pI > 6$) family comprising cytokeratins 1 to 8.[2,4-8] They are also designated as type I or type II cytokeratins, respectively. Individual epithelial cells express a limited selection of two to ten cytokeratins, consisting of "pairs" of one type I and one type II cytokeratin.[2] The basic building block of the cytokeratin filament is indeed a heterotetramer.[3,4,9] The cytokeratin pairs expressed vary according to the type of epithelium: stratified squamous epithelia synthetize, as a rule, cytokeratins of higher molecular weight than simple epithelia, which contain two to four low-molecular weight cytokeratins.[3,4,6-8] Within the stratified squamous epithelia, the cytokeratin expression further varies according to the differentiation program (skin type, corneal type, esophageal type) (for more detailed discussion see References 6–8, 10). Despite the increasing amount of data on their distribution and chemical composition, the precise function of cytokeratins — and of intermediate filaments in general — is still unknown. The still prevailing hypothesis that cytokeratins provide structural strength to epithelial cells does not account for the tremendous diversity within the family. The observation that some cells *in vivo,* like cells of the early embryo and certain neurons, lack intermediate filaments altogether strongly pleads against a housekeeping role.[3,4,11,12] Instead, intermediate filaments and cytokeratins are probably involved in specialized functions related to the differentiation state of cells.[12]

It soon became clear that tumor cells continued to express the intermediate filament type of their (presumed) cell of origin (for reviews see Reference 13–15). Moreover, it appeared that epithelial tumors usually express the same cytokeratins as their presumptive nontumoral counterpart and that this pattern was even retained in metastases.[2,16,17] It was therefore suggested that cytokeratins were not only useful to distinguish epithelial tumors from nonepithelial tumors, but that determining the cytokeratin ''profile'' of a tumor or even a metastasis could help to determine its origin. The advent of monoclonal antibodies directed against single cytokeratin polypeptides made cytokeratin ''fingerprinting'' of tumors possible by immunohistochemical procedures, thus obviating the use of cumbersome gel electrophoretic techniques. Several excellent reviews have been devoted to the use (and pitfalls) of cytokeratin-immunohistochemistry in diagnostic pathology.[18-20]

B. CYTOKERATINS AS MARKERS OF LIVER PARENCHYMAL OR BILE DUCT CELL DIFFERENTIATION

Normal human adult liver parenchymal cells have a very simple cytokeratin composition and express only one cytokeratin pair: cytokeratin 8 (a type II cytokeratin with molecular weight 52 kDa) and cytokeratin 18 (a type I cytokeratin with molecular weight 45 kDa).[2,21-23] The cytokeratins of human, mouse, rat, and even cow liver are grossly similar, although some differences in molecular weight and isoelectric pH values exist.[21,23] The equivalent cytokeratins in rat liver are referred to as cytokeratin A (55 kDa) and cytokeratin D (49 kDa).[21,23] Intrahepatic bile duct cells in adult human liver express, in addition to cytokeratins 8 and 18, also cytokeratin 7 (a type II cytokeratin with molecular weight 54 kDa) and cytokeratin 19 (a type I cytokeratin with molecular weight 40 kDa).[24-28] The equivalents of the latter two cytokeratins in rat liver are not well known, but normal bile duct cells in adult rat liver can be stained with monoclonal antibodies directed against the human cytokeratins 7 and 19.[29,30]

The three-dimensional arrangement of cytokeratin-type intermediate filaments in the liver parenchymal cell has been studied extensively by French and co-workers[31,32] and Katsuma et al.[33] Cytokeratin filaments run in bundles throughout the cytoplasm without terminating and are especially dense around the canaliculus where they form a ''pericanalicular sheath'' enclosing actin filaments. Intermediate filaments attach to the cell membrane, to desmosomes, and to the nuclear lamina.[31,32] They also connect and interact with microtubules and microfilaments.[34] They are thought to mechanically support the contractile actin network around the bile canaliculus.[31-33] In addition, they may provide for the uptake and secretion of dyes and proteins and may be related to the expression of liver-specific genes.[31,35-37] Little is known about the precise organization of cytokeratin filaments in bile duct cells. The ''need'' for an additional two cytokeratins in bile duct cells as opposed to parenchymal cells and the specific function — if any — of cytokeratins 7 and 19 in the former is at present not understood.

This chapter mainly deals with problems related to cell lineage relationships in developing liver, the phenotypic modulation of hepatocytes, and the histogenesis of hepatic tumors. As pointed out by Marceau,[38] at least three classes of markers can be used to trace cell lineage and to define differentiation stages: markers based on metabolic activities of cells, specific surface-exposed components, and components of the cytoskeleton. Examples of the first category include markers as albumin, α-fetoprotein, tyrosine aminotransferase, and γ-glutamyl transpeptidase. Unfortunately, the expression of this class of markers is influenced by changes in the microenvironment.[38] Germain et al.[39] recently described two surface-exposed components recognized by the monoclonal antibodies HES_6 and BDS_7, specific for rat hepatocytes and bile duct cells, respectively. The antigens recognized by these antibodies, however, await chemical characterization and it is not known whether or not they cross react with human liver tissue. The data presented in this chapter were mainly obtained using immunohistochemical staining for individual cytokeratins. Even though the function of cytokeratins in hepatocytes and bile duct cells is poorly understood, they can be regarded — as in other organs and epithelia — as markers for directions of epithelial differentiation.[38] As such, cytokeratins 7 and 19 are within the liver markers for the biliary type of epithelial differentiation.

III. NORMAL DEVELOPMENT OF THE INTRAHEPATIC BILE DUCTS

During embryonic development, the liver arises as a bud of cells, the so-called hepatic diverticulum, growing from the cranioventral edge of the junction of the yolk sac with the foregut.[40,41] In man and in the mouse, the hepatic diverticulum consists of two parts: a pars caudalis or cystica and a pars cranialis or hepatica.[40,42,43] The latter gives off liver precursor cells that grow into the mesenchyme of the septum transversum. According to Elias,[44] cells of the visceral mesothelium also contribute to the liver cell mass. It is generally assumed that the mesenchymal cells of the septum transversum develop into the connective tissue of the liver.[40] The liver sinusoids presumably arise from the capillary plexus between the vitelline veins present in the septum transversum.[40,43] Part of the cephalic end of the diverticulum is reduced to a small tubular structure, the common hepatic duct.[43] Both hepatic ducts and possibly part of the cystic duct are also derived from this part of the diverticulum.[42] The caudal part of the hepatic diverticulum develops into the gallbladder (part of) the cystic duct, and the common bile duct.[43]

It is well established that the development of the intrahepatic bile ducts takes place in close association with the branches of the portal vein. Their precise origin, however, has been debated for decades (for review see References 40, 45, 46). Three main theories have been proposed. The first, or "ductal hypothesis", holds that the intrahepatic bile ducts are derived from an ingrowth of branches of the extrahepatic bile duct into the liver.[47,48] According to the second, "hepatocytogenic" theory, the intrahepatic biliary epithelium arises from primitive hepatocytes around portal vein branches.[40,42,49-59] A third theory combines elements of both theories mentioned above (for review see Reference 40). The majority of authors dealing with the subject favor the second hypothesis. Until recently, this theory derived its main support from routine light microscopic or electron microscopic observations. Several authors described cells with a morphological appearance "intermediate" between hepatocytes and bile duct cells.[49,58,59] Such cells were interpreted as a sign of transformation of hepatocytes into bile duct cells. In a few studies, (immuno)histochemical and (enzyme)histochemical techniques were applied to document the existence of "intermediate" cells by looking for the presence (or absence) of glycogen, glucose-6-phosphatase (both markers of liver parenchymal cells), various other enzymes, and proteins like albumin and α-fetoprotein.[54,55,60] A number of authors failed to find intermediate cells.[47,48,61]

Recently, however, the use of immunohistochemical staining techniques, especially for cytokeratin-type intermediate filaments and cell-surface markers, as well as cell-culturing experiments have provided new insights into the cell lineage relationships in developing liver.[38,45,46,55,56,62-66] The following description of the development of the intrahepatic bile ducts in man is based on two recent cytokeratin-immunohistochemical studies and is schematically represented in Figure 1.[63,67] During the first 7 weeks of human embryonic development, there are no intrahepatic bile ducts.[40,52] The early developing liver appears to be composed of at least bipotential progenitor cells, expressing cytokeratins 8, 18, and 19.[63,67] From about 9 weeks of gestation on, cells located around large portal vein branches close to the liver hilum become

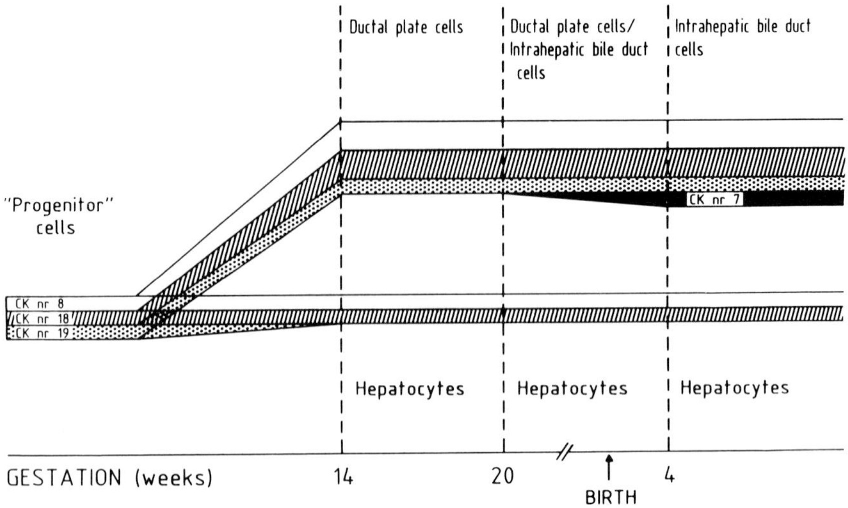

FIGURE 1. Schematic representation of the cell lineage relationships in developing human liver and the development of the intrahepatic bile ducts as determined by cytokeratin-immunohistochemistry.[63,67] For detailed discussion see text. (Reprinted by permission from Desmet et al., *Hepatology,* 12, 1249, 1990, Mosby-Year Book, Inc., St. Louis.)

more strongly positive for the latter three cytokeratins than surrounding cells. This layer of cells ensheathing the portal vein branches as a sleeve, is referred to as the "ductal plate" (Figure 2).[40,47,68,69] Ductal plates are duplicated over a variably long segment of their perimeter. Over the following weeks of gestation, ductal plates also appear around smaller portal vein branches distal to the hilum. In the meantime, cells not forming ductal plates gradually lose cytokeratin 19, and by 10 weeks of gestation they are only immunoreactive for cytokeratins 8 and 18, i.e., the cytokeratin pair normally present in liver parenchymal cells (Figure 2).[21,24] As gestation proceeds, segments of the ductal plates dilate and tubular structures are formed, which are subsequently incorporated into the mesenchyme surrounding the portal vein branches, thus giving rise to individualized bile duct tubes (Figure 3).[67] This "remodelation" of the ductal plates was observed from about 12 weeks of gestation on and first started around large portal vein branches. By 20 weeks of gestation, weak immunoreactivity for cytokeratin 7 was detected in large bile ducts. The intensity of immunoreactivity gradually increased, but only at about 1 month after birth it reached the level observed in adult liver. At birth, the smallest portal vein branches were not yet accompanied by an individual bile duct, but were still surrounded by a layer of cells in ductal plate arrangement.[67] The above findings were essentially confirmed by Shah and Gerber.[56]

Taken together, the studies discussed above provide strong evidence that the early developing human liver is composed of a uniform population of

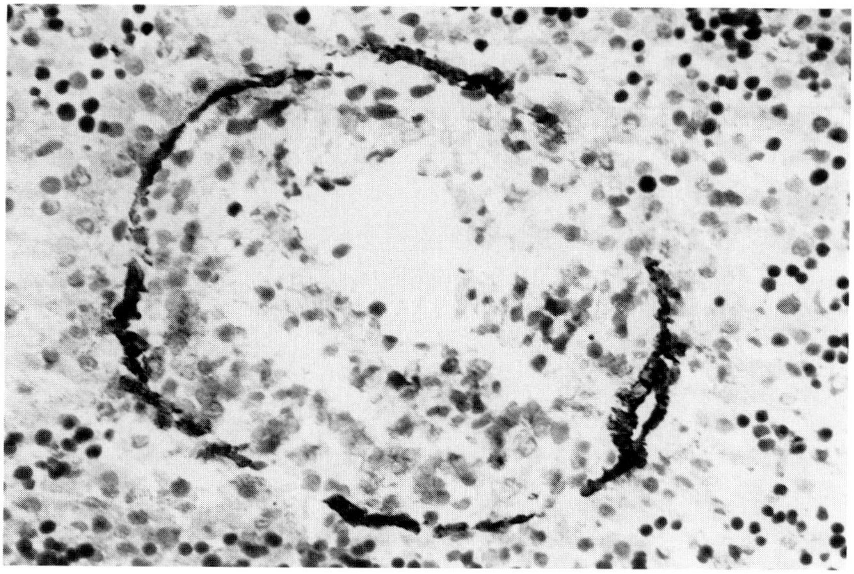

FIGURE 2. Cryostat section of human fetal liver, 20 weeks of gestation, immunostained for cytokeratin 19. A portal vein branch is surrounded by a layer of cells strongly immunoreactive for cytokeratin 19. This layer of cells is termed the "ductal plate". Note the presence of a tubular structure. Surrounding hepatocytes are negative. (Indirect immunoperoxidase, counterstained with Mayer's hemalum; original magnification × 400.)

progenitor cells (characterized by the expression of cytokeratins 8, 18, and 19) that can develop into either liver parenchymal cells (in which case cytokeratin 19 expression is lost) or into bile duct cells (in which case immunoreactivity for cytokeratins 8 and 18 becomes stronger and cytokeratin 7 is acquired). Recent studies using immunohistochemical staining for cytokeratins and cell surface markers of hepatocytes and bile duct cells as well as cell cultures have provided evidence for a similar sequence of events in developing rat liver.[38,39,50,62,65,70] A crucial question that at present goes unexplained, is what determines the developmental fate of the progenitor cells. The hypothesis that the extracellular matrix surrounding the portal vein branches plays a key role in this process is an old one dating back to the first half of the century.[49,52] The observation of Shiojiri[54] that fragments of mouse liver of 12.5 to 13.5 d of gestation cultured under the skin of newborn mice contained far more bile ducts than similar fragments cultured in the testis also points towards a role for the connective tissue in the process of bile duct development. Shah and Gerber[71] recently reported that laminin accompanies the ductal plate cells through all successive phases of differentiation and remodeling. The first attempt at defining the influence of the mesenchyme on bile duct development

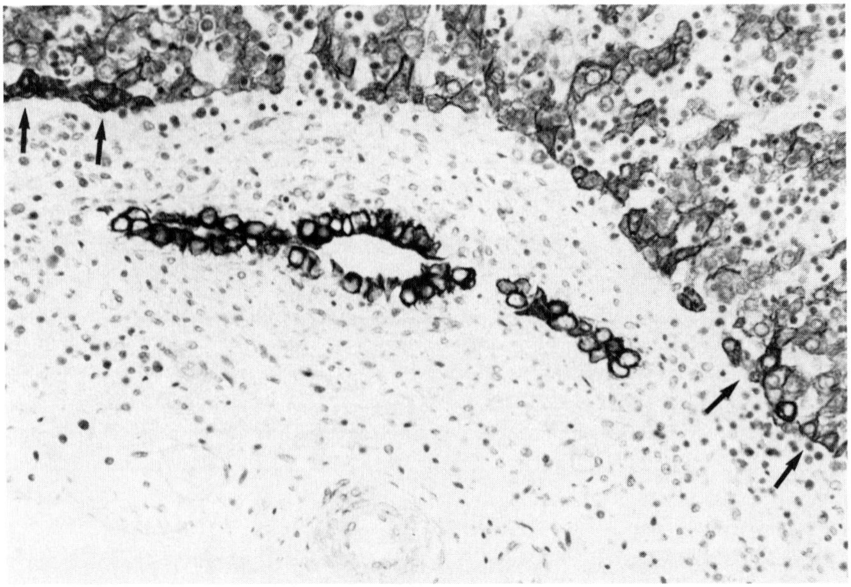

FIGURE 3. Paraffin section of human fetal liver, 16 weeks of gestation, stained with the monoclonal antibody CAM 5.2 reactive with cytokeratins 8, 18, and 19 illustrating the "remodeling" of the ductal plate. Part of the ductal plate is being incorporated into the mesenchyme surrounding the portal vein branch. The remaining part of the ductal plate at the interface with the liver parenchyma is strongly positive for cytokeratins (arrows). (Indirect immunoperoxidase, counterstained with Harris' hematoxylin; original magnification × 400.) (Reprinted by permission from Van Eyken et al., *Hepatology,* 8, 1586, 1988, Mosby-Year Book, Inc., St. Louis.)

by using cell cultures was made by Doljanski and Roulet.[72] New experiments along these lines are urgently needed.

Another important finding is that the bile duct system in both rat and man is still immature at birth.[67,70] This has important implications for the diagnosis of neonatal cholestatic diseases. The bile duct to portal space ratio is the criterion used to diagnose paucity of intrahepatic bile ducts. Kahn et al.[73] reported that this ratio increases with gestational age. This is easily explained by the observation that portal tracts only gradually acquire an individual bile duct.

A recently raised issue of tremendous importance is whether or not a number of bipotential progenitor cells ("stem cells") persist in the adult liver (see also Chapters 5 through 10).[38,39,46,56,62,63,74,75] Shah and Gerber[56,71] observed small ductal plate remnants accompanied by laminin in adult human liver and suggested that they might constitute a source of liver regeneration or ductular proliferation. Stosiek et al.[63] suggested that cytokeratin 19 could be a marker for a stem cell population, but evidence to support this hypothesis is still lacking. It is interesting to note that Stasiak et al.[76] found the expression of cytokeratin 19 in human tissues to be especially characteristic of regions of labile or variable cellular differentiation. They propose that cells may use cytokeratin 19 as a ''neutral'' keratin, which can polymerize (and thus form part of the intermediate filament cytoskeleton), but does not irrevocably commit a cell to a differentiation option.[76]

Several other problems related to the development of the intrahepatic bile ducts remain. It is not clear how the continuity between the different segments of the biliary tree remains or becomes established. In the mouse, it appears that some segments are initially discontinuous and that continuity is established secondarily.[42] During normal bile duct development an excess of structure is formed that subsequently disappears during the ''remodeling'' phase. It is unknown how this process is regulated. Failure of or arrest of remodeling of the ductal plates results in a morphological lesion known as the ''ductal plate malformation''.[40,68,69,77] The latter is a feature of a number of neonatal cholestatic diseases (for detailed discussion see References 69, 77).

IV. TYPICAL AND ATYPICAL DUCTULAR PROLIFERATION AND DUCTULAR METAPLASIA OF HEPATOCYTES

An increased number of ductular structures or ductular ''proliferation'' is a histological feature of various liver diseases. It is most pronounced in cholestatic liver diseases like extrahepatic bile duct obstruction, primary biliary cirrhosis, and primary sclerosing cholangitis, but it can also occur in alcoholic liver disease, (sub)massive hepatic necrosis, and focal nodular hyperplasia.[78-83] Some authors distinguish among different types of ductular proliferation.[79,83-85] The first type, referred to as ''typical'' ductular proliferation, is characterized by the presence of well-formed ductules with a circular profile and a well-formed lumen.[79,85] It is typically observed in acute biliary obstruction. In the second type or ''atypical'' ductular proliferation, the ductules consist of epithelial cells with a small cytoplasmic volume, arranged in anastomosing cords with poorly defined lumina. This type of ductular proliferation is usually seen in chronic cholestatic diseases.[79,83-85] A third type of ductular proliferation is seen in rat liver after the administration of various carcinogenic and noncarcinogenic chemicals and consists of the proliferation of small cells with an oval shape, coined ''oval'' cells.[79,86,87] Oval cells have (until present) not been described in human liver. The origin, nature, and differentiation capacities of oval cells and their role in the development of hepatocellular carcinoma are extensively discussed by Fausto et al. elsewhere in this volume.

Ever since the original description of ductular proliferation by Waldeyer[88] in 1868, the origin of the proliferating ductules has been subject to debate. As will be discussed, the controversy partly stems from the use of different material by different investigators (e.g., liver specimens of rats subjected to bile duct ligation vs. liver specimens from patients with chronic cholestatic diseases like primary biliary cirrhosis). Some authors claim that the ''proliferating'' ductules are derived from multiplication of preexisting ductules.[29,30,89-94] Others state that the ductular increase is at least partly due to a transformation or ''transdifferentiation'' of hepatocytes into bile duct-type cells, a process termed ''ductular metaplasia of hepatocytes''[60,64,80,83,85,97-106] (for a general discussion on ''transdifferentiation'' see References 95 and 96). Methods used to trace the origin of the ductules include routine light and electron microscopy, three-dimensional reconstruction, stereology or preparation of corrosion casts, thymidine labeling and histoautoradiography, enzyme histochemistry, and, more recently, immunohistochemistry.[26,29,30,60,64,83,94-98,97-110] A central issue in all studies is the presence or absence of cells with morphologic or functional features ''intermediate'' between those of hepatocytes and bile duct cells. The presence of such ''intermediate'' cells constitutes strong evidence in favor of the ductular metaplasia concept.

A. EXPERIMENTAL BILE DUCT LIGATION: AN ANIMAL MODEL OF DUCTULAR PROLIFERATION

Most authors performing bile duct ligation in rats claim that the proliferating ductules are derived from extant bile ducts.[29,30,90,92-94,107,109] The ultrastructural appearance of the proliferated bile ducts corresponds to that of normal bile duct cells.[29,91,93] Three-dimensional reconstruction has revealed continuity between the proliferated ductules and the preexistent parts of the biliary tree.[97] This is also supported by volume measurements of the biliary tree in bile duct-ligated rats using [^{3}H]taurocholate as a marker of biliary transit time.[29,30] The proliferated ductules are immunoreactive for cytokeratins 7 and 19 and are also positive for γ-glutamyl transpeptidase.[29,30,92,107,109,111] Intrahepatic bile duct cells isolated from bile duct-ligated rats express high γ-glutamyl transpeptidase and high alkaline phosphatase activities, but are essentially negative for glucose-6-phosphatase activity, glycogen content, and albumin production (the latter three being markers of liver parenchymal cells), indicating that they do not contain a population with ''transitional'' features.[109] It has to be mentioned, however, that Sirica and Cihla[109] reported an occasional bile ductular cell to exhibit a strong cytoplasmic binding of [6,7-^{3}H]estradiol, an indirect measure of α-fetoprotein production. The proliferated ductules retain functional characteristics of normal bile duct cells, since they spontaneously secrete bile and are responsive to secretin.[29,92] Histoautoradiography using ^{3}H-thymidine incorporation provided data consistent with a multiplication of extant ductules.[30,90,93] Slott et al.[30] have clearly shown that bile duct cells divide irrespective of the size of the duct in which they are located and that this proliferation results mainly in elongation, but not in circumferential enlargement or sprouting of side branches. It thus appears that in experimental bile duct ligation in the rat, the increase in ductular structures is mainly due to a multiplication of preexisting ductules retaining their phenotypic and functional characteristics. This assumption is extremely important since several investigators involved in isolating intrahepatic bile duct cells are currently using bile duct ligation as a means of increasing the yield.[107,112-117] Data obtained by this approach can only be applied to normal intrahepatic biliary epithelium if the proliferated cells are indeed entirely comparable to their normal resting counterparts. In this respect it is worth mentioning that Milani et al.[111] found a strong expression of the intermediate filament vimentin in proliferating ductules in bile duct-ligated rats. Since normal liver contains only few weakly vimentin-positive bile duct cells, this finding may point towards subtle differences between ''proliferating'' and ''resting'' ductules. Using cytokeratin immunohistochemistry on livers of bile duct-ligated rats, Gall and Bhathal[94] have demonstrated very small numbers of cells with features intermediate between hepatocytes and bile duct cells, indicating that ductular metaplasia may still play a (very minor) part. The capacity of rat hepatocytes to undergo ductular metaplasia is also demonstrated by a recent experimental study by Hillan et al.[101] Isolated rat hepatocytes were transplanted into the spleen of syngeneic animals which subsequently underwent common bile duct ligation. Light microscopically, the spleen contained numerous ductular structures, and cells with ultrastructural features of both hepatocytes and bile duct cells were found.

Histoautoradiographic experiments using [^{3}H]taurocholate have demonstrated that in extrahepatic cholestasis, bile salts continue to be taken up from the blood by hepatocytes and secreted into bile. It also appears that the proliferated biliary epithelium may play a role in the reabsorption of bile salts from the bile.[118] Ductular proliferation may thus provide the cholestatic liver with a means of getting rid of secreted bile salts.

The factors inducing ductular proliferation following bile duct ligation are only beginning to be elucidated. Shibayama[119] implicated mechanical factors or direct actions of retained bile in the liver. Gall and Bhathal[120] used cultured intact intrahepatic bile ducts to assess putative inducers of biliary epithelial cell hyperplasia including proline, lithocholic acid, and liver extracts from livers of normal rats or rats subjected to bile duct ligation, but failed to find an increase in bile duct cell replication. An elegant study by Slott et al.[30] indicates that an increased biliary pressure rather than retention of irritating and/or stimulating biliary constituents is the initiating factor in cell division. Increasing the biliary pressure to $\sim$65% of bile secretory pressure resulted in thymidine labeling indices virtually the same as those produced by common bile duct ligation.[30] Portal tract infiltration by macrophages apparently does not play a role in the hyperplastic reaction.[30]

The fate of the proliferated ductules has also been studied by several groups.[90,94,121,122] Cameron and Prasad[122] suggested that during recovery after bile duct ligation, ductular cells were partly conversed into liver parenchymal cells. Wilson and Leduc[123] also claimed that during recovery of the mouse liver after dietary injury, ductular cells transformed into hepatocytes. Using ^{3}H-thymidine labeling, Rubin[90] failed to find a transfer of label of ductular cells to hepatocytes or any other cell type. Recent electron-microscopical studies have firmly established that involution of the ductular response takes place by apoptosis.[94,121]

B. DUCTULAR PROLIFERATION IN HUMAN LIVER DISEASE

In chronic cholestatic diseases in man, a large body of evidence is clearly in favor of the ductular metaplasia concept. The geometrical shape of the ductules corresponds to flattened cylinders, and their three-dimensional anatomy is quite similar to that of the hepatic muralia.[108] The elaborate three-dimensional reconstruction studies by Buyssens[97] have also provided arguments in favor of a transformation of hepatocytes into ductules. Sophisticated computer-assisted, three-dimensional reconstructions of the biliary pathways in primary biliary cirrhosis have shown that atypical ductules communicate with liver cell plates, but not with ductules and ducts,[83] suggesting that they arise from the former. Some cells forming cholestatic liver cell rosettes were shown to have a typical hepatocellular arrangement of the nuclear chromatin and to contain alpha particles of glycogen. However, they also contained less organelles than hepatocytes and were resting on a basement membrane.[102] Such cells qualify as ''intermediate'' cells. Similar cells were described by Uchida and Peters[98] in alcoholic liver disease. Wegmann et al.[60] addressed the issue of ductular proliferation using enzyme histochemistry on specimens of cirrhosis of various etiologies. They found cells with the morphological appearances of hepatocytes, but histochemical staining characteristics intermediate between hepatocytes and bile duct cells, as well as cells with the appearance of bile duct cells, but displaying certain histochemical staining characteristics of hepatocytes (e.g., glucose-6-phosphatase activity). Glycogen and glucose-6-phosphatase activity were also demonstrated in ''proliferated ductules'' by Uchida and Peters.[98] In liver diseases associated with ductular proliferation, hepatocytes in acinar zone 1 may become immunoreactive for the S-100 protein, which is (weakly) expressed in normal intrahepatic bile duct cells.[124,125] In alcoholic liver disease and extrahepatic bile duct atresia, positive staining of liver parenchymal cells for tissue polypeptide antigen has been reported.[64,77] The latter antigen is normally restricted to intrahepatic bile duct cells.[64]

Recently, cytokeratin immunohistochemistry was used by several investigators to examine the origin of proliferated ductules. The first studies used antibodies reactive with several cytokeratin polypeptides.[85,99,100,106] Hepatocytes in acinar zone 1 (i.e., the periportal region) as well as hepatocytes forming cholestatic liver cell rosettes were immunoreactive with a polyclonal anticytokeratin antiserum that in the normal liver stains bile duct cells only.[64,85,99,100,106] The availability of monoclonal antibodies directed against the individual cytokeratin polypeptides 7 and 19 opened up the possibility of studying the changes in cytokeratin expression in more detail.[26,104,105,126] Van Eyken et al.[104] studied 38 liver biopsies of cases of primary biliary cirrhosis (stages I to IV), primary sclerosing cholangitis, chronic extrahepatic biliary obstruction, and drug-induced cholestasis. In all specimens, bile ducts and ductules were positive for cytokeratins 7, 8, 18, and 19, whereas hepatocytes were immunoreactive for cytokeratins 8 and 18. In 23 cases, a variable number of liver parenchymal cells were, in addition, positive for the "bile duct type" cytokeratin 7. Cytokeratin 7-positive cells were most numerous in acinar zone 1 and often appeared continuous with ductules at the interface between the connective tissue of portal tracts or septa and the liver parenchyma (Figure 4). Cells forming cholestatic liver cell rosettes were also often stained for cytokeratin 7. The number of parenchymal cells immunoreactive for cytokeratin 7 also increased with the duration of the cholestasis or the stage of the disease (in cases of primary biliary cirrhosis). In three cases of primary sclerosing cholangitis and two cases of primary biliary cirrhosis (one stage III, one stage IV), a small number of hepatocytes were, in addition, positive for cytokeratin 19. These results provide strong arguments in favor of a gradual transformation of hepatocytes into bile duct-type cells as evidenced by the gradual appearance of "bile duct-type" cytokeratins in hepatocytes. Moreover, the data indicate that this process is similar in cholestasis of various etiologies. It appears that during the process of ductular metaplasia cytokeratin 7 appears earlier than cytokeratin 19. Interestingly, this is the opposite of the sequence of events observed during the embryonic development of the intrahepatic bile ducts (see previous section). A recent study by Thung,[103] using the monoclonal antibody AE1, which in the liver is specific for cytokeratin 19, essentially confirmed the findings by Van Eyken et al.[104] In a cytokeratin-immunohistochemical study of alcoholic liver disease, immunoreactivity of hepatocytes for cytokeratin 7 was found in 23 out of 40 cases and cytokeratin 19 was, in addition, demonstrable in 7 cases.[127] However, the localization of cytokeratin 7-positive hepatocytes was different from that observed in cholestatic liver disease. In cases with a histologic picture of alcoholic steatosis, they were found in acinar zone 3. In cases of alcoholic hepatitis, parenchymal cells immunoreactive for cytokeratin 7 were present both in acinar zones 1 and 3, and in alcoholic cirrhosis they were seen both in the center and at the periphery of cirrhotic nodules. Hepatocytes displaying positive staining for both cytokeratins 7 and 19 were only seen in advanced disease (mostly cirrhosis). In cases where hepatocytes positive for "bile duct-type" cytokeratins were present in acinar zone 1 or in a periseptal location, they sometimes appeared in continuity with ductular structures. The difference in localization of cytokeratin 7/19-positive cells (in particular their occurrence in acinar zone

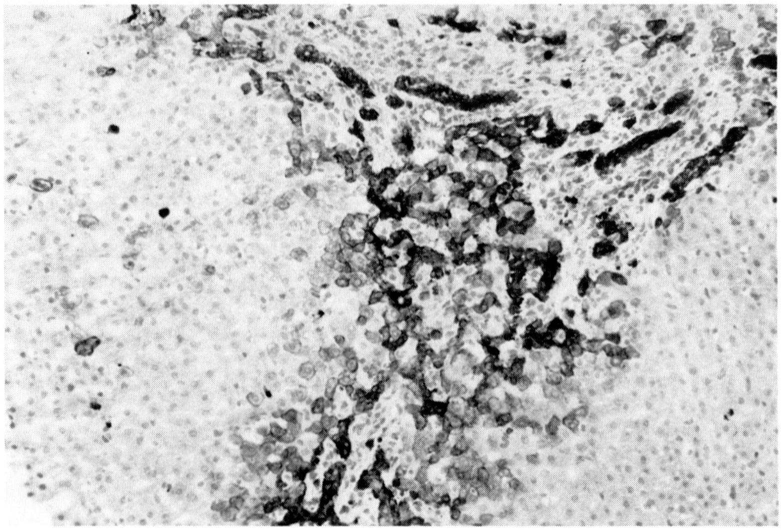

FIGURE 4. Chronic extrahepatic biliary obstruction. Cryostat section stained for cytokeratin 7. Bile ductules are strongly positive. Note the presence of numerous liver parenchymal cells in acinar zone 1 displaying varying degrees of immunoreactivity for cytokeratin 7. Some positive hepatocytes appear to be continuous with ductules. (Indirect immunoperoxidase, counterstained with Mayer's hemalum; original magnification × 160.) (Reprinted by permission from Van Eyken et al., *Histopathology,* 15, 125, 1989, Blackwell Scientific Publications, Oxford.)

3) as compared to cholestatic liver disease is unexplained, but it may be related to a specific effect of alcohol or its metabolites.

The recently reported expression of Le^a and Le^b blood group antigens and of the VLA 2, 3, and 6 by acinar zone 1 hepatocytes in diseases associated with ductular proliferation lends further support to the ductular metaplasia concept, since the former antigens are normally restricted to bile duct cells.[128-130] Proliferating ductules and some acinar zone 1 hepatocytes also express immunoreactivity for type 4 chain H blood group antigens.[131] The appearance of the gamma subunit of the glycolytic enzyme enolase (which is normally expressed in bile duct cells, but not in hepatocytes) in liver parenchymal cells in chronic hepatitis and cirrhosis is a further indication that hepatocytes may acquire phenotypic characteristics of bile duct cells.[132]

Although the increase in the number of ductular structures in chronic cholestatic human liver seems to be largely due to ductular metaplasia, real multiplication of ducts and ductules can also occur. Analogous to the model of bile duct ligation in the rat discussed above, this apparently occurs predominantly in acute extrahepatic bile duct obstruction. The latter condition is histologically characterized by proliferation of "typical" ductules. Nakanuma and Ohta[85] have reported "typical" ductules to be strongly positive for cytokeratin and to exhibit apical immunoreactivity for carcinoembryonic antigen and epithelial membrane antigen, an immunophenotype identical to that of normal bile ducts. Real multiplication of preexisting ductules and ductular metaplasia are by no means mutually exclusive mechanisms. James et al.[133] studied liver specimens of patients with an obstruction of the extrahepatic bile duct in whom the approximate duration of the jaundice was known. Histologically, an increased number of ductules was seen from 2 weeks of icterus on. Immunohistochemically, cytokeratin 7 was detected in a small number of acinar zone 1 hepatocytes from 4 weeks of obstruction on, and their number gradually increased. These findings may indicate that in the acute phase of the obstruction, a real multiplication of ductules takes place with ductular metaplasia becoming increasingly more important as the disease enters a chronic stage.

It is not known whether or not the "atypical" ductules also exhibit functional characteristics of normal biliary epithelium. Electron-microscopic observations by Sasaki et al.[134] suggest that they are also engaged in reabsorption. The presence of lipofuscin and bilirubin in ductular cells also indicates that they can reabsorb biliary constituents.[135] This would again provide the liver parenchymal cells with an efficient mechanism of getting rid of bile acids. Yamada et al.[83] have shown that the "typical" ductules in primary biliary cirrhosis are connected to interlobular bile ducts, and as such may act as a way to bypass obstructed or obliterated segments of the biliary tree. The expression of chromogranin-A (a molecule present in the matrix of neuroendocrine granules) and of the neural cell adhesion molecule N-CAM by bile ductules in cholestatic liver disease shows that these ductules have neuroendocrine features.[136] At the ultrastructural level, they were shown to contain dense cored granules. The contents of the latter are at present unknown, but one may speculate that they may play a role in the process of ductular reaction. Moreover, Roskams et al.[136] found a few hepatocytes in long-standing cholestatic liver disease positive for chromogranin-A, providing further evidence for the ductular metaplasia concept.

The mechanisms inducing a proliferation of preexisting ductules or ductular metaplasia in man are unknown. It is conceivable that in acute extrahepatic obstruction an increased pressure may play a role, as it apparently does in experimental bile duct ligation in the rat. As to ductular metaplasia, there is evidence pointing towards a role for the extracellular matrix in this process. It is well established that the extracellular matrix exerts a profound influence on the differentiated function of cells, and the role of the matrix in the normal development of the intrahepatic bile ducts has already been discussed.[137,138] Schulz et al.[139] and Hahn et al.[140] have reported changes in the composition of the extracellular matrix surrounding hepatocytes undergoing ductular metaplasia, in particular the appearance of collagen type V and laminin around the latter. In addition to the extracellular matrix, humoral factors may play a role.[101] The occurrence of ductular metaplasia in acinar zone 1 and its association with chronic cholestasis had led Desmet[80] to hypothesize that retention of bile constituents may also play a role.

V. CYTOKERATIN EXPRESSION IN HUMAN HEPATIC NEOPLASMS

A. BENIGN LIVER TUMORS AND TUMORLIKE CONDITIONS

1. Adenoma

Published data on the cytokeratin expression of liver cell adenoma are lacking. In the experience of the authors, adenomas are immunoreactive for cytokeratins 8 and 18 and show no immunoreactivity for cytokeratins 7 and 19. This is consistent with the concept that adenomas represent a benign tumoral growth of hepatocytes.

2. Focal Nodular Hyperplasia

Focal nodular hyperplasia is a tumorlike lesion composed of nodules made up of hepatocytes, separated by fibrous septa. The latter contain numerous ductular structures, but genuine interlobular bile ducts are lacking.[99,141,142] The lesion is probably a hyperplastic response of the liver parenchyma to a preexisting vascular malformation.[143] Butron Vila et al.[99] reported staining of hepatocytes close to fibrous septa with a polyclonal anticytokeratin antiserum that in the normal liver stains bile ducts only. Based on this finding, it was suggested that the ductular structures arose by transformation of hepatocytes into bile duct cells.[99] Using monoclonal antibodies specific for individual cytokeratin polypeptides, Van Eyken et al.[105] have shown that the liver cells of the nodules are immunoreactive for cytokeratins 8 and 18, whereas the ductules are positive for cytokeratins 7, 8, 18, and 19. In all specimens analyzed, a variable number of hepatocytes close to fibrous septa or inside the nodules displayed positively for cytokeratin 7. The cytokeratin 7-positive cells close to the septa were often continuous with ductular structures (Figure 5). In four out of eight specimens a few hepatocytes were, in addition, positive for cytokeratin 19. Fischer et al.[144] reported staining of hepatocytes in one case of focal nodular hyperplasia for cytokeratins 7 and 19. These findings are very well comparable to those obtained in cholestatic liver disease (see Section on ductular proliferation). Butron Vila et al.[99] have drawn attention to the fact that hepatocytes inside the lesion are still capable of secreting bile (or some bile constituents). Since they do not have real excretory pathways at their disposal, focal nodular hyperplasia qualifies as a ''cholestatic'' condition. It is thus very well conceivable that the ductular metaplasia arises by the same mechanisms operative in chronic cholestatic liver diseases.

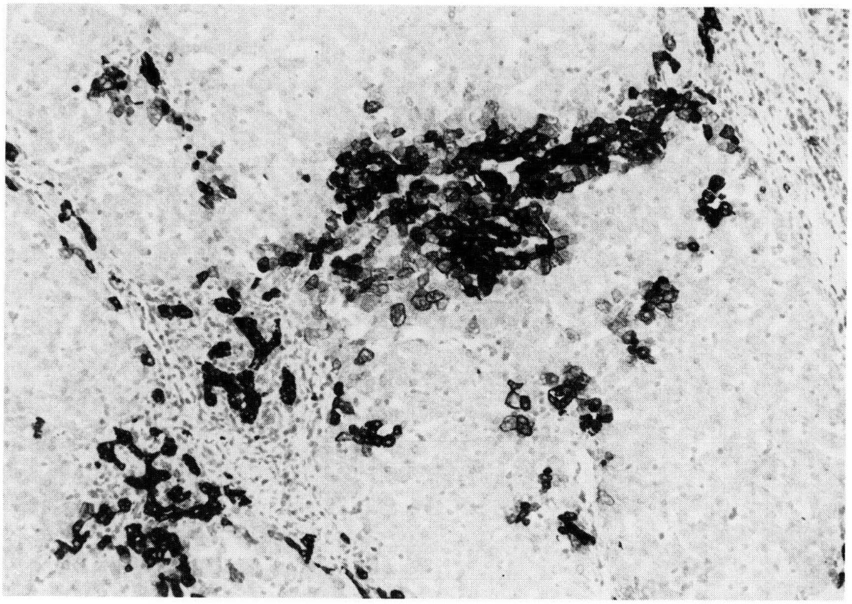

FIGURE 5. Focal nodular hyperplasia. Cryostat section stained for cytokeratin 7. Ductular structures are strongly immunoreactive. Numerous hepatocytes, sometimes in continuity with ductules, are also positive. (Indirect immunoperoxidase, counterstained with Mayer's hemalum; original magnification × 160.)

B. MALIGNANT LIVER TUMORS

1. Hepatocellular Carcinoma

a. Classical Hepatocellular Carcinoma

Denk et al.[22] were the first to show that cytokeratins continue to be expressed when hepatocytes become neoplastic. Moreover, using immunohistochemistry and one- and two-dimensional gel electrophoresis of cytoskeletal extracts, they showed that the hepatocyte-specific pattern of cytokeratin polypeptides is maintained. They analyzed five cases. Moll et al.[2] studied five cases and obtained comparable results. Subsequently, Osborn et al.[28] and Fischer et al.[144] studied two and six cases, respectively, using monoclonal antibodies specific for cytokeratins 7, 8, 18, and 19. Osborn et al.[28] found neoplastic cells to be positive for cytokeratins 8 and 18 in all cases studied. Fischer et al.[144] found focal positivity of tumor cells for cytokeratin 19 in one case. Van Eyken et al.[145] analyzed 34 "classical" hepatocellular carcinomas using monospecific anticytokeratin antibodies. Tumor cells were immunoreactive for cytokeratins 8 and 18 in all cases. An "expected" pattern of reactivity was exhibited in 17 cases in that no staining for cytokeratins 7 or 19 was detected (Figure 6). In 17 cases, however, a variable number of

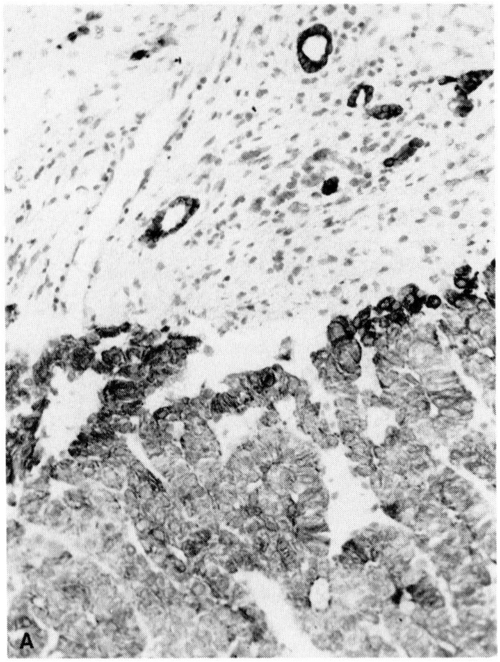

FIGURE 6. Serially cut cryostat sections of a moderately differentiated hepatocellular carcinoma. Tumor cells are strongly positive for cytokeratin 8 (A), but display no immunoreactivity for cytokeratin 19 (B). Note that adjacent nonneoplastic ductules are strongly positive for both cytokeratins 8 and 19. (Indirect immunoperoxidase, counterstained with Mayer's hemalum; original magnification × 200.) (Reprinted by permission from Van Eyken et al., *Hum. Pathol.*, 19, 562, 1988, W. B. Saunders, Orlando, FL.)

tumor cells were positive for cytokeratin 7 (2 cases), cytokeratin 19 (7 cases), or cytokeratins 7 and 19 (8 cases (Figure 7). Interestingly, only 3 out of 11 well-differentiated tumors showed an "unexpected" pattern of immunoreactivity as opposed to 7 out of 7 poorly differentiated tumors. These staining results have in the meantime been confirmed using other monoclonal antibodies specific for cytokeratin 7 (RCK107) or cytokeratin 19 (A53-B/A2) (personal observations).[146]

The data reported by Van Eyken et al. are in conflict with a number of other studies.[110,147,148] Johnson et al.[147] studied 32 hepatocellular carcinomas using the monoclonal antibody AE1 (which, in the liver, reacts with cytokeratin 19) on paraffin sections. Reactivity of tumor cells with this antibody was found in one case. Lai et al.,[110] using the same antibody, failed to find any reactivity in ten hepatocellular carcinomas. Battifora,[18,19] on the other hand, reported staining of 40 to 60% of paraffin-embedded classical hepatocellular carcinomas with AE1. Balaton et al.[148] analyzed 20 hepatocellular

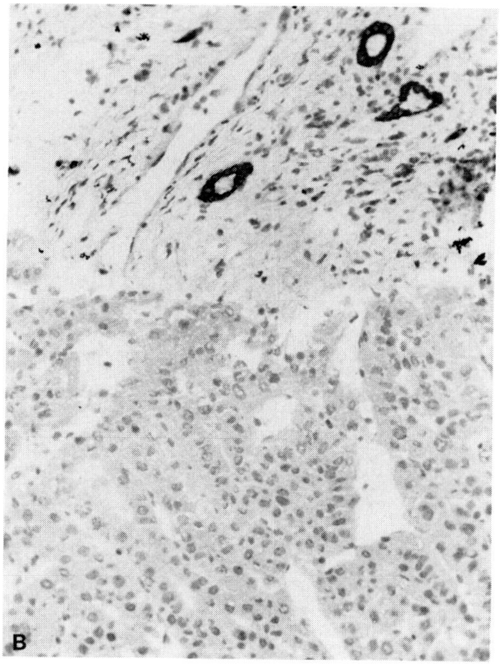

FIGURE 6. *Continued.*

carcinomas on paraffin section using a monoclonal antibody specific for cytokeratin 19 and found none to be reactive. These discrepancies can be explained by several factors. It is well established that fixation and paraffin embedding can destroy or diminish the immunoreactivity of cytokeratin polypeptides.[18,19,24] Monoclonal antibodies reactive with individual cytokeratin polypeptides in particular seem to produce a less reliable staining of paraffin sections.[24] The sensitivity of the immunohistochemical detection procedure and the pretreatment of the paraffin sections with a proteolytic enzyme may further influence the staining results.[149]

b. Fibrolamellar Carcinoma

Fibrolamellar carcinoma is a rare variant of hepatocellular carcinoma, constituting a separate clinicopathologic entity. Its morphological appearance is distinctive. The tumor cells are large, polygonal, and have a peculiar eosinophilic or "oxyphilic" cytoplasm. At the ultrastructural level, the cytoplasm is filled with mitochondria. Large bands or lamellae of fibrous tissue run through the tumor.[141,142,150] A detailed immunohistochemical study by Berman et al.[151] indicates that this tumor is highly differentiated and shows a number of phenotypic characteristics of hepatocytes, including copper-

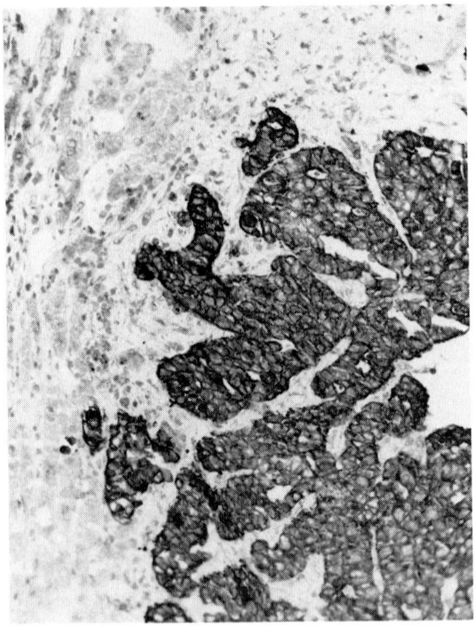

FIGURE 7. Cryostat section of poorly differentiated hepatocellular carcinoma, stained for cytokeratin 19. Tumor cells are strongly positive, but adjacent nontumoral parenchymal cells are negative. (Indirect immunoperoxidase, counterstained with Mayer's hemalum; original magnification × 200.) (Reprinted by permission from Van Eyken et al., *Hum. Pathol.*, 19, 562, 1988, W. B. Saunders, Orlando, FL.)

binding protein, α-1-antitrypsin, fibrinogen, C-reactive protein, and ferritin. Using paraffin-embedded material, Johnson et al.[147] found five fibrolamellar carcinomas to be immunoreactive with the monoclonal antibody CAM 5.2 (reactive with cytokeratins 8, 18, and 19), but negative with the monoclonal antibody AE1 (which in the liver reacts with cytokeratin 19). The expression of cytokeratin 7, however, was not studied. Van Eyken et al.[152] studied two cases of fibrolamellar carcinoma and one lymph node metastasis using immunohistochemical staining with monoclonal antibodies specific for cytokeratins 7, 8, 18, and 19, as well as two-dimensional gel electrophoresis of cytoskeletal extracts. All tumor cells were positive for cytokeratins 8 and 18. Surprisingly, they also expressed cytokeratin 7 (Figure 8). A very small number of tumor cells were also immunoreactive for cytokeratin 19. The presence of cytokeratins 8, 18, and 7 was confirmed using two-dimensional gel electrophoresis and Western blotting of cytoskeletal extracts (Figure 9). The lymph node metastasis of the fibrolamellar carcinoma showed a cytokeratin ''profile'' identical to that of the primary tumor.

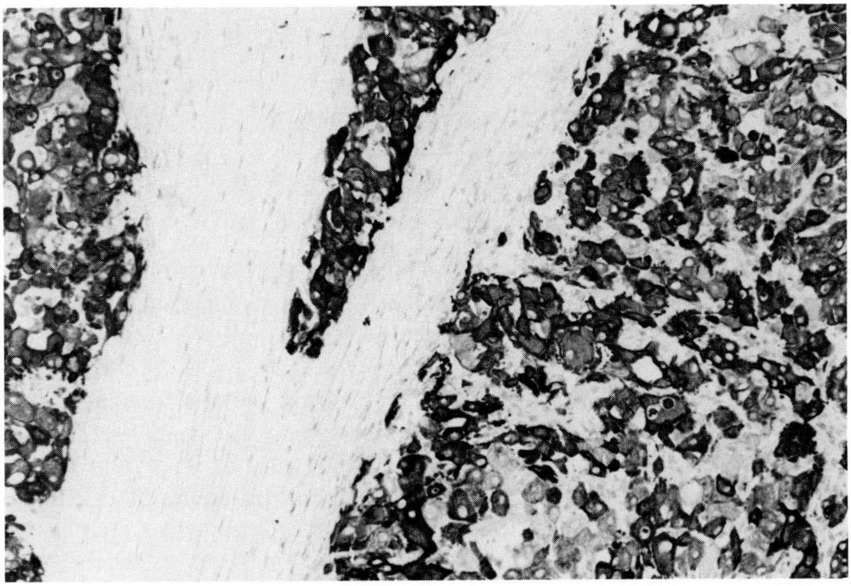

FIGURE 8. Fibrolamellar carcinoma of the liver. Tumor cells display intense immunoreactivity for cytokeratin 7. (Indirect immunoperoxidase on cryostat section, counterstained with Mayer's hemalum; original magnification × 160.)

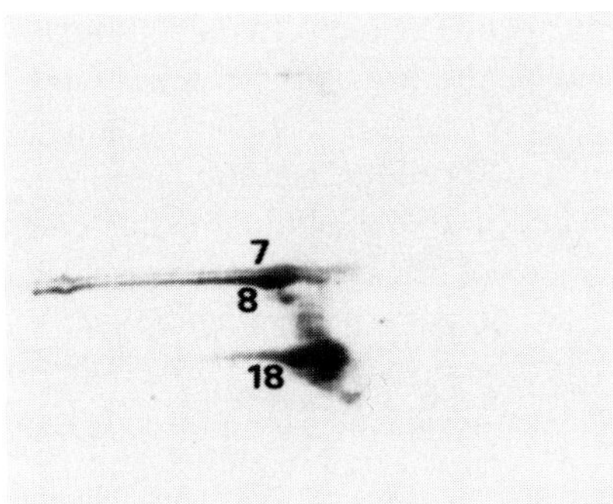

FIGURE 9. Two-dimensional immunoblot of a cytoskeletal preparation of a fibrolamellar carcinoma of the liver metastatic to a lymph node revealing the presence of cytokeratins 7, 8, and 18. (Individual cytokeratins are indicated by their number.) (Reprinted by permission from Van Eyken et al., *Histopathology,* 17, 101, 1990, Blackwell Scientific Publications, Oxford.)

2. Cholangiocarcinoma

Cholangiocarcinomas are tumors thought to arise from intrahepatic bile ducts. Microscopically, they are mucin-secreting adenocarcinomas with a fibrous stroma.[141,142] In all cases analyzed until now with gel electrophoresis or immunohistochemistry, the cytokeratin pattern corresponded to that of normal intrahepatic bile duct cells, i.e., cytokeratins 7, 8, 18, and 19.[2,22,28,110,144,147,148]

3. Combined Hepatocellular-Cholangiocarcinoma

Primary liver tumors displaying both hepatocellular and cholangiocellular features are considered a special entity and are referred to as ''combined hepatocellular-cholangiocarcinoma''. Three histological subtypes have been described.[153] Goodman et al.[153] found the cholangiocellular parts only to be reactive with a polyclonal anticytokeratin antiserum that stains bile ducts in the normal liver. Johnson et al.[147] found immunoreactivity for cytokeratin 19 in areas showing hepatocellular, cholangiocellular, or intermediate features. In the case analyzed by Fischer et al.,[126] hepatocellular parts were positive for cytokeratins 8 and 18 with a few tumor cells adjacent to fibrous septa expressing, in addition, cytokeratins 7 and 19. Adenocarcinomatous cholangiocellular areas were immunoreactive for cytokeratins 7, 8, 18, and 19.

4. Implications of the Pattern of Cytokeratin Expression of Hepatocellular Carcinoma, Cholangiocarcinoma, and Combined Hepatocellular-Cholangiocarcinoma for the Histogenesis and Differential Diagnosis of Primary Liver Tumors

The data discussed in the previous sections have implications for the differential diagnosis of liver tumors using immunohistochemistry as well as for the histogenesis of hepatocellular carcinoma.

As has been discussed in the beginning of this chapter, a fair amount of data indicate that during neoplastic transformation and even during the process of metastasis, the cytokeratin pattern of an epithelium is preserved.[2,16,17] This would imply that a detailed analysis of the cytokeratin profile of a liver tumor would be an excellent tool to differentiate between a primary hepatocellular carcinoma (expected to express cytokeratins 8 and 18 only), cholangiocarcinoma (positive for cytokeratins 7, 8, 18, and 19), and a metastasis of a colorectal adenocarcinoma (positive for cytokeratins 8, 18, and 19). The differential diagnosis between a primary cholangiocarcinoma and a liver metastasis of a pancreatic ductal adenocarcinoma, a ductal carcinoma of the breast, or a follicular thyroid carcinoma would be impossible using cytokeratin immunohistochemistry since these tumors share the same cytokeratin profile.[2,28,144] All available data indicate that in cholangiocarcinoma the pattern of cytokeratin expression of normal intrahepatic biliary epithelium is indeed preserved. Unfortunately, 50% of ''classical'' hepatocellular carcinomas display immunoreactivity for cytokeratins 7 and/or 19.[145] This implies that a differential diagnosis between hepatocellular carcinoma and cholangiocarcinoma (and between primary and secondary liver neoplasms) on the basis of cytokeratin immunohistochemistry may not always be reliable.

The finding of cytokeratin polypeptides which are normally restricted to bile duct cells in tumors thought to originate from liver parenchymal cells raises questions as to the histogenesis of hepatocellular carcinoma. As pointed out by Gould,[154] ''histogenesis'' refers to the putative cell of origin of a neoplasm, whereas differentiation implies the degree to which the tumor cells resemble their normal and mature counterparts. The data on the cytokeratin expression in hepatocellular carcinoma can be interpreted in two ways. Since it is generally assumed that hepatocellular carcinoma originates from liver parenchymal cells, one might conclude that during the process of neoplastic transformation regulation of cytokeratin expression becomes disturbed, resulting in an aberrant cytokeratin pattern. In the section on ''ductular metaplasia'' we have discussed data indicating a gradual transformation or transdifferentiation of hepatocytes into bile duct-type cells. The appearance of bile duct-type cytokeratins in tumoral liver cells may be analogous to the process of ductular metaplasia in the nontumoral liver and may even be brought about by the same mechanisms. In that case, the presence of cytokeratins 7 and/or 19 in primary hepatocellular carcinoma may reflect aberrant differentiation rather than histogenesis.

There is, however, an alternative explanation. If the hypothesis that the cytokeratin pattern of a given epithelium is maintained during carcinogenesis also holds true for the liver, we have to conclude that (at least a proportion of) primary hepatocellular carcinomas may arise from a cell with a bile duct/ductular cytokeratin profile. As discussed elsewhere in this volume (Chapters 5–10), there is an increasing body of evidence pointing towards a stem cell origin of hepatocellular and cholangiocellular carcinoma. The latter evidence has been collected in models of experimental hepatocarcinogenesis.[74,75] It is interesting to note that oval cells (see Chapters 5–8), which are related to — or may themselves be — stem cells, express cytokeratins 7 and 19, indicating that they may belong to the bile duct cell lineage.[155] The existence of stem cells in the human liver is at present entirely speculative, but it is tempting to hypothesize that they do exist and may be the origin of (at least a proportion of) hepatocellular carcinomas. In this respect, the presence of cytokeratins 7 and/or 19 may still reflect histogenesis, i.e., the origin of hepatocellular carcinoma from a cell type (stem cell?) belonging to the bile duct cell lineage.

In the light of the above discussed interpretations, the existence of ''combined hepatocellular-cholangiocellular carcinoma'' can hardly be surprising. These tumors may represent variants of hepatocellular carcinoma in which the appearance of bile duct-type cytokeratins is accompanied by a light-microscopically recognizable differentiation towards bile ductular cells.[126,135,156] Alternatively, the existence of this type of neoplasm can easily be explained by assuming a stem cell origin of primary liver tumors.

5. Hepatoblastoma

Hepatoblastoma is a rare primary malignant liver tumor occurring predominantly in early childhood.[157,158] (see also Chapter 4). Histologically, hepatoblastomas are classified as ''epithelial'' or ''mixed epithelial-mesenchymal''. The former type is composed exclusively of epithelial tumor cells, whereas the latter contains, in addition to an epithelial component, spindle cell mesenchyme or osteoidlike foci.[157,158] Originally, two types of epithelial tumor cells were described: embryonal- and fetal-type cells resembling cells of the normal embryonal and fetal liver, respectively.[158] Schmidt et al.[159] reported positive staining for cytokeratin (not otherwise specified) of neoplastic cells in one case of hepatoblastoma. Abenoza et al.[160] found focal expression of a 54-kDa cytokeratin in embryonal- and fetal-type tumor cells. Van Eyken et al.[161] performed a detailed analysis of the cytokeratin expression of six specimens of hepatoblastoma (five epithelial and one mixed epithelial-mesenchymal). Both embryonal- and fetal-type tumor cells were immunoreactive for cytokeratins 8, 18, and 19 (Figures 10 and 11). Positivity for the latter was more intense in embryonal- than in fetal-type tumor cells. In two cases, some embryonal-type tumor cells adjacent to fibrous septa expressed, in addition, cytokeratin 7 (Figure 12). Focal positivity for vimentin was seen in embryonal areas. Cells in mesenchymal areas expressed vimentin, but no cytokeratins, but cells embedded in osteoidlike foci were, in addition, immunoreactive for cytokeratins 7, 18, and 19. The immunohistochemical staining results were confirmed by gel electrophoresis and immunoblotting.[161]

The cytokeratin patterns detected in the epithelial component of hepatoblastoma bear a striking resemblance to those observed during normal development of the human liver (see section on the development of the intrahepatic bile ducts). Embryonal-type tumor cells have essentially the same cytokeratin profile as the progenitor cells in the early embryonal liver. Fetal-type tumor cells were positive for cytokeratins 8 and 18 and were rather weakly positive to negative for cytokeratin 19. This observation may indicate that the embryonal-type tumor cells — just like the progenitor cells during normal development — can further differentiate into fetal-type tumor cells, thereby gradually losing cytokeratin 19. The gradual ''maturation'' of embryonal- to fetal-type tumor cells is also supported by the immunohistochemical staining results of Abenoza et al.[160] Those authors found a stronger expression of α-1-antitrypsin and ferritin in fetal as compared to embryonal areas. The parallel with normal development is further completed by the observation that embryonal-type tumor cells close to fibrous septa expressed cytokeratin 7 in addition to cytokeratins 8, 18, and 19. As illustrated in Figure

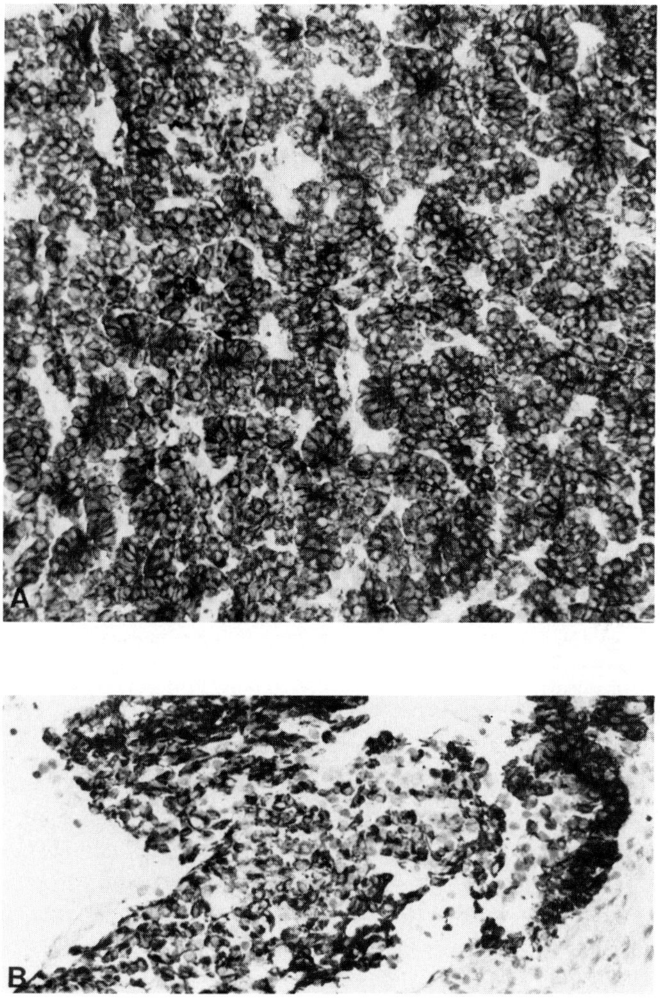

FIGURE 10. Hepatoblastoma. Tumor cells in both fetal (A) and embryonal (B) areas strongly express cytokeratin 18. (Indirect immunoperoxidase, counterstained with Mayer's hemalum; magnification × 156.) (Reprinted by permission from Van Eyken et al., *Hum. Pathol.*, 21, 302, 1990, W. B. Saunders, Orlando, FL.)

12, the resemblance of the layer of tumoral cells adjacent to mesenchymal tissue to a normal ductal plate is striking. Apparently the embryonal-type tumor cells can still embark upon a "bile duct cell" differentiation pathway. The localization of cytokeratin 7-positive tumor cells close to connective tissue may indicate that they are responsive to inductive influences exerted by the mesenchyme. The expression of cytokeratins by cells embedded in osteoidlike

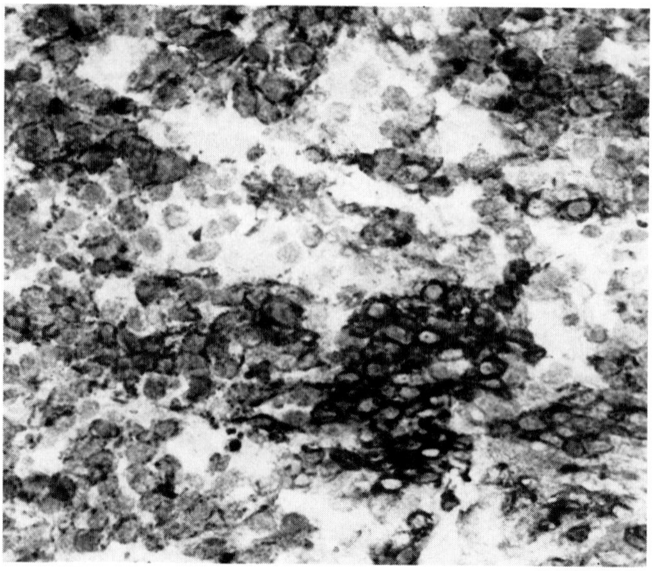

FIGURE 11. Hepatoblastoma. Embryonal-type tumor cells are immunoreactive for cytokeratin 19. (Indirect immunoperoxidase, counterstained with Mayer's hemalum; magnification × 400.) (Reprinted by permission from Van Eyken et al., *Hum. Pathol.*, 21, 302, 1990. W. B. Saunders, Orlando, FL.)

foci, already reported by Abenoza et al.,[160] argued against those cells being mere mesenchymal cells. They may rather correspond to metaplastic epithelial cells. Hence there is no need to invoke a simultaneous neoplastic transformation of a mesenchymal and an epithelial cell to account for the existence of mixed epithelial-mesenchymal hepatoblastoma. Tsao and Grisham[162] have provided unequivocal evidence that a chemically transformed liver epithelial cell line and its single cell-derived clonal subpopulations can produce a variety of tumors, including hepatocellular carcinomas, cholangiocarcinomas, and mixed epithelial-mesenchymal hepatoblastomas in rats. It is worth noting that this epithelial cell line is thought to be derived from bile ductular cells.

6. Malignant Rhabdoid Tumor

Malignant rhabdoid tumor is a rare neoplasm with an extremely aggressive behavior. Most cases arise in the kidney, but the tumor may occur in extrarenal sites, including the liver. Histologically, the neoplasm is characterized by large cells with eosinophilic inclusions and nuclei with prominent nucleoli. Ultrastructurally, whorls of intermediate filaments are present.[163] The authors

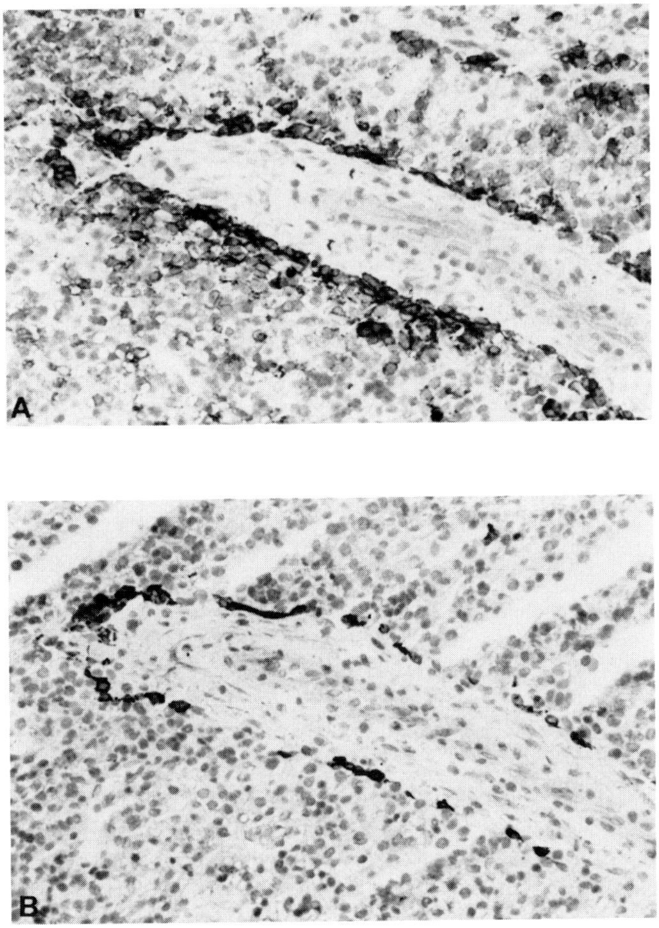

FIGURE 12. Serial sections of hepatoblastoma stained for cytokeratin 19 (A) and cytokeratin 7 (B). Tumor cells close to the connective tissue surrounding a vessel are more strongly positive for cytokeratin 19 and some of them also express cytokeratin 7. (Indirect immunoperoxidase, counterstained with Mayer's hemalum; magnification × 160.) (Reprinted by permission from Van Eyken et al., *Hum. Pathol.*, 21, 302, 1990, W. B. Saunders, Orlando, FL.)

had the opportunity to study one case of primary malignant rhabdoid tumor of the liver.[164] The tumor cells were positive for vimentin and cytokeratin 19 and displayed a variable immunoreactivity for cytokeratins 8 and 18 (Figure 13). A small number of tumor cells were immunoreactive for cytokeratin 7 as well. This cytokeratin profile is again reminiscent of that of the progenitor cells in normal developing human liver.

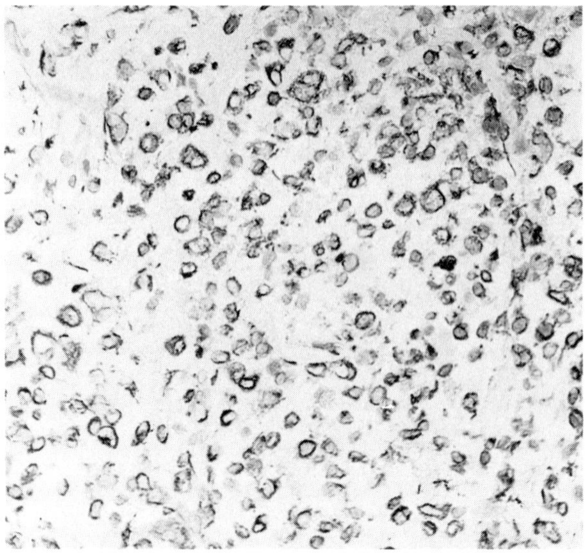

FIGURE 13. Rhabdoid tumor of the liver. Tumor cells display immunoreactivity for cytokeratin 19. (Indirect immunoperoxidase, counterstained with Mayer's hemalum; original magnification × 160.)

VI. CONCLUDING REMARKS

The cell type-specific distribution of cytokeratin polypeptides makes them an extremely useful tool to study cell lineage and differentiation. The use of cytokeratin immunohistochemistry has provided new insights into the cell lineage relationships in developing human liver. Moreover, it has provided additional evidence supporting the concept of "ductular metaplasia of hepatocytes", the mechanism of which remains to be elucidated. A detailed analysis of the cytokeratin profiles of primary liver tumors has yielded some "unexpected" results which have implications for the differential diagnosis of these tumors. More importantly, the findings raise questions concerning the cell of origin of hepatocellular carcinoma and are consistent with the existence of a stem cell in the adult human liver.

Numerous problems remain to be solved. Unfortunately, since the function of cytokeratin filaments is still unknown, it is impossible to infer a functional meaning from changes in cytokeratin expression. At present, cytokeratin polypeptides are merely being used as markers of directions of epithelial differentiation. If ever cell biologists come up with specific functions for individual cytokeratins, changes in their pattern of expression may become more meaningful. The mechanisms involved in the induction of changes in cytokeratin pattern — and therefore in differentiation commitment — both during normal development and in liver disease have to be elucidated. It is likely that several mechanisms are involved. The role of the extracellular matrix or soluble factors may be addressed in cell culture experiments. Finally, the identification of a liver stem cell — if present in the human liver — is a top priority. It is tempting to speculate that the cytokeratin profile of such a stem cell should comprise cytokeratins 8 and 18 as well as cytokeratin 19 (and 7?).

REFERENCES

1. **Alberts, B., Bray, D., Lewis, J., Raff, M., Roberts, K., and Watson, J. D., Eds.,** The cytoskeleton, in *Molecular Biology of the Cell,* 1st ed., Garland Publishing, New York, 1983, 549.
2. **Moll, R., Franke, W. W., Schiller, D. A., Geiger, B., and Krepler, R.,** The catalog of human cytokeratins: patterns of expression in normal epithelia, tumors and cultured cells, *Cell,* 31, 11, 1982.
3. **Steinert, P. M., Stevens, A. C., and Roop, D. R.,** The molecular biology of intermediate filaments, *Cell,* 42, 411, 1985.
4. **Steinert, P. M. and Roop, D. R.,** Molecular and cellular biology of intermediate filaments, *Annu. Rev. Biochem.,* 57, 523, 1988.
5. **Fuchs, E., Tyner, A. L., Giudice, G. J., Marchuk, D., Raychaudhury, A., and Rosenberg, M.,** The human keratin genes and their differential expression, *Curr. Top. Dev. Biol.,* 22, 5, 1987.
6. **Sun, T. T., Eichner, R., Nelson, W. E., Tseng, S. C. G., Weiss, R. A., Jarvinen, M., and Woodcock-Mitchell, J.,** Keratin classes: molecular markers for different types of epithelial differentiation, *J. Invest. Dermatol.,* 81, 109s, 1983.
7. **Sun, T. T., Tseng, S. C. G., Huang, A. J. W., Cooper, D., Schermer, A., Lynch, M. H., Weis, R., and Eichner, R.,** Monoclonal antibody studies of mammalian epithelial keratins: a review, *Ann. N.Y. Acad. Sci.,* 455, 307, 1985.
8. **Cooper, D., Schermer, A., and Sun, T. T.,** Biology of disease. Classification of human epithelia and their neoplasms using monoclonal antibodies to keratin, strategies, applications and limitations, *Lab. Invest.,* 52, 243, 1985.
9. **Quinlan, R. A., Schiller, D. L., Hatzfeld, M., Achtstätter, T., Moll, R., Jorcano, J. L., Magin, T. M., and Franke, W. W.,** Patterns of expression and organization of cytokeratin intermediate filaments, *Ann. N.Y. Acad. Sci.,* 455, 282, 1985.
10. **Eichner, R., Bonitz, P., and Sun, T. T.,** Classification of epidermal keratins according to their immunoreactivity, isoelectric point and mode of expression, *J. Cell. Biol.,* 98, 1388, 1984.
11. **Osborn, M.,** Intermediate filaments as histologic markers: an overview, *J. Invest. Dermatol.,* (Suppl.) 81, 104s, 1983.
12. **Franke, W. W.,** Nuclear lamins and cytoplasmic intermediate filament proteins: a growing multigene family, *Cell,* 48, 3, 1987.
13. **Ramaekers, F. C. S., Puts, J. J. G., Moesker, O., Kant, A., Huijsmans, A., Haag, D., Jap. P. H. K., Herman, C. J., and Vooijs, E. P.,** Antibodies to intermediate filament proteins in the immunohistochemical identification of human tumours. An overview, *Histochem. J.,* 15, 691, 1983.

14. **Osborn, M., Altmannsberger, M., Debus, E., and Weber, K.,** Differentiation of the major human tumor groups using conventional and monoclonal antibodies specific for individual intermediate filament proteins, *Ann. N.Y. Acad. Sci.*, 455, 649, 1984.
15. **Erlandson, R. A.,** Cytoskeletal proteins including myofilaments in human tumors, *Ultrastruct. Pathol.*, 13, 155, 1989.
16. **Moll, R., Krepler, R., and Franke, W. W.,** Complex cytokeratin polypeptide patterns observed in certain human carcinomas, *Differentiation*, 23, 256, 1983.
17. **Moll, R. and Franke, W. W.,** Cytochemical cell typing of metastatic tumors according to their cytoskeletal proteins, in *Biochemistry and Molecular Genetics of Cancer Metastasis*, Liotta, L. A. and Rabson, A. S., Eds., Martinus Nijhoff, The Hague, 1986, 101.
18. **Battifora, H.,** Diagnostic uses of antibodies to keratins: a review and immunohistochemical comparison of seven monoclonal and three polyclonal antibodies, in *Progress in Surgical Pathology*, Vol. 8, Fenoglio-Preiser, C. M., Wolff, M., and Rilke, F., Eds., Field & Wood, Philadelphia, 1988.
19. **Battifora, H.,** The biology of the keratins and their diagnostic applications, in *Advances in Immunohistochemistry*, DeLellis, R. A., Ed., Raven Press, New York, 1988, 191.
20. **Corson, J. M.,** Keratin protein immunohistochemistry in surgical pathology practice, in *Pathology Annual*, Sommers, S. C., Rosen, P. P., and Fechner, R. E., Eds., Appleton-Century-Crofts, Norwalk, CT, 1986.
21. **Franke, W. W., Denk, H., Kalt, R., and Schmid, E.,** Biochemical and immunological identification of cytokeratin proteins present in hepatocytes of mammalian liver tissue, *Exp. Cell Res.*, 131, 299, 1981.
22. **Denk, H., Krepler, R., Lackinger, E., Artlieb, U., and Franke, W. W.,** Biochemical and immunocytochemical analysis of the intermediate filament cytoskeleton in human hepatocellular carcinomas and in hepatic neoplastic nodules of mice, *Lab. Invest.*, 46, 584, 1982.
23. **Denk, H. and Lackinger, E.,** Cytoskeleton in liver diseases, *Semin. Liver Dis.*, 6, 199, 1986.
24. **Van Eyken, P., Sciot, R., Van Damme, B., De Wolf-Peeters, C., and Desmet, V. J.,** Keratin-immunohistochemistry in normal human liver. Cytokeratin pattern of hepatocytes, bile ducts and acinar gradient, *Virchows Arch. A*, 412, 63, 1987.
25. **Ramaekers, F., Huysmans, A., Schaart, G., Moesker, O., and Vooijs, P.,** Tissue distribution of keratin 7 as monitored by a monoclonal antibody, *Exp. Cell Res.*, 170, 235, 1987.
26. **Stosiek, P., Karsten, V., and Kasper, M.,** Monoclonal anti-cytokeratin antibodies reveal bile duct origin of cirrhotic ductular proliferations. A case report, *Virchows Arch. A*, 407, 233, 1985.
27. **Kasper, M., Stosiek, P., Typlt, H., and Karsten, U.,** Histological evaluation of three new monoclonal anti-cytokeratin antibodies. 1. Normal Tissues, *Eur. J. Cancer Clin. Oncol.*, 23, 137, 1987.
28. **Osborn, M., van Lessen, G., Weber, K., Klöppel, G., and Altmannsberger, M.,** Methods in laboratory investigations. Differential diagnosis of gastrointestinal carcinomas by using monoclonal antibodies specific for individual keratin polypeptides, *Lab. Invest.*, 55, 497, 1986.
29. **Alpini, G., Lenzi, R., Zhai, W.-R., Slott, P. A., Liu, M. H., Sarkozi, L., and Tavoloni, N.,** Bile secretory function of intrahepatic biliary epithelium in the rat, *Am. J. Physiol.*, 257, G124, 1989.
30. **Slott, P. A., Liu, M. H., and Tavoloni, N.,** Origin, pattern and mechanism of bile duct proliferation following biliary obstruction in the rat, *Gastroenterology*, 99, 466, 1990.
31. **French, S. W., Kawahara, H., Katsuma, Y., Ohta, M., and Swierenga, S. H. H.,** Interaction of intermediate filaments with nuclear lamina and cell periphery, *Electron Microsc. Rev.*, 2, 17, 1989.

32. **French, S. W., Okanoue, T., Swierenga, S. H. H., and Marceau, N.,** The cytoskeleton of hepatocytes in health and disease, in *Pathogenesis of Liver Diseases,* Farber, E., Phillips, M. J., and Kaufman, N., Eds., Williams & Wilkins, Baltimore, 1987, 95.
33. **Katsuma, Y., Marceau, N., Ohta, M., and French, S. W.,** Cytokeratin intermediate filaments of rat hepatocytes: different cytoskeletal domains and their three-dimensional structure, *Hepatology,* 8, 559, 1988.
34. **Feldmann, G.,** The cytoskeleton of the hepatocyte. Structure and functions, *J. Hepatol.,* 8, 380, 1989.
35. **Leroux-Nicollet, I., Noël, M., Baribault, H., Goyette, R., and Marceau, N.,** Selective increase in cytokeratin synthesis in cultured rat hepatocytes in response to hormonal stimulation, *Biochem. Biophys. Res. Commun.,* 114, 556, 1983.
36. **Ben Ze'ev, A., Robinson, G. S., Bucher, N. L. R., and Farmer, S. R.,** Cell-cell and cell-matrix interactions differentially regulate the expression of hepatic and cytoskeletal genes in primary cultures of rat hepatocytes, *Proc. Natl. Acad. Sci. U.S.A.,* 85, 2161, 1988.
37. **Marceau, N., Baribault, H., and Leroux-Nicollet, I.,** Dexamethasone can modulate the synthesis and organization of cytokeratins in cultured differentiating rat hepatocytes, *Can. J. Biochem. Cell Biol.,* 63, 448, 1985.
38. **Marceau, N.,** Cell lineages and differentiation programs in epidermal, urothelial and hepatic tissues and their neoplasms, *Lab. Invest.,* 63, 4, 1990.
39. **Germain, L., Blouin, M. J., and Marceau, N.,** Biliary epithelial and hepatocyte cell lineage relationships in embryonic rat liver as determined by the differential expression of cytokeratins, alpha-fetoprotein, albumin and cell surface-exposed components, *Cancer Res.,* 48, 4909, 1988.
40. **Jørgensen, M. J.,** The ductal plate malformation. A study of the intrahepatic bile duct lesion in infantile polycystic disease and congenital hepatic fibrosis, *Acta Pathol. Microbiol. Scand. (Suppl.),* 257, 1, 1976.
41. **Desmet, V. J.,** Embryology of the liver and intrahepatic biliary tract and an overview of malformations of the bile duct, in *The Oxford Textbook of Clinical Hepatology,* McIntyre, N., Benhamou, J.-P., Bircher, J., Rizzetto, M., and Rodes, J., Eds., Oxford University Press, 1991, 497.
42. **Shiojiri, N. and Katayama, H.,** Secondary joining of the bile ducts during the hepatogenesis of the mouse embryo, *Anat. Embryol.,* 177, 153, 1987.
43. **Severn, C. B.,** A morphological study of the development of the human liver. II. Establishment of liver parenchyma, extrahepatic ducts and associated venous channels, *Am. J. Anat.,* 133, 85, 1972.
44. **Elias, H.,** Origin and early development of the liver in various vertebrates, *Acta Hepatol.,* 3, 1, 1955.
45. **Van Eyken, P.,** Phenotypic Modulation of Hepatocytes: a Cytokeratin-Immunohistochemical Study, Ph.D. thesis, Katholieke Universiteit Leuven, Acco, Leuven, 1990.
46. **Desmet, V. J. and Van Eyken, P., and Sciot, R.,** Cytokeratins for probing cell lineage relationships in developing liver, *Hepatology,* 12, 1249, 1990.
47. **Hammar, J. A.,** Über die erste Entstehung der nicht kapillaren intrahepatischen Gallengänge beim Menschen, *Z. Mikrosk. Anat. Forsch.,* 5, 59, 1926.
48. **Koga, A.,** Morphogenesis of intrahepatic bile ducts of the human fetus. Light and electron microscopic study, *Z. Anat. Entwicklungsgesch.,* 135, 156, 1971.
49. **Horstmann, E.,** Entwicklung und Entwicklungsbeidingungen des intrahepatischen Gallengangsystems, *Wilhelm Roux' Arch. Entwicklungsmech. Org.,* 139, 363, 1939.
50. **Gall, J. A. M. and Bhathal, P. S.,** Morphological and immunohistochemical assessment of intrahepatic bile duct development in the rat, *J. Gastroenterol. Hepatol.,* 4, 241, 1989.
51. **Aron, M.,** Sur le développement des voies biliaires intrahépatiques et l'établissement de la fonction biliaire du foie, *C R Mém. Soc. Biol.,* 85, 110, 1921.

52. **Bloom, W.,** The embryogenesis of human bile capillaries and ducts, *Am. J. Anat.*, 36, 451, 1926.
53. **Enzan, H., Ohkita, T., Fujita, H., and Iijima, S.,** Light and electron microscopic studies on the development of periportal bile ducts of the human embryo, *Acta Pathol. Jpn.*, 24, 427, 1974.
54. **Shiojiri, N.,** The origin of intrahepatic bile duct cells in the mouse, *J. Embryol. Exp. Morphol.*, 79, 25, 1984.
55. **Shiojiri, N.,** Enzymo- and immunocytochemical analyses of the differentiation of liver cells in the prenatal mouse, *J. Embryol. Exp. Morphol.*, 62, 139, 1981.
56. **Shah, K. and Gerber, M. A.,** Development of intrahepatic bile ducts in humans. Immunohistochemical study using monoclonal cytokeratin antibodies, *Arch. Pathol. Lab. Med.*, 113, 1135, 1989.
57. **Wood, R. L.,** An electron microscope study of developing bile canaliculi in the rat, *Anat. Rec.*, 151, 507, 1965.
58. **Picardi, R., Gardiol, D., and Gautier, A.,** Etude de la cholangiogénèse chez le foetus humain. I. Aspects ultrastructuraux et classification des divers types cellulaires épitheliaux rencontrés dans les régions periportales, *Z. Zellforsch. Mikrosk. Anat.*, 84, 311, 1968.
59. **Picardi, R., Gardiol, D., and Gautier, A.,** Etude de la cholangiogénèse chez le foetus human. II. Aspects histologiques et ultrastructuraux de la génèse des canaux biliaires intrahépatiques, *Z. Zellforsch. Mikrosk. Anat.*, 84, 319, 1968.
60. **Wegmann, R., Corcos, V., and Caroli, J.,** Histoenzymologie des ductules biliaries chez l'embryon humain normal et au cours des cirrhoses humaines, *Arch. Mol. Appr. Dig.*, 54, 215, 1965.
61. **Werin, W. K.,** Zur Histogenese der Gewebe der menschlichen Leber, *Z. Mikrosk. Anat. Forsch.*, 76, 425, 1987.
62. **Marceau, N., Blouin, M. J., Germain, L., and Noël, M.,** Role of different epithelial cell types in liver ontogenesis, regeneration and neoplasia, *In Vitro Cell Dev. Biol.*, 25, 336, 1989.
63. **Stosiek, P., Kasper, M., and Karsten, U.,** Expression of cytokeratin 19 during human liver organogenesis, *Liver,* 10, 59, 1990.
64. **Burt, A. D., Stewart, J. A., Aitchison, M., and MacSween, R. N. M.,** Expression of tissue polypeptide antigen (TPA) in fetal and adult liver: changes in liver disease, *J. Clin. Pathol.*, 40, 719, 1987.
65. **Carthew, P., Edwards, R. E., Hill, R. J., and Evans, J. G.,** Cytokeratin expression in cells of the rodent bile duct developing under normal and pathological conditions, *Br. J. Exp. Pathol.*, 70, 717, 1989.
66. **Ruebner, B. H., Blankenberg, T. A., Burrows, D. A., Soohoo, W., and Lund, J. K.,** Development and transformation of the human hepatocytic-ductal plate, *Lab. Invest.*, 60, 7P, 1989.
67. **Van Eyken, P., Sciot, R., Callea, F., Van der Steen, K., Moerman, P., and Desmet, V. J.,** The development of the intrahepatic bile ducts in man: a keratin-immunohistochemical study, *Hepatology,* 8, 1586, 1988.
68. **Desmet, V. J.,** Intrahepatic bile ducts under the lens, *J. Hepatol.*, 1, 445, 1985.
69. **Desmet, V. J.,** Cholangiopathies: past, present and future, *Semin. Liver Dis.*, 7, 67, 1987.
70. **Van Eyken, P., Sciot, R., and Desmet, V.,** Intrahepatic bile duct development in the rat: a cytokeratin-immunohistochemical study, *Lab. Invest.*, 59, 52, 1988.
71. **Shah, K. D. and Gerber, M. A.,** Development of intrahepatic bile ducts in humans. Possible role of laminin, *Arch. Pathol. Lab. Med.*, 114, 597, 1990.
72. **Doljanski, L. and Roulet, F.,** Uber die gestaltende Wechselwirkung zwischen dem Epithel und dem Mesenchym, zugleich ein Beitrag zur Histogenese der sogenannten "Gallengangserwucherungen", *Virchows Arch. Pathol. Anat. Physiol.*, 292, 256, 1934.

73. **Kahn, E., Markowitz, J., Aiges, H., and Daum, F.,** Human ontogeny of the bile duct to portal space ratio, *Hepatology,* 10, 21, 1989.
74. **Sell, S.,** Is there a liver stem cell?, *Cancer Res.,* 50, 3811, 1990.
75. **Sell, S. and Dunsford, H. A.,** Evidence for the stem cell origin of hepatocellular carcinoma and cholangiocarcinoma, *Am. J. Pathol.,* 134, 1347, 1989.
76. **Stasiak, P. C., Purkis, P. E., Leigh, J. M., and Lane, E. B.,** Keratin 19: predicted amino acid sequence and broad tissue distribution suggest it evolved from keratinocyte keratins, *J. Invest. Dermatol.,* 92, 707, 1989.
77. **Desmet, V. J. and Callea, F.,** Cholestatic syndromes of infancy and childhood, in *Hepatology. A Textbook of Liver Disease,* Vol. 2, 2nd ed., Zakim, D. and Boyer, T. D., Eds., W. B. Saunders, Philadelphia, 1990, 1355.
78. **Desmet, V. J.,** Cholestasis: extrahepatic obstruction and secondary biliary cirrhosis, in *Pathology of the Liver,* 2nd ed., MacSween, R. N. M., Anthony, P. P., and Scheuer, P. J., Eds., Churchill Livingstone, Edinburgh, 1987, 364.
79. **Tavoloni, N.,** The intrahepatic biliary epithelium: an area of growing interest in hepatology, *Semin. Liver Dis.,* 7, 280, 1987.
80. **Desmet, V. J.,** Modulation of biliary epithelium, in *Modulation of Liver Cell Expression,* Reutter, W., Popper, H., Arias, I. M., Heinrich, P. C., Keppler, D., and Landmann, L., Eds., MTP Press, Lancaster, England, 1987, 195.
81. **Gerber, M. A., Thung, S. N., Shen, S., Stromeyer, W., and Ishak, K. G.,** Phenotypic characterization of hepatic proliferation. Antigenic expression by proliferating epithelial cells in fetal liver, massive hepatic necrosis and nodular transformation of the liver, *Am. J. Pathol.,* 110, 70, 1983.
82. **International Group,** Histopathology of the intrahepatic biliary tree, *Liver,* 3, 161, 1983.
83. **Yamada, S., Howe, S., and Scheuer, P. J.,** Three-dimensional reconstruction of biliary pathways in primary biliary cirrhosis: a computer-assisted study, *J. Pathol.,* 152, 317, 1987.
84. **Rubin, E., Schaffner, F., and Popper, H.,** Primary biliary cirrhosis: chronic nonsuppurative destructive cholangitis, *Am. J. Pathol.,* 46, 387, 1965.
85. **Nakanuma, Y. and Ohta, G.,** Immunohistochemical study on bile ductular proliferation in various hepatobiliary diseases, *Liver,* 6, 205, 1986.
86. **Desmet, V.,** Experimentele Levercarcinogenese. Histochemische Studie, Ph.D. thesis, Arscia, Brussels, 1963.
87. **Farber, E.,** Similarities in the sequence of early histological changes induced in the liver of the rat by ethionine, 2-acetylaminofluorene and 3′ methyl-4-dimethylaminoazobenzene, *Cancer Res.,* 16, 142, 1956.
88. **Waldeyer, H. W. G.,** Bacteriencolonien mit Pseudomelanose in der Leber (Acute Atrophie), *Virchows Arch.,* 43, 533, 1868.
89. **Masuko, K., Rubin, E., and Popper, H.,** Proliferation of bile ducts in cirrhosis, *Arch. Pathol.,* 78, 421, 1964.
90. **Rubin, E.,** The origin and fate of proliferated bile ductular cells, *Exp. Mol. Pathol.,* 3, 279, 1964.
91. **Schaffner, F., and Popper, H.,** Electron microscopic studies of normal and proliferated bile ductules, *Am. J. Pathol.,* 38, 393, 1961.
92. **Alpini, G., Lenzi, R., Sarkozi, L., and Tavoloni, N.,** Biliary physiology in rats with bile ductular cell hyperplasia. Evidence for a secretory function of proliferated bile ductules, *J. Clin. Invest.,* 81, 569, 1988.
93. **Grisham, J. W. and Porta, E. A.,** Origin and fate of proliferated hepatic ductal cells in the rat: electron microscopic and autoradiographic studies, *Exp. Mol. Pathol.,* 3, 242, 1964.
94. **Gall, J. A. M. and Bhathal, P. S.,** Origin and involution of hyperplastic bile ductules following total biliary obstruction, *Liver,* 10, 106, 1990.

95. **Okada, T. S.,** Transdifferentiation in animal cells: fact or artifact?, *Dev. Growth Differ.,* 28, 213, 1986.
96. **Scarpelli, D. G.,** Multipotent developmental capacity of cells in the adult animal, *Lab. Invest.,* 52, 331, 1985.
97. **Buyssens, N.,** La prolifération des canaux biliaires et la formation de structures tubulaires au cours des affections hépatiques en pathologie humaine et expérimentale, *Rev. Belge Pathol. Med. Exp.,* 29, 7, 1962.
98. **Uchida, T. and Peters, R. L.,** The nature and origin of proliferated bile ductules in alcoholic liver disease, *Am. J. Clin. Pathol.,* 79, 326, 1983.
99. **Butron Vila, M. M., Haot, J., and Desmet, V. J.,** Cholestatic features in focal nodular hyperplasia of the liver, *Liver,* 4, 387, 1984.
100. **Vanstapel, M. J., Butron, M., and Desmet, V. J.,** Keratin immunohistochemistry in normal and pathological human liver, *Hepatology,* 4, 1037, 1984.
101. **Hillan, K. J., Burt, A. D., George, W. D., MacSween, R. N. M., Griffiths, M. R., and Bradley, J. A.,** Intrasplenic hepatocyte transplantation in rats with experimental liver injury: morphological and morphometric studies, *J. Pathol.,* 159, 67, 1989.
102. **Nagore, N., Howe, S., Boxer, L., and Scheuer, P. J.,** Liver cell rosettes: structural differences in cholestasis and hepatitis, *Liver,* 9, 43, 1989.
103. **Thung, S. N.,** The development of proliferating ductular structures in liver disease, *Arch. Pathol. Lab. Med.,* 114, 407, 1990.
104. **Van Eyken, P., Sciot, R., and Desmet, V. J.,** A cytokeratin immunohistochemical study of cholestatic liver disease: evidence that hepatocytes can express "bile duct-type" cytokeratins, *Histopathology,* 15, 125, 1989.
105. **Van Eyken, P., Sciot, R., Callea, F., and Desmet, V. J.,** A cytokeratin-immunohistochemical study of focal nodular hyperplasia of the liver: further evidence that ductular metaplasia of hepatocytes contributes to ductular "proliferation", *Liver,* 9, 372, 1989.
106. **Sciot, R., Van Damme, B., and Desmet, V. J.,** Cholestatic features in hepatitis A, *J. Hepatol.,* 3, 172, 1986.
107. **Sirica, A. E., Sattler, C. A., and Cihla, H. B.,** Characterization of a primary bile ductular cell culture from the livers of rats during extrahepatic cholestasis, *Am. J. Pathol.,* 120, 67, 1985.
108. **Jorgensen, M.,** A stereological study of intrahepatic bile ducts. 2. Bile duct proliferation in some pathological conditions, *Acta Pathol. Microbiol. Scand. (A),* 81, 663, 1973.
109. **Sirica, A. E. and Cihla, H. P.,** Isolation and partial characterizations of oval and hyperplastic bile ductular-enriched populations from the livers of carcinogen and non-carcinogen treated rats, *Cancer Res.,* 44, 3454, 1984.
110. **Lai, Y.-S., Thung, S. N., Gerber, M. A., Chen, M.-L., and Schaffner, F.,** Expression of cytokeratins in normal and diseased livers and in primary liver carcinomas, *Arch. Pathol. Lab. Med.,* 113, 134, 1989.
111. **Milani, S., Herbst, H., Schuppan, D., Niedobitek, G., Kim, K. Y., and Stein, H.,** Vimentin expression of newly formed rat bile duct epithelial cells in secondary biliary fibrosis, *Virchows Arch. A.,* 415, 237, 1989.
112. **Grant, A. G. and Billing, B. H.,** The isolation and characterization of a bile ductule cell population from normal and bile-duct ligated rat livers, *Br. J. Exp. Pathol.,* 58, 301, 1977.
113. **Marceau, N., Germain, L., Goyette, R., Noël, M., and Gourdeau, H.,** Cell of origin of distinct cultured rat liver epithelial cells, as typed by cytokeratin and surface component selective expression, *Biochem. Cell Biol.,* 64, 788, 1986.
114. **Mathis, G. A., Walls, S. A., and Sirica, A. E.,** Biochemical characteristics of hyperplastic rat bile ductular epithelial cells cultured "on top" and "inside" different extracellular matrix substitutes, *Cancer Res.,* 48, 6145, 1988.

115. **Mathis, G. A., Walls, S. A., D'Amico, P., Gengo, T. F., and Sirica, A. E.,** Enzyme profile of rat bile ductular epithelial cells in reference to the resistance phenotype in hepatocarcinogenesis, *Hepatology,* 9, 477, 1989.
116. **Ledda, G. M., Sells, M. A., Yokoyama, S., and Lombardi, B.,** Metabolic properties of isolated rat liver cell preparations enriched in epithelial cells other than hepatocytes, *Int. J. Cancer,* 31, 231, 1983.
117. **Alpini, G., Lenzi, R., Zhai, W.-R., Liu, M. H., Slott, P. A., Paronetto, F., and Tavoloni, N.,** Isolation of a nonparenchymal liver cell fraction enriched in cells with biliary epithelial phenotypes, *Gastroenterology,* 97, 1248, 1989.
118. **Buscher, H.-P., Miltenberger, C., MacNelly, G., and Gerok, W.,** The histoautoradiographic localization of taurocholate in rat liver after bile duct ligation, *J. Hepatol.,* 8, 181, 1989.
119. **Shibayama, Y.,** Factors producing bile infarction and bile duct proliferation in biliary obstruction, *J. Pathol.,* 160, 57, 1990.
120. **Gall, J. A. M. and Bhathal, P. S.,** Isolation and culture of intrahepatic bile ducts and its application in assessing putative inducers of biliary epithelial cell hyperplasia, *Br. J. Exp. Pathol.,* 68, 501, 1987.
121. **Bhathal, P. S. and Gall, J. A. M.,** Deletion of hyperplastic biliary epithelial cells by apoptosis following removal of the proliferative stimulus, *Liver,* 5, 311, 1985.
122. **Cameron, G. R. and Prasad, L. B. M.,** Recovery from biliary obstruction after spontaneous restoration of the obstructed common bile duct, *J. Pathol. Bacteriol.,* 80, 127, 1960.
123. **Wilson, J. W. and Leduc, E. H.,** Role of cholangioles in restoration of the liver of the mouse after dietary injury, *J. Pathol. Bacteriol.,* 76, 441, 1958.
124. **Vanstapel, M. J., Gatter, K. C., De Wolf-Peeters, C., Mason, D. Y., and Desmet, V. J.,** New sites of human S-100 immunoreactivity detected with monoclonal antibodies, *Am. J. Clin. Pathol.,* 85, 160, 1986.
125. **Vanstapel, M. J., Mason, D. Y., and Desmet, V. J.,** Immunohistochemical detection of S-100 protein in hepatocyte derived proliferating bile ducts, *Hepatology,* 4, 784, 1984.
126. **Fischer, H.-P., Doppl. W., Osborn, M., and Altmannsberger, M.,** Evidence for a hepatocellular lineage in a combined hepatocellular-cholangiocarcinoma of transitional type, *Virchows Arch.* [*Cell Pathol.*], 56, 71, 1988.
127. **Van Eyken, P., Sciot, R., and Desmet, V. J.,** A cytokeratin-immunohistochemical study of alcoholic liver disease: evidence that hepatocytes can express "bile duct type" cytokeratins, *Histopathology,* 13, 605, 1988.
128. **Nakanuma, Y. and Sasaki, M.,** Expression of blood group-related antigens in the intrahepatic biliary tree and hepatocytes in normal livers and various hepatobiliary diseases, *Hepatology,* 10, 174, 1989.
129. **Volpes, R., Van den Oord, J. J., and Desmet, V. J.,** Distribution of the VLA family of integrins in normal and pathological human liver tissue, *Gastroenterology,* 101, 200, 1991.
130. **Volpes, R., Van den Oord, J. J., and Desmet, V. J.,** Expression of VLA adhesion molecules in normal and pathological human liver tissue, *J. Hepatol.,* 11 (Suppl. 2), S65, 1990.
131. **Okada, Y., Colnaghi, M. I., and Tsuji, T.,** Type 4 chain H expression by bile ductules and hepatocytes in cirrhosis, *J. Pathol.,* 157, 329, 1989.
132. **Fukuda, Y., Miyazawa, Y., Imoto, M., Koyama, Y., Nakano, I., Nagura, H., and Kato, K.,** In situ distribution of enolase isozymes in chronic liver disease, *Am. J. Gastroenterol.,* 84, 601, 1989.
133. **James, J., Lygidakis, N. J., Van Eyken, P., Tanaka, A. K. F., Bosch, K. S., Ramaekers, F. C. S., and Desmet, V.,** Application of keratin immunocytochemistry and Sirius red staining in evaluating intrahepatic changes with acute extrahepatic cholestatis due to hepatic duct carcinoma, *Hepatogastroenterology,* 36, 151, 1989.

134. **Sasaki, H., Schaffner, F., and Popper, H.,** Bile ductules in cholestasis: morphologic evidence for secretion and absorption in man, *Lab. Invest.*, 16, 84, 1967.
135. **Desmet, V.,** Morphologic and histochemical aspects of cholestasis, in *Progress in Liver Diseases,* Vol. 4, Popper, H. and Schaffner, F., Eds., Grune & Stratton, New York, 1972, 97.
136. **Roskams, T., Van den Oord, J. J., De Vos, R., and Desmet, V. J.,** Neuro-endocrine features of reactive bile ductules in cholestatic liver disease, *Am. J. Pathol.*, 137, 1019, 1990.
137. **Bissell, D. M. and Choun, M. O.,** The role of extracellular matrix in normal liver, *Scand. J. Gastroenterol. (Suppl.),* 151, 1, 1988.
138. **Bissell, M. J., Hall, H. G.,and Parry, G.,** How does the extracellular matrix direct gene expression, *J. Theor. Biol.*, 99, 31, 1982.
139. **Schulz, R., Schuppan, D., and Hahn, E. G.,** Connective tissue and ductular metaplasia of hepatocytes, in *Mesenchymal-epithelial Interactions in Neural Development,* Wolff, J. R., Sievers, J., and Berry, M., Eds., Springer-Verlag, Berlin, 1987, 119.
140. **Hahn, E. G., Kirchner, J., and Schuppan, D.,** Perisinusoidal fibrogenesis and ductular metaplasia of hepatocytes in focal nodular hyperplasia of the liver, *J. Hepatol.*, 5, S32, 1987.
141. **Anthony, P. P.,** Tumours and tumour-like lesions of the liver and biliary tract, in *Pathology of the Liver,* 2nd ed., MacSween, R. N. M., Anthony, P. P., and Scheuer, P. J., Eds., Churchill Livingstone, Edinburgh, 1987, 574.
142. **Anthony, P. P.,** Liver tumours: an update, in *Recent Advances in Histopathology,* Vol. 14, Anthony, P. P. and MacSween, R. N. M., Eds., Churchill Livingstone, Edinburgh, 1989, 185.
143. **Wanless, J. R., Mawdsley, C., and Adams, R.,** On the pathogenesis of focal nodular hyperplasia of the liver, *Hepatology,* 5, 1194, 1985.
144. **Fischer, H. P., Altmannsberger, M., Weber, K., and Osborn, M.,** Keratin polypeptides in malignant epithelial liver tumors, *Am. J. Pathol.*, 127, 530, 1987.
145. **Van Eyken, P., Sciot, R., Paterson, A., Callea, F., Kew, M. C., and Desmet, V. J.,** Cytokeratin expression in hepatocellular carcinoma: an immunohistochemical study, *Hum. Pathol.*, 19, 562, 1988.
146. **Ramaekers, F., Van Niekerk, C., Poels, L., Schaefsma, E., Hiijsmans, A., Robben, H., Schaart, G., and Vooijs, P.,** Use of monoclonal antibodies to keratin 7 in the differential diagnosis of adenocarcinomas, *Am. J. Pathol.*, 136, 641, 1990.
147. **Johnson, D. E., Herndier, B. G., Medeiros, L. J., Warnke, R. A., and Rouse, R. V.,** The diagnostic utility of the keratin profiles of hepatocellular carcinoma and cholangiocarcinoma, *Am. J. Surg. Pathol.*, 12, 187, 1988.
148. **Balaton, A. J., Nehama-Sibony, M., Gotheil, C., Callard, P., and Baviera, E. E.,** Distinction between hepatocellular carcinoma based on immunohistochemical staining for carcinoembryonic antigen and for cytokeratin 19 on paraffin sections, *J. Pathol.*, 156, 305, 1988.
149. **Battifora, H. and Kopinsky, M.,** The influence of protease digestion and duration of fixation on the immunostaining of keratins. A comparison of formalin and alcohol fixation, *J. Histochem. Cytochem.*, 34, 1095, 1986.
150. **Rolfes, D. B.,** Fibrolamellar carcinoma of the liver, in *Neoplasms of the Liver,* Okuda, K. and Ishak, K. G., Eds., Springer-Verlag, Tokyo, 1987, 137.
151. **Berman, M. A., Burnham, J. A., and Sheahan, D. G.,** Fibrolamellar carcinoma of the liver: an immunohistochemical study of nineteen cases and a review of the literature, *Hum. Pathol.*, 19, 784, 1988.
152. **Van Eyken, P., Sciot, R., Brock, P., Casteels-Van Daele, M., Schaart, G., Ramaekers, F. C. S., and Desmet, V. J.,** Abundant expression of cytokeratin nr 7 in fibrolamellar carcinoma of the liver, *Histopathology,* 17, 101, 1990.

153. **Goodman, Z. D., Ishak, K. G., Langloss, J. M., Sesterhenn, I. A., and Sobin, L.,** Combined hepatocellular-cholangiocarcinoma: a histologic and immunohistochemical study, *Cancer,* 55, 124, 1985.
154. **Gould, V. E.,** Current topics. Histogenesis and differentiation: a re-evaluation of these concepts as criteria for the classification of tumors, *Hum. Pathol.,* 17, 212, 1986.
155. **Liu, M. H., Han, T., Lenzi, R., Slott, P. A., and Tavoloni, N.,** Immunohistochemical characteristics of bile-duct like (oval) cells proliferating during chemical hepatocarcinogenesis, *Hepatology,* 10, 621, 1989.
156. **Okada, Y., Jinno, K., Moriwaki, S., Moriwaki, S., Torigoe, S. I., Arima, T., Nagashima, H., and Koprowski, H.,** Expression of ABH and Lewis blood group antigens in combined hepatocellular-cholangiocarcinoma. Possible evidence for the hepatocellular origin of combined hepatocellular-cholangiocarcinoma, *Cancer,* 60, 345, 1987.
157. **Dehner, L. P.,** Tumors and tumorlike lesions, in *Pediatric Surgical Pathology,* Dehner, L. P., Ed., Williams & Wilkins, Baltimore, 1987, 478.
158. **Ishak, K. G. and Glunz, P. R.,** Hepatoblastoma and hepatocarcinoma in infancy and childhood. A report of 47 cases, *Cancer,* 20, 396, 1966.
159. **Schmidt, D., Harms, D., and Lang, W.,** Primary malignant hepatic tumors in childhood, *Virchows Arch. A,* 407, 387, 1985.
160. **Abenoza, P., Minivel, J. C., Wick, M. R., Hagen, K., and Dehner, L. P.,** Hepatoblastoma: an immunohistochemical and ultrastructural study, *Hum. Pathol.,* 18, 1025, 1987.
161. **Van Eyken, P., Sciot, R., Callea, F., Ramaekers, F., Schaart, G., and Desmet, V. J.,** A cytokeratin-immunohistochemical study of hepatoblastoma, *Hum. Pathol.,* 21, 302, 1990.
162. **Tsao, M. S. and Grisham, J. W.,** Hepatocarcinomas, cholangiocarcinomas and hepatoblastomas produced by chemically transformed cultured rat liver epithelial cells. A light and electron-microscopic analysis, *Am. J. Pathol.,* 127, 168, 1987.
163. **Weeks, D. A., Beckwith, J. B., Mierau, G. W., and Luckey, D. W.,** Rhabdoid tumor of kidney. A report of 111 cases from the National Wilm's Tumor Study Pathology Center, *Am. J. Surg. Pathol.,* 13, 439, 1989.
164. **Foschini, M. P., Van Eyken, P., Brock, P., Casteels-Van Daele, M., De Vos, R., and Desmet, V. J.,** Malignant rhabdoid tumour of the liver. A case report, *Histopathology,* 20, 157, 1992.

Chapter 12

ROLE OF VIRAL AND CELLULAR ONCOGENES AND GROWTH FACTORS IN HEPATOCARCINOGENESIS IN CULTURE AND *IN VIVO*

H. C. Isom and S. C. Strom

TABLE OF CONTENTS

ISBN 0-8493-4746-7
© 1992 by CRC Press, Inc.

I. *IN VITRO* IMMORTALIZATION AND TRANSFORMATION OF LIVER CELLS BY VIRAL AND CELLULAR ONCOGENES

A. TYPES OF LIVER CELLS USED FOR IMMORTALIZATION AND TRANSFORMATION

In reviewing studies that have been carried out to generate liver cell lines, it is important to note the source of cells that have been used, the culture conditions, and the growth properties of the cells. Fetal and adult hepatocytes from normal and transgenic animals, oval cells, and rat liver epithelial cells (RLE) are the predominant cell types that have been successfully used to analyze the ability of specific genes to immortalize or transform liver cells. Successful transformation of normal intrahepatic biliary epithelial cells by viral genes or oncogenes has not been accomplished. There is one preliminary report indicating that rat intrahepatic biliary epithelial cells can be immortalized by simian virus 40 (SV40) large T antigen.[1] When adult hepatocytes are isolated from normal animals and placed in culture, only a low percentage of the cells will synthesize DNA unless induced by the presence of specific mitogens. Adult hepatocytes can enter into two to three rounds of cell replication,[2-4] but will not continue to grow to form a cell line that can be subcultured. Therefore, to generate a cell line from an adult parenchymal hepatocyte, it is necessary to introduce a gene which has the capacity to enable the cells to replicate and become immortalized. The resulting replicating epithelial cells can then be transformed, yielding cells that have the capacity to form a tumor in a test animal. Fetal hepatocytes from normal animals synthesize DNA and undergo limited replication *in vitro* in the presence of mitogens.[5] Neonatal hepatocytes can replicate in culture in medium supplemented with epidermal growth factor (EGF).[6] The *in vitro* growth properties of fetal and adult hepatocytes isolated from transgenic animals depends upon what the transgene is and whether or not it is being expressed. Oval cells are often derived from the liver of an adult animal that has been treated with a hepatocarcinogen and fed a choline-deficient diet, and differ from normal adult hepatocytes in that they replicate and can spontaneously immortalize.[7] RLE cells can be derived by trypsin digestion of liver from 7-d-old,[8] 10- to 12-d-old,[9] and adult[10] Fischer rats. This results in the outgrowth of replicating epithelial cells that can be passaged, are diploid, do not grow in soft agar, and do not form tumors in Fischer rats or nude mice. These cells are usually used for transformation experiments at reasonably low passages (below passage 20). Cloned cell lines derived from RLE cells, such as RLE ϕ 13,[11] RLE-228,[8] and C61,[8] have been used for transformation experiments. The origin of RLE cells is unknown, but it has been suggested that they may be derived from terminal bile ductular epithelium.[12] However, some cell lines may have been derived from the mesothelium of Glisson's capsule (see Chapter 7). Untransformed rat liver-derived epithelial cell lines which spontaneously arose from primary rat hepatocyte cultures also have been used for transformation studies.[13-15] Since the cellular origin of liver tumors *in vivo* has not been established, use of any of these various cell types for immortalization and transformation studies has merit.

B. IMMORTALIZATION OF PRIMARY HEPATOCYTES BY VIRAL ONCOGENES

Primary rat hepatocytes isolated from adult rats have been immortalized by transfection and/or infection with SV40 or adenovirus genes (Table 1).[16-19] Isolated adult mouse hepatocytes have also been immortalized by SV40.[20] Viral genetic information can be introduced into hepatocytes by infection with viruses or by transfection or electroporation of DNA. Isom et al.[16] demonstrated that SV40 genetic information could transform cells in primary rat hepatocyte cultures, and that there was no background in the assay (that is, no colonies arose spontaneously). The assay was also quantitative.[17] In addition, it was demonstrated that transformation could be accomplished using hepatocytes maintained in a chemically defined medium[17,18] and the transformation frequency was increased if the cells were placed in a medium supplemented with EGF or if the cells were isolated from an animal that had been previously subjected to partial hepatectomy.[18] When primary rat hepatocytes maintained on a rat tail-coated collagen dish in a chemically defined medium were transfected (using calcium phosphate precipitation) with plasmid DNA containing the complete SV40 genome, colonies of epithelial cells appeared by 4 to 6 weeks posttransfection. The timing was delayed to 8 to 10 weeks if dimethylsulfoxide (DMSO) was included in the culture medium. When the colonies were analyzed using an immunooverlay procedure, 20 to 50% of the colonies contained cells that secreted albumin. In general, the colonies that contained albumin-secreting cells were smaller and had more regular borders than the colonies containing cells that did not secrete albumin. The percentage of colonies containing albumin-secreting cells was higher when the transfection was carried out using hepatocytes maintained in a medium supplemented with DMSO.

Several cell lines were successfully prepared from individual colonies of replicating albumin-secreting epithelial cells and were designated SV40-immortalized hepatocyte cell lines.[21] These cell lines differed from each other morphologically and with regard to the genes the cells expressed. Most of the cell lines continued to express albumin, but the level and mechanism of expression varied. The SV40-immortalized hepatocyte cell lines, at low cell passage number, were either nontumorigenic or weakly tumorigenic when inoculated subcutaneously into newborn syngeneic rats.[22] It was also not possible to successfully transplant CWSV1, an albumin-producing SV40-

TABLE 1
Examples of Immortalization and Transformation of Liver Cells by Viral and Cellular Oncogenes

Cell type	Agent used for immortalization or transformation	Growth properties of cell line	State of differentiation of cell line	Ref.
Adult rat hepatocytes	SV40 DNA	Immortalization	Liverlike alb levels; AFP^-; alb production regulated by glucocorticoid; production of other liver-specific genes	18, 21, 22, 64
	SV40 ts mutant virus infection	Immortalization (40°C); transformation (33°C); growth in soft agar at permissive temperature	Low alb; alb regulated by glucocorticoid	26
	Adenovirus; E1A + E1B	Immortalization	Low alb or alb^-	19
	v-*ras* infection	Transformation; tumors in nude mice; growth in soft agarose	Undifferentiated hepatocellular carcinoma	34
Adult mouse hepatocytes	SV40 ts mutant virus infection	Immortalization	Low alb	20
Fetal rat hepatocytes	SV40 ts mutant virus infection	Immortalization (40°C); transformation (33°C); growth in soft agar at permissive temperature	Low alb; AFP^+; $transferrin^+$; glucagon receptors	25, 27, 28
Neonatal rat hepatocytes	Polyoma virus early region DNA: LT, MT, and ST	Transformation; loss of EGF requirement; tumors in nude mice	Moderate alb; AFP^+	33
Neonatal mouse hepatocytes	SV40 virus infection	Selection in arg-free medium; immortalization; loss of EGF requirement	Moderate alb; AFP^+; TAT^+	29

Neonatal hepatocytes from albino deletion mutant mice	SV40 virus infection	Selection in arg-free medium; immortalization; loss of EGF requirement	Moderate alb; AFP^+; TAT^-; lacks tyrosinase encoding sequences	29
Newborn hepatocytes from albino deletion mutant mice	SV40 ts mutant virus infection	Ability to overgrow nontransformed human fibroblasts at 33°C	Lower levels of TAT and PEPCK mRNA than wild-type cell line	30
Newborn mouse hepatocytes	SV40 ts mutant virus infection	Ability to overgrow nontransformed human fibroblasts at 33°C	Alb^+; $transferrin^+$; $PEPCK^+$, TAT^+	30
Oval cells	Activated c-Ha-*ras* DNA	Transformation; tumors in nude mice	Well-differentiated hepatocellular carcinoma	7
SV40-immortalized hepatocyte cell lines (CWSV series)	*In vitro* passage	Transformation; tumors in syngeneic hosts	Variable expression of alb and AFP; well-differentiated hepatocellular tumors	22
SV40-immortalized hepatocyte cell line (CWSV1)	Activated c-Ha-*ras* DNA	Transformation; tumors in syngeneic hosts; growth in soft agarose	Liverlike alb levels; AFP^-; undifferentiated tumors	31
RLE	Infection with retrovirus containing v-H-*ras*	Transformation; tumors in nude mice and syngeneic hosts	Alb, AFP, and p450d same as in control RLE	39
	Infection with retrovirus containing v-*raf*	Transformation; tumors in nude mice and syngeneic hosts; no growth in soft agar	Alb, AFP, and p450d RNA same as in control RLE; tumors are sarcomalike mixed epithelial-mesenchymal tumors	39–41
RLEϕ13: clonal cell line derived from early passage RLE cells	Infection with retrovirus containing v-*raf* and v-*myc*	Transformation; tumors in nude mice, growth in soft agar	Tumors resemble hepatocellular carcinoma; altered expression of fibronectin and collagen I	11, 41, 42
RLE	c-*myc* DNA	Higher saturation density	Not examined	43
	Activated c-Ha-*ras* DNA	Transformation; tumors in nude mice; growth in soft agar	Not examined	43

TABLE 1 (continued)
Examples of Immortalization and Transformation of Liver Cells by Viral and Cellular Oncogenes

Cell type	Agent used for immortalization or transformation	Growth properties of cell line	State of differentiation of cell line	Ref.
RLE (*continued*)	Activated c-*myc* and c-Ha-*ras* DNA	Transformation; tumors in nude mice; growth in soft agar	Not examined	43
RLE-228 cells	Activated c-Ha-*ras* DNA driven by metallothionein promoter	Transformation; tumors in immunocompetent syngeneic hosts; growth in soft agar	Not examined	8
RLE	Activated c-Ha-*ras* DNA driven by MMTV promoter	Transformation; tumors in nude mice; growth in soft agar	Undifferentiated tumors; *in vitro* growth kinetics altered by dexamethasone; altered glucose transport	38
BL8: rat liver epithelial cell line derived from adult rat hepatocyte cultures	Activated c-Ha-*ras* DNA	Transformation; tumors in nude mice; growth in soft agar	Not examined	13–15
BL8	N-*ras* DNA	Transformation; tumors in nude mice; growth in soft agar	Not examined	13–15
Fetal hepatocytes from transgenic mouse	SV40 driven by the mouse metallothionein promoter	Immortalization; requires insulin for growth	Fetal liver levels of alb and AFP	47
SV40-immortalized fetal hepatocyte cell line from transgenic mouse	*In vitro* passage	Transformation; growth in soft agar; tumors in nude mice	Not examined	47

	Head to tail dimer of HBV DNA	Transformation; growth in soft agar; tumors in nude mice	Well-differentiated hepatocellular carcinoma	62
Adult preneoplastic liver cells from transgenic mouse	SV40 T antigen driven by alb enhancer/promoter	Immortalization	Not examined	63
Hepatocellular carcinoma from transgenic mouse	SV40 T antigen driven by alb enhancer/promoter	Transformation; growth in soft agar; tumors in nude mice	Not examined	63
c-*myc* hepatocellular carcinoma from transgenic mouse	Mouse c-*myc* and human A_1AT driven by human A_1AT promoter	Transformation	Mouse alb, A_1AT and transferrin; human A_1AT	52

Note: alb, Albumin; AFP, α-fetoprotein; EGF, epidermal growth factor; arg, arginine; TAT, tyrosine aminotransferase; PEPCK, phosphoenolpyruvate carboxykinase; A_1AT, α_1-antitrypsin; −, negative expression; +, positive expression.

immortalized hepatocyte cell line that had been derived from a single colony, to the spleen of an adult syngeneic rat.[23] In contrast, when cultures of SV40-transfected hepatocytes containing multiple albumin-positive and albumin-negative colonies were harvested at approximately 6 weeks posttransfection and inoculated into the spleen of syngeneic animals, SV40 T antigen-positive cells that formed well-organized ducts were observed in the spleens at 1 month after transplantation to spleen. SV40-immortalized or -transformed cells are highly immunogenic and are often rapidly rejected by an immunocompetent host. The ability of a cell to survive suggests that it can replicate rapidly enough to overcome rejection or in such a way as to not be killed by the immune defense mechanisms of the host. The results of the spleen implant studies suggest that the heterogeneous population of transfected cells contained T antigen-positive liver cells. These most likely differed in origin from the cells that became the CWSV1 cell line and were able to survive and replicate at least temporarily in the spleens of recipient hosts. In contrast, the clonally derived albumin-positive CWSV1 cells, which have a slow growth rate, were eliminated by the immune response of the host.

In control cultures transfected with hepatocyte carrier DNA, no colonies of replicating epithelial cells appeared throughout the duration of the transformation assays.[18] If the hepatocytes were maintained in a chemically defined medium with or without EGF, the initial primary hepatocytes died by approximately 4 weeks. If the transfections were carried out using hepatocytes maintained in medium supplemented with DMSO, the colonies of replicating epithelial cells arose amidst a background of nondividing primary hepatocytes. In control transfected DMSO cultures, no colonies of replicating cells arose during the duration of the transformation assay. When control DMSO cultures were maintained for 9 months to 1 year, the spontaneous outgrowth of replicating cells that morphologically resemble hepatocytes was observed on rare occasions and cell lines were successfully prepared from these cells.[24] These cell lines have not been characterized for tumorigenicity, growth properties, or gene expression.

Isolated adult and fetal rat hepatocytes have also been immortalized by infection with temperature-sensitive mutants of SV40 (Table 1).[25-28] The cell lines established were immortalized at the nonpermissive temperature of 40°C and did not form colonies in soft agar at this temperature. At the permissive temperature, 33°C, the cells behaved more like transformed cells in that they demonstrated anchorage-independent growth. One fetal liver cell line had a cloning efficiency in soft agar of 0.0015% at the restrictive temperature and 0.25% at the permissive temperature.[25] One adult liver cell line had a cloning efficiency of 0.001% at the restrictive temperature and 0.5% at the permissive temperature.[26]

Adult rat hepatocytes have also been immortalized by infection with adenovirus type 5 or by transfection with adenovirus virion DNA or recombinant plasmid DNA containing the E1A and E1B genes.[19] Only two albumin-producing adenovirus-immortalized cell lines were obtained and both were derived from cultures transfected with a plasmid containing the E1A and E1B genes. Both cell lines only secreted 7 to 14% of the amount of albumin secreted by the SV40-immortalized hepatocyte cell line CWSV1, and albumin secretion by one of the two cell lines tested was undetectable by passage 9. Five of the adenovirus-immortalized cell lines, tested at passages below 10, were not tumorigenic in newborn syngeneic animals.

Neonatal hepatocytes from normal and albino deletion mutant mice have been immortalized by infection with SV40 followed by selection in arginine-deficient medium.[29] One cell line from normal hepatocytes and five cell lines from mutant hepatocytes were derived. These cell lines, in contrast to the parental neonatal hepatocytes, did not require EGF for growth. The authors referred to the cell lines as immortalized and did not provide information as to whether these cell lines grow in soft agar or are tumorigenic. Neonatal hepatocytes from normal mice and mice heterozygous and homozygous for the albino deletion were immortalized by infection with a temperature-sensitive mutant of SV40.[30] At the permissive temperature (33°C) the cell lines behaved as transformed cells and overgrew nontransformed human fibroblasts. The ability of the cell lines to overgrow was greatly diminished when they were assayed at the nonpermissive temperature (39.5°C). The cell lines were not tested for growth in soft agar or tumorigenicity in animals.

Primary hepatocytes from 6-week-old mice have also been immortalized by infection with an SV40 temperature-sensitive mutant.[20] A cell line, H2.35, which was derived to study liver-specific gene expression, was not tested for tumorigenicity or growth properties associated with transformation.

C. TRANSFORMATION OF IMMORTALIZED PRIMARY HEPATOCYTES AS A RESULT OF *IN VITRO* PASSAGE OR INTRODUCTION OF ONCOGENES AND VIRAL GENES

Transformation of primary hepatocytes in *in vitro* model systems has been observed to occur as a one-step or multistep process depending upon the source of the cells and/or the transforming agent (Table 1). Transformation of adult rat hepatocytes with SV40 is a multistep process; it requires either *in vitro* passage, which apparently induces secondary cellular changes, or the addition of the oncogene *ras*.[12,31,32] Five SV40-immortalized hepatocyte cell lines of the CWSV series at various *in vitro* passage levels were tested for tumorigenicity after subcutaneous transplantation into syngeneic newborn rats.[22] At low passage levels (12 to 22), three of the five cell lines were not tumorigenic and two were weakly tumorigenic, indicating that low passage cells demonstrated a low tumor incidence. Low passage cells also produced predominantly tumors that developed and then regressed. At higher *in vitro* passage levels (33 to 61), all five cell lines became tumorigenic. One, the CWSV1 cell line, was only weakly tumorigenic. The other four cell lines demonstrated tumor incidences ranging from 48 to 78% and had a high tendency to produce progressive tumors. Histological examination of tissue sections from tumors from two of the cell lines (CWSV2 and CWSV14) indicated that the tumors were well-differentiated hepatocellular carcinomas. The tumors were trabecular and in one tumor, well-differentiated ducts composed of cuboidal epithelium were also observed. Tumor cell lines derived from three of the tumors had a tumor incidence of 80 to 100% and produced predominantly progressive tumors.

Comparison of gene expression in high compared to low passage CWSV cell lines revealed that Ha-*ras* RNA levels were higher in the high passage cells.[22] For this reason, the activated c-Ha-*ras* gene was introduced by transfection into low passage immortalized CWSV1 cells to determine whether or not these cells would transform.[31,32] The CWSV1 cells were chosen for these studies became they had essentially undetectable levels of Ha-*ras* RNA at low passage and lowest tumorigenic potential at late passage of the five cell lines examined. CWSV1 cells were cotransfected with pT24-C3 (activated c-Ha-*ras*) and pSV2-*neo* or with pSV2-*neo* alone as a control. Colonies that survived G418 treatment were developed into cell lines. Two cell lines, NR3 and NR4, transfected with *ras* + *neo*, expressed reasonably high levels of *ras*. The NR4 cell line expressed higher levels of *ras* than NR3. Both NR3 and NR4 cells formed tumors in syngeneic newborn rats with a tumor incidence of 50% or greater and with a latency period of 5 to 7 weeks. The CWSV1 cells were essentially negative for growth in soft agarose (0.015% cloning efficiency). Both the NR3 and NR4 cells grew reasonably well in soft agarose. The control cell lines were not tumorigenic in syngeneic animals and also did not demonstrate anchorage-independent growth in soft agarose. Analysis of growth curves showed that both the parent CWSV1 cells and one *neo* cell line, N1, grew slowly. The NR4 cells grew at a more rapid rate and achieved a significantly higher saturation density. Four tumors that arose after inoculation of NR4 cells and one that arose after inoculation of NR3 cells into syngeneic newborn rats were undifferentiated; the cells in the tumors were epithelial, but did not form trabecular patterns and did not resemble hepatocytes. The results indicate that a well-differentiated SV40-immortalized hepatocyte cell line was able to be transformed by activated c-Ha-*ras*. In addition, the intentional introduction of the activated c-Ha-*ras* gene yielded cell lines that produced undifferentiated tumors while transformation that resulted from *in vitro* passage (and was accompanied by increased levels of *ras* RNA) yielded two cell lines that formed well-differentiated hepatocellular carcinomas.

Transformation of neonatal rat hepatocytes with polyoma virus early region DNA occurs as a one-step process.[33] This may be due to the fact that neonatal rat hepatocytes actively replicate and therefore were partially immortalized before being transfected with polyoma virus early region DNA, or it may be due to the properties of polyoma small, middle, and large T expression. When SV40 infection or transfection of adult primary rat or mouse hepatocytes was carried out, colonies of immortalized replicating cells were established into cell lines without interference from the parental cells, since the adult hepatocytes either died (hepatocytes in the absence of DMSO) or did not divide (DMSO-treated hepatocytes). In contrast, it was necessary to separate polyoma-transformed neonatal hepatocytes from a background of replicating normal neonatal hepatocytes. Therefore, neonatal rat hepatocytes were transfected with pSV2-*neo* as well as polyoma DNA and transformed cells were selected with G418. In order to enrich for cells retaining the capacity to synthesize arginine, the transfected cultures were also fed an arginine-deficient medium. Several polyoma-transformed neonatal hepatocyte cell lines were derived and were reported to grow in soft agar and produce tumors in nude mice. Transfection of neonatal hepatocytes with pSV2-*neo* also did not result in colony formation.

Transformation of primary adult rat hepatocytes has been accomplished in a single step by infection with an Ha-*ras* sarcoma virus (Harvey MSV/Mo) pseudotyped with Friend MuLV.[34] Colonies of replicating refractile, rounded cells were apparent 1 week after infection. The v-*ras*-transformed hepatocytes grew in soft agar and were tumorigenic in nude mice. The v-*ras*-transformed hepatocytes and the activated c-Ha-*ras*-transformed SV40 hepatocytes were similar in that both demonstrated anchorage-independent growth, formed tumors in test animals, and yielded undifferentiated tumors. In contrast, it has not been possible to transform primary adult rat hepatocytes by transfection with the pT24-C3 (activated c-Ha-*ras*) plasmid alone or in combination with c-*myc*.[35] The difference in the outcome of these two sets of experiments is most likely due to the fact that infection with v-*ras* is more efficient than transfection; that is, the number of cells taking up and expressing the *ras* gene may be critical in successfully achieving transformation.

D. TRANSFORMATION OF OVAL CELLS BY ONCOGENES

Oval cells are small, nonhepatocyte epithelial cells (see Chapters 1 and 5–8). Oval cells proliferate in livers of animals treated with hepatocarcinogenic regimens. They differ from hepatocytes in that they are absent or present at essentially undetectable levels in normal liver. They also differ from adult hepatocytes in that they proliferate and occasionally will spontaneously develop into immortalized cell lines. With continued *in vitro* passage, oval cell lines can be generated which will grow in soft agar in the presence of EGF. It is also possible to derive spontaneously transformed oval cell lines by passaging the cells infrequently during *in vitro* cultivation.[36] These spontaneously transformed oval cell lines produced moderate to well-differentiated hepatocellular carcinoma in nude mice.

Early passage immortalized oval cell lines have been transformed by transfection with activated c-Ha-*ras*.[7] The *ras*-transformed oval cell lines produced well-differentiated hepatocellular carcinomas when inoculated into nude mice.

E. TRANSFORMATION OF RAT LIVER EPITHELIAL CELLS BY ONCOGENES

The origin of RLE is not clearly established, but some are most likely derived from bile ductular epithelium and not hepatocytes.[12] Support for studying oncogene transformation of RLE cells as an *in vitro* model for hepatocarcinogenesis comes from the knowledge that WB-F344 RLE cells transformed *in vitro* by exposure to *N*-methyl-*N'*-nitro-*N*-nitrosoguanidine (MNNG) when inoculated subcutaneously or intraperitoneally into 1-d-old syngeneic F344 rats can yield tumors that are morphologically indistinguishable from hepatocellular carcinomas.[37] The MNNG-transformed WB-F344 cell line and 18 clonal subpopulations derived from this cell line produced hepatocellular carcinomas, cholangiocarcinomas, hepatoblastomas, well-differentiated adenocarcinomas demonstrating features of intestinal epithelial differentiation, mixed epithelial-mesenchymal tumors, undifferentiated sarcomas, etc., when the cells were inoculated subcutaneously or intraperitoneally into 1-d-old syngeneic rats. This study showed that RLE cells are multipotent cells that can produce tumors upon transplantation that reproduce the types of malignant neoplasms that can develop *in vivo* in liver.

Continuously dividing, nontransformed RLE cells have been transformed by activated c-Ha-*ras*, v-*ras*, N-*ras*, v-*raf*, and activated c-Ha-*ras* in combination with activated c-*myc* and v-*raf* in combination with v-*myc*.[8,13,14,34,38-43] Oncogenes were introduced into RLE cell lines by transfection or infection with retroviruses containing the specific genes. Introduction of c-*myc* into RLE cells yielded cells that grew at a higher saturation density than the parental RLE cells, but did not form tumors in animals and did not grow in soft agar.[34,43] RLE cells have been transformed by the activated c-Ha-*ras* oncogene driven by its own promoter yielding cells that constitutively express *ras*,[13,34,43] and they have also been transformed by *ras* driven by the regulatable mouse mammary tumor virus (MMTV)[38] and metallothionein[8] promoters. In all cases, transformation with *raf* or *ras* and *raf* or *ras* in combination with *myc* yielded cell lines that formed tumors in nude mice. Cells transformed by v-*ras* or v-*raf* also produced tumors in syngeneic hosts.[39] An RLE cell line transformed by activated c-Ha-*ras* was capable of producing tumors in immunocompetent syngeneic hosts.[8] The ability of many of the oncogene-transformed RLE cells to form tumors in syngeneic hosts was not measured. It is interesting to note that although v-*raf*-transformed cells were able to produce tumors in test animals, they grew poorly in soft agar without EGF.[42] Similarly, although six v-*raf* tumor cell lines (derived from a tumor produced by v-*raf*-transformed

RLE cells) yielded tumors in 100% of the nude mice inoculated, none of these cell lines grew in soft agar without EGF and only three of the cell lines grew in soft agar supplemented with EGF.

A series of v-*raf* and v-*raf*/v-*myc*-transformed RLE cell lines that differed in properties associated with transformation were generated.[41,42] RLE cells transformed by v-*raf* alone were weakly tumorigenic; when athymic nude mice were inoculated subcutaneously with 10^6 v-*raf*-transformed cells, tumors developed in 17% of the animals. Six v-*raf* tumor cell lines developed from single-cell clones derived from a tumor produced by the *raf*-transformed cells.[42] Tumors developed in 100% of the animals that received v-*raf* tumor cells or v-*raf*/v-*myc*-transformed cells. v-*raf* tumor cells were also more tumorigenic than v-*raf*-transformed cells in that the latency period for tumor formation decreased threefold to sevenfold and the growth rate of the tumors increased by fourfold to ninefold. Cotransformation with both v-*raf* and v-*myc* yielded cells that were the most tumorigenic of the series; these cell lines formed tumors at a rate of growth two to three times that of the tumors produced by the v-*raf* tumor cell lines.

RLE cells transformed with *ras* driven by the MMTV long terminal repeat (LTR) promoter were used to measure the effect of increased *ras* expression on growth kinetics and anchorage-independent growth.[38] RLE cells were infected with a retroviral vector containing the activated c-Ha-*ras* gene under the control of the dexamethasone inducible LTR of MMTV. The *ras*-infected cells produced tumors in nude mice. Control cells infected with a virus containing only the *neo* gene driven by the MMTV LTR did not form tumors. The tumors were reestablished into tissue culture as tumor cell lines. One *ras* tumor cell line was studied. In the absence of dexamethasone, the *ras* tumor cell line expressed higher levels of *ras* than the parental untransformed REL cells. Treatment with dexamethasone increased the *ras* mRNA levels in the tumor cell line by approximately 15-fold. In the absence of dexamethasone, the *ras* tumor cell line grew in soft agar with a cloning efficiency of 5.3%. In the presence of dexamethasone, the cloning efficiency increased to an average of 8.4%, but the difference was not statistically different. The control REL cells and *neo* cells did not form colonies in soft agar in the absence or presence of dexamethasone. A significant effect of dexamethasone treatment and the subsequent induction of *ras* was observed when the kinetics of anchorage-dependent growth were measured. REL cells and *neo* cells in the absence and presence of dexamethasone demonstrated approximately a 4-d lag period in defined medium until logarithmic growth started. In the absence of dexamethasone, the *ras* tumor cells had a shorter lag phase than the control cells. In the presence of dexamethasone, the tumor cells demonstrated an even greater alteration in growth properties in that they entered logarithmic growth without a lag phase.

The morphology of tumors induced by RLE cells transformed by different oncogenes varied. Tumors produced by *ras*-transformed cells were undifferentiated carcinomas.[34] Tumors produced by v-*raf*-transformed cells were sarcomalike mixed epithelial mesenchymal tumors.[40] Tumors produced by v-*raf*/v-*myc*-transformed cells were hepatocellular carcinomas.[40]

II. DERIVATION OF IMMORTALIZED AND TRANSFORMED LIVER CELL LINES FROM THE LIVERS OF TRANSGENIC ANIMALS

Transgenic animals represent a powerful tool for studying immortalization and transformation of liver cells by oncogenes, viral genes, and growth factors for two reasons (see Chapter 13). First, the transgenic animal can be used to determine whether or not introduction of specific genetic information under the control of a promoter that will be expressed in a hepatocyte can alter the growth properties of hepatocytes in the liver leading to the development of hepatocellular carcinoma. Second, the liver from an animal containing such a transgene can be used for preparation of immortalized or transformed liver cells that will grow in culture. These cells can be prepared from livers at any stage of development of the animal and hence also from livers with no lesions, preneoplastic lesions, or carcinoma.

A variety of different transgenes directed by various promoters have been used successfully to generate hepatocellular carcinoma in transgenic animals. Some of these include SV40 T antigen driven by a segment of the human α-1-antitrypsin (A_1AT) promoter,[44,45] SV40 early region and control region driven by the mouse metallothionein promoter,[46] SV40 early region lacking the SV40 enhancer driven by the mouse metallothionein promoter.[46-48] the SV40 early region lacking the SV40 enhancer and 21-bp repeats driven by the mouse metallothionein promoter,[46] SV40 T antigen driven by the mouse albumin enhancer/promoter,[49] SV40 T antigen driven by the mouse major urinary protein,[50,51] activated c-Ha-*ras* driven by the mouse albumin enhancer/promoter,[49] c-*myc* driven by the mouse albumin enhancer/promoter,[49] c-*myc* and human A_1AT driven by the human A_1AT promoter,[52] human transforming growth factor-α (TGF-α) driven by the mouse metallothionein promoter,[53] the hepatitis B virus (HBV) *BGl*II-A fragment driven by the mouse albumin enhancer/promoter,[54-58] and the HBV X gene driven by its own regulatory elements.[59]

SV40 T antigen, activated c-Ha-*ras*, c-*myc*, TGF-α, HBV surface antigen (HBsAg), and HBV X antigen when expressed *in vivo* in liver can lead to the development of hepatocellular carcinoma. The time course for transformation and the type of pathogenicity observed vary markedly depending upon the transgene. A slow time course for development of hepatocellular carcinoma in many of the animals indicates that secondary events, in addition to expression of the transgene, are important in the hepatocarcinogenic process. The most rapid and dramatic effects were observed with the *ras* transgene.[49] Out of 19 alb-*ras* transgenic founder mice, 14 were born with enlarged, hyperplastic livers and died within several days of birth. The remaining five animals developed focal areas of bile duct hyperplasia and dysplasia and diffuse hepatocellular dysplasia. Two of the five developed hepatocellular carcinomas and all five animals died by 5 months from lung adenomas. These studies suggest that when *ras* expression exceeds a specific threshold in neonatal liver, it induces uncontrolled cell division, apparently in a single step, resulting in death. Induction of hepatocellular carcinoma by SV40 T antigen was considerably slower and underwent distinct steps. Hyperplastic hepatocytes appeared, followed by diffuse dysplasia. By 2 to 7 months of age the animals developed multifocal hepatic tumors against a background of dysplasia. Hepatic neoplasms in SV40 T antigen transgenics included hepatocellular carcinomas, adenomas, and cholangiomas and cholangiocarcinomas. In c-*myc* transgenic mice, only dysplasia was seen in the founder mice by 1 year of age; well-differentiated hepatocellular adenomas developed in three offspring between 15 and 18 months of age.

The effect of coexpression of oncogene pairs using *ras, myc,* and SV40 T antigen has also been studied by mating *ras, myc,* or SV40 mice.[49] Each combination resulted in accelerated tumor development. It is interesting to note that mice carrying the dual transgene of alb-SV40 and alb-*ras* by 1 week of age had enlarged livers that histologically displayed a prominent ductular pattern reminiscent of severe bile duct hyperplasia. These data suggest that biliary epithelium is particularly susceptible to transformation by these oncogenes or that these oncogenes can convert progenitor cells or hepatocytes to proliferating bile duct epithelium. It was also observed in the same study that bile duct tumors did form in one founder mouse containing only the SV40 T antigen.

In TGF-α mice, the morphological alterations observed in the development of hepatocellular carcinoma were remarkably similar to those seen in SV40 mice, although in TGF-α mice the process was considerably slower.[53] Hepatocellular adenomas and carcinomas appeared against a background of normal and dysplastic hepatocytes between 10 and 15 months of age.

Transgenic mice, which produce the HBV large and major enveloped polypeptides at equimolar ratios and generate long nonsecretable HBsAg filaments that accumulate in the endoplasmic reticulum of the hepatocyte, develop hepatocellular carcinoma at 18 to 21 months of age.[57] Two lineages have been established in which tumors develop. The HBsAg mice differ from the SV40 and TGF-α animals in that the development of neoplasia in HBsAg mice is preceded by cell injury. In two different mouse lineages, palpable liver tumors were apparent approximately 9 to 12 months after onset of liver injury. The tumors were histologically heterogeneous, including hepatocellular adenomas and carcinomas. Some of the carcinomas were trabecular and well differentiated while others were poorly differentiated. It is interesting to note that the c-*myc*, c-*fos*, Ha-*ras*, and K-*ras* RNA levels in liver tumors from HBsAg mice were not increased relative to the level in nontumorous transgenic mice;[57] however, the multidrug resistance gene (*mdr3*) was overexpressed in these tumors.[60]

There is also recent evidence using transgenic animals that HBV X gene can transform hepatocytes *in vivo*.[59] Two independent studies have been carried out using transgenic animals to determine whether or not expression of the HBV X gene leads to hepatocellular carcinoma.[59,61] In the first study,[61] the transgene was the X gene open reading frame under the control of the human A_1AT regulatory region. The logic behind this study was that transgenic mice carrying the SV40 T antigen gene controlled by the same A_1AT gene regulatory region reproducibly developed liver tumors. In contrast, transgenic mice carrying the X gene did not develop liver tumors. Three of the lines of mice expressed X gene mRNA and the expression was at low levels. The X protein was expressed maximally during the first month after birth and became undetectable by 2 months of age. None of the transgenic animals died prematurely, compared to the littermates. The most common disorder was mild focal lobular hepatitis. It is possible that pathology of the liver in some of the animals was due to unrelated factors such as sporadic infection by mouse hepatitis virus. One animal had a nodule that resembled an adenoma, but progeny from this did not develop similar hyperplastic nodules. The authors were not concerned about the fact that the X gene was no longer expressed by 2 months, since the same results had been observed for T antigen expression driven by the A_1AT promoter, and the T antigen animals still developed liver tumors.

In the second study, the HBV X gene was under the control of its own regulatory elements.[59] Of three founder mouse cell lines developed that contained at least one copy of the HBV X transgene, all developed hepatocellular carcinoma. The males died with hepatocellular carcinoma between 11 and 15 months of age and the females between 17 and 21 months. The histopathological changes in the liver were progressive, leading first to the appearance of benign adenomas and then to the development of malignant carcinomas. A comparison of the two studies indicates that persistence of expression of X gene and magnitude of X gene product accumulation may be critical in determining whether or not X protein can induce the development of hepatocellular carcinoma. The studies also suggest that SV40 and X gene transform hepatocytes by different mechanisms.

As indicated by the above studies, a variety of different oncogenes, viral genes, and growth factor genes directed to the liver by liver-specific promoters have been used to develop transgenic models for hepatocarcinogenesis. The stepwise pathological changes observed during chemically induced and spontaneous hepatocarcinogenesis have been reproduced in many of the transgenic models. Livers from transgenic fetuses and animals of various ages should be able to be used to derive cell lines that maintain the *in vivo* phenotypes and may provide *in vitro* systems for analyzing molecular and cellular changes during hepatocarcinogenesis. Several liver cell lines have been derived from transgenic animals and characterized (Table 1).

An SV40-immortalized fetal hepatocyte cell line, designated FMH-202, was derived from animals containing SV40 in the absence of its own enhancer and driven by the mouse metallothionein enhancer.[47] The goal of using hepatocytes from a transgenic mouse fetus instead of an adult was to develop a cell line which was immortalized, but not transformed by the SV40 transgene. The FMH-202 cell line was derived from primary day 19 fetal mouse liver cell cultures maintained in an arginine-free, serum-free medium. FMH-202 cells contained integrated SV40 sequences, expressed SV40 T antigen, did not grow in soft agar, did not form tumors in nude mice, and were growth arrested in the absence of insulin. When FMH-202 cells were triggered with insulin to reenter the cell cycle, the cells transiently expressed the protooncogenes c-*fos*, c-*myc*, and c-Ha-*ras* prior to entering S phase.[62] FMH-202 cells expressed albumin and α-fetoprotein (AFP), but did not express tyrosine aminotransferase (TAT) or phosphoenolpyruvate carboxykinase (PEPCK). The cells could not be cloned initially. At passage 26, clonal cell lines were derived and these cells grew in soft agar and were tumorigenic.

It has recently been demonstrated that immortalized FMH-202 cells can be transformed by HBV.[62] FMH-202 cells were transfected with a shuttle vector containing a head-to-tail dimer of cloned HBV DNA downstream of the SV40 enhancer and early promoter and a plasmid containing the *neo* gene. Control cell lines were generated by transfecting the fetal hepatocytes with the shuttle vector alone and plasmid containing the *neo* gene. Four of the clones that arose from the HBV-transfected cells and that secreted HBsAg and HBeAg antigens and viral DNA (suggesting that HBV particles were being produced) were analyzed further. All four cell lines grew well in soft agar and formed tumors in nude mice. All of the tumors were identified as hepatocellular carcinomas. One cell line produced tumors with a trabecular pattern typical of a well-differentiated heptocellular carcinoma. The control transfected cell lines did not grow in soft agar and were not tumorigenic in nuce mice. Two of the cell lines, tumors produced by these cell lines, and tumor cell lines derived from the tumors were analyzed for viral expression at the RNA and protein level. The expression pattern of viral sequences in the tumors and tumor cell lines were highly variable. In one tumor, viral gene expression was virtually absent, while in the more highly differentiated trabecular tumor viral gene expression was only partially blocked.

The two cell lines expressed four transcripts (0.9, 2.1, 3.6, and 4.0 kb) which hybridized to HBV DNA. The authors state that the 2.1-kb transcript most likely corresponds to mRNA, which expresses the small and middle-sized HBsAg polypeptides. The 2.1 transcript was expressed in the cell lines, was virtually absent in RNA from the tumors, and was weakly reexpressed in cell lines derived from two mouse tumors. The 0.9-kb transcript, which hybridized with the X probe, was expressed in the cell lines and in tumors from one cell line, but not in tumors from the other cell line. No HBsAg was detected in the tumors or in the sera of the tumor-bearing mice. HBV X protein was weakly detected in purified nuclei from one of the cell lines. Southern blot analysis of DNA from tumors and tumor-derived cell lines indicated that extensive rearrangement of HBV DNA had occurred.

Liver cell lines were also established from HBsAg transgenic mice.[57] The cell lines were derived from two micronodular livers displaying early histological features of malignant transformation.

SV40-immortalized ("preneoplastic") and transformed (hepatocellular carcinoma) cell lines were also derived from mice harboring an SV40 T antigen transgene driven by the albumin enhancer/promoter.[63] The ten immortalized cell lines, which were established from 3- to 12-week-old animals, were unable to grow in soft agar or form tumors in nude mice at early passage or late passage. Nine cell lines were isolated from hepatocellular carcinomas in livers of 24-week-old animals. Six of these cell lines grew in soft agar and nude mice at early passage; three did not grow in soft agar or nude mice at early passage, but acquired these properties at later passage. These cell lines were characterized for expression of activated c-Ha-*ras* as discussed below. These cell lines were not characterized for expression of growth factor genes, liver-specific genes associated with differentiation, or other oncogenes.

Mouse hepatoma cell lines that produce human A_1AT have been established from tumors in transgenic mice coexpressing the mouse c-*myc* oncogene and the human A_1AT gene under the control of the human A_1AT promoter.[52] Initial attempts at establishing cell lines from the hepatocellular carcinomas yielded cells that rapidly dedifferentiated. In order to maintain cells with a hepatocyte morphology, it was necessary to isolate the cells from the tumor with collagenase, culture within a Primaria® flask, and feed the cells Williams medium supplemented with dialyzed fetal calf serum (FCS) supplemented with hydrocortisone, transferrin, EGF, and insulin. After five passages, the cells could be grown in the absence of FCS. The cells produced rat albumin, A_1AT, and transferrin. The cells also secreted the 55-kDa glycosylated form of human A_1AT encoded by the transgene. These cells maintained a differentiated hepatic phenotype and continued to secrete human A_1AT for at least 40 generations.

III. EFFECT OF IMMORTALIZATION AND TRANSFORMATION ON EXPRESSION OF GENES ASSOCIATED WITH DIFFERENTIATION

Expression of liver-specific genes, predominantly albumin, and other genes associated with the differentiated phenotype is not automatically maintained during *in vitro* generation of immortalized and/or transformed liver cell lines, but it can be. Not all oncogene and viral gene immortalized and/or transformed liver cell lines derived *in vitro* or from transgenic livers or tumors have been selected for maintenance of expression of differentiated functions, and few have been characterized rigorously or at a quantitative level for such expression. In addition, only a limited number of studies involving liver cell lines have been directed at determining the effect of transformation on expression of genes associated with differentiation.

The gene product associated with differentiation in the liver that has been most extensively measured in liver cell lines derived from hepatocytes is albumin. It is possible to immortalize cell lines from adult and neonatal hepatocytes which continue to express albumin (Table 1). The amount of albumin produced by these cell lines varies and is often difficult to evaluate relative to albumin produced by normal liver. Maintenance of albumin expression in several of the cell lines requires glucocorticoid (usually dexamethasone) in the medium.[21,26] Albumin expression can also be regulated by cell density[21] and by cytokines that induce the acute-phase response.[64]

The SV40-immortalized hepatocyte cell line, CWSV1, expressed albumin mRNA at steady-state levels equal to or greater than liver[22] and transcribed albumin as measured in a nascient chain extension assay at 145% the level of liver.[65] The CWSV2 and CWSV14 cell lines transcribed albumin at approximately 140 and 70% the level of liver, respectively.[65] To our knowledge, these are the only cell lines which transcribe albumin at this high level. Transfection experiments have also demonstrated that the mouse albumin enhancer and promoter can function to drive the expression of a CAT reporter gene in CWSV1 and CWSV2 cells,[65] suggesting that these cells retain many of the appropriate transcription factors for albumin expression. The CWSV cells used for the above studies on albumin expression and albumin enhancer function were cultured on plastic cell culture dishes and fed a chemically defined medium.

H2.35 cells, an adult mouse hepatocyte cell line immortalized by infection with a temperature-sensitive mutant of SV40, plated on collagen gels at 39°C, and cultured in a serum-free media (differentiating conditions) transcribed albumin at only 5% the level of liver, but were able to support the function of the albumin enhancer.[20] When H2.35 cells were cultured on plastic dishes at 33°C in serum-supplemented media (dedifferentiating conditions), albumin mRNA levels decreased to less than 1% what they were under differentiating conditions. The H2.35 cells have also been used to study cell-extracellular matrix interactions important for transcriptional activation, in particular with regard to the interaction of specific transcription factors and the albumin enhancer.[66,67]

RALA255-10g cells, adult rat hepatocytes immortalized by infection with a temperature-sensitive mutant of SV40, produced albumin at 40°C at levels ranging from 0.4 to 0.6 pg per cell per 24 h.[26] In comparison, CWSV1 cells which expressed albumin mRNA at 100% or greater than liver, secreted albumin at levels ranging from 20 to 40 pg per cell per 24 h[21] or approximately 50-fold higher than RALA255-10g cells. Temperature altered the growth characteristics of RALA255-10g cells and their ability to express albumin. At a permissive temperature (33°C), RALA255-10g cells secreted extremely low levels of albumin (0.05 to 0.1 pg per cell per 24 h). The studies with RALA255-10g cells indicate that conversion of these cells from an immortalized phenotype to a transformed phenotype as a result of growth temperature and hence stability of the SV40 T antigen resulted in approximately a tenfold decrease in albumin secretion. However, it must be realized that at both permissive and nonpermissive temperatures, albumin production by RALA255-10g cells was low relative to liver. Also, the effect of temperature on albumin expression independent of effects on T antigen and growth properties was not measured. In addition, it is doubtful that the presence of a stable SV40 T antigen (at permissive temperature) directly decreases albumin expression since CWSV1 cells express stable SV40 T antigen and produce albumin at liverlike levels.

FMH-202 cells which were derived from the fetuses of transgenic mice harboring SV40 early region driven by the mouse metallothionein enhancer have the characteristics of an immortalized cell line and express albumin RNA at approximately 50% the level in fetal liver as determined by northern blot analysis.[47] FMH-202 cells transform spontaneously with *in vitro* passage; the authors reported that the transformed clonally derived cell lines display hepatocyte-specific functions similar to low passage FMH-202 cells, but no data with regard to albumin levels were provided.[47] FMH-202 cells have been transformed by HBV, but albumin levels in the HBV-transformed cells were not reported.[62]

Studies have been carried out to measure albumin expression in CWSV cell lines that transform spontaneously[22] and CWSV1 cells that transform as a result of introduction of the activated c-Ha-*ras* gene.[31,31] CWSV2 cells at high *in vitro* passage, transformed, produced tumors in syngeneic hosts and continued to produce albumin RNA at the same level as low passage immortalized CWSV2 cells. AFP RNA was not detected in low or high passage CWSV2 cells. CWSV14 cells were derived in the same manner as CWSV2 cells, and at high *in vitro* passage transformed and produced tumors in syngeneic hosts. However, in contrast to CWSV2 cells, CWSV14 cells at high passage transcribed albumin at markedly reduced levels (less than 10% compared to 60 to 70% for low passage CWSV14 cells) and began to express low levels of AFP RNA. NR4 cells, which were derived by transfection of low passage immortalized CWSV1 cells with activated c-Ha-*ras,* were transformed and expressed albumin RNA at levels equal to or greater than liver and did not express AFP RNA. NR4T-1 cells, which are tumor cells derived from undifferentiated, rapidly proliferating carcinomas produced by NR4 cells, expressed albumin at twofold the level of liver and also did not express AFP RNA. We conclude from these studies that transformation of hepatocyte-derived cell lines can be, but is not necessarily, accompanied by loss of albumin expression.

Several immortalized liver cell lines derived from hepatocytes have also been shown to express a variety of liver-specific genes in addition to albumin. Fetal rat hepatocytes immortalized by infection with a temperature-sensitive mutant of SV40 expressed transferrin and AFP.[28] The transgenic mouse fetal line FMH-202 expressed AFP, but did not express PEPCK or TAT.[47] A fetal mouse cell line derived by infection with temperature-sensitive mutants of SV40 expressed TAT, PEPCK, and transferrin RNA at approximately 4, 3, and 3% relative to adult mouse liver, respectively.[30] SV40-immortalized neonatal mouse hepatocyte[29] and polyoma virus-transformed neonatal rat hepatocyte[33] cell lines expressed AFP.

Analysis of immortalized fetal mouse hepatocytes from albino deletion mutant mice showed that the cell lines expressed reduced levels of PEPCK and several other genes expressed at lower levels in mutant mice.[30] In contrast, the metallothionein gene, which is expressed in lower levels in mutant as compared to wild type mice, was expressed equally in mutant and wild-type mouse cell lines. In general, gene expression in the immortalized cell lines largely mimicked the effects of the deletion in newborn mutant mice.[29,30]

Characterization of the SV40-immortalized adult rat hepatocyte (CWSV) cell lines showed that the cell lines secreted transferrin, the third component of complement, and hemopexin, but did not produce AFP.[21] The cell lines differed from each other with regard to the level of production of each of these serum proteins. Northern blot analyses confirmed that the cell lines produce transferrin and also demonstrated that they produce PEPCK, A_1AT, and glutathione *S*-transferase-P (GST-P).[22] In a separate study, it was shown that the CWSV1 and CWSV16 cells express acid 1-glycoprotein, haptoglobin, α- β-, and γ-fibrinogen, serum amyloid A, and major acute-phase protein RNA and that expression of these acute-phase gene transcripts was increased when the cells were treated with conditioned medium from stimulated lymphocytes.[64]

Limited studies have been carried out using cell lines immortalized by oncogenes and viral genes to study the effects of transformation on expression of liver-specific genes. *Ras* transformation of SV40-immortalized CWSV1 cells not only did not diminish the RNA levels for albumin in the transformed cells, but also did not reduce transferrin, A_1AT, or PEPCK RNA levels.[32]

The effect of transformation on gene expression by RLE cells has not involved measuring effects on albumin because albumin is either not expressed or poorly expressed by RLE cells. Instead, attempts have been made in various RLE systems to determine whether or not expression of γ-glutamyl transpeptidase and GST-P[8,13,14,39] and more recently production of extracellular matrix proteins[41,42] correlates with tumorigenicity. Interaction of cells with extracellular matrix and production of extracellular matrix proteins are important for maintenance of differentiation and regulation of growth control. Using v-*raf* and v-*raf*/v-*myc*-immortalized and -transformed REL cells, it has been demonstrated that tumorigenic potential was correlated with decreased expression of collagen I, fibronectin,[42] and two polypeptides designated RP2 and RP3.[41]

IV. EXPRESSION OF ONCOGENES AND GROWTH FACTOR GENES IN IMMORTALIZED AND TRANSFORMED LIVER CELL LINES: ASSOCIATION WITH TUMORIGENICITY

We have discussed above the ability of viral genes, oncogenes, and growth factor genes, alone and in combination, to immortalize and/or transform liver cells in culture. In order to understand the involvement of oncogene expression in hepatocarcinogenesis, it is also important to analyze immortalized and transformed cell lines of differing tumorigenic potential for expression of oncogenes. Cell lines derived by chemical, viral, and oncogene transformation *in vitro* and cell lines derived from transgenic animals will be reviewed. We will not include discussions of oncogene expression in hepatoma cell lines derived from spontaneous or chemically induced liver tumors.

A. RELATIONSHIP BETWEEN INCREASED *ras* OR *myc* EXPRESSION AND TUMORIGENICITY

As indicated above, SV40-immortalized adult rat hepatocytes acquire tumorigenic potential with *in vitro* passage. Northern blot analysis showed that nontumorigenic low passage CWSV1 and CWSV2 cell lines expressed undetectable levels of Ha-*ras* RNA.[22] When these cell lines were examined at a higher passage level at which they were tumorigenic, Ha-*ras* RNA levels were markedly higher. This increase appeared to be specific in that analysis of the same RNA samples for *raf* RNA showed no change in expression with passage. These findings suggested that SV40 expression alone was not sufficient for transformation. Completion of the carcinogenic process may involve increased expression of Ha-*ras*. A similar result was found when hepatocyte cell lines derived from animals harboring an albumin promoter-driven

SV40 T antigen transgene were examined with *in vitro* passage.[63] Abnormal hepatocytes appeared in these animals at 4 to 7 weeks of age, followed by a multistep progression resulting in hepatocellular carcinoma at 24 weeks of age. Ten ''preneoplastic'' cell lines were established from animals 3 to 12 weeks of age and nine hepatocellular carcinoma cell lines were established from 24-week-old transgenic mice. These cell lines were examined for properties associated with transformation and for Ha-*ras* point mutation by polymerase chain reaction. The ten preneoplastic cell lines did not grow in soft agar or produce tumors in nude mice. An Ha-*ras* point mutation was detected in one cell line, but not until it was at the 38th generation. Four of the hepatocellular carcinoma cell lines grew in soft agar, formed tumors in animals, and had an Ha-*ras* point mutation at time of establishment. All of the five remaining cell lines acquired anchorage-independent growth and the ability to form tumors in animals with *in vitro* passage, and in three out of the five cell lines Ha-*ras* activation at position 61 also became apparent. The two transformed hepatocellular carcinoma cell lines that were negative for Ha-*ras* codon 61 mutations may have possessed a different *ras* mutation or expressed higher levels of *ras*. Alternatively, completion of the carcinogenic process in these two cell lines may not involve altered expression of *ras*. We conclude from these two studies and the above cited report that SV40-immortalized hepatocyte cell lines can be transformed by transfection with activated c-Ha-*ras* and that Ha-*ras* activation may be an event occurring at a relatively late stage of hepatocarcinogenesis.

Enhanced expression of c-*myc* has been correlated with tumorigenic potential in numerous, but not all transformed liver cell lines. The relationship between c-*myc* expression and tumorigenicity has been analyzed predominantly in transformed cell lines derived from RLE cells. It has been reported recently using multiple cell lines clonally derived from chemically transformed WB-F344 RLE cell lines that overexpression of the c-*myc* gene cosegregated with tumorigenicity.[68] In this study, 16 clonally derived, transformed liver epithelial cell lines that varied in tumorigenicity from 0 to 97% were analyzed for expression of TGF-α, EGF receptor (EGFR), c-*myc,* c-Ha-*ras,* and c-K-*ras.*[69] Only the expression of c-*myc* correlated with tumorigenicity. Elevated expression of TGF-α did not correlate with tumorigenicity among all the clones, but did correlate with tumorigenicity among clones that expressed the c-*myc* gene above the median level for all 16 clones. Clones which simultaneously overexpressed c-*myc* and TGF-α were highly tumorigenic. In a separate study, it was shown that REL cells transformed *in vitro* by MNNG and selected for resistance to growth inhibition by TGF-β1 also overexpress c-*myc.*[70]

Increased expression of c-*myc* RNA was also observed in REL cells transformed by the oncogene v-*raf.*[42] REL cells transformed by infection with v-*raf* and cells derived from tumors produced by v-*raf*-transformed cells expressed high levels of c-*myc* message. However, REL cells transformed by v-*raf* and v-*myc* in combination expressed only low levels of c-*myc* compared to the v-*raf* transformants and tumor cells. c-*Myc* expression in the v-*raf*/v-*myc*-transformed cells was higher than the control helper virus-infected REL cells. These studies indicate that c-*myc* expression is increased by v-*raf* transformation of REL cells. However, since the v-*raf*/v-*myc*-transformed cells were considerably more tumorigenic than v-*raf*-transformed REL cells, the level of c-*myc* expression did not correlate with tumorigenicity in the v-*raf in vitro* model of tumor progression.

Oval cells that transform spontaneously as a result of infrequent subculture[36] or are transformed by *ras*[71] lose cell cycle control of *myc* expression; that is, the levels of c-*myc* RNA are the same in growth-arrested and proliferating cells. The *myc* gene was not amplified or rearranged in these cells.

Studies to evaluate *myc* expression in liver cell lines derived from hepatocytes are limited. Examination of one *ras*-transformed SV40 hepatocyte cell line that expresses reasonably high levels of *ras* RNA indicated that the c-*myc* RNA levels were enhanced compared to the control cells.[72]

Experiments using the transformed BL8 rat liver-derived epithelial cell line suggest that immortalization yields dividing liver cells with enhanced c-*myc* expression, but that transformation does not further increase *myc* expression and may even decrease it.[15] The immortalized BL8 cell line had higher levels of expression of c-*myc* than freshly isolated hepatocytes. *Ras*-transformed BL8 cells expressed similar levels of c-*myc* to those found in the parental BL8 cells. BL8 cells transformed *in vitro* by activated aflatoxin B_1 had lower levels of *myc* than nontransformed BL8 cells. In attempting to compare the results with BL8 cells to the above studies with RLE cells, it is important to realize that c-*myc* expression in the BL8 study was determined by measuring c-*myc* protein using flow cytometry instead of c-*myc* RNA. In addition, the BL8 cell line was derived from cells that immortalized spontaneously from cultures of isolated rat hepatocytes. The origin of BL8 cells is unknown and may differ from that of other RLE cells.

B. RELATIONSHIP BETWEEN ALTERED TGF-α AND TGF-β1 EXPRESSION AND TUMORIGENICITY

The relationship between the expression of the growth factors TGF-α and TGF-β1 and tumorigenicity has been studied predominantly in chemically and oncogene-transformed RLE cells and oncogene-transformed oval cells. The set of cloned transformed cell lines derived by MNNG transformation of the normal WB-F344 RLE cell line and cell lines derived from tumors produced by these transformed cell lines have been used to measure the relationship between TGF-α production and tumorigenic potential.[73] The normal WB-F344 cells did not secrete TGF-α as measured by the ability of conditioned medium from these cells to induce NRK-c49 cells to form colonies in agar in the absence of EGF. Some, but not all, of the transformed cell lines secreted TGF-α-like activity. Similarly, the tumor cell lines also varied in their ability to produce TGF-α. Analysis of four of these cell lines was carried out at the mRNA level.[74] The normal WB-F344 cells did not produce detectable levels of TGF-α mRNA. One of the clonally derived transformed cell lines, GP6, produced low levels of TGF-α RNA which was only detectable if poly A^+ RNA was probed. The GP6ac cell line, which was derived by passing GP6 cells in soft agar, and the GP6TB tumor cell line, which was derived from a GP6 tumor, both produced TGF-α message that could be detected using blots of total RNA. The level of TGF-α mRNA in GP6TB cells was sevenfold greater than in GP6ac cells. In addition, it was demonstrated that treatment of GP6, GP6ac, and GP6TB with 12-*O*-tetradecanoylphorbol-13-acetate (TPA) increased the TGF-α mRNA levels. TPA treatment had no effect on TGF-α mRNA levels in WB-F344 cells. Recently, these same investigators analyzed expression of TGF-α RNA and tumorigenicity in 16 clonally derived MNNG-transformed WB-F344 cells, including G6P,[69] and showed that TGF-α RNA expression was not correlated with tumorigenicity among all the clones. However, as discussed above, TGF-α RNA expression was highly correlated with tumorigenicity among clones that expressed the c-*myc* gene above the medium level for all 16 clones. Transformation of RLE cells with a mutated H-*ras* gene induces the expression and secretion of TGF-α.[43,75] Expression of TGF-α was nondetectable in control cells transfected with the selectable marker gene alone, or in cells transfected with an activated c-*myc* gene. Control parental cells, those transfected with the selectable marker gene alone, and lines transfected with c-*myc* were not tumorigenic when inoculated into nude mice. Tumorigenicity in nude mice and TGF-α expression was only observed in cells transfected with a mutated H-*ras*. The production and secretion of TGF-α, coupled to the presence of EGFR on the cells, leaves open the possibility that an autocrine stimulatory pathway contributes to the transformation of RLE by the H-*ras* oncogene.

Association between TGF-α RNA expression and tumorigenicity was also measured in the v-*raf,* v-*raf*/v-*myc*-transformed RLE model.[42] These studies showed that there was a direct correlation between TGF-α mRNA levels and tumor aggressiveness. TGF-α RNA was not expressed in the normal RLE cells, the v-*raf*-transformed clone, and some of the tumor cell lines derived from tumors produced by the v-*raf*-transformed cells. Two of the tumor cell lines and the cell line infected with v-*raf* and v-*myc* in combination expressed TGF-α RNA at high levels, and these cell lines also had the fastest growth rate and shortest latency period for tumor formation in animals.

Studies have been carried out to determine whether or not a relationship exists between TGF-β1 production and tumorigenicity by liver cell lines at least with regard to transformed RLE and oval cells. In the WB-F344 system, normal, chemically transformed, and tumor cell lines produced TGF-β1, but only the tumor cell lines activated TGF-β1.[73] In the v-*raf*, v-*raf*/v-*myc* RLE system, the same direct correlation with tumorigenicity observed for TGF-α RNA was also reported for TGF-β1 RNA.[42] The only difference was that all the cell lines, including normal RLE, produced detectable levels of TGF-β1 RNA and the magnitude of expression varied with tumorigenicity. Oval cells in culture (the LE/6 cell line), in contrast with RLE cells, produce TGF-β1 RNA. Transformation of LE/6 cells with *ras* produced a cell line, LE/6EJ, which expressed TGF-β1 RNA at a 40% lower level than the parent LE/6 cells.[76,77] Further analysis showed that subcloning of LE/6 cells yielded six out of seven cell lines with TGF-β1 RNA levels similar to those in LE/6, while seven tumorigenic clonal lines derived from an EJ *(ras)*-transformed tumor cell line produced variable levels of TGF-β1 RNA. The results indicate that *ras* transformation of oval cells yields cells that express decreased levels of TGF-β1 RNA. However, cell lines that continue to produce high levels of TGF-β1 RNA can be tumorigenic.

The expression and secretion of TGF-β1 by normal or transformed cells is not the only mechanism by which TGF-β1 could play a role in transformation. While some mesenchymal cells are stimulated to grow by TGF-β1, it is a most potent inhibitor of growth for most epithelial cells.[78] The resistance of many transformed cell types to the growth inhibitory effects of TGF-β1 supports the hypothesis that escape from negative growth control might be an important first step during carcinogenesis.[78-82] EGF-stimulated DNA synthesis in primary cultures of hepatocytes is inhibited by picomolar concentrations of TGF-β1.[83-85] As TGF-α has been proposed to function as an autocrine stimulator of liver regeneration, TGF-β1 has been proposed to function in a paracrine manner to terminate the regenerative response, since its expression is induced late in liver regeneration in nonparenchymal liver cells[86] and authentic TGF-β1 injected into animals reversibly inhibits liver regeneration.[87] Liver epithelial cells and parenchymal hepatocytes transformed with a mutated H-*ras* oncogene are completely resistant to the growth inhibitory effects of TGF-β1.[34] Many chemically or spontaneously transformed liver cells are resistant to TGF-β1-induced growth inhibition.[34,70,77,88] If, as proposed, TGF-β1 performs a negative regulatory function on the growth of normal liver,[86] resistance to the growth inhibitory effects of TGF-β1 might be a crucial first step in the development of liver neoplasms. One study of cells from tumors and enzyme-altered nodules induced in Fischer 344 rats by exposure to the liver carcinogen 2-acetylaminofluorene indicated that primary tumors in the rat liver may still be responsive to the growth inhibitory effects of TGF-β1.[89] Although no other data exist on the sensitivity of other primary rat or mouse liver tumors to TGF-β1, resistance to TGF-β1-induced growth inhibition is a mechanism which would favor the clonal expansion of cells expressing this phenotype. Because of a possible role for TGF-β1 as a negative regulator of normal liver growth and the number of transformed liver cell types which are resistant to the growth inhibitory effects of TGF-β1, it is important to determine the role that resistance to TGF-β1 may play in hepatocarcinogenesis *in vivo*.

More studies need to be carried out with liver cells before conclusions can be drawn about the relationship between the production of TGF-α, TGF-β1, and immortalization and transformation of various liver cell types. In particular, studies need to be carried out using cell lines derived from hepatocytes. That TGF-α expression is related to transformation of liver cells and may play a role in the process is supported by the *in vitro* studies detailed above, a report that TGF-α transgenic mice develop hepatocellular carcinoma, and the clinical observations that patients with hepatocellular carcinoma excrete elevated amounts of TGF-α in their urine.[90] With regard to TGF-β1, the question is far more complex, involving not only measuring the amount of TGF-β1 produced, but also determining whether or not the cells can activate it and if they remain sensitive to its growth inhibitory effects. Although TGF-β1 may play a role in hepatocarcinogenesis, it is not possible at this time to define that role. However, TGF-β1 is known to modulate the expression of extracellular matrix proteins and has demonstrable angiogenic activity;[78-81] properties which would be expected to favor *in vivo* growth and metastasis, but would not necessarily be easily detected in *in vitro* studies.

V. THE ROLE OF VIRAL GENES AND GROWTH FACTORS IN HUMAN HEPATOCARCINOGENESIS

Transgenic animals expressing viral SV40 T or HBV genes develop hepatocellular carcinomas, indicating that these genes are effective in the transformation of hepatocytes.[44-49] Particularly in the case of HBV, tumors develop in these animals after a considerable latent period.[54-57] The long latent period suggests that secondary changes or genetic alterations are required in addition to the expression of the viral transgene to induce full malignant transformation. The nature of the secondary events which may work in concert with the viral gene in cellular transformation is under investigation in many laboratories, but data from human patients with hepatocellular carcinomas associated with HBV may provide some insight. Mutations in the cellular p53 gene have been detected in 50% of hepatocellular carcinomas with a clustering of point mutations at the third base of codon 249.[91,92] Transversions from G → T were the most frequently identified genetic change. The areas of Africa and eastern Asia where the patients were selected is known to have a high incidence of hepatocellular carcinoma in the population. HBV infection and the exposure to the naturally occurring carcinogen, aflatoxin B_1, are suspected to play important roles in the initiation of hepatocellular carcinoma in these patients. Mutations in the p53 gene at the guanine residue at codon 249 were thought to be significant because of the predilection for N-7 guanine binding observed for aflatoxin B_1. In other studies, cell lines derived from hepatocellular carcinomas from African patients contained the codon 249 mutations, while four other hepatoma cell lines from non-African patients expressed wild type p53.[92] Thus, aflatoxin-induced mutations in p53 genes may be crucial to the further development of hepatocellular carcinoma in approximately 50% of the cases.

The mechanism by which p53 mutations cooperate with HBV in the transformation of liver cells is not known; however, p53 is thought to act as a tumor-suppressor gene, and that its genetic loss and/or mutational inactivation favors the clonal outgrowth of transformed cells[93,94] (see Chapter 13).

Patients with hepatocellular carcinomas show an average fourfold elevation in the amount of TGF-α in their urine.[90] Urinary TGF-α levels may be a better diagnostic test for hepatocellular cancers than serum AFP, and may be of significant value when used as a complementary tumor marker in patients with low AFP.[90] Transcripts for insulinlike growth factor-II (IGF-II) have been detected in primary liver tumors and in surrounding cirrhotic areas.[95-97] In addition to the major adult transcript for IGF-II, reexpression of fetal isozyme IGF-II was observed in 9 of 16 patient specimens. Normal liver expresses the adult transcript, and overall 22% of primary liver tumors express higher than normal levels of IGF-II. Interestingly, HBV was never coexpressed in tissues expressing AFP or fetal isozymes of IGF-II, suggesting a developmentally regulated expression of this growth factor.

A pattern of genetic changes is developing which is characteristic of human hepatocellular carcinomas. High-level expression of certain growth factors, such as TGF-α or IGF-II, are common. Allelic deletion of one copy of the p53 gene and mutational inactivation of the normal function of the p53 protein expressed from the remaining gene is also frequently observed in human hepatocellular cancers. More than 65% of hepatocellular carcinomas produce TGF-α, and 50% of the tumors in some regions of the world contain p53 gene mutations. A very high percentage of all hepatocellular carcinomas in these same regions develop in patients infected with HBV. Like the experimental programs conducted with laboratory animals discussed earlier in this chapter, viral genes cooperate with environmental factors and secondary genetic changes in human liver carcinogenesis. Key cellular targets in the transformation process involving growth factor pathways (i.e., TGF-α or IGF-II) and cell cycle regulatory/tumor-suppressor proteins have been identified in a large proportion of the tumors studied. A major challenge for future work will be to identify how the pathways already identified contribute and cooperate in transformation and to identify other pathways which are causal in the neoplastic process in the 50% or more of hepatocellular tumors which express normal p53 function.

ACKNOWLEDGMENTS

The Authors' research described in this chapter was supported by Public Health Service grant CA 23931 from the National Cancer Institute (H. C. Isom) and American Cancer Society grant CN-43B (S. C. Strom).

REFERENCES

1. **Dahl, N. K., Grubman, S. A., Cox, L. J., and Jefferson, D. M.,** Immortalization of intrahepatic bile duct epithelium: a model for biliary function, *Hepatology,* 12, 888, 1990.
2. **Michalopoulos, G., Cianciulli, H. D., Novotny, A. R., Kligerman, A. D., Strom, S. C., and Jirtle, R. L.,** Liver regeneration studies with rat hepatocytes in primary culture, *Cancer Res.,* 42, 4673, 1982.
3. **Eckl, P. M., Whitcomb, W. R., Michalopoulos, G., and Jirtle, R. L.,** Effects of EGF and calcium on adult parenchymal hepatocyte proliferation, *J. Cell. Physiol.,* 132, 363, 1987.
4. **Mitaka, T., Sattler, C. A., Sattler, G. L., Sargent, L. M., and Pitot, H. C.,** Multiple cell cycles occur in rat hepatocytes cultured in the presence of nicotinamide and epidermal growth factor, *Hepatology,* 13, 21, 1991.
5. **Leffert, H. L. and Paul, D.,** Studies on primary cultures of differentiated fetal liver cells, *J. Cell Biol.,* 52, 559, 1972.
6. **Draghi, E., Armato, U., and Andreis, P. G.,** The stimulation by epidermal growth factor (urogastrone) of the growth of neonatal rat hepatocytes in primary tissue culture and its modulation by serum and associated pancreatic hormones, *J. Cell. Physiol.,* 103, 129, 1980.
7. **Braun, L., Goyette, M., Yaswen, P., Thompson, N. L., and Fausto, N.,** Growth in culture and tumorigenicity after transfection with the *ras* oncogene of liver epithelial cells from carcinogen-treated rats, *Cancer Res.,* 49, 4116, 1987.
8. **Li, Y., Seyama, T., Godwin, A. K., Winokur, T. S., Lebovitz, R. M., and Lieberman, M. W.,** MT*ras*T24, a metallothionein-*ras* fusion gene, modulates expression in cultured rat liver cells of two genes associated with *in vivo* liver cancer, *Proc. Natl. Acad. Sci. U.S.A.,* 85, 344, 1988.
9. **Herring, A. S., Raychaudhuri, R., Kelley, S. P., and Iype, P. T.,** Repeated establishment of diploid epithelial cell cultures from normal and partially hepatectomized rats, *In Vitro,* 19, 576, 1983.
10. **Tsao, M.-S., Smith, J. D., Nelson, K. G., and Grisham, J. W.,** A diploid epithelial cell line from normal adult rat liver with phenotypic properties of 'oval' cells, *Exp. Cell Res.,* 154, 38, 1984.
11. **Hugget, A. C., Hampton, L. L., Ford, C. P., Wirth, P. J., and Thorgeirsson, S. S.,** Altered responsiveness of rat liver epithelial cells to transforming growth factor-β1 following their transformation with v-*raf, Cancer Res.,* 50, 7468, 1990.
12. **Grisham, J. W.,** Cell types in long-term propagable cultures of rat liver, *Ann. N.Y. Acad. Sci.,* 349, 128, 1980.
13. **Sinha, S., Marshall, C. J., and Neal, G. E.,** Gamma-glutamyltranspeptidase and the *ras*-induced transformation of a rat liver cell line, *Cancer Res.,* 46, 1440, 1986.
14. **Power, C., Sinha, S., Webber, C., Manson, M. M., and Neal, G. E.,** Transformation related expression of glutathione-S-transferase P in rat liver cells, *Carcinogenesis,* 8, 797, 1987.

15. **Sinha, S., Neal, G. E., Legg, R. F., Watson, J. V., and Pearson, C.,** The expression of c-*myc* related to the proliferation and transformation of rat liver-derived epithelial cells, *Br. J. Cancer,* 59, 674, 1989.
16. **Isom, H. C., Tevethia, M. J., and Taylor, J. M.,** Transformation of isolated rat hepatocytes with simian virus 40, *J. Cell Biol.,* 85, 651, 1980.
17. **Isom, H. C. and Georgoff, I.,** Quantitative assay for albumin-producing liver cells after simian virus 40 transformation of rat hepatocytes maintained in chemically defined medium, *Proc. Nat. Acad. Sci. U.S.A.,* 81, 6378, 1984.
18. **Woodworth, C., Secott, T., and Isom, H. C.,** Transformation of rat hepatocytes by transfection with simian virus 40 DNA to yield proliferating differentiated cells, *Cancer Res.,* 46, 4018, 1986.
19. **Woodworth, C. D. and Isom, H. C.,** Transformation of differentiated rat hepatocytes with adenovirus and adenovirus DNA, *J. Virol.,* 61, 3570, 1987.
20. **Zaret, K. S., DiPersio, C. M., Jackson, D. A., Montigny, W. J., and Weinstat, D. L.,** Conditional enhancement of liver-specific gene transcription, *Proc. Natl. Acad. Sci. U.S.A.,* 85, 9076, 1988.
21. **Woodworth, C. D. and Isom, H. C.,** Regulation of albumin gene expression in a series of rat hepatocyte cell lines immortalized by simian virus 40 and maintained in chemically defined medium, *Mol. Cell. Biol.,* 7, 3740, 1987.
22. **Woodworth, C. D., Kreider, J. W., Mengel, L., Miller, T., Meng, Y., and Isom, H. C.,** Tumorigenicity of simian virus 40-hepatocyte cell lines: effect of *in vitro* and *in vivo* passage on expression of liver-specific genes and oncogenes, *Mol. Cell. Biol.,* 8, 4492, 1988.
23. **Isom, H. C. and Woodworth, C. D.,** unpublished data.
24. **Isom, H. C. and Miller, T.,** unpublished data.
25. **Schlegel-Haueter, S. E., Schlegel, W., and Chou, J. Y.,** Establishment of a fetal rat liver cell line that retains differentiated liver functions, *Proc. Natl. Acad. Sci. U.S.A.,* 5, 2731, 1980.
26. **Chou, J. Y.,** Temperature-sensitive adult liver cell line dependent on glucocorticoid for differentiation, *Mol. Cell. Biol.,* 3, 1013, 1983.
27. **Chou, J. Y. and Ito, K.,** Temperature-sensitive fetal liver cells: a model for hepatocellular differentiation, *Ann. N.Y. Acad. Sci.,* 417, 308, 1983.
28. **Chou, J. Y.,** Establishment of rat fetal liver lines and characterization of their metabolic and hormonal properties: use of temperature-sensitive SV40 virus, *Meth. Enzymol.,* 109, 385, 1985.
29. **Paul, D., Kwon, B. S., Höhne, M., Tönjes, R., Haq, A. K., and Hoffmann, B.,** Establishment and partial characterization of SV40 virus-immortalized hepatocyte lines of normal and lethal mutant mice carrying a deletion on chromosome 7, *J. Cell. Physiol.,* 139, 599, 1989.
30. **Chou, J. Y., Ruppert, S., Shelly, L. L., and Pan, C.-J.,** Isolation and characterization of mouse hepatocyte lines carrying a lethal albino deletion, *J. Biol. Chem.,* 266, 5716, 1991.
31. **Meng, Y., Woodworth, C., Miller, T., Mengel, L., and Isom, H. C.,** Conversion of simian virus 40-immortalized hepatocyte cell line into a tumorigenic cell line by the *ras* oncogene, *Proc. Am. Assoc. Cancer Res.,* 29, 1836, 1988.
32. **Isom, H. C., Woodworth, C. D., Meng, Y., Kreider, J., Miller, T., and Mengel, L.,** unpublished data.
33. **Höhne, H., Piasecki, A., Ummelmann, E., and Paul, D.,** Transformation of differentiated neonatal rat hepatocytes in primary culture by polyoma virus early region sequences, *Oncogene,* 1, 337, 1987.
34. **Houck, K. A., Michalopoulos, G. K., and Strom, S. C.,** Introduction of a Ha-*ras* oncogene into rat liver epithelial cells and parenchymal hepatocytes confers resistance to the growth inhibitory effects of TGF-β, *Oncogene,* 4, 19, 1989.

35. **Woodworth, C. D. and Isom, H. C.,** unpublished data.
36. **Braun, L., Mikumo, R., and Fausto, N.,** Production of hepatocellular carcinoma by oval cells: cell cycle expression of c-*myc* and p53 at different stages of oval cell transformation, *Cancer Res.,* 49, 1554, 1989.
37. **Tsao, M.-S. and Grisham, J. W.,** Hepatocarcinomas, cholangiocarcinomas, and hepatoblastomas produced by chemically transformed cultured rat liver epithelial cells. A light- and electron-microscopic analysis, *Am. J. Pathol.,* 127, 168, 1987.
38. **Huber, B. E. and Cordingley, M. G.,** Expression and phenotypic alterations caused by an inducible transforming *ras* oncogene introduced into rat liver epithelial cells, *Oncogene,* 3, 245, 1988.
39. **Burt, R. K., Garfield, S., Johnson, K., and Thorgeirsson, S. S.,** Transformation of rat liver epithelial cells with v-H-*ras* or v-*raf* causes expression of MDR-1, glutathione-S-transferase-P and increased resistance to cytotoxic chemicals, *Carcinogenesis,* 9, 2329, 1988.
40. **Garfield, S., Huber, B. E., Nagy, P., Cordingley, M. G., and Thorgeirsson, S. S.,** Neoplastic transformation and lineage switching of rat liver epithelial cells by retrovirus-associated oncogenes, *Mol. Carcinogenesis,* 1, 189, 1988.
41. **Worland, P. J., Hampton, L. L., Thorgeirsson, S. S., and Huggett, A. C.,** Development of an *in vitro* model of tumor progression using v-*raf* and v-*raf*/v-*myc* transformed rat liver epithelial cells: correlation of tumorigenicity with the down-regulation of specific proteins, *Mol. Carcinogenesis,* 3, 20, 1990.
42. **Hampton, L. L., Worland, P. J., Yu, B., Thorgeirsson, S. S., and Huggett, A. C.,** Expression of growth-related genes during tumor progression in v-*raf*-transformed rat liver epithelial cells, *Cancer Res.,* 50, 7460, 1990.
43. **Strom, S. C., Faust, J. B., Cappelluti, E., Harris, R. B., and Lalwani, N. D.,** Characterization of liver epithelial cells transfected with *myc* and/or *ras* oncogenes, *Dig. Dis. Sci.,* 36, 642, 1991.
44. **Sepulveda, A. R., Finegold, M. J., Smith, B., Slagele, B. L., DeMayo, J. L., Shen, R.-F., Woo, S. L. C., and Butel, J. S.,** Development of a transgenic mouse system for the analysis of stages in liver carcinogenesis using tissue-specific expression of SV40 large T-antigen controlled by regulatory elements of the human α-1-antitrypsin gene, *Cancer Res.,* 49, 6108, 1989.
45. **Butel, J. S., Sepulveda, A. R., Finegold, M. J., and Woo, S. L. C.,** SV40 large T antigen directed by regulatory elements of the human alpha-1-antitrypsin gene, *Intervirology,* 31, 85, 1990.
46. **Palmiter, R. D., Chen, H. Y., Messing, A., and Brinster, R. L.,** SV40 enhancer and large-T antigen are instrumental in development of choroid plexus tumours in transgenic mice, *Nature,* 316, 457, 1985.
47. **Paul, D., Höhne, M., Pinkert, C., Piasecki, A., Ummelmann, E., and Brinster, R. L.,** Immortalized differentiated hepatocyte lines derived from transgenic mice harboring SV40 T-antigen genes, *Exp. Cell Res.,* 175, 354, 1988.
48. **Messing, A., Chen, H. Y., Palmiter, R. D., and Brinster, R. L.,** Peripheral neuropathies, hepatocellular carcinomas and islet cell adenomas in transgenic mice, *Nature,* 316, 461, 1985.
49. **Sandgren, E. P., Quaife, C. J., Pinkert, C. A., Palmiter, R. D., and Brinster, R. L.,** Oncogene-induced liver neoplasia in transgenic mice, *Oncogene,* 4, 715, 1989.
50. **Held, W. A., Mullins, J. J., Kuhn, N. J., Gallagher, J. F., Gu, G. D., and Gross, K. W.,** T Antigen expression and tumorigenesis in transgenic mice containing a mouse major urinary protein/SV40 T antigen hybrid gene, *EMBO J.,* 8, 183, 1989.
51. **Schirmacher, P., Held, W. A., Yang, D., Biempica, L., and Rogler, C. E.,** Selective amplification of periportal transitional cells precedes formation of hepatocellular carcinoma in SV40 large tag transgenic mice, *Am. J. Pathol.,* 139, 231, 1991.

52. **Dalemans, W., Perraud, F., Le Meur, M., Gerlinger, P., Courtney, M., and Pavirani, A.,** Heterologous protein expression by transimmortalized differentiated liver cell lines derived from transgenic mice (hepatomas/α_1 antitrypsin/*ONC* mouse), *Biologicals,* 18, 191, 1990.
53. **Jhappan, C., Stahle, C., Harkins, R. N., Fausto, N., Smith, G. H., and Merlino, G. T.,** TGF-α overexpression in transgenic mice induces liver neoplasia and abnormal development of the mammary gland and pancreas, *Cell,* 61, 1137, 1990.
54. **Chisari, F. V., Pinkert, C. A., Milich, D. R., Filippi, P., McLachlan, A., Palmiter, R. D., and Brinster, R. L.,** A transgenic mouse model of the chronic hepatitis B surface antigen carrier state, *Science,* 230, 1157, 1985.
55. **Chisari, F. V., Filippi, P., McLachlan, A., Milich, D. R., Riggs, M., Lee, S., Palmiter, R. D., Pinkert, C. A., and Brinster, R. L.,** Expression of hepatitis B virus large envelope polypeptide inhibits hepatitis B surface antigen secretion in transgenic mice, *J. Virol.,* 60, 880, 1986.
56. **Chisari, F. V., Filippi, P., Buras, J., McLachlan, A., Popper, H., Pinkert, C. A., Palmiter, R. D., and Brinster, R. L.,** Structural and pathological effects of synthesis of hepatitis B virus large envelope polypeptide in transgenic mice, *Proc. Natl. Acad. Sci. U.S.A.,* 84, 6909, 1987.
57. **Chisari, F. V., Klopchin, K., Moriyama, T., Pasquinelli, C., Dunsford, H. A., Sell, S., Pinkert, C. A., Brinster, R. L., and Palmiter, R. D.,** Molecular pathogenesis of hepatocellular carcinoma in hepatitis B virus transgenic mice, *Cell,* 59, 1145, 1989.
58. **Dunsford, H. A., Sell, S., and Chisari, F. V.,** Hepatocarcinogenesis due to chronic liver cell injury in hepatitis B virus transgenic mice, *Cancer Res.,* 50, 3400, 1990.
59. **Kim, C.-M., Koike, K., Saito, I., Miyamura, T., and Jay, G.,** *HBx* gene of hepatitis B virus induces liver cancer in transgenic mice, *Nature,* 351, 317, 1991.
60. **Teeter, L. D., Becker, F. F., Chisari, F. V., Li, D., and Kuo, M. T.,** Overexpression of the multidrug resistance gene *mdr3* in spontaneous and chemically induced mouse hepatocellular carcinomas, *Mol. Cell. Biol.,* 10, 5728, 1990.
61. **Lee, T.-H., Finegold, M. J., Shen, R.-F., DeMayo, J. L., Woo, S. L. C., and Butel, J. S.,** Hepatitis B virus transactivator X protein is not tumorigenic in transgenic mice, *J. Virol.,* 64, 5939, 1990.
62. **Höhne, M., Schaefer, S., Seifer, M., Feitelson, M. A., Paul, D., and Gerlich, W. H.,** Malignant transformation of immortalized transgenic hepatocytes after transfection with hepatitis B virus DNA, *EMBO J.,* 9, 1137, 1990.
63. **Li, H., Liu, J., Nomura, K., Lee, G.-H., Hino, O., Ohtake, K., Aizawa, S., and Kitagawa, T.,** *In vitro* progression-associated c-H-*ras* activation in neoplastic hepatocyte lines established from SV40-T antigen gene-harboring transgenic mice, *Jpn. J. Cancer Res.,* 82, 4, 1991.
64. **Liao, W. S., Ma, K.-T., Woodworth, C. D., Mengel, L., and Isom, H. C.,** Stimulation of the acute-phase response in simian virus 40-hepatocyte cell lines, *Mol. Cell. Biol.,* 9, 2779, 1989.
65. **Hu, J.-M., Camper, S. A., Tilghman, S. M., Miller, T., Georgoff, I., Serra, R., and Isom, H. C.,** unpublished data.
66. **Liu, J.-K., DiPersio, M., and Zaret, K. S.,** Extracellular signals that regulate liver transcription factors during hepatic differentiation *in vitro, Mol. Cell. Biol.,* 11, 773, 1991.
67. **DiPersio, C. M., Jackson, D. A., and Zaret, K. S.,** The extracellular matrix coordinately modulates liver transcription factors and hepatocyte morphology, *Mol. Cell. Biol.,* 11, 4405, 1991.
68. **Lee, L. W., Harris, J., Lee, D. C., Earp, H. S., and Grisham, J. W.,** Relationship between expression of H-*ras,* K-*ras, myc,* α-transforming growth factor (TGF-α), and epidermal growth factor (EGF)-receptor mRNAs and tumorigenicity among carcinogen-transformed rat liver epithelial cells, *Proc. Am. Assoc. Cancer Res.,* 30, 457, 1989.

69. **Lee, L. W., Raymond, V. W., Tsao, M.-S., Lee, D. C., Earp, H. S., and Grisham, J. W.,** Clonal cosegregation of tumorigenicity with overexpression of c-*myc* and transforming growth factor-α genes in chemically transformed rat liver epithelial cells, *Cancer Res.,* 51, 5235, 1991.
70. **Chapekar, M. S., Huggett, A. C., Cheng, C. C., Hampton, L. L., Lin, K.-H., and Thorgeirsson, S. S.,** Isolation and characterization of a rat liver epithelial cell line resistant to the antiproliferative effects of transforming growth factor-β (type 1), *Cancer Res.,* 50, 3600, 1990.
71. **Fausto N.,** personal communication.
72. **Isom, H. C. and Mengel, L.,** unpublished data.
73. **Liu, C., Tsao, M.-S., and Grisham, J. W.,** Transforming growth factors produced by normal and neoplastically transformed rat liver epithelial cells in culture, *Cancer Res.,* 48, 850, 1988.
74. **Raymond, V. W., Lee, D. C., Grisham, J. W., and Earp, H. S.,** Regulation of transforming growth factor-α messenger RNA expression in a chemically transformed rat hepatic epithelial cell line by phorbol ester and hormones, *Cancer Res.,* 49, 3608, 1989.
75. **Faust, J. B., Rogers, K. T., Lee, D. C., and Srom, S. C.,** Production and secretion of transforming growth factors in oncogene transfected liver epithelial cells, *Proc. Am. Assoc. Cancer Res.,* 29, 485, 1988.
76. **Fausto, N., Mead, J. E., Gruppuso, P. A., and Braun, L.,** TGF-β1 in liver development, regeneration, and carcinogenesis, *Ann. N.Y. Acad. Sci.,* 593, 231, 1990.
77. **Braun, L., Gruppuso, P., Mikumo, R., and Fausto, N.,** Transforming growth factor-β1 in liver carcinogenesis: messenger RNA expression and growth effects, *Cell Growth Differ.,* 1, 103, 1990.
78. **Lyons, R. and Moses, H. L.,** Transforming growth factors and the regulation of cell proliferation, *Eur. J. Biochem.,* 187, 467, 1990.
79. **Goustin, A. S., Leof, E. B., Shipley, G. D., and Moses, H. L.,** Growth factors and cancer, *Cancer Res.,* 46, 1015, 1986.
80. **Roberts, A. B., Thompson, N. L., Heine, U., Flanders, C., and Sporn, M. B.,** Tansforming growth factor-β: possible roles in carcinogenesis, *Br. J. Cancer,* 57, 594, 1988.
81. **Sporn, M. B., Roberts, A. B., Wakefield, L. M., and Crombrugghe, B.,** Some recent advances in the chemistry and biology of TGF-β, *J. Cell Biol.,* 105, 1039, 1987.
82. **Bascom, C. C., Sipes, N. J., Coffey, R. J., and Moses, H. L.,** Regulation of epithelial cell proliferation by transforming growth factors, *J. Cell Biochem.,* 39, 25, 1989.
83. **Nakamura, T., Tomita, Y., Hirai, R., Yamaoka, K., Kaji, K., and Ichihara, A.,** Inhibitory effect of TGF-β on DNA synthesis of adult rat hepatocytes in primary culture, *Biochem. Biophys. Res. Commun.,* 133, 1042, 1985.
84. **Carr, B. I., Hayashi, I., Branum, E. L., and Moses, H. L.,** Inhibition of DNA synthesis in rat hepatocytes by platelet-derived type B transforming growth factor, *Cancer Res.,* 46, 2330, 1986.
85. **McMahon, J. B., Richards, W. L., del Campo, A. A., Song, M. K., and Thorgeirsson, S. S.,** Differential effects of TGF-β on proliferation of normal and malignant rat liver epithelial cells in culture, *Cancer Res.,* 46, 4665, 1986.
86. **Braun, L., Mead, J., Panzica, M., Mikumo, R., Bell, C. I., and Fausto, N.,** TGF-β mRNA increases during liver regeneration: a possible paracrine mechanism of growth regulation, *Proc. Natl. Acad. Sci. U.S.A.,* 85, 4665, 1986.
87. **Russell, W. E., Coffee, R. J., Ouellette, A. J., and Moses, H. L.,** Type β TGF reversibly inhibits the early proliferative response to partial hepatectomy in the rat, *Proc. Natl. Acad. Sci. U.S.A.,* 85, 5126, 1988.

88. **Lin, P., Liu, C., Tsao, M. S., and Grisham, J. W.,** Inhibition of proliferation of cultured liver epithelial cells at specific cell cycle stages by TGF-β, *Biochem. Biophys. Res. Commun.*, 143, 26, 1987.
89. **Wollenberg, G. K., Semple, E., Quinn, B. A., and Hayes, M. A.,** Inhibition of proliferation of normal preneoplastic and neoplastic rat hepatocytes by TGF-β, *Cancer Res.*, 47, 6595, 1987.
90. **Yeh, Y. C., Tsai, J.-F., Chuang, L.-Y., Yeh, H.-W., Tsai, J.-H., Florine, D. L., and Tam, J. P.,** Elevation of transforming growth factor-α and its relationship to the epidermal growth factor and α-fetoprotein levels in patients with hepatocellular cancer, *Cancer Res.*, 47, 896, 1987.
91. **Hsu, I. C., Metcalf, R. A., Sun, T., Welsh, J. A., Wang, N. J., and Harris, C. C.,** Mutational hotspot in the p53 gene in human hepatocellular carcinomas, *Nature*, 355, 427, 1991.
92. **Bressac, B., Few, M., Wands, J., and Ozturk, M.,** Selective G to T mutations of p53 gene in hepatocellular carcinoma from southern Africa, *Nature*, 350, 429, 1991.
93. **Chen, P. L., Chen, Y., Brookstein, R., and Lee, W.-H.,** Genetic mechanisms of tumor suppression by the human p53 gene, *Science*, 250, 1576, 1990.
94. **Diller, L., Kassel, J., Nelson, C. E., Gryka, M. A., Litwak, G., Gebhardt, M., Bressac, B., Ozturk, M., Baker, S. J., Vogelstein, B., and Friend, S. H.,** p53 Functions as a cell cycle control protein in osteosarcomas, *Mol. Cell. Biol.*, 10, 5772, 1990.
95. **Cariani, E., Lasserre, C., Seurin, D., Hamelin, B., Kemeny, F., Franco, D., Czech, M. P., Ullrich, A., and Brechot, C.,** Differential expression of insulin-like growth factor II mRNA in human primary liver cancers, benign liver tumors and liver cirrhosis, *Cancer Res.*, 48, 6844, 1988.
96. **Cariani, E., Seurin, D., Lasserre, C., Franco, D., Binoux, M., and Brechot, C.,** Expression of insulin-like growth factor II (IGF-II) in human primary liver cancer: mRNA and protein analysis, *J. Hepatol.*, 11, 226, 1990.
97. **Cariani, E., Lasserre, C., Kemeny, F., Franco, D., and Brechot, C.,** Expression of insulin-like growth factor II, α-fetoprotein, and hepatitis B virus transcripts in human primary liver cancer, *Hepatology*, 31, 644, 1991.

Chapter 13

TRANSGENIC MOUSE MODELS OF HEPATOCARCINOGENESIS

S. Sell and B. Knoll

TABLE OF CONTENTS

ISBN 0-8493-4746-7
© 1992 by CRC Press, Inc.

I. INTRODUCTION

The insertion of genes into transgenic animals allows control of expression and determination of the function of specific genes in selected tissues of a living animal. Using techniques for hybrid gene construction to engineer genes and gene-controlling elements, and transgenic animals as assay systems, the effect of gene activation during development and in specific adult tissues can be determined. In addition, specific genetic defects caused by missing or nonfunctioning genes can be restored and specific genes can be deleted, allowing the evaluation of the effect of their loss. Application of these approaches to the study of cancer promises to revolutionize our understanding of the carcinogenic process. Transgenic mice carrying hybrid genes not only permit study of the significance of the genes themselves in causing or preventing cancer (oncogenes and suppressor genes), but also, by controlling the production of hormones or enzymes, can provide models which allow analysis of ''epigenetic'' factors that determine the carcinogenic potential of chemical carcinogens.

II. TRANSGENIC MICE

Transgenic mice are generated by microinjection of selected segments of DNA into one-cell mouse embryos (fertilized eggs), which are implanted into the oviduct of a female mouse to allow development into fully formed fetal mice.[1,2] Usually about 20% of the implanted embryos will grow into a live mouse; and about 10% of these (2% overall) will carry the injected DNA. The injected DNA usually integrates into a random chromosomal location as a head to tail concatenation of 2 to 50 copies of the gene; there may be only one or several sites of integration per mouse. The transgenic locus usually is stable and heritable, so that when the transgenic ''founder'' mouse is bred, the transgene will be inherited in classic Mendelian fashion.[3-6] In the first generation, half of the F1 animals can be expected to carry the transgene. The F1 animals may be tested for the presence of transgenes by determining if the transgene is present in DNA from tail scrapings shortly after birth, and then the cohort divided into groups of transgenic or nontransgenic littermates for controlled experiments. Expression of a transgene during fetal development is often fatal, so that many attempts may be needed to obtain founder mice that can be bred to establish viable transgenic lineages that express the transgene.

In addition, preexisting genes can be replaced by similar genes through homologous recombination.[7] DNA with homology to a cellular gene is introduced into cultures of embryonic stem cells *in vitro*. The infrequent cells that integrate the DNA by homologous recombination with the cognate chromosomal gene are selected and introduced into mouse blastocysts. In the blastocyst, the cells carrying the modified genes may be incorporated into the developing mouse embryo to produce a chimeric mouse, composed of normal cells and modified cells. In some animals the modified stem cells will differentiate into germ cells, permitting transmission of the modified genes to heterozygous offspring through sexual reproduction. These offspring can then be crossbred to produce mice that are homozygous for the modified gene. This approach has the potential to complement the use of transgenic mice not only to disrupt functional genes, but also to replace nonfunctional or defective genes with normal-functioning genes.

The DNA inserted may be a natural gene or a construct that employs highly refined recombinant DNA technology to produce a "hybrid" gene containing a tissue-specific promoter linked to the coding sequence (structural gene) of interest. For example, in the pioneering work of Brinster and Palmiter,[3] a transgene was constructed using the mouse metallothioneine (MT) promoter fused to the rat growth hormone (rGH) structural gene. The MT promoter is activated by dietary zinc, which permits control of expression of the transgene in the transgenic animals through the feeding of zinc-poor (low expression) or zinc-enriched diets (high expression). The transgenic MT-rGH mice fed the zinc-enriched diet have elevated serum levels of growth hormone and a marked increase in growth rate as compared to nontransgenic littermates. This result was extended by restitution of the growth of genetically defective, growth hormone-deficient mice by incorporation of the same MT-rGH construct into the germline of the deficient mice.[8] Transgenic animals of other species, such as rabbits, sheep, cattle, and pigs, have been constructed and a variety of normal functions and pathologic conditions, in addition to neoplasia, have been studied using transgenic models.[9,10] For instance, understanding of the role of the major histocompatibility complex in tolerance induction[11] and in the production of human disease states[12] has been greatly enhanced through the use of transgenic mice and rats.

In this review, the use of transgenic animals to examine the process of hepatocarcinogenesis will be reviewed. Most of the studies using transgenic mice have addressed the role of oncogenes in liver cancer.

III. ONCOGENES IN LIVER CANCER

Oncogenes are transforming genetic elements, originally discovered by their ability to confer increased growth potential on cell lines *in vitro*.[13-15] The viral oncogenes, designated by v-, are altered versions of normal mammalian cellular regulatory genes, known as protooncogenes or cellular oncogenes, designated by the prefex c-.[16-18] In the process of cellular transformation protooncogenes may mutate to produce a transforming oncogene (i.e., a gene that can transform cells *in vitro*), or retroviruses may provide exogenous oncogenes that become incorporated into host DNA. For instance, the *ras* oncogene family consists of a number of genes coding for a distinct transforming p21 protein. Two of the major *ras* genes are from rat retroviruses; Harvey murine sarcoma virus (*v-Ha-ras*) and Kirsten murine sarcoma virus (*v-Ki-ras*). Cellular homologs of the viral *ras* genes are *c-Ha-ras* and *c-Ki-ras*.[19,20] A related gene, *N-ras*, was discovered as a transforming oncogene in a neuroblastoma line.[21] *Ras* protooncogenes acquire transforming capacity by a single-point mutation in their coding sequence, often at codons 12, 13, and 61, in human and rodent tumors. The activated *ras* genes were detected after transfection of tumor DNA into NIH/3T3 cells, followed by assay for the ability of these transfected cells to form morphologically transformed foci in culture or tumors in mice.

The expression of the *ras* family of oncogenes as well as other protooncogenes in the liver during carcinogenesis, as well as in liver tumors, has been used as evidence that protooncogene activation or expression may be an important event in hepatocarcinogenesis (see Tables 1 and 2). However, although oncogene expression may be greater than normal and activated forms of protooncogenes have been reported in human and rodent tumors as well as in chemically induced "premalignant" lesions, a clear sequence of events has not been established. A listing of the relevant observations follows. (1) There is increased expression of cellular oncogenes in chemically induced rat hepatomas as well as in "preneoplastic" liver cells. However, only 2/10 aflatoxin-[26] and 1/18 dibutylnitrosamine[27] (DBN)-induced rat hepatomas examined contained activated *ras* genes. (2) Increased expression of *c-ras, c-myc,* and other cellular protooncogenes is seen during fetal liver development, and during restitutive liver cell proliferation following partial hepatectomy, as well as in cultures of normal liver cells that are stimulated to proliferate. (3) Increased expression of *c-ras* and *c-myc* may be seen in "precancerous" livers during exposure of rats to chemical carcinogens. (4) Activated (mutated) *ras* genes are also found in mouse hepatomas, but at very different percentages. Although 70% of spontaneous hepatomas in B6C3Fl mice[40] and 24 of 25 chemically induced tumors in these carcinogen-sensitive mice had a mutation at codon 61 of the *c-Ha-ras* gene,[41] no such mutations were found in 31 liver tumors of carcinogen-resistant C57BL/6j and BALB/c mice. (5) Only 1 of 30 human hepatomas examined had a mutation at codon 61 of the *ras* gene.[46]

The significance of the expression and mutation of cellular oncogenes during carcinogenesis and in liver tumors is not clear. Since many tumors do not have activated oncogenes, it is difficult to conclude that activation is required for cancer. Similarly, even though greater than normal expression of protooncogenes may be seen in premalignant lesions following exposure to chemical hepatocarcinogens and in hepatomas, elevation of the expression of cellular oncogenes is also observed during proliferation of normal liver cells. Similar, but less extensive, results have been reported for *c-myc, c-fos,* and with other oncogenes associated with liver cancer.

TABLE 1
Expression of Oncogenes in Rat Hepatomas, During Normal Liver Cell Proliferation, and After Exposure to Chemical Hepatocarcinogens

Oncogenes	Observation	Ref.
	Rat hepatomas	
c-myc	Overexpressed in DAB-induced rat hepatomas	22
c-ras	Elevation of mRNA for c-Ha-ras, N-ras, and c-Ki-ras in DEN-induced hepatomas and preneoplastic rat liver cells	23
ras^H, c-raf	H-ras and activated c-raf found in DNA from chemically induced hepatomas after transfection	24, 25
c-Ki-ras	Activated in 2/10 aflatoxin-induced rat hepatomas	26
N-ras	Activated in 1/18 DBN-induced rat hepatomas	27
ras^K, ras^H, ras^N	Elevated in rat hepatomas induced by choline-devoid diet	28
ras	Mutations rare in DMN-induced rat tumors, although DNA transforms NIH 3T3 cells.	29
	Rat, proliferating liver cells	
c-ras, c-myc	c-myc mRNA elevated before DNA synthesis; c-Ki-ras elevated during DNA synthesis in normal liver cells	30, 31
c-fos, ras^H	Early increase in c-fos expression after partial hepatectomy or stimulation of liver cell culture proliferation *in vitro;* later increase in ras^H	32
N-myc, c-myc	Elevation of both N-myc and c-myc during liver regeneration and after DEN exposure	33
myc	Continuously dividing normal liver cells *in vitro* express high levels of c-myc; transformation by AFB_1 produced lower expression	34
	Rat, chemical hepatocarcinogenesis	
c-ras, c-myc	Elevation of c-Ki-ras, c-Ha-ras, and c-myc in fetal liver; early elevations during chemical hepatocarcinogenesis, high c-Ki-ras and c-myc expression in "oval cells"	35
ras^H, c-myc	Transcription not elevated in GGT-positive hepatocytes, but may be elevated or decreased in DEN-induced rat hepatomas	36
mdr	Increased expression during regeneration, during AAF exposure, and in rat hepatomas	37
ras	Hypomethylation of c-H-ras and c-Ki-ras in tumors and preneoplastic livers of DEN-exposed rats	38
c-myc	Elevated in basophilic foci, oval cells, and nodules by *in situ* hybridization	39

Note: Abbreviations: DAB, 3′-methyl-4-dimethylaminobenzene; DEN, diethylnitrosamine; DBN, dibutylnitrosamine; DMN, dimethylnitrosamine; AFB_1, aflatoxin B_1; GGT, gamma glutamyl transpeptidase.

TABLE 2
Expression of Oncogenes in Mouse and Human Hepatomas

Oncogene	Species	Observation	Ref.
ras	Mouse	Activated c-Ha-ras (usually codon 61) in spontaneous hepatomas (70%) and benign liver neoplasms of B6C3F1 mice	40
ras	Mouse	24 of 25 chemically induced hepatomas in B6C3F1 mice have mutation at codon 61 of c-Ha-ras	41
ras	Mouse	Spontaneous and chemically induced hepatomas have similar activated ras genes, but chemically induced tumors have mutations at positions other than codon 61 or in K-ras (approximately 50%)	42
c-Ha-ras	Mouse	Mutations at codon 61 or 12 seen in hepatomas in mouse strains highly susceptible to chemical hepatocarcinogens; mutations not seen in tumors from resistant strains	43
myc, fos ras	Mouse	Early elevations of c-fos and c-myc seen in both male and female mice after partial hepatectomy, but later reelevation only seen in males; c-myc and c-fos elevated in preneoplastic lesions and tumors, c-Ha-ras unchanged	44
c-myc	Man	Reduction in c-myc transcription in 2/7 human hepatomas as compared to normal liver	45
ras	Man	Only 2 of 30 human hepatomas have mutated ras, codon 12 of c-Ki-ras and codon 61 of N-ras	46

IV. THE p53 GENE IN LIVER CANCER

p53 is a 375-amino acid nuclear phosphoprotein that was first identified as a cellular protein that complexed with the simian virus 40 (SV40) large T antigen and coimmunoprecipitated with T antigen (T-Ag) in extracts of SV40-transformed cells.[47] It is present at low levels in most tissues.[48] Recently, mutations at the third base position of codon 249 resulting in a G→T or a G→C transversion have been reported in 8/16 human hepatocellular carcinomas in one study from China[49] and in 4 hepatocellular carcinomas from Africa.[50] This mutation site contrasts with the p53 mutations previously reported in carcinomas and sarcoma of human lung, colon, esophagus, and breast which are usually scattered over four of the five evolutionarily conserved domains, involving codons 130 to 234.[51-55] The amino acid conversions found in these liver cancers are consistent with mutations caused by aflatoxin in mutagenesis experiments.[56,57]

The role of the coding region for p53 as both an oncogene and a tumor-suppressor gene has received considerable attention. It was first thought that expression of wild-type p53 behaved like *c-myc* as a cooperating oncogene.[58-60] However, it has now been shown that mutant forms of the p53 gene behave as cooperating oncogenes: they cooperate with the activated human *c-Ha-ras* oncogene in the transformation of rat embryo fibroblasts.[61,62] Wild-type forms of the p53 gene can suppress the transformed phenotype of some tumor cell lines.[63] Recent data indicate that point mutations at multiple sites in the p53 gene remove the ability of the p53 product to function as a tumor suppressor. The original cloned p53 gene contained a point mutation. It has now been shown by a number of studies that only p53 genes that carry mutations act as cooperating oncogenes and that the wild-type p53 gene serves to suppress transformation. The human p53 gene, mapped to the chromosome 17p13 region, was found to be deleted in colorectal cancer.[64] Subsequently, one copy of the p53 gene was found to be deleted in all colorectal cancers and the remaining p53 allele contained at least one point mutation resulting in an amino acid substitution. Thus, p53 fits the description of a recessive tumor-suppressor gene.

Transgenic mice carrying mutant murine p53 genomic fragments controlled by the natural p53 gene promoter have high levels of the p53 protein and a high incidence of lung adenocarcinomas, osteosarcomas, and lymphomas.[65] There were no tumors in the livers of these mice. This could be explained either because the p53 transgene was not sufficiently expressed in the liver, or because the liver is protected by production of the natural product. It is also possible that, in the liver, other events in addition to the expression of mutant p53 must occur to bring about transformation.

It appears that normal p53 functions to control cell proliferation and that mutated p53 may fail to be able to inhibit cell-cycle stimulating proteins of oncoviruses. p53 has been shown to be a substrate of the cdc2 protein kinase.[66] A serine at residue 315 is phosphorylated by both p60-cdc2 and cyclin B-cdc2, and is cell-cycle dependent. p53 protein accumulates during the G1 phase of the cell cycle. The mutant forms of the p53 protein fail to bind SV40 T antigen as well as does wild-type p53 protein. Viral products which bind and presumably neutralize the normal cell-cycle regulatory functions of p53 protein release cells from normal cell cycle control and in doing so use the cell for virus replication. The binding of p53 protein to the transforming proteins of various DNA tumor viruses could thereby inhibit viral transformation, and mutations in p53 could abrogate this regulatory function.

The wild-type (but not the mutant) p53 protein has a potent transcription-activating domain. This activity becomes apparent when the p53 protein is attached to a sequence-specific DNA-binding domain and then assayed for the activation of a promoter sequence adjacent to the DNA-binding site of the DNA-binding domain.[67,68] A cellular gene that normally is regulatable by p53 has not yet been described; however, it is proposed that normal p53 protein activates the transcription of genes that in turn suppress transformation.[68] Using immunoprecipitation of DNA-protein complexes, Kern et al.[69] described specific sites bound by p53 as characteristic of DNA replication origins. Mutant p53 proteins did not bind the DNA fragments. Thus, the p53 protein could be involved in both transcription and replication of DNA.

V. ACCUMULATING ONCOGENIC MUTATIONS DURING CARCINOGENESIS

Although in retinoblastoma and in Wilms' tumor there are indications that mutation of a single tumor-suppressor gene could lead to cancer, investigations of more common tumors indicate that multiple loci producing an accumulation or a sequential series of changes is more likely to culminate in neoplasia.[70] For example, the model developed by Fearon et al.[71] and Vogelstein et al.[72,73] for colon carcinoma postulates that tumorigenesis proceeds through a series of genetic alterations involving tumor-suppressor genes and oncogenes. The model is as follows: Normal Epithelium (5q mutation or loss)→Hyperplastic Epithelium (DNA hypomethylation)→Early Adenoma (12q mutation *K-ras*)→Intermediate Adenoma (18q loss)→Late Adenoma (17 p loss p53)→ Carcinoma (other alterations)→ Metastases.

If neoplasia is the result of the accumulation of four mutations, each occurring at the frequency of 10^{-3}, then the process of A 10^{-3} + B 10^{-3} + C 10^{-3} + D 10^{-3} would occur with a frequency of 10^{-12}, a not unreasonable incidence for the cellular origin of cancer. However, in a rapidly turning over cell population, the classic model of initiation and promotion argues that the order of the mutations or events may be critical. Thus in skin and liver chemical carcinogenesis, promoting agents are not effective unless administered after initiating agents (for reviews see References 74 to 79). Using hepatocarcinogenesis as an example of sequential interplay of oncogene expression, a chemical carcinogen and a promoting event: A might be a carcinogen, such as aflatoxin; B activation of an oncogene, such as *c-Ha-ras;* C a mutation of a tumor-suppressor gene, such as p53; and D a promoting event, such as liver cell proliferation induced by CCl_4, or phenobarbital administration.

VI. ONCOGENES IN TRANSGENIC MICE

Transgenic mice bearing oncogenes linked to the regulatory regions of tissue-specific genes provide a means of determining the significance of expression of an oncogene in various tissues (see reviews by Palmiter and Brinster,[4] Cory and Adams,[80] Jaenisch,[5] Hanahan,[6] and Pattengale et al.[10]) The expression pattern of the genetically engineered hybrid gene usually conforms to that of the gene from which the regulatory region was derived. Sometimes novel patterns are observed, perhaps because of the creation of a recombinant enhancer and/or the influence of flanking DNA. Collaboration between oncogenes can be studied by breeding two transgenic lines bearing different oncogenes.[5,81] An abbreviated list of some of the oncogenes studied in other systems is presented in Table 3.

TABLE 3
Some Selected Oncogenes and their Putative Mechanism of Action

Protein (gene) product	(Onco) gene	Activity of product
Cellular protein	SV40 T-Ag	Binds to antioncogene protein p53, modified p53 allows transformation of cells
Nuclear proteins	myc, myb, fos	Binds to nuclear DNA, presumably activates cell cycle
Membrane-associated protein		
Inner	ras	GTP-binding protein that affects signal transduction
Outer	src	Enzyme that phosphorylates tyrosine residues on proteins
Growth factor(s)	sis	β-Polypeptide chain of platelet-derived growth factor
Growth factor receptor	erb-B/erb-B2 (neu)	Truncated epidermal growth factor receptor

Modified from Hanahan, *Annu. Rev. Genet.*, 22, 479, 1988; and Pattengale et al., *Am. J. Pathol.*, 135, 39, 1989.

TABLE 4
Organ Expression and Tumor Development in Transgenic Mice with the MMTV Promoter Directing Oncogene Expression

Oncogene	Tissue expression	Tumors
myc	Breast, salivary gland	Adenocarcinoma of breast
myc	Breast, salivary gland, testis, pancreas, lung, spleen, liver, heart, muscle, thymus	Adenocarcinoma of breast, testicular and mast cell tumors, and lymphomas
ras	Breast, salivary gland, lacrimal gland, thymus, spleen, ovary, seminal vesicle	Adenosquamous carcinomas of breast, salivary gland tumors, lymphomas
myc/ras	Breast, salivary gland, lacrimal gland, spleen, lung, seminal vesicle	Same as above plus seminal vesicle tumors; many animals with multiple tumors
erb-B2[neu]	Breast, salivary gland, lacrimal gland, epididymis	Adenocarcinomas of breast

Modified from Pattengale et al., *Am. J. Pathol.*, 135, 39, 1989.

As an example of the tissue site-directed effect of a promoter on expression of an oncogene and tumor development, the studies of Pattengale et al.[10] show that the mouse mammary tumor virus (MMTV) promoter directs expression mainly to the breast and lacrimal gland, but also to other tissues (Table 4), and malignant tumors arise in the organs that express the oncogene. Several generalities can be concluded from such studies. First, the inheritance pattern

TABLE 5
Selected Studies on Expression of SV40 T-Antigen and Tumors in Transgenic Mice

Promoter	Effect	Ref.
Metallothionein	Tumors of choroid plexis	82, 83
MT-SVΔe	Deletion of SV40 enhancer leads to liver tumors	84
Inverse MT	Zinc-dependent islet cell adenomas and hepatocellular carcinomas	85
MUP[a]	Liver "hyperplasia" preceding hepatomas	86
α-Amylase (early)	Metastatic hibernomas	87
α-Amylase (late)	Osteosarcomas	88
Insulin	Pancreatic β-cell tumors	89
Insulin	Intestinal tumors metastatic to liver	90
Albumin	Liver dysplasia, adenomas, and hepatomas	91
Albumin	Hepatomas contain activated c-H-*ras*	92
α-1-Antitrypsin	Hepatoma preceded by dysplasia and hyperplasia	93, 93a

[a] Mouse major urinary protein.

of oncogene expression in transgenic mice is autosomal dominant with variable expression. Second, tumors do not always develop in tissues that express the oncogene. Some lineages that expressed the *myc* oncogene only in the salivary gland and not in the breast did not develop salivary gland tumors. Third, tumor tissue usually expresses a high level of the mRNA of the oncogene, but normal nonneoplastic cells may also express mRNA. It is concluded that events in addition to the overexpression of the *myc* gene are required to induce malignant transformation. Finally, tumors that develop in transgenic mice appear to be the result of a timed series of random (stochastic) events. In both the lacrimal gland and the breast, polyclonal hyperplasia appears to precede the appearance of a monoclonal malignant tumor, typically occurring over a 6- to 12-month period.[81]

A. SV40 T ANTIGEN IN TRANSGENIC MICE

The results of studies on the expression of the SV40 T-Ag and development of tumors in transgenic mice are tabulated in Table 5. The SV40 T-Ag is included as the model of oncogene-produced tumors because it is a potent oncogene, causing both immortalization and transformation of many cell types *in vitro*.[86,94] In addition, *in vivo* tumorigenicity of SV40 T-Ag has been demonstrated in transgenic mice containing the gene for and expressing the SV40 T-Ag. This was first seen in the choroid plexus (SV40 T-Ag gene driven by its own promoter),[82,83] and later in the liver (MT promoter),[84,85] the pancreatic islets (insulin promoter),[89] the exocrine pancreas (elastase promoter),[94] and other tissues.[95,96]

Liver cancer develops following stages of liver dysplasia and hyperplasia in transgenic mouse lines with the T-Ag gene under the control of the regulatory elements of the human albumin *(alb)*[89] or α-1-antitrypsin (AAT)

TABLE 6
Some Effects of Other Genes Expressed in the Livers of Transgenic Mice

Gene	Promoter	Effect	Ref.
hGH	MT	Organomegaly, hypertrophy, and hyperplasia of liver	98
IGF	MT	No organomegaly or liver change	
H-ras	alb	Neonatal death with liver dysplasia or death at 1 to 5 months with liver dysplasia and lung tumors in founder mice, lung adenomas and hepatomas in lineages	91, 97
c-myc	alb	No tumors in founder mice; hepatic dysplasia, adenomas at 15 to 18 months in lineages	91
ras/myc	alb	Hepatic dysplasia, hepatomas, and lung adenomas in dual-expressing hybrid lines	91
TGF-α	MT	Hyperplasia of liver, pancreas, stomach, intestine, coagulation glands; breast carcinoma in one lineage	99
TGF-α	MT	Hepatic dysplasia, hepatomas, mammary, and pancreatic dysplasia in one lineage	100
TGF-α	MMTV	Breast hyperplasia and carcinoma	101
α_1-Pi	α_1-Pi	Accumulation of mutant (Z allele) protein in hepatocytes causes neonatal hepatitis	102, 103
μPA	alb	μPA-expressing cells die and are replaced by transgene-deficient cells that proliferate from a few clones	104

Note: hGH, human growth hormone; MT, metallothionein; IGF, insulinlike growth factor; alb, albumin; TGF-α, transforming growth factor alpha; MMTV, mouse mammary tumor virus; α_1-Pi, α_1-antitrypsin; μPA, urokinase-type plasminogen activator.

gene.[93,93a] With the albumin promoter, T-Ag founder mice died at 3 to 7 months with liver tumors. Transgenic lineages demonstrated severe hepatic dysplasia, altered hepatocytic foci, and had multiple hepatic tumors at 3 to 5 months of age. Transgenic mice bearing the AAT promoter/T-Ag construct died within 12 to 14 weeks of age after birth with a progressive series of morphologic developments. These included (1) no recognizable change during embryonic and fetal life; (2) hyperplastic hepatocytes with no dysplasia or nuclear alterations, neonatal to 2 weeks of age; (3) diffuse liver cell dysplasia without nodules, 3 to 8 weeks; and (4) hepatocellular carcinoma in a background of dysplasia and normal liver at 8 to 12 weeks. In these mice, the SV40 T-Ag is expressed in day 18 fetuses and later when under control of the albumin promoter and throughout fetal and adult life when under control of the AAT promoter. Activated *c-Ha-ras* genes were found in hepatocellular carcinomas developing in transgenic mice harboring the T-Ag gene driven by the *alb* gene promoter.[92] Thus mutation of *c-Ha-ras* may be a critical step in the development of liver cancer (see Chapter 12).

A partial listing of other genes expressed in the livers of transgenic mice using different promoters is shown in Table 6. Expression of a mutated *c-*

Ha-ras gene produces neonatal death with liver dysplasia in some founder mice and later death with liver dysplasia and lung tumors in other founder mice. Some lines derived from the latter founders have hepatomas.[91,97] Founder mice expressing *c-myc* have no abnormalities, but lineages may develop hepatic dysplasia and adenomas after 15 to 18 months.[91] After cross breeding to obtain dual transgenics expressing both *c-Ha-ras* and *c-myc,* hepatic dysplasia and hepatomas were seen in 10 of 30 mice between 1 and 4 months of age. Dual transgenics with *alb*-SV40-T-Ag and *alb-ras* died within 1 week of birth with severe hepatobiliary dysplasia, whereas those with *alb*-T-Ag and *alb-myc* developed hepatomas and died within 3 to 6 weeks of age. Thus dual expression accelerates the developmental abnormalities of tumors seen in single-expressing transgenic mice.[91] Expression of the growth hormone gene under control of the MT promoter causes hypertrophy and hyperplasia of the liver, but no tumors.[98] Expression of the transforming growth factor-α (TGF-α) gene produces hyperplasia and dysplasia in the liver,[99] with hepatomas seen in one lineage.[100] A TGF-α gene under control of the MMTV promoter consistently produces adenocarcinomas of the breast, but no liver lesions.[101] Accumulation of Z-type mutant AAT produces hepatocyte death associated with the accumulation of the mutant protein in the liver cells,[102,103] and liver cells in transgenic mice expressing urokinase-type plasminogen activator die, but the liver is replaced by a few clones of transgene-deficient cells.[104] These studies demonstrate the variability of results that can be expected in transgenic mice expressing foreign genes in the liver. In general, expression of transforming oncogenes leads to developmental abnormalities and/or tumor formation, but at different rates. Expression of growth factor genes leads to hyperplasia and hypertrophy, and the accumulation of toxic proteins in the cells leads to cell death. Further studies using different promoters, different oncogenes, and different combinations of oncogenes may lead to insights into the role of oncogenes and growth factors in the carcinogenic process (see below).

B. HEPATITIS B TRANSGENIC MICE

Transgenic mice containing the coding region for the HBV surface antigen (HBsAg) and pre-S and X antigens, as well as most of the p coding region, but with deletions in the core region (PSX), were produced using either the normal HBV promoter[105,106] or the MT promoter (MT-PSX).[105] The founder mice produced high serum concentrations of HBsAg and high cellular concentrations were found in liver, with less in other organs. The mice were found to be immunologically tolerant to the HBsAg, and little or no tissue pathology was noted, supporting the hypothesis that cellular injury in hepatitis B is immunologically mediated. However, using albumin promoter-HBV transgenic mice (*alb*-PSX) overproduction of the HBV large envelope polypeptide inhibited the secretion of the surface antigen.[107] When the large and

TABLE 7
SAlaAT,[a] AFP, HBsAg Expression, Liver Damage, and Hepatoma Formation in alb-PSX Transgenic Mice

Lineage	HBsAg expression	Hepatonecrosis	SAlaAT[a]	AFP	Dysplasia	Tumors
45-3	+	0	0	0	0	0
45-2	+ +	+ +	+ +	+	+	+
50-4	+ + + +	+ + + +	+ + + +	+ + +	+ + +	+ + +

[a] Serum alanine aminotransferase, formerly SGPT.

major envelope polypeptides are produced by hepatocytes at roughly equimolar ratios, long branching filamentous HBsAg particles are formed that become entrapped within the endoplasmic reticulum of the hepatocytes, forming 22-nm diameter HBsAg particles that are not efficiently secreted.[108] The accumulation leads to focal hepatocellular degeneration and necrosis, inflammation, and hyperplasia,[109] which in turn progress to aneuploidy and the development of hepatocellular carcinomas.[110] The degree of liver cell damage (SAlaAT levels), restitutive proliferation (α-fetoprotein [AFP] elevation), dysplasia, and the eventual appearance of hepatomas is related to the amount of expression of the large polypeptide antigen in different lineages (Table 7).

The process of storage (present at birth), injury (beginning at 3 months), hyperplasia (beginning at 6 months), tumor development (beginning at 15 months), and death (beginning at 19 months) occurs with fair uniformity in male mice of the 50-4 lineage.[108] The process is slower in female 50-4 transgenic alb-PSX mice, but by 20 months all mice of this lineage developed hepatocellular carcinoma, whereas none of the age- and sex-matched nontransgenic littermates had any histopathologic changes.[110] The tumors contain neither viral envelope proteins nor HBV X-specific transcripts. In addition, the levels of *c-myc, c-fos, H-ras,* and *K-ras* transcripts are not elevated.[109] It is concluded that overproduction of the HBV large envelope polypeptide initiates liver cell injury and inflammation, followed by restitutive hyperplasia, which places large numbers of hepatocytes at risk for development of transforming mutations, and progression to hepatocellular carcinoma.

However, an alternative which does not include liver injury in hepatitis B-related liver cancer is provided by transgenic mice carrying a DNA fragment spanning positions 707 to 1856 in the HBV genome.[111] This segment contains the coding region for the *HBx* gene as well as the transcriptional enhancer reported to upregulate other viral genes.[112,113] The livers of transgenic lineages expressing the HBx protein show accumulation of the protein in altered foci of hepatocytes at 4 months of age, followed by the development of basophilic adenomas and some clear cell adenomas at 8 to 10 months of age, and death of most of the male transgenics at 11 to 15 months of age with clear cell hepatocellular carcinomas.[111] Thus in this study the expression of the *HBx* gene alone results in progressive cellular changes in the liver of transgenic mice, leading to hepatocellular carcinoma. However, in other studies in which the *HBx* gene was used to produce transgenic mice that expressed the HBx transactivator protein, liver tumors were not found.[114-116] This was true even when the *HBx* gene was placed under the control of the human AAT regulatory region and transgenic mice produced HBx protein in patterns similar to that observed for SV40 T-Ag in transgenic mice with the SV40 T-Ag gene controlled by the same promoter.[116]

If liver injury increases the chance for mutations in regenerating hepatocytes, then it may be hypothesized that chemical hepatocarcinogens may have more effect in transgenic mice with HBsAg-related injury. This hypothesis is supported by the conclusion based on epidemiologic data that diets which stimulate liver cell proliferation increase the incidence of liver cancer in hepatitis B+ populations.[117] Although data from this same study do not indicate an association between aflatoxin and liver cancer in hepatitis B endemic areas,[117] other studies have indicated that aflatoxin in the diet is a high-risk factor in such populations.[118-120] In the transgenic mice described in Reference 114, exposure to DEN or diaminoazobenzene, known chemical hepatocarcinogens, was associated with the development of liver nodules, adenomas, and hepatocellular carcinomas.[121] However, in these lineages, nontransgenic littermates have a similar development of neoplastic lesions. The study suggested a ''weak'' enhancement of the effect of the chemicals in transgenic mice. For instance, 9 adenomas were found in 23 HBV-transgenic male mice exposed to DEN (39%), whereas 5 adenomas were found in 19 HBV nontransgenic male controls (26%). In the alb-PSX transgenic mice described in Reference 107, exposure to DEN produced nine adenomas and two hepatomas after 18 months in eight female transgenic mice; ten adenomas and two carcinomas were found in eight transgenic female mice receiving 6 μg/g of aflatoxin B1 at 18 months (Figure 1); no adenomas or carcinomas were seen in nine nontransgenic female littermates of the same age.[122] These results support the hypothesis that liver proliferation induced in these transgenic mice predisposes them to the carcinogenic effects of chemicals. Epidemiologic data indicate that dietary factors that enhance liver proliferation are a ''necessary and sufficient cause'' for development of hepatocellular cancer in HBsAg+ carriers.[117] The question of cause and effect arises when considering the proliferation of hepatocytes in HBsAg+ individuals. Does HBV infection induce restitutive proliferation that increases the effectiveness of chemical carcinogens, or do dietary factors enhance proliferation, allowing expression of the carcinogenic effect of HBV?

VII. FUTURE STUDIES

The ability to construct transgenic mice to produce different gene products allows dissection of the events in chemical carcinogenesis in ways not pre-

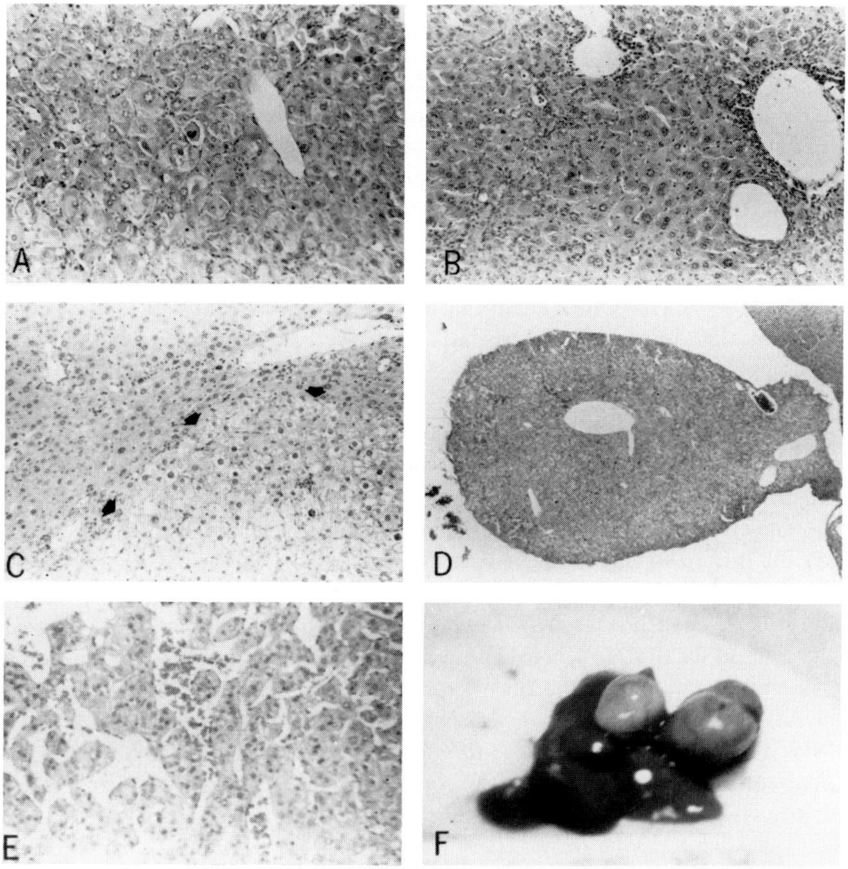

FIGURE 1. Photomicrographs (A through E) and gross photo (F) of livers of female lineage 50-4 hepatitis B transgenic mice. (A) Untreated; (B) through (F) treated with aflatoxin B1. In both (A) and (B) there are swollen "ground-glass" cells and dysplastic changes. There is a mild periportal round cell infiltrate in (B). The arrows in (C) outline the edge of a nodule of large, clear hepatocytes. A pedunculated dysplastic nodular zone (?adenoma) is shown in (D). A trabecular hepatocellular carcinoma is shown in (E). Gross nodules on a finely nodular background is shown in (F). (A, B, C, and E, original magnification × 200; D, original magnification × 100.)

viously possible. The role of oncogenes in chemical carcinogenesis can be studied by determining the effects of chemical carcinogens or promoting agents in transgenic mice expressing a given oncogene. Selective expression of an oncogene may be accomplished using transcriptional promoters that can be controlled by diet, such as the zinc-inducible MT gene promoter, or promoters turned on by other events, such as the AFP gene-controlling region, which is activated by liver injury or restitutive proliferation,[123] or the C-reactive protein promoter, which is controlled by inflammatory stimuli.[124] In addition, the effect of fetal control of oncogenes may be determined using the AFP promoter, as this gene normally is expressed at high levels during fetal development, but drops to very low levels after 3 to 5 weeks of age in mice and rats.[123] Using combinations of the AFP and *alb* promoters, the effect of early expression of one gene and the continuous expression of another could be determined in dual transgenic cross-bred lines. This type of approach is now under way in our laboratory.

As with the HBV transgenic mouse work described above, the interaction of chemical or promoting agents in transgenic mice expressing other genes could be determined. Thus, the effect of overexpressing a cooperating or weak oncogene such as *c-myc* on the induction of liver cancer could be examined. The effect of ''antioncogenes'' such as wild type p53 and ''anti-antioncogenes'' such as mutated p53 could be determined either in combination with oncogenes in transgenic animals or as modifiers of exposure to chemicals.

The effect of overexpression of carcinogen-metabolizing enzymes or other host modifiers of carcinogenesis may be studied. For instance, transgenic mice expressing the bacterial[125] or human[126] 06-alkylguanine-DNA alkyltransferase have been produced that express a total alkyltransferase activity that is increased fourfold to fivefold above background in the livers of transgenic mice. Similarly, transgenic mice expressing the bacterial 06 methylguanine DNA methyltransferase under control of the MT promoter produce three times the activity found in nontransgenic littermates.[127] If these mice are resistant to carcinogens, such as *N*-nitroso compounds, it would indicate that DNA repair can block the effect of the carcinogen.

The stochastic nature of chemical carcinogenesis may be examined in transgenic mice by various combinations of chemicals, promoters, controlling genes, and oncogenes. For example, transgenic mice may be used to investigate the effects of enhancing a weak oncogene (i.e., *c-myc*) by a tumor promoter vs. suppressing a strong oncogene (i.e., SV40 T-Ag) by an antioncogene on sequential events in hepatocarcinogenesis. The availability of different liver-specific constitutive and inducible promoters, different strong, weak, and cooperating oncogenes, strong and weak carcinogens, different promoting events, different antioncogenes, and different modifying enzymes makes the possible combinations and permutations staggering, but provides ways of sorting out the importance of key events in the carcinogenic process.

REFERENCES

1. **Gordon, J. W. and Ruddle, F. H.,** Gene transfer into mouse embryos: production of transgenic mice by pronuclear injection, *Methods Enzymol.,* 101, 411, 1983.
2. **Hogan, B. L. M., Costantini, F., and Lacy, E.,** *Manipulation of the Mouse Embryo: A Laboratory Manual,* Cold Spring Harbor Laboratory, New York, 1986.
3. **Brinster, R. L. and Palmiter, R. D.,** Introduction of genes into germ line of animals, *Harvey Lect.,* 80, 1, 1985.
4. **Palmiter, R. D. and Brinster, R. L.,** Germline transformation of mice, *Annu. Rev. Genet.,* 20, 465, 1986.
5. **Jaenisch, R.,** Transgenic animals, *Science,* 240, 1468, 1988.
6. **Hanahan, D.,** Transgenic mice as probes into complex systems, *Science,* 246, 1265, 1989.
7. **Capecchi, M. R.,** Altering the genome by homologous recombination, *Science,* 244, 1288, 1989.
8. **Hammer, R. E., Palmiter, R. D., and Brinster, F. L.,** Partial correction of murine hereditary growth disorder by germline incorporation of a new gene, *Nature,* 311, 65, 1985.
9. **Cuthbertson, R. A. and Klintworth, G. K.,** Transgenic mice: a gold mine for furthering knowledge in pathobiology, *Lab. Invest.,* 58, 484, 1988.
10. **Pattengale, P. K., Stewart, T. A., Leder, A., Sinn, E., Muller, W., Tepler, I., Schmidt, E., and Leder, P.,** Animal models of human disease. Pathology and molecular biology of spontaneous neoplasms occurring in transgenic mice carrying and expressing activated cellular oncogenes, *Am. J. Pathol.,* 135, 39, 1989.
11. **Hanahan, D.,** Transgenic mouse models of self-tolerance and autoreactivity by the immune system, *Annu. Rev. Cell Biol.,* 6, 493, 1990.
12. **Hammer, R. E., Maika, S. D., Richardson, J. A., Tang, J.-P., and Taurog, J. D.,** Spontaneous inflammatory disease in transgenic rats expressing HLA-B27 and human β2m: an animal model of HLA-B27-associated human disorders, *Cell,* 63, 1099, 1990.
13. **Bishop, J. M.,** Viral oncogenes, *Cell,* 42, 23, 1985.
14. **Land, H., Parda, L., and Weinberg, R. A.,** Cellular oncogenes and multistep carcinogenesis, *Science,* 222, 771, 1983.
15. **Weinberg, R. A.,** The action of oncogenes in the cytoplasm and nucleus, *Science,* 230, 770, 1985.
16. **Hunter, T. and Cooper, J. A.,** Protein-tyrosine kinases, *Annu. Rev. Biochem.,* 54, 897, 1985.
17. **Burgess, A.,** Growth factors and oncogenes, *Immunol. Today,* 6, 107, 1985.
18. **Muller, R. and Verma, I.,** Expression of cellular oncogenes, *Curr. Top. Microbiol. Immunol.,* 112, 73, 1984.
19. **Norton, J. D. and Avery, R. J.,** Genetic organization and cloning of Kirsten murine sarcoma virus DNA, *Biochem. Biophys. Res. Commun.,* 108, 1631, 1982.
20. **Eva, A., Tronick, S., Gol, R. A., Pierce, J. H., and Aaronson, S. A.,** Transforming genes of human hematopoietic tumors: frequent detection of *ras*-related oncogenes whose activation appears to be independent of tumor phenotype, *Proc. Natl. Acad. Sci. U.S.A.,* 80, 4926, 1983.
21. **Shimizu, K., Goldfarb, M., Perucho, M., and Wigler, M.,** Isolation and preliminary characterization of the transforming gene of a human neuroblastoma cell line, *Proc. Natl. Acad. Sci. U.S.A.,* 80, 383, 1983.
22. **Makino, R., Hayashi, K., Sato, S., and Sugimura, T.,** Expression of the *c-Ha-ras* and *c-myc* genes in rat liver tumors, *Biochem. Biophys. Res. Commun.,* 119, 1096, 1984.

23. **Corcos, D., Defer, N., Raymondjean, M., Paris, B., Corral, M., Tichonicky, L., Kruh, J., Glaise, D., Saulnier, A., and Guguen-Guillouzo, C.,** Correlated increase of the expression of the *c-ras* genes in chemically induced hepatocarcinomas, *Biochem. Biophys. Res. Commun.*, 122, 259, 1984.
24. **Ishikawa, F., Takakau, F., Nagao, M., Ochiai, M., Hayashi, K., Takayama, S., and Sugimura, T.,** Activated oncogenes in a rat hepatocellular carcinoma induced by 2-amino-3-methyl-imidazo[4,5-f]quinoline, *Jpn. J. Cancer Res.*, 76, 425, 1985.
25. **Ishikawa, F., Takaku, F., Ochiai, M., Hayashi, K., Hirohashi, S., Terada, M., Takayama, S., Nagao, M., and Sugimura, T.,** Activated *c-raf* gene in a rat hepatocellular carcinoma induced by 2-amino-3-methylimidazo[4,5-f]quinoline, *Biochem. Biophys. Res. Commun.*, 132, 186, 1985.
26. **McMahon, G., Hanson, L., Lee, J.-J., and Wogan, G. N.,** Identification of an activated *c-Ki-ras* oncogene in rat liver tumors induced by aflatoxin B1, *Proc. Natl. Acad. Sci. U.S.A.*, 83, 9418, 1986.
27. **Funato, T., Yokota, J., Sakamoto, H., Kameya, T., Fukushima, S., Ito, N., Terada, M., and Surimura T.,** Activation of *N-ras* gene in a rat hepatocellular carcinoma induced by dibutylnitrosamine and butylated hydroxytolune, *Jpn. J. Cancer Res.*, 78, 689, 1987.
28. **Chandar, N., Lombardi, B., Schultz, W., and Locker, J.,** Analysis of *Ras* genes and linked viral sequences in rat hepatocarcinogenesis, *Am. J. Pathol.*, 129, 241, 1987.
29. **Goyette, M., Dolan, M., Kaufmann, W., Kaufmann, D., Shank, P. R., and Fausto, N.,** Transforming activity of DNA from rat liver tumors induced by carcinogen methyl(acetoxymethyl)nitrosamine, *Mol. Carcinogenesis*, 1, 26, 1988.
30. **Goyette, M., Petropoulous, M. C., Shank, P. R., and Fausto, N.,** Expression of a cellular oncogene during liver regeneration, *Science*, 219, 510, 1983.
31. **Goyette, M., Petropoulous, C. J., Shank, P. R., and Fausto, N.,** Regulated transcription of *c-Ki-ras* and *c-myc* during compensatory growth of rat liver, *Mol. Cell. Biol.*, 4, 1493, 1984.
32. **Kruijer, W., Skelly, H., Botteri, F., van der Putten, H., Barber, J. R., Berma, I. M., and Leffert, H. L.,** Proto-oncogene expression in regenerating liver is simulated in cultures of adult rat hepatocytes, *J. Biol. Chem.*, 261, 7929, 1986.
33. **Corral, M., Paris, B., Guguen-Guillouzo, C., Corcos, D., Kruh, J., and Defer, N.,** Increased expression of the *N-myc* gene during normal and neoplastic rat liver growth, *Exp. Cell Res.*, 174, 107, 1988.
34. **Sinha, S., Neal, G. E., Legg, R. F., Watson, J. V., and Pearson, C.,** The expression of *c-myc* related to the proliferation and transformation of rat liver-derived epithelial cells, *Br. J. Cancer*, 59, 674, 1989.
35. **Yaswen, P., Goyette, M., Shank, P. B., and Fausto, N.,** Expression of *c-Ki-ras, c-Ha-ras* and *c-myc* in specific cell types during hepatocarcinogenesis, *Mol. Cell. Biol.*, 5, 780, 1985.
36. **Beer, D. G., Schwarz, M., Sawada, N., and Pitot, H. C.,** Expression of *H-ras* and *c-myc* protooncogenes in isolated γ-glutamyl transpeptidase-positive rat hepatocytes and in hepatocellular carcinomas induced by diethylnitrosamine, *Cancer Res.*, 46, 2435, 1986.
37. **Fairchild, C. R., Ivy, S. P., Rushmore, T., Lee, G., Koo, P., Goldsmith, M. E., Myers, C. E., Farber, E., and Cowan, K. H.,** Carcinogen-induced *mdr* overexpression is associated with xenobiotic resistance in rat preneoplastic liver nodules and hepatocellular carcinomas, *Proc. Natl. Acad. Sci. U.S.A.*, 84, 7701, 1987.
38. **Bhave, M. R., Wilson, M. J., and Poirier, L. A.,** *c-H-ras* and *c-K-ras* gene hypomethylation in the livers and hepatomas of rats fed methyl-deficient, amino acid-defined diets, *Carcinogenesis*, 9, 343, 1988.
39. **Nagy, P., Evarts, R. P., Marsden, E., Roach, J., and Thorgeirsson, S. S.,** Cellular distribution of *c-myc* transcripts during chemical hepatocarcinogenesis in rats, *Cancer Res.*, 48, 5522, 1988.

40. **Reynolds, S. H., Stowers, S. J., Maronpot, R. R., Anderson, M., and Aaronson, S. A.,** Detection and identification of activated oncogenes in spontaneously occurring benign and malignant hepatocellular tumors of B6C3F1 mouse, *Proc. Natl. Acad. Sci. U.S.A.,* 83, 33, 1986.
41. **Wiseman, R. W., Stowers, S. J., Miller, E. C., Anderson, M. W., and Miller, J. A.,** Activating mutations of the *c-Ha-ras* proto-oncogene in chemically induced hepatomas of the male B6C3F1 mouse, *Proc. Natl. Acad. Sci. U.S.A.,* 83, 5285, 1986.
42. **Reynolds, S. H., Stowers, S. J., Patterson, R. M., Maronpot, R. R., Aaronson, S. A., and Anderson, M. W.,** Activated oncogenes in B6C3F1 mouse liver tumors: implications for risk assessment, *Science,* 237, 1309, 1987.
43. **Buchman, A., Bauer-Hofmann, R., Mahr, J., Drinkwater, N. R., Luz, A., and Schwarz, M.,** Mutational activation of the *c-Ha-ras* gene in liver tumors of different rodent strains: correlation with susceptibility to hepatocarcinogenesis, *Proc. Natl. Acad. Sci. U.S.A.,* 88, 911, 1991.
44. **Porsch-Hallstrom, I., Blanck, A., Ericksson, L. C., and Gustafsson, J.-A.,** Expression of the *c-myc, c-fos* and *c-ras*^{Ha} protooncogenes during sex-differentiated rat liver carcinogenesis in the resistant hepatocyte model, *Carcinogenesis,* 10, 1793, 1989.
45. **Su, T.-S., Lin, L.-H., Lui, W.-Y., Chang, C., Chou, C.-K., Timg, L.-P., Hu, C.-P., Han, S.-H., and P'eng, F.-K.,** Expression of *c-myc* gene in human hepatoma, *Biochem. Biophys. Res. Commun.,* 132, 264, 1985.
46. **Tsuda, H., Hirohashi, S., Shimosato, Y., Yoshida, T., and Terada, M.,** Low incidence of point mutation of *c-Ki-ras* and *N-ras* in human hepatocellular carcinoma, *Jpn. J. Cancer Res.,* 80, 196, 1989.
47. **Lane, D. R. and Crawford, L. V.,** T-Antigen is bound to host protein in SV40-transformed cells, *Nature (London),* 278, 261, 1979.
48. **Levine, A. J., Momans, J., and Finlay, C. A.,** The p53 tumor suppressor gene, *Nature (London),* 351, 453, 1991.
49. **Hu, I. C., Metcalf, R. A., Sun, T., Welsh, J. A., Wang, N. J., and Harris, C. C.,** Mutational hotspot in the p53 gene in human hepatocellular carcinomas, *Nature (London),* 350, 427, 1991.
50. **Bressac, B., Kew, M., Wands, J., and Ozturk, M.,** Selective G to T mutations of p53 gene in hepatocellular carcinoma from southern Africa, *Nature (London),* 350, 429, 1991.
51. **Nigro, J. M., Baker, S. J., Preisinger, A. C., Jessup, J. M., Hostetter, R. et al.,** Mutations in p53 gene occur in diverse human tumor types, *Nature (London),* 342, 705, 1989.
52. **Takahashi, T., Nau, M. M., Chiba, I., Birrer, M. J., Rosenberg, R. K., et al.,** p53: A frequent target for genetic abnormalities in lung cancer, *Science,* 246, 491, 1989.
53. **Iggo, R., Gatter, K., Bartek, J., Lane, D., and Harris, A. L.,** Increased expression of mutant forms of p53 oncogene in primary lung cancer, *Lancet,* 335, 675, 1990.
54. **Hollstein, M., Metcal, R. A., Welsh, J., Montisane, R., and Harris, C.,** Frequent mutation of the p53 gene in human esophageal cancer, *Proc. Natl. Acad. Sci. U.S.A.,* 87, 9958, 1990.
55. **Mulligan, L. M., Matalshewski, G., Scrable, H. J., and Cavenee, W. K.,** Mechanisms of p53 loss in human sarcomas, *Proc. Natl. Acad. Sci. U.S.A.,* 87, 5863, 1990.
56. **McMahon, G., Davis, E. F., Huber, L. J., Kim, Y., and Wogan, G. N.,** Characterization of *c-Ki-ras* and *N-ras* oncogenes in aflatoxin B1 induced rat tumors, *Proc. Natl. Acad. Sci. U.S.A.,* 87, 1104, 1990.
57. **Sambamurti, K., Callahan, J., Luo, X., Perkins, C. P., Jacobsen, J. S., and Humayun, M. Z.,** Mechanisms of mutagenesis by a bulky DNA lesion of the guanine n7 position, *Genetics,* 120, 863, 1988.

58. **Crawford, L. V., Pim, D. C., and Lamb, P.,** The cellular protein p53 in human tumors, *Mol. Biol. Med.,* 2, 261, 1984.
59. **Robiniski, B. and Benchimol, S.,** Immortalization of rat embryo fibroblasts by the cellular p53 oncogene, *Oncogene,* 2, 445, 1988.
60. **Land, H., Parada, L. F., and Weinberg, R. A.,** Tumorigenic conversion of primary embryo fibroblasts requires at least two cooperating oncogenes, *Nature (London),* 304, 596, 1983.
61. **Eliyahu, D., Raz, A., Gruss, P., Givol, D., and Oren, M.,** Participation of p53 cellular tumor antigen in transformation of normal embryonic cells, *Nature (London),* 312, 646, 1984.
62. **Parada, L. F., Land, H., Weinberg, R. A., Wolf, D., and Rotter, V.,** Cooperation between gene encoding p53 tumor antigen and *ras* in cellular transformation, *Nature (London),* 312, 649, 1984.
63. **Chen, P.-L., Chen, Y., Bookstein, R., and Lee, W.-H.,** Genetic mechanisms of tumor suppression by the human p53 gene, *Science,* 250, 1576, 1990.
64. **Baker, S. J., Fearon, E. R., Nigro, J. M., Preisinger, A. C. et al.,** Chromosome deletions and p53 mutations in colorectal carcinomas, *Science,* 244, 217, 1989.
65. **Lavigueur, A., Maltby, V., Mock, D., Rossant, J., Pawson, T., and Bernstein, A.,** High incidence of lung, bone, and lymphoid tumors in transgenic mice overexpressing mutant alleles of the p53 oncogene, *Mol. Cell. Biol.,* 9, 3982, 1989.
66. **Bischoff, J. R., Friedman, P. N., Marshak, D. R., Prives, C., and Beash, D.,** Human p53 is phosphorylated by p60-cdc2 and cyclin B-cdc2, *Proc. Natl. Acad. Sci. U.S.A.,* 87, 4766, 1990.
67. **Fields, S. and Jang, S. K.,** Presence of a potent transcription activating sequence in the p53 protein, *Science,* 249, 1046, 1990.
68. **Raycroft, L., Wu, H., and Lozano, G.,** Transcriptional activation by wild-type but not transforming mutants of the p53 anti-oncogene, *Science,* 249, 1049, 1990.
69. **Kern, S. E., Kinzler, K. W., Bruskin, A., Jarosz, D., Friedman, P., Prives, C., and Vogelstein, B.,** Identification of p53 as a sequence-specific DNA-binding protein, *Science,* 252, 1708, 1991.
70. **Stanbridge, E. J.,** Human tumor suppressor genes, *Annu. Rev. Genet.,* 24, 615, 1990.
71. **Fearon, E. R., Hamilton, S. R., and Vogelstein, B.,** Clonal analysis of human colorectal tumors, *Science,* 238, 193, 1987.
72. **Vogelstein, B., Fearon, E. R., Hamilton, S. R., Kern, S. E., Preisinger, A. C., et al.,** Genetic alterations during colorectal tumor development, *N. Engl. J. Med.,* 319, 525, 1988.
73. **Vogelstein, B., Fearon, B., Kern, S. E., Hamilton, S. R., Preisinger, A. C., et al.,** Allotype of colorectal carcinomas, *Science,* 244, 207, 1989.
74. **Berenblum, I.,** A speculative review: the probable nature of promoting action and its significance in the understanding of the mechanism of carcinogenesis, *Cancer Res.,* 14, 471, 1954.
75. **Argyris, T. S.,** Regeneration and the mechanism of epidermal tumor promotion, *CRC Crit. Rev. Toxicol.,* 14, 211, 1985.
76. **Bannasch, P.,** Sequential cellular changes during chemical carcinogenesis, *J. Cancer Res. Clin. Oncol.,* 108, 11, 1984.
77. **Farber, E. and Sarma, D. S. R.,** Biology of disease: hepatocarcinogenesis: a dynamic cellular perspective, *Lab. Invest.,* 56, 4, 1987.
78. **Sell, S., Hunt, J. M., Knoll, B. J., and Dunsford, H. A.,** Cellular events during hepatocarcinogenesis in rats and the question of premalignancy, *Adv. Cancer Res.,* 48, 37, 1987.
79. **Yuspa, S. H. and Poirier, M. C.,** Chemical carcinogenesis: from animal models to molecular models in one decade, *Adv. Cancer Res.,* 50, 25, 1989.

80. **Cory, S. and Adams, A. M.,** Transgenic mice and oncogenesis, *Annu. Rev. Immunol.,* 6, 25, 1988.
81. **Sinn, E., Muller, W., Pattengale, P. K., Tepler, I., Wallace, R., and Leder, P.,** Coexpression of MMTV/*v-Ha-ras* and MMTV/*c-myc* in transgenic mice: synergistic action of oncogenes in vivo, *Cell,* 49, 465, 1987.
82. **Brinster, R. L., Chen, H. R., Messing, A., van Dyke, T., Levine, A. J., and Palmiter, R. D.,** Transgenic mice harboring SV40 T-antigen genes develop characteristic brain tumors, *Cell,* 37, 367, 1984.
83. **Palmiter, R. D., Chen, H. Y., Messing, A., and Brinster, R. L.,** SV40 enhancer and large-T antigen are instrumental in development of choroid plexus tumors in transgenic mice, *Nature (London),* 316, 457, 1985.
84. **Messing, A., Chen, H. Y., Palmiter, R. D., and Brinster, R. L.,** Peripheral neuropathies, hepatocellular carcinomas and islet cell adenomas in transgenic mice, *Nature (London),* 316, 461, 1985.
85. **Dyer, K. R. and Messing, A.,** Metal-inducible pathology in the liver, pancreas and kidney of transgenic mice expressing SV40 early region genes, *Am. J. Pathol.,* 135, 401, 1989.
86. **Held, W. A., Mullins, J. J., Kuhn, N. J., Gallagher, J. F., Gu, G. D., and Gross, K. W.,** T Antigen expression and tumorigenesis in transgenic mice containing a mouse major urinary protein/SV40 T antigen hybrid gene, *EMBO J.,* 8, 183, 1989.
87. **Fox, N., Crooke, R., Hwang, L.-H., Schibler, U., Knowles, B. B., and Solter, D.,** Metastatic hibernomas in transgenic mice expressing an α-amylase hybrid gene, *Science,* 244, 460, 1989.
88. **Knowles, B. B., McCarrick, J., Fos, N., Solter, D., and Damjanov, I.,** Osteosarcomas in transgenic mice expressing an α-amylase-SV40 T-antigen hybrid gene, *Am. J. Pathol.,* 137, 259, 1990.
89. **Hanahan, D.,** Heritable formation of pancreatic B-cell tumors in transgenic mice expressing recombinant insulin simian virus 40 oncogenes, *Nature (London),* 315, 115, 1985.
90. **Rindi, G., Grant, S. G. N., Yiangos, Y., Ghatei, M. A., Bloom, S. R., Bautch, V. L., Solcia, E., and Polak, J. M.,** Development of neuroendocrine tumors in the gastrointestinal tract of transgenic mice, *Am. J. Pathol.,* 136, 1349, 1990.
91. **Sandgren, E. P., Quaife, C. J., Pinkert, C. A., Palmiter, R. D., and Brinster, R. L.,** Oncogene-induced liver neoplasia in transgenic mice, *Oncogene,* 4, 715, 1989.
92. **Lee, G.-H., Li, H., Ohtake, K., Nomura, K., Hino, O., Furuta, Y., Aizawa, S., and Kitagawa, T.,** Detection of activated *c-H-ras* oncogene in hepatocellular carcinomas developing in transgenic mice harboring albumin promoter-regulated simian virus 40 gene, *Carcinogenesis,* 11, 1145, 1990.
93. **Sepulveda, A. R., Finegold, M. J., Smith, B., Slagle, B. E., DeMayo, J. L., Shen, R.-F., Woo, S. L. C., and Butel, J. S.,** Development of a transgenic mouse system for the analysis of stages in liver carcinogenesis using tissue-specific expression of the SV40 large T-antigen controlled by regulatory elements of the human alpha-1-antitrypsin gene, *Cancer Res.,* 49, 6108, 1989.

93a. **Butel, J. S., Sepulveda, A. R., Finegold, M. J., and Woo, S. L. C.,** SV40 large T antigen directed by regulatory elements of the human alpha-1-antitrypsin gene. A transgenic mouse system that exhibits stages in liver carcinogenesis, *Intervirology,* 31, 85, 1990.

94. **Ornitz, D. M., Hammer, R. E., Messing, A., Palmiter, R. D., and Brinster, R. L.,** Pancreatic neoplasia induced by SV40 T-antigen expression in acinar cells of transgenic mice, *Science,* 238, 188, 1987.
95. **Field, L. J.,** Atrial natriuretic factor-SV40 T antigen transgenes produce tumors and cardiac arrhythmias in mice, *Science,* 239, 1029, 1988.

96. **Reynolds, R. K., Hokezema, G. S., Vogel, J., Hinrichs, S. H., and Jay, G.,** Multiple endocrine neoplasia induced by the promiscuous expression of a viral oncogene, *Proc. Natl. Acad. Sci. U.S.A.*, 85, 3135, 1988.
97. **Wilkinson, W. O., Sandgren, E. P., Palmiter, R. D., Brinster, R. L., and Bell, R. M.,** Elevation of 1,2-diacylglycerol in *ras*-transformed neonatal liver and pancreas of transgenic mice, *Oncogene*, 4, 625, 1989.
98. **Quaife, C. J., Mathews, L. S., Pinkert, C. A., Hammer, R. E., Brinster, R. L., and Palmiter, R. D.,** Histopathology associated with elevated levels of growth hormone and insulin-like growth factor I in transgenic mice, *Endocrinology*, 124, 40, 1989.
99. **Sandgren, E. P., Luetteke, N. C., Palmiter, R. D., Brinster, R. L., and Lee, D. C.,** Overexpression of TGFα in transgenic mice: induction of epithelial hyperplasia, pancreatic metaplasia and carcinoma of the breast, *Cell*, 61, 1121, 1990.
100. **Jhappan, Cl., Stahle, C., Harkins, R. N., Fausto, N., Smith, G. H., and Merlino, G. T.,** TGFα overexpression in transgenic mice induces liver neoplasia and abnormal development of the mammary gland and pancreas, *Cell*, 61, 1137, 1990.
101. **Matsui, Y., Halter, S. A., Holt, J. T., Hogan, B. L. M., and Coffey, R. J.,** Development of mammary hyperplasia and neoplasia in MMTV-TGFα transgenic mice, *Cell*, 61, 1147, 1990.
102. **Dycaico, M. J., Grant, S. G. N., Felts, K., Nichols, W. S., Geller, S. A., Hager, J. H., Pollard, A. J., Kohler, S. W., Short, H. P., Jirik, F. R., Hanahan, D., and Sorge, J. A.,** Neonatal hepatitis induced by α_1-antitrypsin: a transgenic mouse model, *Science*, 242, 1409, 1988.
103. **Carlson, J. A., Rogers, B. B., Sifers, R. N., Finegold, M. J., Clift, S. M., DeMayo, F. J., Bullock, D. W., and Woo, S. L. C.,** Accumulation of PiZ α_1-antitrypsin causes liver damage in transgenic mice, *J. Clin. Invest.*, 83, 1183, 1989.
104. **Sandgren, E. P., Palmiter, R. D., Heckel, J. L., Daugherty, C. C., Brinster, R. L., and Degen, J. L.,** Complete hepatic regeneration after somatic deletion of an albumin-plasminogen activator transgene, *Cell*, 66, 245, 1991.
105. **Chisari, F. V., Pinkert, C. A., Milish, D. R., Filippi, P., McLachlan, A., Palmiter, R. D., and Brinster, R. L.,** A transgenic mouse model of the chronic hepatitis B surface antigen carrier state, *Science*, 230, 1157, 1985.
106. **Babinet, C., Farza, H., Morello, D., Hadchouel, M., and Pourcel, C.,** Surface expression of hepatitis B surface antigen (HBsAg) in transgenic mice, *Science*, 230, 1160, 1985.
107. **Chisari, F. V., Pilippi, P., McLachlan, A., Milich, D. R., Riggs, M., Lee, S., Palmiter, R. D., Pinkert, C. A., and Brinster, R. L.,** Expression of hepatitis B virus large envelope polypeptide inhibits hepatitis B surface antigen secretion in transgenic mice, *J. Virol.*, 60, 880, 1986.
108. **Chisari, F. V., Filippi, P., Buras, J., McLachlan, A., Popper, H., Pinkert, C. A., Palmiter, R. D., and Brinster, R. L.,** Structural and pathological effects of synthesis of hepatitis B virus large envelope polypeptide in transgenic mice, *Proc. Natl. Acad. Sci. U.S.A.*, 84, 6909, 1987.
109. **Chisari, F. V., Klopchin, K., Moriyama, T., Pasquinelli, C., Dunsford, H. A., Sell, S., Pinkert, C. A., Brinster, R. L., and Palmiter, R. D.,** Molecular pathogenesis of hepatocellular carcinoma in hepatitis B virus transgenic mice, *Cell*, 59, 1145, 1989.
110. **Dunsford, H. A., Sell, S., and Chisari, F. V.,** Hepatocarcinogenesis due to chronic liver cell injury in hepatitis B virus transgenic mice, *Cancer Res.*, 50, 3400, 1990.
111. **Kim, C.-M., Koike, K., Saito, I., Miyamura, T., and Jay, G.,** *HBx* gene of hepatitis B virus induces liver cancer in transgenic mice, *Nature (London)*, 351, 317, 1991.
112. **Twu, J.-S. and Schloemer, R. H.,** Transcriptional trans-activating function of hepatitis B virus, *J. Virol.*, 61, 3448, 1987.

113. **Seto, E., Mitchell, P. J., and Yen, T. S. B.,** Transactivation by the hepatitis B virus X protein depends on AP-2 and other transcription factors, *Nature (London),* 344, 72, 1990.
114. **Farza, H., Hadchouel, M., Scotto, J., Tiollais, P., Babinet, C., and Pourcel, C.,** Replication and gene expression of hepatitis B virus in a transgenic mouse that contains the complete viral genome, *J. Virol.,* 62, 4144, 1988.
115. **Araki, K., Miyazaki, J.-I., Hino, O., Tomita, N., Chisaka, O., Matsubara, K., and Yamamura, K.-I.,** Expression and replication of hepatitis B virus genome in transgenic mice, *Proc. Natl. Acad. Sci. U.S.A.,* 86, 207, 1989.
116. **Lee, T.-H., Finegold, M. J., Shen, R.-F., DeMayo, J. L., Woo, S. L. C., and Butel, J. S.,** Hepatitis B virus transactivator X protein is not tumorigenic in transgenic mice, *J. Virol.,* 64, 5939, 1990.
117. **Campbell, T. C., Chen, J., Liu, C., Li, J., and Parpia, B.,** Nonassociation of aflatoxin with primary liver cancer in a cross-sectional ecological survey in the People's Republic of China, *Cancer Res.,* 50, 6882, 1990.
118. **Szmuness, W.,** Hepatocellular carcinoma and the hepatitis B virus: evidence for a causal association, *Prog. Med. Virol.,* 24, 40, 1978.
119. **Beasley, R. P.,** Hepatitis B virus as the etiologic agent in hepatocellular carcinoma: epidemiologic considerations, *Hepatology,* 2, 21S, 1982.
120. **Waterhouse, J., Muir, C., Shanmugaratnam, K., and Powell, J., Eds.,** *Cancer Incidence in Five Continents,* Vol. 4, IARC Scientific Pub., Lyon, 1982.
121. **Dragani, T. A., Manenti, G., Farza, H., Della Porta, G., Tiollais, P., and Pourcel, C.,** Transgenic mice containing hepatitis B virus sequences are more susceptible to carcinogen-induced hepatocarcinogenesis, *Carcinogenesis,* 11, 953, 1989.
122. **Sell, S., Hunt, J. M., Dunsford, H. A., and Chisari, F. V.,** Synergy between hepatitis B virus expression and chemical hepatocarcinogens in transgenic mice, *Cancer Res.,* 51, 1278, 1991.
123. **Sell, S., Sala-Trepat, J., Sargent, T., Thomas, K., Nahon, J.-L., Goodman, T., and Bonner, J.,** Molecular mechanisms of control of albumin and alphafetoprotein production: a system to study the early effects of chemical hepatocarcinogens, *Cell Biol. Int. Rep.,* 4, 235, 1980.
124. **Toniatti, C., Arcone, R., Majello, B., Anter, U., Arpaia, G., and Gilberto, G.,** Regulation of the human C-reactive protein gene, a major marker in inflammation and cancer, *Mol. Biol. Med.,* 7, 699, 1990.
125. **Lim, I. K., Dumenco, L. L., Yun, J., Conovan, C., Warman, B., Gorodetzkaya, N., Wagner, G., Clapp, D. W., Lim, I. K. T., Hanson, R. W., and Gerson, S. L.,** High level, regulated expression of the chimeric p-enolpyruvate carboxykinase (GTP)-bacterial 06-alkylguanine-DNA alkyltansferase (ADA) gene in transgenic mice, *Cancer Res.,* 50, 1701, 1990.
126. **Fan, C. Y., Potter, P. M., Rafferty, J., Watson, A. J., Cakewell, L., Searle, P. F., O'Connor, P. J., and Margison, G. P.,** Expression of a human 06-alkylguanine-DNA alkyltransferase cDNA in human cells and transgenic mice, *Nucleic Acids Res.,* 18, 5723, 1990.
127. **Matsukuma, S., Nakatsuru, Y., Nakagawa, K., Utakoji, T., Surano, H., Kataoka, H., Sekiguchi, M., and Ishikawa, T.,** Enhanced 06-methyltransferase activity in transgenic mice containing an integrated E. coli ADA repair gene, *Mutat. Res.,* 218, 197, 1989.

Chapter 14

HEPATOCARCINOGENESIS IN EXPERIMENTAL WOODCHUCK HEPATITIS VIRUS INFECTION

B. C. Tennant

TABLE OF CONTENTS

ISBN 0-8493-4746-7
© 1992 by CRC Press, Inc.

I. INTRODUCTION

Demonstration of the Australia antigen by Blumberg and colleagues[1,2] and establishment of the relationship between the Australia antigen and the virus of hepatitis B virus (HBV)[3-5] are among the most important medical discoveries of this century and have resulted in major advances in understanding the natural history of viral hepatitis. It is now known that following HBV infection, a small percentage of adults and a much higher proportion of infants and young children develop persistent infection, becoming chronic carriers and potential shedders of HBV. Chronic HBV carriers often develop chronic hepatitis which may progress to cirrhosis,[6-9] and there now is convincing epidemiological evidence that HBV has an etiologic role in the pathogenesis of primary hepatocellular carcinoma (HCC).[10-19]

There are three lines of evidence that indicate HBV has an important role in human hepatocarcinogenesis. Several types of epidemiological data support this view. On a worldwide basis, there is a generally close geographical relationship between the incidence of HBV and HCC.[12,16] Individuals with chronic HBV infection have a significantly higher rate of HCC than case controls both in Africa[11,12,15] and Asia,[12,20-23] where HBV infection is hyperendemic, and in the U.S. and U.K., where the incidence of HBV infection and HCC are relatively low.[24-26] Perhaps the most compelling seroepidemiologic evidence for an etiologic role of HBV in human hepatocarcinogenesis has come from the prospective study of HBV carriers that has been conducted in Taiwan by Beasley and colleagues.[27-30] They have shown that the relative risk of developing HCC is two orders of magnitude greater in chronic HBV carriers than in controls. A comparably high relative risk of HCC among HBV carriers also exists in the U.S.[31]

The second type of evidence suggesting that HBV has a role in hepatocarcinogenesis has come from studies conducted at the molecular level demonstrating covalent integration of HBV DNA sequences into the host genome.[32-34] Such integrations have been demonstrated not only in hepatic tumor tissue, but also in adjacent nontumorous tissue. Integrated HBV DNA has been found in many human hepatocellular carcinomas and in virtually all in which HBV infection was present.[33,34] These observations, together with data indicating that integration precedes tumor formation, suggest an oncogenic role for HBV similar to that of other tumor-producing viruses in which the presence of integrated viral DNA or, in the case of retroviruses, proviral DNA, have been associated with malignant transformation.[33,35]

A third line of research that supports the hypothesis that HBV is a carcinogen involves the study of naturally occurring HCC in animals. At least three members of the family Sciuridae now have been described in which infection with HBV-like viruses has been associated with the development of HCC.[36-43] The objectives of this chapter are to summarize the seroepidemiologic and experimental evidence that indicates one of these agents, the woodchuck hepatitis virus (WHV), has an etiologic role in hepatocarcinogenesis, and to review some of the possible oncogenic mechanisms that are currently under investigation.

II. NATURALLY OCCURRING HEPADNAVIRUS INFECTIONS OF ANIMALS

A. HEPATOCELLULAR CARCINOMA IN WOODCHUCKS

The initial description of hepatic tumors in the woodchuck may be that of Fox[44] who, in 1912, reported multiple ''adenomas'' of the liver ranging in size from 0.3 to 1.0 cm in diameter in a woodchuck from the Philadelphia Zoological Gardens. A similar case (possibly identical to that described by Fox) was cited in a comprehensive report of tumors of captive animals in 1933 by Ratcliffe.[45] Habermann et al.[46] subsequently reported two woodchucks with hepatoma, one from the Washington Zoological Park and the other captured in Bethesda, MD.

Snyder and Ratcliffe[37,38] were the first to recognize the unusually high frequency of HCC in woodchucks trapped in the native habitat in Pennsylvania and maintained in the Penrose Research Laboratory of the Philadelphia Zoological Gardens. Dietary mycotoxins tentatively were ruled out as a cause of HCC in woodchucks because similar hepatic tumors were not observed in other animal species fed the same diet, and Snyder hypothesized a viral etiology. Bond[47] subsequently reported a 4-year-old woodchuck from New York State that had been maintained in captivity for $3^1/_2$ years prior to recognition of hepatoma, and Long et al.[48] described three woodchucks with hepatoma and a fourth with hepatic nodular regeneration from a group of 15 woodchucks that had been trapped in Frederick County, MD, prior to being maintained in captivity.

B. WOODCHUCK HEPATITIS VIRUS

In 1978, Summers et al.[36] described a virus related to HBV in woodchucks from the Penrose Research Laboratory of the Philadelphia Zoological Gardens. Woodchucks from this colony had experienced a remarkably high rate of chronic hepatitis for a period of 18 years, and 22.5% were found at necropsy to have HCC. Sera from 15% of the woodchucks contained DNA polymerase activity and, in such sera, it was possible to demonstrate spherical and filamentous particles that were morphologically similar to HBV. Virion DNA was similar in size and structure to HBV DNA, and they concluded that the WHV belonged to the same unique class of viruses as HBV.[36]

WHV now is classified as a member of the group of *hepa*totropic *DNA* viruses known as hepadnaviruses (family Hepadnaviridae),[49-52] and HBV is the generally accepted prototype. Woodchucks infected with WHV have large numbers of virus particles present in the blood.[36,53] Three types of viral

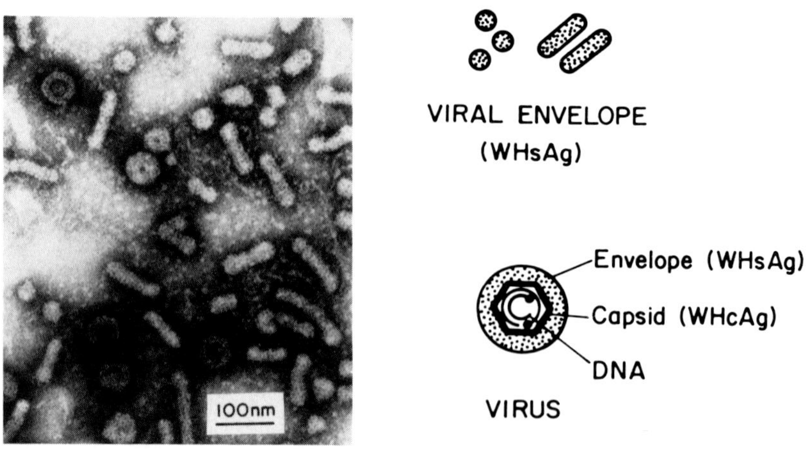

FIGURE 1. Electronphotomicrograph of WHV prepared from serum of a chronic carrier. The virion particles are approximately 50 nm in diameter. The smaller 22- to 25-nm spheres and filaments are composed of the envelope protein or surface antigen (WHsAg).

particles can be identified morphologically (Figure 1). Numerous 22-nm diameter filaments and spherical particles are present. These represent incomplete viral particles that contain the envelope protein or surface antigen (WHsAg). The 42- to 45-nm diameter virions are composed of the exterior envelope containing WHsAg, the nucleocapsid that is composed of the core antigen (WHcAg), and contained within the nucleocapsid is the DNA genome.[53-55]

The genome of WHV is a circular, partially double-stranded DNA molecule approximately 3.3 kb in length. The complete minus strand contains four large open reading frames that code for the two structured proteins, WHsAg and WHcAg, for viral DNA polymerase, and for the X protein, a transactivator that may regulate expression of viral and/or cellular genes during viral replication and that could have an important role in oncogenesis.[56] The polymerase region of the viral genome overlaps the other three viral genes, providing efficient use of the otherwise limited genetic coding capacity.[53-55]

The initial steps in viral replication involve entry of the virion into the hepatocyte and removal of the surface envelope.[53-55] This is followed by translocation of the nucleocapsid to the nucleus. After completion of plus-strand synthesis, viral DNA is converted to covalently closed, circular DNA which serves as a template for the synthesis of viral mRNA and for the RNA pregenome.[56,57] In the cytoplasm, pregenomic RNA and viral polymerase are packaged within core particles where synthesis of minus-strand viral DNA is completed by reverse transcription with concomitant degradation of RNA. Minus-strand DNA serves as the template for subsequent synthesis of the incomplete plus strand.[56,57]

Cyclic transition from DNA to RNA during the life cycle is a characteristic of both hepadnaviruses and retroviruses.[54,57,58] In the case of retroviruses, integration of proviral DNA into the host genome is an essential step in replication. Integration is not necessary for replication of hepadnaviruses; however, replicative intermediates may serve as substrate for integration of hepadnaviral nucleic acid sequences.[58]

C. SEROEPIDEMIOLOGY OF WHV INFECTION

The natural habitat of the woodchuck ranges from northern Georgia, Alabama, and Mississippi on the south, to Quebec and Labrador on the north, west to Oklahoma, Kansas, Nebraska, and North and South Dakota, across Canada to British Columbia and the Yukon Territory, and includes a small area of southeastern Alaska (Figure 2). A comprehensive study of the prevalence of WHV infection in this large population has not been completed. The woodchucks from which WHV was originally isolated by Summers et al.[36] originated in Pennsylvania. Tyler et al.[59] subsequently conducted a serologic study of woodchucks in Pennsylvania, New Jersey, and Maryland. They found that 23% of adult woodchucks were positive for WHsAg, and 36% were WHs antibody positive, for an overall rate of WHV infection of 59%. Similar WHV seroprevalence data from the mid-Atlantic states, including Delaware, have been reported by Wong et al.[60]

The hyperendemic status of WHV infection in the mid-Atlantic states is in marked contrast to that observed in New York State and New England. In New York, the rate of WHV infection, based on detection of anti-WHs antibody, was estimated to be approximately 2%,[60] and the rate of confirmed, persistent, WHsAg-positive infection is 0.2%.[61] Although the data are limited, no evidence of WHV infection has been found in Vermont, Massachusetts, or Iowa.[62-64] Most of the initial research on the relationship between WHV infection and HCC utilized wild woodchucks trapped in the mid-Atlantic states where WHV infection is hyperendemic. Woodchucks from the low-prevalence area of central New York have been valuable for development of breeding colonies to produce laboratory-born and -reared experimental animals.

D. RELATIONSHIP OF NATURALLY ACQUIRED WHV INFECTION AND HCC

The HCC described in woodchucks by Snyder and Summers[65] that were associated with WHV infection were variable in size and appeared multicentric in origin. Histologically, the trabecular form of hepatocyte arrangement was the most common, but the acinar or pseudoglandular form also was observed. The tumors typically were composed of rather well-differentiated hepatocytes containing hyperchromatic nuclei with large, prominent nucleoli. In nonneoplastic hepatic tissue, degenerative and regenerative changes were common and there were varying degrees of hepatitis. In mild portal hepatitis, inflam-

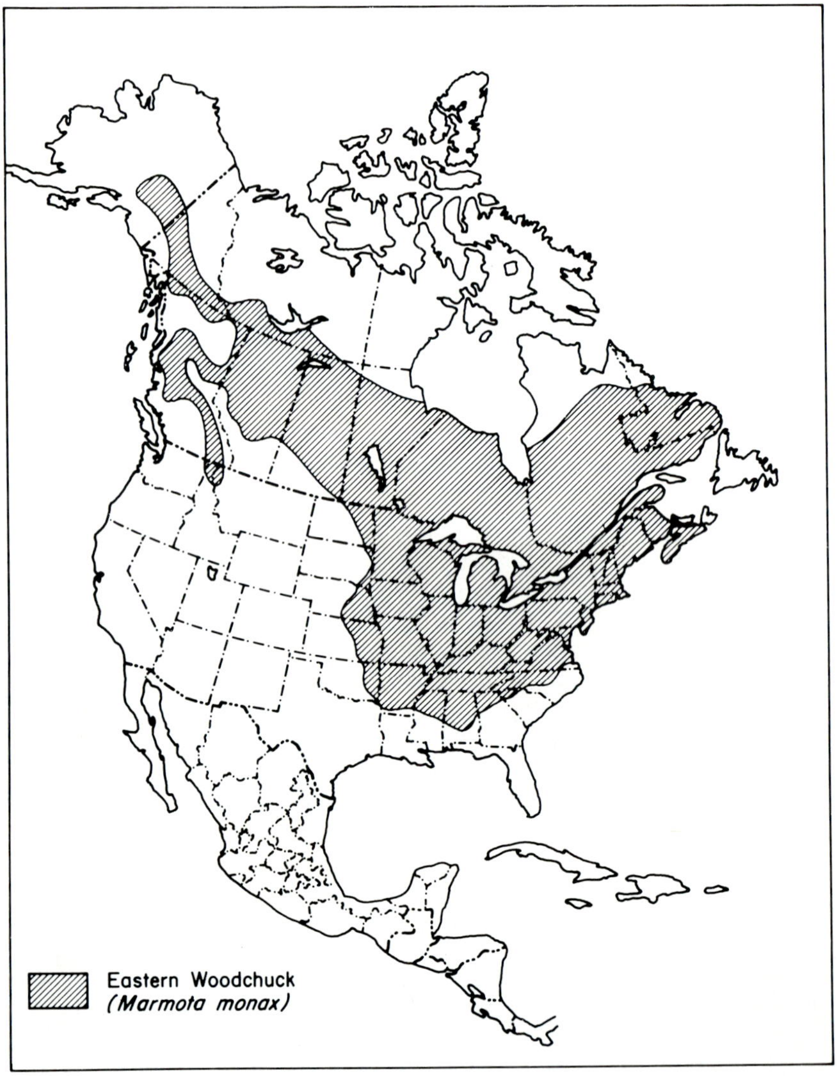

FIGURE 2. Geographic distribution of the Eastern woodchuck *(Marmota monax)*. WHV infection is hyperendemic in the mid-Atlantic states, but is either not present or rare in the other regions so far studied. (Reprinted from Hall, E. R., *The Mammals of North America*, John Wiley & Sons, New York, 1981. With permission.)

matory cells were located within the limiting plate, hepatocyte necrosis was absent, and bile duct proliferation and fibroplasia were minimal. In 75% of cases, a more severe form of chronic active hepatitis was present. In this group, abundant mononuclear cell inflammation extended beyond the limiting plate, and there were scattered necrosis, bile duct proliferation, and evidence of early fibrosis. Of the cases of chronic active hepatitis, 87% had histologic evidence of WHsAg in liver cells based on orcein (Shikata)-positive staining and 53% had HCC.[65]

The close association between the development of HCC in woodchucks and the presence of naturally acquired WHV infection has been confirmed by several groups. Popper et al.[66] reported serologic and histopathologic studies of 33 woodchucks that had been trapped in Maryland, Pennsylvania, and Delaware and maintained in laboratory animal facilities for up to 2 years. Of these, 14 had serologic evidence of WHV infection, and 10 were chronic carriers. Of the ten carriers, six developed histologically confirmed HCC similar to those reported by Snyder and Ratcliffe.[37,38] Popper et al.[66] also recorded lesions of moderate to severe portal hepatitis with varying degrees of bile duct proliferation and perilobular fibrosis. They also reported progression of neoplasia from small nodules of altered hepatocytes to frank carcinoma and the presence of significant numbers of hematopoietic cells infiltrating HCC.[66]

Subsequent reports have largely confirmed these original observations (Figure 3A and B).[67-69] Metastasis outside the liver, which occurs with some frequency in man and other experimental animals with HCC, has not been reported by most investigators, but Roth et al.[69] have reported one woodchuck with pulmonary metastases. This individual had had a full-thickness liver biopsy with partial resection of a hepatic tumor performed 10 weeks prior to postmortem examination, and it is possible that surgical manipulation may have predisposed to hematogenous metastasis. We have observed spontaneous pulmonary metastases in other WHV carriers with HCC, but the rate of such metastases appears to be less than 5%.[61]

E. OTHER MAMMALIAN AND AVIAN HEPADNAVIRUSES

Since description of WHV,[36] hepadnaviruses have been described in several other mammalian and avian species. These include the ground squirrel hepatitis virus (GSHV) of California ground squirrels *(Spermophilus beecheyi),*[39-41] the duck hepatitis B virus (DHBV) of Pekin ducks *(Anas domesticus),*[70,71] and the heron hepatitis B virus (HHBV) of German gray herons *(Ardea cinerea).*[72] Evidence also has been provided for hepadnavirus infections in Eastern gray squirrels *(Spermophilus carolinensis)* from Pennsylvania,[73,74] and for Richardson's ground squirrels *(S. richardsonii)* originating in Alberta.[42,43]

HCC has been observed in California ground squirrels with naturally acquired GSHV infection. Hepatic tumors in California ground squirrels were reported to occur somewhat less frequently and characteristically develop at an older age than in woodchucks infected with WHV.[41] The chronic hepatitis and HCC observed in Richardson's ground squirrels, however, appear to be

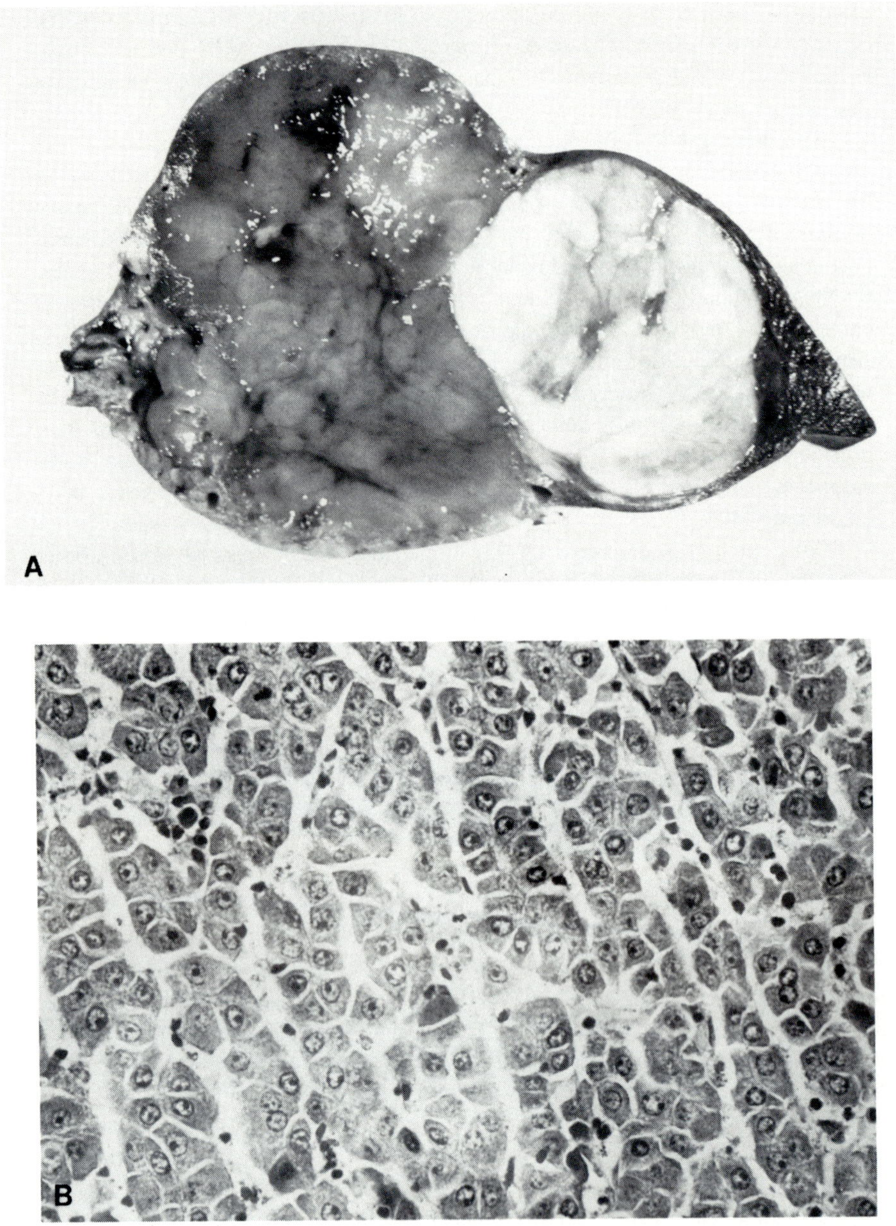

FIGURE 3. (A) Two large hepatocellular neoplasms that have virtually replaced the normal hepatic tissue of the liver lobe. (B) Photomicrograph of typical well-differentiated HCC.

virtually identical to the disease process observed in woodchucks.[43] The association of HCC with DHBV infection is uncommon,[71] and the role of DHBV in duck hepatocarcinogenesis remains problematic. Hepatic neoplasms have not been reported in association with infection of either the GSHV[73,74] or HHBV infections.[72]

III. EXPERIMENTAL WOODCHUCK HEPATITIS VIRUS INFECTION AND HEPATOCELLULAR CARCINOMA

A. WHV INFECTION OF ADULT WOODCHUCKS

Initial research on the relationship between WHV and HCC utilized woodchucks trapped in the native habitat in which WHV infection had been acquired naturally. Wild animals proved to be valuable sources of virus and of hepatic tissue for histologic and molecular analysis, but the limitations of such animals for experimental purposes were soon recognized. It was impossible, for example, to known when and for how long woodchucks trapped in the wild had been infected with WHV or to be certain about nutritional and/or environmental factors that could influence the natural course of WHV infection.

Experimental infection with WHV was reported initially by Summers et al.,[75] who injected susceptible 4- to 8-month-old woodchucks with serum from chronic carriers. Although it was concluded that acute infection had occurred in all animals, infection was self-limiting, and none became chronic carriers. Others subsequently have experimentally infected juvenile or adult woodchucks, and acute infection is regularly observed.[60,76-78] With one exception,[77] however, chronic infection has not been established.

B. WHV INFECTION OF NEONATAL WOODCHUCKS

When adult humans are infected with HBV, the chronic carrier state develops in less than 5%.[79-81] Perinatal infection or infection early in life, however, results in high rates of persistent HBV infection.[13,16,18,54] Because of the very high rates of chronic WHV infection in woodchucks from hyperendemic areas,[59,66] it was suspected that, under natural conditions, infection of woodchucks with WHV occurred early in life and that, as in man, vertical transmission may be important in maintaining the high rate of chronic natural infection. If this were so, experimental infection of woodchucks during the neonatal period should result in a high rate of persistent WHV infection.

To fully exploit the woodchuck as an animal model, laboratory raised experimental animals were required that were genetically and nutritionally defined, and that were free of WHV infection and other parasitic infections known to be endemic in wild woodchuck populations. To meet this need, a breeding colony of WHV-negative woodchucks was established at Cornell University using breeding stock trapped in central New York. Such animals now serve as a source of woodchucks for experimental purposes.

Woodchucks born in the laboratory were inoculated at 1-d of age with diluted serum from a standardized infectious pool obtained from a woodchuck with chronic WHV infection that ultimately developed HCC.[82] Following infection, woodchucks were monitored, and tests performed for woodchuck-specific markers of WHsAg,[83] anti-WH core,[84,85] and anti-WH surface antibodies.[60] Of 43 woodchucks initially infected with WHV at birth, 28 developed serologic evidence of self-limited infection, i.e., they cleared WHsAg and developed anti-WHs antibody, 13 became chronic carriers and, after 48 months, 12 (92%) had developed histologically confirmed HCC identical to that observed in woodchucks with naturally acquired WHV infection.[86] During the same period, 4 of the 28 woodchucks that appeared to recover serologically from acute WHV infection developed HCC, and the neoplasms in the anti-WHs-positive group were similar in appearance to those of chronic carriers. Over 100 uninfected, laboratory born control woodchucks have been raised in the laboratory and monitored clinically and biochemically for evidence of HCC, and none so far have developed HCC. These studies have been confirmed and extended[87] and now provide direct experimental evidence that WHV (and, by analogy, other hepadnaviruses) has an etiologic role in the pathogenesis of HCC.

The rate of HCC in woodchucks with experimentally induced, chronic WHV infection was fully comparable to that observed in woodchucks with naturally acquired, chronic infection.[65,66,69] HCC also has been observed in both experimentally infected woodchucks and woodchucks with naturally acquired WHV infection that have recovered and serologically converted to anti-WHs. While the rate of HCC in serologically recovered woodchucks was significantly lower than that of carriers, the risk of HCC was significantly higher than that observed in controls.

C. COMPARISON OF THE ONCOGENIC POTENTIAL OF WHV AND GROUND SQUIRREL HEPATITIS VIRUS

Studies of the molecular virology of hepadnaviruses have been impeded by the lack of readily available tissue culture systems. Fortunately, however, the cloned genomes of HBV,[88] WHV,[89] DHBV,[90] and GSHV[91,92] have proven to be infectious when injected directly into the parenchyma of the appropriate host species. During a series of transfection experiments designed to determine the viral gene or genes responsible for the presumed restriction of host range between WHV and GSHV,[92] we discovered that woodchucks could be infected with GSHV.[93] This had been unexpected, although GSHV earlier had been shown to infect chipmunks (*Eutamias* sp.).[94]

The frequency of HCC in woodchucks[65-69,82,87] infected with WHV is significantly higher than that reported in California ground squirrels infected with GSHV,[40,41] and HCC develops at a much older age in California ground

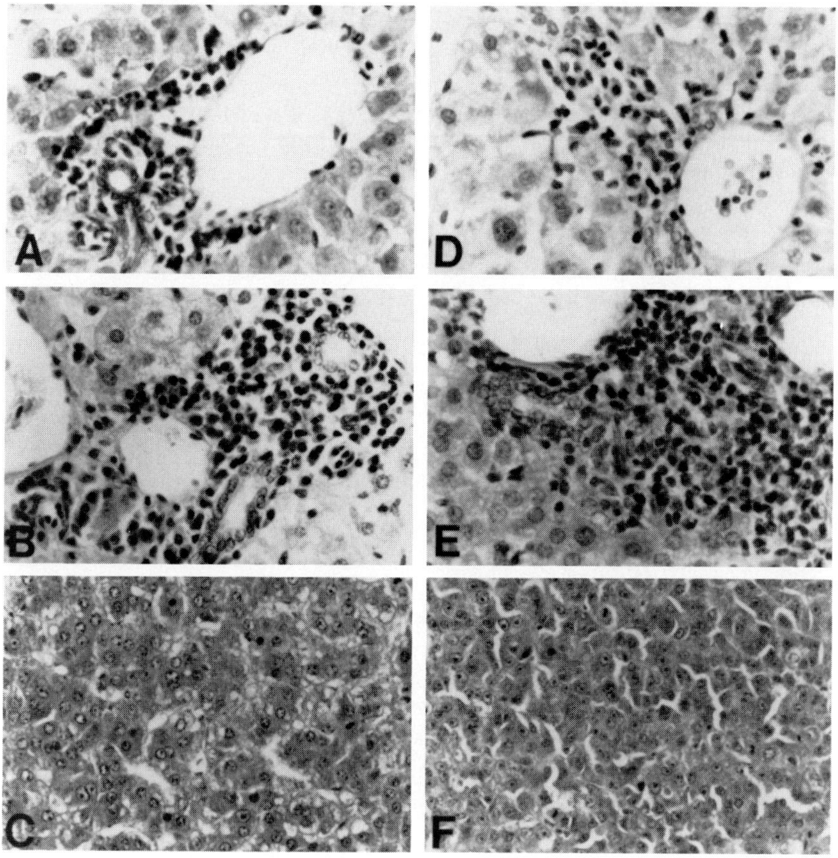

FIGURE 4. Mild portal hepatitis at 1 year of age (A), more severe hepatitis at 2 years of age (B), and typical trabecular HCC (C) in woodchucks that are chronic carriers of WHV. The hepatitis at 1 year of age (D) and at 2 years of age (E) in woodchucks that are chronic GSHV carriers if fully comparable to that associated with WHV infection. A small (5-mm diameter) neoplastic nodule in a chronic GSHV carrier woodchuck (F).

squirrels than in woodchucks. Because both hepadnaviruses could infect woodchucks, it became possible experimentally to determine if the apparent differences in oncogenic potential of GSHV and WHV were the result of differences in mammalian host response or were caused by genetic differences between the two viruses.

When neonatal woodchucks were experimentally infected with WHV or GSHV, the rates of chronic infection for each virus were similar[93] as was the severity of chronic hepatitis at 1 and 2 years following infection (Figure 4). Hepatic neoplasms were observed in 13 of 16 chronic WHV carriers 2 years after infection, while 1 of 16 chronic carriers of GSHV had developed a tumor nodule (5 mm diameter). These results suggest that the difference in hepatocarcinogenicity observed between GSHV and WHV in their natural hosts is due primarily to genetic differences between the viruses, although host factors cannot be entirely excluded on the basis of results obtained thus far. The reciprocal experiment using California ground squirrels infected with WHV may be impossible. Attempts to infect California ground squirrels with WHV by *in situ* transfection have not been successful.[92] It should, however, be possible to assess differences in the viral genetic determinants of oncogenicity by transfection of woodchucks with chimeric hepadnavirus constructs similar to those originally designed to examine host range restriction.[92]

IV. CHEMICAL VS. VIRAL HEPATOCARCINOGENESIS

There are several exceptions to the generally close association between HBV infection and HCC. Studies of native Eskimo populations in Greenland indicate that, despite a high prevalence of HBV infection, the incidence of HCC may be no greater than that in the U.S. or Northern Europe.[95] The incidence of both HBV infection and HCC are relatively high in Alaskan Eskimos,[96] but notable variations in the incidence of HCC have been observed between communities that have similar rates of chronic HBV infection.[97] There also are reports of clusters of HCC cases in which no serologic evidence of HBV involvement can be found.[98] These data, taken together with other observations described below, suggest that factors other than HBV (genetic, environmental, and/or dietary) may also be important in the pathogenesis of HCC,[95-98] and aflatoxin has frequently been suggested as such a factor.[99-102]

Aflatoxins frequently contaminate the food supply of man. Soon after publication of the original animal studies describing aflatoxicosis, aflatoxin was suspected of having a role in hepatocarcinogenesis.[103] Aflatoxins have been linked epidemiologically to the increased prevalence of HCC in Africa[104-109] and Asia.[110,111] Outbreaks of severe human liver disease have been reported following high dietary intake of aflatoxin,[112] and the presence of aflatoxin B_1 (AFB_1) in human liver, including that of patients with HCC, also has been demonstrated.[113-117] These data are sufficient to cause concern that aflatoxin, a potent hepatocarcinogen in animals, may play a role in the pathogenesis of human HCC.

The prospective study in Taiwan demonstrating that the relative risk of HCC in HBsAg-positive men is two orders of magnitude greater than that of seronegative controls[29] leaves little doubt about the decisive role HBV can have in the pathogenesis of HCC.[118,119] This study has gone perhaps as far as is possible using epidemiological techniques to establish that HBV is an important oncogenic agent. How other environmental or dietary factors may interact with HBV remains to be established, but such factors might be directly related to the chronic HBV carrier state.[29]

The observation that the relative risk of HCC in HBsAg carriers in the U.S. is similar to that of carriers in Taiwan has been interpreted to mean that environmental or dietary factors are relatively insignificant in the pathogenesis of HCC.[31] An alternate interpretation is that a dietary factor such as aflatoxin (known to be in relative abundance in areas of the world where HBV infection is hyperendemic) may actually influence the rate of persistence of HBV infection by affecting the immune response.[99] Aflatoxin has been shown to increase susceptibility of poultry to bacterial,[120,121] mycotic,[122] protozoal,[123,124] and viral infections,[125,126] including at least one oncogenic virus.[124] Aflatoxin-induced immunosuppression is associated with decreased immunoglobulin synthesis[127,128] and decreased specific antibody titers during the primary immune response.[129] Aflatoxin also causes impaired phagocytosis[130-132] and has been reported to impair lymphocyte blastogenesis.[133]

Chemically induced hepatic injury significantly increases the hepatocarcinogenicity of aflatoxin.[134,135] A variety of nutritional, drug, and other xenobiotic treatments, and partial hepatectomy, all induce liver cell regeneration and increase the hepatocarcinogenic response to aflatoxin.[136-140] In the marmoset, experimental hepatitis A infection has been shown to increase the rate of aflatoxin-induced HCC.[141] It is reasonable to suspect that the chronic active hepatitis caused by persistent hepadnavirus infection might increase the hepatotoxic and/or hepatocarcinogenic effects of aflatoxin.

Until recently, most epidemiological studies of human HCC have assessed the effects of HBV infection or of aflatoxin independently, but not their possible interaction. Even studies reported from the same city and supported by the same research support process dealt independently with aflatoxin exposure[142,143] and with HBV effects[144] without suggesting possible interrelationships. In a series of studies in Kenya comparing two populations with differences in the prevalence of HCC, the high-HCC population had significantly greater aflatoxin exposure than the lower-HCC population, but the prevalence of HBV infection in the two populations was similar.[104,105]

To investigate experimentally the influence of AFB_1 on WHV infection, neonatal woodchucks were inoculated with WHV and, after the acute phase of infection had begun, were treated for periods of 3 to 6 months with AFB_1.[145] The principals (AFB_1+, WHV+) were compared to woodchucks receiving AFB_1 alone, WHV alone, or to controls that received no experimental treatment. The rate of chronic WHV infection in the AFB_1+, WHV+ group was similar to woodchucks infected with WHV alone, indicating no effect of AFB_1 on the WHV immune response. HCC developed slightly more often in woodchucks with WHV alone than in the AFB_1+, WHV+ group. There was an increased number of nonhepatic tumors in both AFB_1-treated groups. There were no significant differences between the WHV alone and WHV+, AFB_1 groups in number of hepatic tumors observed in individual woodchucks or in median time to tumor. Of 21 woodchucks, 2 (10%) treated with AFB_1 alone developed hepatic neoplasms.

Under the conditions of these studies, AFB_1 was shown to be mildly hepatocarcinogenic in woodchucks. AFB_1 did not, however, alter the frequency of persistent WHV infection, an observation inconsistent with the hypothesis that aflatoxin may influence development of HCC by altering the cell-mediated immune response to HBV infection.[99] Development of HCC in WHV-infected woodchucks did not appear to be accelerated by AFB_1 treatment. The dose of AFB_1 used in these studies resulted in a high rate of hepatotoxicity, and it is possible that the process of viral hepatocarcinogenesis may have been altered by AFB_1. The results in woodchucks, however, are similar to those reported in ducks in which no significant interaction between AFB_1 treatment and chronic DHBV infection in hepatocarcinogenesis has been observed.[146-149] The experimental animal data also are consistent with a recent comprehensive epidemiologic study conducted in China in which it was concluded that aflatoxin had an insignificant role relative to HBV in human hepatocarcinogenesis.[149] Other human epidemiologic studies have produced data leading to the opposite conclusion,[104,105,150] and further experimental studies are needed to answer the questions that remain concerning possible interactions between AFB_1 (and other hepatotoxins) and hepadnaviruses in hepatocarcinogenesis.

V. MOLECULAR ETIOLOGY OF HEPADNAVIRUS-INDUCED HEPATOCELLULAR CARCINOMA

Current evidence strongly suggests that WHV is the proximate cause of HCC in woodchucks *(vida supra),* but how hepadnaviruses *cause* HCC is not completely understood. Two alternate hypotheses have been considered. In one hypothesis, a direct molecular role in hepatocarcinogenesis is proposed for hepadnaviruses. Integration of viral nucleic acid into the host genome is seen as the essential event leading to alteration in the expression of cellular genes (protooncogenes or tumor-suppressor genes) that ultimately results in neoplastic transformation of hepatocytes. Using terminology from the multistage model of chemical hepatocarcinogenesis,[151,152] hepadnaviruses are seen to be responsible for *initiation* of transforming events. In the alternate hypothesis, the principal role of hepadnaviruses is considered to be production of chronic hepatitis and regeneration of liver cells. Progressive hepatic injury associated with increased hepatocyte turnover leads to increased frequency and/or increased fixation of spontaneous mutations, rearrangements, or chromosomal translocations that alter expression of the cellular genes responsible for control of cell replication and growth. In the second hypothesis, the role of hepadnaviruses in hepatocarcinogenesis is comparable to the processes identified in the multistage model as *promotion* or *progression.*[151,152] It is necessary to note that these two hypotheses are not mutually exclusive, and it is possible that both mechanisms may be involved in hepadnavirus-induced hepatocarcinogenesis.

The molecular hypothesis was suggested first by the observation that HBV or WHV nucleic acid sequences frequently can be found covalently integrated into the cellular DNA of hepatic tumors from individuals with hepadnavirus infection.[32-34] Integration of viral nucleic acids is known to be a critical event in neoplastic transformation associated with several types of oncogenic viruses. Retroviruses, for example, when integrated in certain cellular DNA sites, appear to directly influence expression of cellular protooncogenes by insertional mutagenesis, and similar integrations of viral nucleic acids have been observed in other mammalian cancers caused by DNA tumor viruses.[153]

Several reports have described the occurrence of integrated WHV sequences in HCC from woodchucks. Typically, only restricted portions of the viral genome are found to be integrated, and these are often rearranged.[54,55] Molecular cloning of WHV DNA and flanking cellular DNA sequences indicates that integrations occur at multiple sites within the host genome.[154-160] Typically, multiple integrations are seen in the same tumor. Rogler[161] has summarized two possible models to explain the presence of multiple integrations. The first is that several viral integrations may accumulate in an individual hepatocyte and that these are clonally propagated after transformation of the hepatocyte. The alternate model is that the multistage process of hepatocarcinogenesis is associated with instability of the genome that favors integration of viral DNA and secondary rearrangements of integrated viral DNA sequences.

Integrated WHV DNA has often been reported in nontumorous portions of the liver.[154-157,159,160] This observation, combined with the presence of unique patterns of integration in most individual hepatic tumors, has been interpreted to mean that integrations occur early in the process of hepatocarcinogenesis, prior to clonal expansion of individual tumors. Such molecular lesions have been reported in the early neoplastic nodules that precede development of HCC[161-163] and that may be precursors of HCC.

Hepadnavirus genomes contain no known oncogene sequences similar to those of transforming retroviruses.[53-55] Screening of HCC RNA from woodchucks infected with WHV by dot blot and Northern blot hybridization failed to demonstrate increased levels of transcription of almost all of the known protooncogenes.[164] However, enhanced expression of the c-myc gene (5- to 50-fold) associated with truncation and rearrangement of the gene was demonstrated in three of nine HCC examined.[164-166] (Recent evidence suggests the rate of c-myc activation is approximately 10%.)[167] In one tumor, WHV DNA was not integrated near the c-myc locus, and no direct linkage between integration of viral nucleic acid sequences and c-myc activation was observed. The rearrangements and activation of the c-myc gene appeared to be similar to the chromosomal translocations observed in Burkitt's lymphoma and in acute B and T leukemias.[164] Molecular cloning of the other two integrations demonstrated that activation was the result of insertional mutagenesis of the c-myc gene by WHV DNA.[166] In these two tumors, viral insertion interrupted different regions of the c-myc locus. The position and orientation of WHV and c-myc sequences and the use of the c-myc promoter appeared to exclude involvement of the viral promoter (promoter insertion) in c-myc activation, but in both cases WHV sequences analogous to the HBV enhancer were identified, suggesting an enhancer insertion mechanism.

Recently, Fourel et al.[168] have examined 30 HCC from woodchucks, and 6 (20%) had insertions of WHV nucleic acid sequences adjacent to cellular N-myc sequences. In 18 tumors (60%), there was overexpression of N-myc messenger RNA which is normally not detectable in woodchuck liver. Woodchucks were reported to have two N-myc loci, one homologous to the known mammalian N-myc genes, and the other, based on nucleotide sequence analysis, had a structure typical of a retroposon. These data, combined with those related to c-myc, indicate that activation of myc genes by insertion of hepadnavirus DNA is a relatively common event in the pathogenesis of HCC in woodchucks.[168]

Results similar to those describing enhanced expression of rearranged myc protooncogenes have not been reported in human HCC. Analysis of human HCC for restriction fragment length polymorphisms has demonstrated loss of heterozygosity in human HCC and the possibility that loss of tumor-suppressor genes may be involved in hepatocarcinogenesis.[169,170] Recently, molecular analyses of human hepatic tumors from China[171] and Africa[172] were performed, and mutation of the p53 tumor-suppressor gene was demonstrated. In both studies, a high frequency of point mutations was observed in the third base position of codon 249. G→T and G→C transversions were the types of mutations identified, suggesting possible etiologic involvement of AFB_1 which causes specific G→T transversions in mutation assays. While these findings are interesting, it was not possible to determine whether or not the p53 mutations alone were sufficient to induce neoplastic transformation or whether or not they were secondary events (see Chapters 12 and 13). Hayward et al.[173] have recently analyzed 16 hepatic tumors from Australians thought to have minimal exposure to AFB_1 and found no mutations in codon 249 of p53. Experimental studies with animal models should be particularly valuable in assessing the role of tumor-suppressor genes and possible AFB_1-induced mutations of p53 in hepatocarcinogenesis.

Although no known oncogene sequences are present in the hepadnavirus genome, evidence is accumulating that specific viral gene sequences may have important roles in oncogenesis. The X gene of HBV has been shown in transfection experiments to induce growth stimulation and tumorigenic cell transformation of NIH3T3 cells.[174] The X gene product is a potent transactivator of both viral and cellular genes[54,55,175,176] and has been shown to have protein kinase activity.[177] In transgenic mice in which X is the transgene, the rate of HCC is significantly increased over control mice.[178]

The finding that the integrated HBV X open reading frame encodes transactivating activity[179] is consistent with the idea that integrated hepadnavirus sequences play a role in hepatocarcinogenesis by transcriptional transactivation of cellular genes.[179-181] Not all HCC, however, have integrated X gene sequences. Kekule et al.[182] have reported integration of truncated preS2/S sequences with transactivating capacity and demonstrated experimentally that these sequences were capable of transactivating the c-myc promoter.

The second general hypothesis concerning the role of hepadnaviruses in hepatocarcinogenesis proposes that by inducing hepatic inflammation, with increased hepatocellular necrosis and regeneration, hepadnaviruses enhance the acquisition and/or fixation of spontaneous genetic events responsible for neoplastic transformation. Support for this hypothesis comes from the observation that chronic liver injury, including cirrhosis of the liver, characteristically precedes development of HCC in humans with HBV infection.[183,184] A significant increase in the rate of HCC also has been observed in inherited forms of chronic liver disease in humans, such as hemochromatosis and α-1-antitrypsin deficiency. In both of these diseases, cirrhosis almost always precedes development of HCC.[185]

Important experimental support for hepatic injury as a key event in hepadnavirus-associated hepatocarcinogenesis has come from the work of Chisari et al.[186] and Moriyama et al.[187] with transgenic mice. They have studied lineages of transgenic mice that vary in their capacity to express the large envelope protein of HBV and have demonstrated a close relationship between the severity of liver cell injury in these lines of mice and the frequency of HCC. They have concluded that it is unlikely that insertional mutagenesis of the currently known oncogenes is involved. Recent observations suggesting increased free-radical formation in woodchucks with persistent WHV infection and the possibility that there may be associated increases in endogenous production of hepatocarcinogens is consistent with the view that protracted inflammation may be important in viral hepatocarcinogenesis.[188]

VI. CONCLUSIONS

Although there is now convincing evidence of the importance of hepadnaviruses in hepatocarcinogenesis, the oncogenic mechanism remains to be fully understood. Evidence has been presented that supports both a direct molecular role of the hepadnavirus and a role for chronic hepatic inflammation and liver cell injury caused by viral infection. It seems plausible that hepadnaviruses under certain conditions may have the capacity to act as complete carcinogens and under other conditions to interact with chemical carcinogens as cofactors or promoters of hepatocarcinogenesis. Regardless of the nature of the oncogenic process, the possibility exists that the control of one known risk factor (HBV) might independently result in significant reduction in the prevalence of human HCC. Safe and effective vaccines are now available that are capable of preventing neonatal and juvenile infection with HBV, and it seems probable that vaccination may be a useful method of preventing one of mankind's most serious forms of cancer.

The woodchuck appears to be of great potential value as an experimental animal model for studies of the pathogenesis of viral hepatitis and viral hepatocarcinogenesis. Such a model should be ideally suited for investigation of the suspected immunopathological mechanisms of liver injury in acute and chronic hepatitis, in the assessment of the oncogenic role of the virus, and the possible interactions between hepadnaviruses, diet, and/or other environmental factors in hepatocarcinogenesis and in developing and evaluating new and improved antiviral strategies to treat chronic liver disease and prevent HCC.

REFERENCES

1. **Blumberg, B. S., Alter, H. J., and Visnich, S.,** A new antigen in leukemia sera, *J. Am. Med. Assoc.*, 191, 541, 1965.
2. **Blumberg, B. S., Gerstley, B. J. S., Hungerford, D. A., London, W. T., and Sutnick, A. I.,** A serum antigen (Australia antigen) in Down's syndrome, leukemia and hepatitis, *Ann. Int. Med.*, 66, 924, 1967.
3. **Prince, A. M.,** An antigen detected in the blood during the incubation period of serum hepatitis, *Proc. Natl. Acad. Sci. U.S.A.*, 60, 814, 1968.
4. **Okochi, K. and Murakami, S.,** Observations on Australia antigen in Japanese, *Vox Sang.*, 15, 374, 1968.
5. **Dane, D. S., Cameron, C. H., and Briggs, M.,** Virus-like particles in serum of patients with Australia-antigen associated hepatitis, *Lancet*, 1, 695, 1970.
6. **Prince, A. M., Leblanc, L., Krohn, K., Masseyeff, R., and Alpert, M. E.,** SH antigen and chronic liver disease, *Lancet*, 2, 717, 1970.
7. **Dudley, F. J., Scheuer, P. J., and Sherlock, S.,** Natural history of hepatitis-associated antigen-positive chronic liver disease, *Lancet*, 2, 1388, 1972.
8. **Alberti, A., Diana, S., Scullard, G. H., Eddleston, A. L. W. F., and Williams, R.,** Full and empty Dane particles in chronic hepatitis B virus infection: relation to hepatitis B antigen and presence of liver damage, *Gastroenterology*, 75, 869, 1978.
9. **Alberti, A., Diana, S., Eddleston, A. L. W. F., and Williams, R.,** Changes in hepatitis B virus DNA polymerase in relation to the outcome of acute hepatitis type B, *Gut*, 20, 190, 1979.
10. **Sherlock, S., Fox, R. A., Niazi, S. P., and Scheuer, P. J.,** Chronic liver disease and primary liver-cell cancer with hepatitis associated (Australia) antigen in serum, *Lancet*, 1, 1243, 1970.
11. **Blumberg, B. S., Larouze, B., London, W. T., Werner, B., Hessen, J. E., Millman, I., Saimot, G., and Payet, M.,** The relation of infection with hepatitis B agent to primary hepatic carcinoma, *Am. J. Pathol.*, 81, 669, 1975.
12. **Maupas, P., Larouze, B., London, W. T., Werner, B., Millman, I., O'Connell, A., Blumberg, B., Saimot, G., and Payet, M.,** Antibody to hepatitis-B core antigen in patients with primary hepatic carcinoma, *Lancet*, 2, 9, 1975.

13. **Maupas, P., Goudeau, A., Coursaget, P., Chiron, J.-P., Drucker, J., Barin, F., Perrin, J., Denis, F., Mar, I. D., and Summers, J.,** Hepatitis B virus infection and primary hepatocellular carcinoma: epidemiological, clinical, and virological studies in Senegal from the perspective of prevention by active immunization, in *Viruses in Naturally Occurring Cancers,* Book A, Essex, M., Todaro, G., and zur Hausen, H., Eds., Cold Spring Harbor Conferences on Cell Proliferation, Vol. 7, Cold Spring Harbor Laboratory, Cold Spring Harbor, NY, 1980, 481.
14. **Macnab, G. M., Urbanowicz, J. M., Geddes, E. W., and Kew, M. C.,** Hepatitis-B surface antigen and antibody in Bantu patients with primary hepatocellular cancer, *Br. J. Cancer,* 33, 544, 1976.
15. **Prince, A. M., Szmuness, W., Michon, J., Demaille, J., Diebolt, G., Linhard, J., Quenum, C., and Sankale, M.,** A case/control study of the association between primary liver cancer and hepatitis B infection in Senegal, *Int. J. Cancer,* 16, 376, 1975.
16. **Szmuness, W.,** Hepatocellular carcinoma and the hepatitis B virus: evidence for a causal association, *Prog. Med. Virol.,* 24, 40, 1978.
17. **Hadziyannis, S. J.,** Primary liver cancer and its relationship to chronic infection with the hepatitis B virus, *Springer Semin. Immunopathol.,* 3, 473, 1981.
18. **Blumberg, B. S. and London, W. T.,** Hepatitis B virus: pathogenesis and prevention of primary cancer of the liver, *Cancer,* 50, 2657, 1982.
19. **Popper, H., Gerber, M. A., and Thung, S. N.,** The relationship of hepatocellular carcinoma to infection with hepatitis B and related viruses in man and animals, *Hepatology,* 2, 1S, 1982.
20. **Chien, M.-C., Tong, M. J., Lo, K.-J. L., Lee, J.-K. L., Milich, D. R., Vyas, G. N., and Murphy, B. L.,** Hepatitis B viral markers in patients with primary hepatocellular carcinoma in Taiwan, *J. Natl. Cancer Inst.,* 66, 475, 1981.
21. **Okuda, K., Nakashima, T., Sakamoto, K., Ikari, T., Hidaka, H., Kubo, Y., Sakuma, K., Motoike, Y., Okuda, H., and Obata, H.,** Hepatocellular carcinoma arising in noncirrhotic and highly cirrhotic livers: a comparative study of histopathology and frequency of hepatitis B markers, *Cancer,* 49, 450, 1982.
22. **Asano, M., Kato, H., Yoshimoto, K., Seyama, S., Itakura, H., Hamada, T., and Iijima, S.,** Primary liver carcinoma and liver cirrhosis in atomic bomb survivors, Hiroshima and Nagasaki, 1961–75, with special reference to hepatitis B surface antigen, *J. Natl. Cancer Inst.,* 69, 1221, 1982.
23. **Hann, H.-W. L., Kim, C. Y., London, W. T., Whitford, P., and Blumberg, B. S.,** Hepatitis B virus and primary hepatocellular carcinoma: family studies in Korea, *Int. J. Cancer,* 30, 47, 1982.
24. **Tabor, E., Gerety, R. S., Voegel, C. L., Bayley, A. C., Anthony, P. P., Chan, C. H., and Barker, L. F.,** Hepatitis B virus infection and primary hepatocellular carcinoma, *J. Natl. Cancer Inst.,* 58, 1197, 1977.
25. **Omata, M., Ashcavai, M., Liew, C.-T., and Peters, R. L.,** Hepatocellular carcinoma in the USA, etiologic considerations. Localization of hepatitis B antigens, *Gastroenterology,* 76, 279, 1979.
26. **Viola, L. A., Coleman, J. C., Fluker, J. L., Barrison, I. G., Paradinas, F. J., Evans, B. A., and Murray-Lyon, I. M.,** Natural history of liver disease in chronic hepatitis B surface antigen carriers, *Lancet,* 2, 1156, 1981.
27. **Beasley, R. P. and Lin, C.-C.,** Hepatoma risk among HBsAg carriers, *Am. J. Epidemiol.,* 108, 247, 1978.
28. **Beasley, R. P., Lin, C.-C., Hwang, L.-Y., and Chien, C.-S.,** Hepatocellular carcinoma and hepatitis B virus. A prospective study of 22,707 men in Taiwan, *Lancet,* 2, 1129, 1981.
29. **Beasley, R. P.,** Hepatitis B virus as the etiologic agent in hepatocellular carcinoma — epidemiologic considerations, *Hepatology,* 2, 21S, 1982.

30. **Beasley, R. P.,** Hepatitis B virus. The major etiology of hepatocellular carcinoma, *Cancer,* 61, 1942, 1987.
31. **Prince, A. M. and Alcabes, P.,** The risk of development of hepatocellular carcinoma in hepatitis B virus carriers in New York: a preliminary estimate using death-records matching, *Hepatology,* 2, 15S, 1982.
32. **Brechot, C., Hadchouel, M., Scotto, J., Fonck, M., Potet, F., Vyas, G. N., and Tiollais, P.,** State of hepatitis B virus DNA in hepatocytes of patients with hepatitis B surface antigen-positive and negative liver diseases, *Proc. Natl. Acad. Sci. U.S.A.,* 78, 3906, 1981.
33. **Shafritz, D. A.,** Hepatitis B virus DNA molecules in the liver of hepatitis B surface antigen carriers: mechanistic considerations in the pathogenesis of hepatocellular carcinoma, *Hepatology,* 2, 35S, 1982.
34. **Chen, D.-S., Hoyer, B. H., Nelson, J., Purcell, R. H., and Gerin, J. L.,** Detection and properties of hepatitis B viral DNA in liver tissues from patients with hepatocellular carcinoma, *Hepatology,* 2, 42S, 1982.
35. **Summers, J. and Mason, W. S.,** Replication of the genome of a hepatitis B-like virus by reverse transcription of an RNA intermediate, *Cell,* 29, 403, 1982.
36. **Summers, J., Smolec, J. M., and Snyder, R.,** A virus similar to human hepatitis B virus associated with hepatitis and hepatoma in woodchucks, *Proc. Natl. Acad. Sci. U.S.A.,* 75, 4533, 1978.
37. **Snyder, R. L.,** Hepatomas of captive woodchucks, *Am. J. Pathol.,* 52, 32a, 1968.
38. **Snyder, R. L. and Ratcliffe, H. L.,** *Marmota monax:* a model for studies of cardiovascular, cerebrovascular and neoplastic disease, *Acta Zool. Pathol. Antverp.,* 48, 265, 1969.
39. **Marion, P. L., Oshiro, L. S., Regner, D. C., Scullard, G. H., and Robinson, W. S.,** A virus in Beechey ground squirrels that is related to hepatitis B virus of humans, *Proc. Natl. Acad. Sci. U.S.A.,* 77, 2941, 1980.
40. **Marion, P. L., Knight, S. S., Salazar, F. H., Popper, H., and Robinson, W. S.,** Ground squirrel hepatitis virus infection, *Hepatology,* 3, 519, 1983.
41. **Marion, P. L., Van Davelaar, M. J., Knight, S. S., Salazar, F. H., Garcia, G., Popper, H., and Robinson, W. S.,** Hepatocellular carcinoma in ground squirrels persistently infected with ground squirrel hepatitis virus, *Proc. Natl. Acad. Sci. U.S.A.,* 83, 4543, 1986.
42. **Minuk, G. Y., Shaffer, E. A., Hoar, D. I., and Kelly, J.,** Ground squirrel hepatitis virus (GSHV) infection and hepatocellular carcinoma in the Canadian Richardson ground squirrel *(Spermophilus richardsonii), Liver,* 6, 350, 1986.
43. **Tennant, B. C., Mrosovsky, N., McLean, K., Cote, P. J., Korba, B. E., Engle, R. E., Gerin, J. L., Wright, J., Michener, G. R., Uhl, E., and King, J. M.,** Hepatocellular carcinoma in Richardson's ground squirels *(Spermophilus richardsonii):* evidence for association with hepatitis B-like virus infection, *Hepatology,* 13, 1215, 1991.
44. **Fox, H.,** Observations upon neoplasms in wild animals in the Philadelphia Zoological Garden, *J. Pathol. Bacteriol.,* 17, 217, 1912.
45. **Ratcliffe, H. L.,** Incidence and nature of tumors in captive mammals and birds, *Am. J. Cancer.,* 17, 116, 1933.
46. **Habermann, R. T., Williams, F. P., and Eyestone, W. H.,** Spontaneous hepatomas in two woodchucks and a carcinoma of the testis in a badger, *J. Am. Vet. Med. Assoc.,* 125, 295, 1954.
47. **Bond, E.,** Hepatoma and arteriosclerosis in a woodchuck, *J. Wildl. Dis.,* 6, 418, 1970.
48. **Long, G. G., Terrell, T. G., and Stockey, J. L.,** Hepatomas in a group of captured woodchucks, *J. Am. Vet. Med. Assoc.,* 167, 589, 1975.
49. **Robinson, W. S.,** Genetic variation among hepatitis B and related viruses, *Ann. N.Y. Acad. Sci.,* 354, 371, 1980.

50. **Robinson, W. S., Marion, P. L., Feitelson, M., and Siddiqui, A.,** The hepadnavirus group: hepatitis B and related viruses, in *Viral Hepatitis,* Szmuness, W., Alter, H. J., and Maynard, J. E., Eds., Franklin Institute Press, Philadelphia, 1982, 57.
51. **Melnick, J. L.,** Classification of hepatitis A virus as enterovirus type 72 and hepatitis B virus as hepadnavirus type 1, *Intervirology,* 18, 105, 1982.
52. **Gust, I. D., Burrell, C. J., Coulepis, A. G., Robinson, W. S., and Zuckerman, A. J.,** Taxonomic classification of human hepatitis B virus, *Intervirology,* 25, 14, 1986.
53. **Tiollais, P., Pourcel, C., and Dejean, A.,** The hepatitis B virus, *Nature,* 317, 489, 1985.
54. **Ganem, D. and Varmus, H. E.** The molecular biology of the hepatitis B viruses, *Annu. Rev. Biochem.,* 56, 651, 1987.
55. **Schodel, F., Sprengel, R., Weimer, T., Fernholz, D., Schneider, R., and Will, H.,** Animal hepatitis B viruses, in *Viral Oncology,* Vol. 8, Klein, G., Ed., Raven Press, New York, 1989, 73.
56. **Seeger, C., Ganem, D., and Varmus, H. E.,** Biochemical and genetic evidence for the hepatitis B virus replication strategy, *Science,* 232, 477, 1986.
57. **Summers, J.,** The replication cycle of hepatitis B viruses, *Cancer,* 61, 1957, 1987.
58. **Schroder, C. H. and Zentgraf, H.,** Hepatitis B virus related hepatocellular carcinoma: chronicity of infection — the opening to different pathways of malignant transformation, *Biochim. Biophys. Acta,* 1032, 137, 1990.
59. **Tyler, G. V., Summers, J., and Snyder, R.,** Woodchuck hepatitis virus in natural woodchuck population, *J. Wildl. Dis.,* 17, 297, 1981.
60. **Wong, D. C., Shih, J. W.-K., Purcell, R. H., Gerin, J. L., and London, W. T.,** Natural and experimental infection of woodchucks with woodchuck hepatitis virus as measured by new specific assays for woodchuck surface antigen and antibody, *J. Clin. Microbiol.,* 15, 484, 1982.
61. **Tennant, B. C.,** unpublished data.
62. **Lutwick, L. I., Hebert, M. B., and Sywassink, J. M.,** Cross-reactions between the hepatitis B and woodchuck hepatitis viruses, in *Viral Hepatitis,* Szmuness, W., Alter, H. J., and Maynard, J. E., Eds., The Franklin Institute Press, Philadelphia, 1982, 711.
63. **Summers, J.,** Three recently described animal virus models for human hepatitis B, *Hepatology,* 1, 179, 1981.
64. **Young, R. A. and Sims, E. A. H.,** The woodchuck, *Marmota monax,* as a laboratory animal, *Lab. Anim. Sci.,* 29, 770, 1979.
65. **Snyder, R. L. and Summers, J.,** Woodchuck hepatitis virus and hepatocellular carcinoma, in *Viruses in Naturally Occurring Cancers,* Book A, Essex, M., Todaro, G., and zur Hausen, H., Eds., Cold Spring Harbor Conferences on Cell Proliferation, Vol. 7., Cold Spring Habor Laboratory, Cold Spring Harbor, NY, 1980, 447.
66. **Popper, H., Shih, J. W.-K., Gerin, J. L., Wong, D. C., Hoyer, B. H., London, W. T., Sly, D. L., and Purcell, R. H.,** Woodchuck hepatitis and hepatocellular carcinoma: correlation of histologic with virologic observations, *Hepatology,* 1, 91, 1981.
67. **Snyder, R. L., Tyler, G., and Summers, J.,** Chronic hepatitis and hepatocellular carcinoma associated with woodchuck hepatitis virus, *Am. J. Pathol.,* 107, 422, 1982.
68. **Chomel, B., Duclos, P. H., Gilbert, J. M., Trepo, C., and Joubert, L.,** Animal models for studying hepadna human virus disease, *Rev. Med. Vet.,* 134, 425, 1983.
69. **Roth, L., King, J. M., Hornbuckle, W. E., Harvey, H. J., and Tennant, B. C.,** Chronic hepatitis and hepatocellular carcinoma associated with persistent woodchuck hepatitis virus infection, *Vet. Pathol.,* 22, 338, 1985.
70. **Mason, W. S., Seal, G., and Summers, J.,** Virus of Pekin ducks with structural and biological relatedness to human hepatitis B virus, *J. Virol.,* 36, 829, 1980.
71. **Omata, M., Uchiumi, K., Ito, Y., Yokosuka, O., Mori, J., Terao, K., Wei-Fa, Y., O'Connell, A. P., London, W. T., and Okuda, K.,** Duck hepatitis B virus and liver disease, *Gastroenterology,* 85, 260, 1983.

72. **Sprengel, R., Kaleta, E. F., and Will, H.,** Isolation and characterization of a hepatitis B virus endemic in herons, *J. Virol.,* 62, 3832, 1988.
73. **Feitelson, M. A., Millman, I., Halbherr, T., Simmons, H., and Blumberg, B. S.,** A newly identified hepatitis B type virus in tree squirrels, *Proc. Natl. Acad. Sci. U.S.A.,* 83, 2233, 1986.
74. **Feitelson, M. A., Millman, I., and Blumberg, B. S.,** Tree squirrel hepatitis B virus: antigenic and structural characterization, *Proc. Natl. Acad. Sci. U.S.A.,* 83, 2994, 1986.
75. **Summers, J., Smolec, J. M., Werner, B. G., Kelly, T. J., Jr., Tyler, G. V., and Snyder, R. L.,** Hepatitis B virus and woodchuck hepatitis virus are members of a novel class of DNA viruses, in *Viruses in Naturally Occurring Cancers,* Book A, Essex, M., Todaro, G., and zur Hausen, H., Eds., Cold Spring Harbor Conferences on Cell Proliferation, Vol. 7, Cold Spring Harbor Laboratory, Cold Spring Harbor, NY, 1980, 459.
76. **Millman, I., Southam, L., Halbherr, T., Simmons, H., and Kang, C. M.,** Woodchuck hepatitis virus: experimental infection and natural occurrence, *Hepatology,* 4, 817, 1984.
77. **Lindberg, J., Pichoud, C., Hantz, O., Vitvitski, L., Grimaud, J. A., Gilbert, J. M., Joubert, L., Frommel, D., and Trepo, C.,** Woodchuck hepatitis virus infection: serologic and histopathologic course and outcome, *Eur. J. Clin. Microbiol.,* 4, 59, 1985.
78. **Tyler, G. V., Snyder, R. L., and Summers, J.,** Experimental infection of the woodchuck *(Marmota monax monax)* with woodchuck hepatitis virus, *Lab. Invest.,* 55, 51, 1986.
79. **Bortolotti, F., Bertaggia, A., Crivellaro, C., Armigliato, M., Alberti, A., Pontisso, P., Chemello, C., and Realdi, G.,** Chronic evolution of acute hepatitis type B: prevalence and predictive markers, *Infection,* 14, 64, 1986.
80. **Seeff, L. B., Beebe, G. W., Hoffnagle, J. H., Norman, J. E., Buskell-Bales, Z., Waggoner, J. G., Kaplowitz, N., Koff, R. S., Petrini, J. L., Jr., Schiff, E. R., Shorey, J., and Stanley, M. M.,** A serologic follow-up of the 1942 epidemic of post-vaccination hepatitis in the United States Army, *N. Engl. J. Med.,* 316, 965, 1987.
81. **Tassopoulos, N. C., Papaevangelou, G. J., Sjogren, M. H., Roumeliotou-Karayannis, A., Gerin, J. L., and Purcell, R. H.,** Natural history of acute hepatitis B surface antigen-positive hepatitis in Greek adults, *Gastroenterology,* 92, 1844, 1987.
82. **Popper, H., Roth, L., Purcell, R. H., Tennant, B. C., and Gerin, J. L.,** Hepatocarcinogenicity of the woodchuck hepatitis virus, *Proc. Natl. Acad. Sci. U.S.A.,* 84, 866, 1987.
83. **Cote, P. J., Engle, R. E., Langer, C. A., Ponzetto, A., and Gerin, J. L.,** Antigenic analysis of woodchuck hepatitis surface antigen with site-specific radioimmunoassays, *J. Virol.,* 49, 701, 1984.
84. **Ponzetto, A., Cote, P. J., Ford, E. C., Purcell, R. H., and Gerin, J. L.,** Core antigen and antibody in woodchucks after infection with woodchuck hepatitis virus, *J. Virol.,* 52, 70, 1984.
85. **Ponzetto, A., Cote, P. J., Ford, E. C., Engle, R., Cicmanec, J., Shapiro, M., Purcell, R. H., and Gerin, J. L.,** Radioimmunoassay and characterization of woodchuck hepatitis core antigen and antibody, *Virus Res.,* 2, 301, 1985.
86. **Tennant, B., Hornbuckle, W. E., Baldwin, B. H., King, J. M., Cote, P., Popper, H., Purcell, R. H., and Gerin, J. L.,** Influence of age on the response to experimental woodchuck hepatitis virus infection, in *Viral Hepatitis and Liver Disease,* Zuckerman, A. J., Ed., Alan R. Liss, New York, 1988, 462.
87. **Gerin, J. L., Cote, P. J., Korba, B. E., and Tennant, B. C.,** Hepadnavirus-induced liver cancer in woodchucks, *Cancer Detect. Prevent.,* 14, 227, 1989.
88. **Will, H., Cattaneo, R., Koch, H.-G., Darai, G., Schaller, H., Schellekens, H., van Eerd, P. M. C. A., and Deinhardt, F.,** Cloned HBV DNA causes hepatitis in chimpanzees, *Nature,* 299, 740, 1982.
89. **Miller, R. H., Girones, R., Cote, P. J., Hornbuckle, W. E., Chestnut, T., Baldwin, B. H., Korba, B. E., Tennant, B. C., Gerin, J. L., and Purcell, R. H.,** Evidence against a requisite role for defective virus in the estabishment of persistent hepadnavirus infections, *Proc. Natl. Acad. Sci. U.S.A.,* 87, 9329, 1990.

90. **Sprengel, R., Kuhn, C., Manso, C., and Will, H.,** Cloned duck hepatitis B virus DNA is infectious in Pekin ducks, *J. Virol.,* 52, 932, 1984.
91. **Seeger, C., Ganem, D., and Varmus, H. E.,** Nucleotide sequence of an infectious molecularly cloned genome of ground squirrel hepatitis virus, *J. Virol.,* 51, 367, 1984.
92. **Seeger, C., Marion, P. L., Ganem, D., and Varmus, H. E.,** In vitro recombinants of ground squirrel and woodchuck hepatitis viral DNAs produce infectious virus in squirrels, *J. Virol.,* 61, 3241, 1987.
93. **Seeger, C., Baldwin, B., Hornbuckle, W. E., Yeager, A. E., Tennant, B. C., Cote, P., Ferrell, L., Ganem, D., and Varmus, H. E.,** Woodchuck hepatitis virus is a more efficient oncogenic agent than ground squirrel hepatitis virus in a common host, *J. Virol.,* 65, 1673, 1991.
94. **Trueba, D., Phelan, M., Nelson, J., Beck, F., Pecha, B. S., Brown, R. J., Varmus, H. E., and Ganem, D.,** Transmission of ground squirrel hepatitis virus to homologous and heterologous hosts, *Hepatology,* 5, 435, 1985.
95. **Skinhoj, P., Hart Hansen, J. P., Nielsen, N. H., and Mikkelsen, F.,** Occurrence of cirrhosis and primary liver cancer in an Eskimo population hyperendemically infected with hepatitis B virus, *Am. J. Epidemiol.,* 108, 121, 1978.
96. **Heyward, W. L., Lanier, A. P., Bender, T. R., Hardison, H. H., Dohan, P. H., McMahon, B. J., and Francis, D. P.,** Primary hepatocellular carcinoma in Alaskan natives, 1969–1979, *Int. J. Cancer,* 28, 47, 1981.
97. **Boss, L. P., Bender, T. R., Schreeder, M. T., Lanier, A. P., Hardison, H. H., and Maynard, J. E.,** Hepatitis B testing in the families and villages of five young Eskimos with primary hepatocellular carcinoma, *Am. J. Epidemiol.,* 114, 95, 1981.
98. **Thomas, G. E., Wicks, A. C. B., Clain, D. J., Loon, N., Seggie, J., and Bramstom, B.,** Hepatocellular carcinoma in the Rhodesian African, *Am. J. Dig. Dis.,* 22, 573, 1977.
99. **Lutwick, L. I.,** Relation between aflatoxin, hepatitis-B virus, and hepatocellular carcinoma, *Lancet,* 1, 755, 1979.
100. **Gibson, J. B., Wu, P.-C., Ho, J. C. I., and Lauder, I. J.,** Hepatitis B surface antigen, hepatocellular carcinoma and cirrhosis in Hong Kong: a necropsy study: 1963–1976, *Br. J. Cancer,* 42, 370, 1980.
101. **Joishy, S. K., Bennett, J. M., Balasegaram, M., MacIntyre, J. M., Falkson, G., Moertel, C., and Carbone, P. P.,** Clinical and chemotherapeutic study of hepatocellular carcinoma in Malaysia. A comparison with African and American patients, *Cancer,* 50, 1065, 1982.
102. **Lam, K. C., Yu, M. C., Leung, J. W. C., and Henderson, B. E.,** Hepatitis B virus and cigarette smoking: risk factors for hepatocellular carcinoma in Hong Kong, *Cancer Res.,* 42, 5246, 1982.
103. **Oettle, A. F.,** The aetiology of primary carcinoma of the liver in Africa: a critical appraisal of previous ideas with an outline of the mycotoxin hypothesis, *S. Afr. Med. J.,* 39, 817, 1965.
104. **Alpert, M. E., Hutt, M. S. R., Wogan, G. N., and Davidson, C. S.,** Association between aflatoxin content of food and hepatoma frequency in Uganda, *Cancer,* 28, 253, 1971.
105. **Peers, F. G. and Linsell, C. A.,** Dietary aflatoxins and liver cancer — a population based study in Kenya, *Br. J. Cancer,* 27, 473, 1973.
106. **Bagshawe, A. F., Gacengi, D. M., Cameron, C. H., Dorman, J., and Dane, D. S.,** Hepatitis Bs antigen and liver cancer: a population based study in Kenya, *Br. J. Cancer,* 31, 581, 1975.
107. **Perrs, F. G., Gilman, G. A., and Linsell, C. A.,** Dietary aflatoxin and human liver cancer: a study in Swaziland, *Int. J. Cancer,* 17, 167, 1976.
108. **Peers, F. G. and Linsell, C. A.,** Dietary aflatoxins and human primary liver cancer, *Ann. Nutr. Aliment.,* 31, 1005, 1977.

109. **Van Rensburg, S. J., Kirsipuu, A., Coutinho, L. P., and Van Der Watt, J. J.,** Circumstances associated with the contamination of food by aflatoxin in a high primary liver cancer area, *S. Afr. Med. J.*, 49, 877, 1975.
110. **Shank, R. C., Bhamarapravati, N., Gordon, J. E., and Wogan, G. N.,** Dietary aflatoxins and human liver cancer. IV. Incidence of primary liver cancer in two municipal populations of Thailand, *Food Cosmet. Toxicol.*, 10, 171, 1972.
111. **Wogan, G. N.,** Dietary factors and special epidemiological situations of liver cancer in Thailand and Africa, *Cancer Res.*, 35, 3499, 1975.
112. **Anon.,** Epidemic of hepatitis in man due to aflatoxicosis, *Nutr. Rev.*, 34, 45, 1976.
113. **Phillips, D. L., Yourtee, D. M., and Searles, S.,** Presence of aflatoxin B_1 in human liver in the United States, *Toxicol. Appl. Pharmacol.*, 36, 403, 1976.
114. **Wray, B. B. and Hayes, A. W.,** Aflatoxin B_1 in the serum of a patient with primary hepatic carcinoma, *Environ. Res.*, 22, 400, 1980.
115. **Akintonwa, D. A. A.,** Some aspects of the human carcinogenicity of aflatoxin B_1 and 3,4 benz(a)pyrene, *Int. J. Epidemiol.*, 11, 293, 1982.
116. **Onyemelukwe, G. C., Ogbadu, G. H., and Salifu, A.,** Aflatoxins B_1, B_2, G_1, G_2 in primary liver cell carcinoma, *Toxicology Lett.*, 10, 309, 1982.
117. **Siraj, M. Y., Hayes, A. W., Unger, P. D., Hogan, G. R., Ryan, N. J., and Wray, B. B.,** Analysis of aflatoxin B_1 in human tissues with high-pressure liquid chromatography, *Toxicol. Appl. Pharmacol.*, 58, 422, 1981.
118. **Trichopoulos, D., Kremastinou, J., and Tzonou, A.,** Does hepatitis B virus cause hepatocellular carcinoma?, *IARC Sci. Pub.*, 39, 317, 1982.
119. **London, W. T. and Blumberg, B. S.,** A cellular model of the role of hepatitis B virus in the pathogenesis of primary hepatocellular carcinoma, *Hepatology*, 2, 10, 1982.
120. **Boonchuvit, B. and Hamilton, P. B.,** Interaction of aflatoxin and paratyphoid infection in broiler chickens, *Poultry Sci.*, 54, 1567, 1975.
121. **Wyatt, R. D. and Hamilton, P. B.,** Interaction between aflatoxicosis and a natural infection of chickens with Salmonella, *Appl. Microbiol.*, 30, 870, 1975.
122. **Hamilton, P. B. and Harris, J. R.** Interaction of aflatoxicosis with *Candida albicans* infections and other stresses in the chicken, *Poultry Sci.*, 50, 906, 1971.
123. **Wyatt, R. D., Ruff, M. D., and Page, R. K.,** Interaction of aflatoxin with *Eimeria tenella* infection and monensin in young broiler chickens, *Avian Dis.*, 19, 730, 1975.
124. **Edds, G. T., Nair, K. P. C., and Simpson, C. F.,** Effect of aflatoxin B_1 on resistance in poultry against cecal coccidiosis and Marek's disease, *Am. J. Vet. Res.*, 34, 819, 1973.
125. **Giambrone, J. J., Partadiredja, M., Eidson, C. S., Kleven, S. H., and Wyatt, R. D.,** Interaction of aflatoxin with infectious bursal disease virus infection in young chickens, *Avian Dis.*, 22, 431, 1978.
126. **Chang, C.-F. and Hamilton, P. B.,** Increased severity and new symptoms of infectious bursal disease during aflatoxicosis in broiler chickens, *Poultry Sci.*, 61, 1061, 1982.
127. **Giambrone, J. J., Ewert, D. L., Wyatt, R. D., and Eidson, C. S.,** Effects of aflatoxin on the humoral and cell-mediated immune systems of the chicken, *Am. J. Vet. Res.*, 39, 305, 1978.
128. **Tung, H. T., Wyatt, R. D., Thaxton, P., and Hamilton, P. B.,** Concentrations of serum proteins during aflatoxicosis, *Toxicol. Appl. Pharmacol.*, 34, 320, 1975.
129. **Thaxton, J. P., Tung, H. T., and Hamilton, P. B.,** Immunosuppression in chickens by aflatoxin, *Poultry Sci.*, 53, 721, 1974.
130. **Michael, G. Y., Thaxton, P., and Hamilton, P. B.,** Impairment of the reticuloendothelial system of chickens during aflatoxicosis, *Poultry Sci.*, 52, 1206, 1973.
131. **Chang, C. F. and Hamilton, P. B.,** Impairment of phagocytosis in chicken monocytes during aflatoxicosis, *Poultry Sci.*, 58, 562, 1979.
132. **Chang, C. F. and Hamilton, P. B.,** Impaired phagocytosis by heterophils from chickens during aflatoxicosis, *Toxicol. Appl. Pharmacol.*, 48, 459, 1979.

133. **Pier, A. C., Richard, J. L., and Thurston, J. R.,** The influence of mycotoxins on resistance and immunity, in *Interactions of Mycotoxins in Animal Production,* National Academy of Science, Washington, D.C., 1979, 56.
134. **Newberne, P. M., Harrington, D. H., and Wogan, G. N.,** Effects of cirrhosis and other liver insults on induction of liver tumors by aflatoxin in rats, *Lab. Invest.,* 15, 962, 1966.
135. **Cardeilhac, P. T. and Nair, K. P.,** Inhibition by castration of aflatoxin induced hepatoma in carbon tetrachloride treated rats, *Toxicol. Appl. Pharmacol.,* 26, 393, 1973.
136. **Miller, E. C.,** Some current perspectives on chemical carcinogenesis in humans and experimental animals: presidential address, *Cancer Res.,* 38, 1479, 1978.
137. **Pitot, H. C. and Sirica, A. E.,** The stages of initiation and promotion in hepatocarcinogenesis, *Biochim. Biophys. Acta,* 605, 191, 1980.
138. **Becker, F. F.,** Recent concepts of initiation and promotion in carcinogenesis, *Am. J. Pathol.,* 105, 3, 1981.
139. **Farber, E.,** Chemical carcinogenesis: a biologic perspective, *Am. J. Pathol.,* 106, 271, 1982.
140. **Sell, S. and Leffert, H. L.,** An evaluation of cellular lineages in the pathogenesis of experimental hepatocellular carcinoma, *Hepatology,* 2, 77, 1982.
141. **Lin, J. J., Liu, C., and Svoboda, D. J.,** Long term effects of aflatoxin B_1 and viral hepatitis on marmoset liver. A preliminary report, *Lab. Invest.,* 30, 267, 1974.
142. **Bulatao-Jayme, J., Almero, E. M., Castro, M. C. A., Salamat, L. A., and Velandria, F. V.,** Dietary aflatoxin and hepatocellular carcinoma in the Philippines, *Philipp. J. Intern. Med.,* 19, 95, 1981.
143. **Bulatao-Jayme, J., Almero, E. M., Castro, M. C. A., Jardeleza, M. T. R., and Salamat, L. A.,** A case-control dietary study of primary liver cancer risk from aflatoxin exposure, *Int. J. Epidemiol.,* 11, 112, 1982.
144. **Lingao, A. L., Domingo, E. O., and Nishioka, K.,** Hepatitis B virus profile of hepatocellular carcinoma in the Philippines, *Cancer,* 48, 1590, 1981.
145. **Tennant, B. C., Hornbuckle, W. E., Yeager, A. E., Baldwin, B. H., Sherman, W. K., Anderson, W. I., Steinberg, H., Cote, P. J., Korba, B. E., and Gerin, J. L.,** Effects of aflatoxin B_1 on experimental woodchuck hepatitis virus infection and hepatocellular carcinoma, in *Viral Hepatitis and Liver Disease,* Hollinger, F. B., Lemon, S. M., and Margolis, H., Eds., Williams & Wilkins, Baltimore, 1991, 599.
146. **Uchida, T., Suzuki, K., Esumi, M., Arii, M., and Shikata, T.,** Influence of aflatoxin B_1 intoxication on duck livers with duck hepatitis B virus infection, *Cancer Res.,* 48, 1559, 1988.
147. **Cova, L., Wild, C. P., Mehrotra, R., Turusov, V., Shirai, T., Lambert, V., Jacquet, C., Tomatis, L., Trepo, C., and Montesano, R.,** Contribution of aflatoxin B_1 and hepatitis B virus infection in the induction of liver tumors in ducks, *Cancer Res.,* 50, 2156, 1990.
148. **Cullen, J. M., Marion, P. L., Sherman, G. J., Hong, X., and Newbold, J. E.,** Hepatic neoplasms in aflatoxin B_1-treated, congenital duck hepatitis B virus-infected, and virus-free Pekin ducks, *Cancer Res.,* 50, 4072, 1990.
149. **Campbell, T. C., Chen, J., Liu, C., Li, J., and Parpia, B.,** Nonassociation of aflatoxin with primary liver cancer in a cross-sectional ecological survey in the People's Republic of China, *Cancer Res.,* 50, 6882, 1990.
150. **Yeh, F.-S., Yu, M. C., Mo, C.-C., Luo, S., Tong, M. J., and Henderson, B. E.,** Hepatitis B virus, aflatoxins, and hepatocellular carcinoma in Southern Guangxi, China, *Cancer Res.,* 49, 2506, 1989.
151. **Farber, E. and Sarma, D. S. R.,** Biology of disease — hepatocarcinogenesis: a dynamic cellular perspective, *Lab. Invest.,* 56, 4, 1987.
152. **Pitot, H. C. and Dragan, Y. P.,** Facts and theories concerning the mechanisms of carcinogenesis, *FASEB J.,* 5, 2280, 1991.

153. **Johnson, P. J. and Williams, R.,** Of woodchucks and men: the continuing story of hepatitis B and hepatocellular carcinoma, *Br. Med. J.,* 284, 1586, 1982.
154. **Ogston, C. W., Jonak, G. J., Rogler, C. E., Astrin, S. M., and Summers, J.,** Cloning and structural analysis of integrated woodchuck hepatitis virus sequences from hepatocellular carcinomas of woodchucks, *Cell,* 29, 385, 1982.
155. **Mitamura, K., Hoyer, B. H., Ponzetto, A., Nelson, J., Purcell, R. H., and Gerin, J. L.,** Woodchuck hepatitis virus DNA in woodchuck liver tissues, *Hepatology,* 2, 47S, 1982.
156. **Rogler, C. E. and Summers, J.,** Cloning and structural analysis of integrated woodchuck hepatitis virus sequences from a chronically infected liver, *J. Virol.,* 50, 832, 1984.
157. **Rogler, C. E., Hino, O., and Su, C.-Y.,** Molecular aspects of persistent woodchuck hepatitis virus and hepatitis B virus infection and hepatocellular carcinoma, *Hepatology,* 7, 74S, 1987.
158. **Kaneko, S., Oshima, T., Kodama, K., Aoyama, S., Yoshikawa, H., Unoura, M., Fukuoka, K., Matsushita, F., Morimoto, H., Kobayashi, K., Hattori, N., and Murakami, S.,** Stable integration of woodchuck hepatitis virus DNA in transplanted tumors and established tissue culture cells derived from a woodchuck primary hepatocellular carcinoma, *Cancer Res.,* 46, 3608, 1986.
159. **Korba, B. E., Wells, F. V., Baldwin, B., Cote, P. J., Tennant, B. C., Popper, H., and Gerin, J. L.,** Hepatocellular carcinoma in woodchuck hepatitis virus-infected woodchucks: presence of viral DNA in tumor tissue from chronic carriers and animals serologically recovered from acute infections, *Hepatology,* 9, 461, 1989.
160. **Fuchs, K., Heberger, C., Weimer, T., and Roggendorf, M.,** Characterization of woodchuck hepatitis virus DNA and RNA in the hepatocellular carcinomas of woodchucks, *Hepatology,* 10, 215, 1989.
161. **Rogler, C. E.,** Cellular and molecular mechanisms of hepatocarcinogenesis associated with hepadnavirus infection, in *Current Topics in Microbiology and Immunology. Hepadnaviruses. Molecular Biology and Pathogenesis,* Mason, W. S. and Seeger, C., Eds., Springer-Verlag, New York, 1991, 103.
162. **Fu, X.-X., Su, C. Y., Lee, Y., Hintz, R., Biempica, L., Snyder, R., and Rogler, C. E.,** Insulinlike growth factor II expression and oval cell proliferation associated with hepatocarcinogenesis in woodchuck hepatitis virus carriers, *J. Virol.,* 62, 3422, 1988.
163. **Yang, D. and Rogler, C. E.,** Analysis of insulin-like growth factor II (IGF-II) expression in neoplastic nodules and hepatocellular carcinomas of woodchucks utilizing *in situ* hybridization and immunocytochemistry, *Carcinogenesis,* 12, 1893, 1991.
164. **Moroy, T., Marchio, A., Etiemble, J., Trepo, C., Tiollais, P., and Buendia, M.-A.,** Rearrangement and enhanced expression of *c-myc* in hepatocellular carcinoma of hepatitis virus infected woodchucks, *Nature,* 324, 276, 1986.
165. **Moroy, T., Etiemble, J., Bougueleret, L., Hadchouel, M., Tiollais, P., and Buendia, M.-A.,** Structure and expression of *her,* a locus rearranged with *c-myc* in a woodchuck hepatocellular carcinoma, *Oncogene,* 4, 59, 1989.
166. **Hsu, T.-Y., Moroy, T., Etiemble, J., Louise, A., Trepo, C., Tiollais, P., and Buendia, M.-A.,** Activation of *c-myc* by woodchuck hepatitis virus insertion in hepatocellular carcinoma, *Cell,* 55, 627, 1988.
167. **Dejean, A. and de The, H.,** Hepatitis B virus as an insertional mutagene in a human hepatocellular carcinoma, *Mol. Biol. Med.,* 7, 213, 1990.
168. **Fourel, G., Trepo, C., Bougueleret, L., Henglein, B., Ponzetto, A., Tiollais, P., and Buendia, M.-A.,** Frequent activation of N-*myc* genes by hepadnavirus insertion in woodchuck liver tumors, *Nature,* 347, 294, 1990.
169. **Buetow, K. H., Murray, J. C., Israel, J. L., London, W. T., Smith, M., Kew, M., Blanquet, V., Brechot, C., Redeker, A., and Govindarajah, S.,** Loss of heterozygosity suggests tumor suppressor gene responsible for primary hepatocellular carcinoma, *Proc. Natl. Acad. Sci. U.S.A.,* 86, 8852, 1989.

170. **Walker, G. J., Hayward, N. K., Falvey, S., and Cooksley, W. G. E.,** Loss of somatic heterozygosity in hepatocellular carcinoma, *Cancer Res.,* 51, 4367, 1991.
171. **Hsu, I. C., Metcalf, R. A., Sun, T., Welsh, J. A., Wang, N. J., and Harris, C. C.,** Mutational hotspot in the p53 gene in human hepatocellular carcinomas, *Nature,* 350, 427, 1991.
172. **Bressac, B., Kew, M., Wands, J., and Ozturk, M.,** Selective G to T mutations of p53 gene in hepatocellular carcinoma from southern Africa, *Nature,* 350, 429, 1991.
173. **Hayward, N. K., Walker, G. J., Graham, W., and Cooksley, E.,** Hepatocellular carcinoma mutation, *Nature,* 352, 764, 1991.
174. **Shirakata, Y., Kawada, M., Fujiki, Y., Sano, H., Oda, M., Yaginuma, K., Kobayashi, M., and Koike, K.,** The X gene of hepatitis B virus induced growth stimulation and tumorigenic transformation of mouse NIH3T3 cells, *Jpn. J. Cancer Res.,* 80, 617, 1989.
175. **Seto, E., Mitchell, P. J., and Yen, T. S. B.,** Transactivation by the hepatitis B virus X protein depends on AP-2 and other transcription factors, *Nature,* 344, 72, 1990.
176. **Ritter, S. E., Whitten, T. M., Quets, A. T., and Schloemer, R. H.,** An internal domain of the hepatitis B virus X antigen is necessary for transactivating activity, *Virology,* 182, 841, 1991.
177. **Wu, J. Y., Zhou, Z.-Y., Judd, A., Cartwright, C. A., and Robinson, W. S.,** The hepatitis B virus-encoded transcriptional trans-activator hbx appears to be a novel protein serine/threonine kinase, *Cell,* 63, 687, 1990.
178. **Kim, C.-M., Koike, K., Saito, I., Miyamura, T., and Jay, G.,** HBx gene of hepatitis B virus induces liver cancer in transgenic mice, *Nature,* 351, 317, 1991.
179. **Wollersheim, M., Debelka, U., and Hofschneider, P. H.,** A transactivating function encoded in the hepatitis B virus X gene is conserved in the integrated state, *Oncogene,* 3, 545, 1988.
180. **Twu, J. S. and Schloemer, R.,** Transcriptional trans-activating function of hepatitis B virus, *J. Virol.,* 61, 3448, 1987.
181. **Zahm, P., Hofschneider, P. H., and Koshy, R.,** The HBV X-ORF encodes a transactivator: a potential factor in viral hepatocarcinogenesis, *Oncogene,* 3, 169, 1988.
182. **Kekule, A. S., Lauer, U., Meyer, M., Caselmann, W. H., Hofschneider, P. H., and Koshy, R.,** The preS2/S region of integrated hepatitis B virus DNA encodes a transcriptional transactivator, *Nature,* 343, 457, 1990.
183. **Zaman, S. N., Melia, W. M., Johnson, R. D., Portmann, B. C., Johnson, P. J., and Williams, R.,** Risk factors in development of hepatocellular carcinoma in cirrhosis: prospective study of 613 patients, *Lancet,* 1, 1357, 1985.
184. **Kalayci, C., Johnson, P. J., Davies, S. E., and Williams, R.,** Hepatitis B virus related hepatocellular carcinoma in the non-cirrhotic liver, *J. Hepatol.,* 12, 54, 1991.
185. **Johnson, P. J. and Williams, R.,** Cirrhosis and the aetiology of hepatocellular carcinoma, *J. Hepatol.,* 4, 140, 1987.
186. **Chisari, F. V., Klopchin, K., Moriyama, T., Pasquinelli, C., Dunsford, H. A., Sell, S., Pinkert, C. A., Brinster, R. L., and Palmiter, R. D.,** Molecular pathogenesis of hepatocellular carcinoma in hepatitis B virus transgenic mice, *Cell,* 59, 1145, 1989.
187. **Moriyama, T., Guilhot, S., Klopchin, K., Moss, B., Pinkert, C. A., Palmiter, R. D., Brinster, R. L., Kanagawa, O., and Chisari, F. V.,** Immunobiology and pathogenesis of hepatocellular injury in hepatitis B virus transgenic mice, *Science,* 248, 361, 1990.
188. **Liu, R. H., Baldwin, B., Tennant, B. C., and Hotchkiss, J. H.,** Elevated formation of nitrate and N-nitrosodimethylamine in woodchucks *(Marmota monax)* associated with chronic woodchuck hepatitis virus infection, *Cancer Res.,* 51, 3925, 1991.

INDEX

A

B

C

D

H

N

O

P

R

S